THE ELEMENTS

	IB	IIB	IIIA	IVA	VA	VIA	VIIA	O
							1 H 1.00797 ±0.00001	2 He 4.0026 ±0.00005
			5 B 10.811 ±0.003	6 C 12.01115 ±0.00005	7 N 14.0067 ±0.00005	8 O 15.9994 ±0.0001	9 F 18.9984 ±0.00005	10 Ne 20.183 ±0.0005
			13 Al 26.9815 ±0.00005	14 Si 28.086 ±0.001	15 P 30.9738 ±0.00005	16 S 32.064 ±0.003	17 Cl 35.453 ±0.001	18 Ar 39.948 ±0.0005
28 Ni 58.71 ±0.005	29 Cu 63.54 ±0.005	30 Zn 65.37 ±0.005	31 Ga 69.72 ±0.005	32 Ge 72.59 ±0.005	33 As 74.9216 ±0.00005	34 Se 78.96 ±0.005	35 Br 79.909 ±0.002	36 Kr 83.80 ±0.005
46 Pd 106.4 ±0.05	47 Ag 107.870 ±0.003	48 Cd 112.40 ±0.005	49 In 114.82 ±0.005	50 Sn 118.69 ±0.005	51 Sb 121.75 ±0.005	52 Te 127.60 ±0.005	53 I 126.9044 ±0.00005	54 Xe 131.30 ±0.005
78 Pt 195.09 ±0.005	79 Au 196.967 ±0.0005	80 Hg 200.59 ±0.005	81 Tl 204.37 ±0.005	82 Pb 207.19 ±0.005	83 Bi 208.980 ±0.0005	84 Po (210)	85 At (210)	86 Rn (222)

63	64	65	66	67	68	69	70	71
Eu 151.96 ±0.005	Gd 157.25 ±0.005	Tb 158.924 ±0.0005	Dy 162.50 ±0.005	Ho 164.930 ±0.0005	Er 167.26 ±0.005	Tm 168.934 ±0.0005	Yb 173.04 +0.005	Lu 174.97 ±0.005

95	96	97	98	99	100	101	102	103
Am (243)	Cm (247)	Bk (247)	Cf (249)	Es (254)	Fm (253)	Md (256)	No (253)	Lw (257)

Atomic Weights are based on C^{12}—12.0000 and Conform to the 1961 Values

Printed in U. S. A.

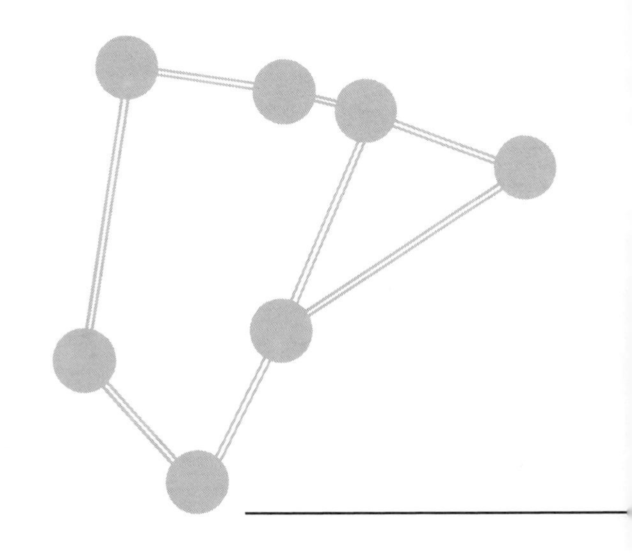

1971 W. B. SAUNDERS COMPANY • PHILADELPHIA • LONDON • TORONTO

A BRIEF INTRODUCTION TO GENERAL, ORGANIC AND BIOCHEMISTRY

JOSEPH I. ROUTH, Ph.D.
Professor, Department of Biochemistry,
College of Medicine, The University of Iowa

DARRELL P. EYMAN, Ph.D.
Associate Professor of Chemistry,
The University of Iowa

DONALD J. BURTON, Ph.D.
Professor of Chemistry,
The University of Iowa

W. B. Saunders Company: West Washington Square
Philadelphia, Pa. 19105

12 Dyott Street
London WC1A 1DB

1835 Yonge Street
Toronto 7, Ontario

A Brief Introduction to General, Organic and Biochemistry SBN 0-7216-7768-1

Print Number: 9 8 7 6 5 4 3 2 1

PREFACE

Trends in education exhibit interesting changes over the years. From the 1920's through the 1950's higher education consisted essentially of four year college programs, with a premium placed on B.A. and B.S. degrees. Graduate studies leading to an M.S. or Ph.D. were necessary requisites for college teachers and research scientists. Beginning in the 1960's and receiving added emphasis in the 1970's were programs designed to provide college training of a more practical nature extending in many instances for only two years beyond a high school education. Community and junior colleges are rapidly satisfying a need and demand for practical higher education.

An important question that faces educators at this level is the type of chemistry course that should be provided for their students. The authors have accumulated many years of experience teaching a practical approach to chemistry to beginning students in nursing, home economics, liberal arts, and medical technology. They are convinced that a chemistry course for these students and for students enrolled in community and junior colleges should consider aspects of general, inorganic, organic, and biochemistry. In this age of relevance, students must be prepared to understand current developments in medical research, drug research, plastics, insecticides, herbicides, nuclear power plants, the application of radioactive isotopes, and structural metal alloys.

To meet this need, the authors have designed a text that covers the essentials of general, inorganic, organic, and biochemistry. The material is chosen to fit a one semester or two quarter course, depending on the weekly class schedule. For a more detailed coverage serving the needs of the student in a one year course, we suggest the larger edition, *Essentials of General, Organic and Biochemistry* (1969). For colleges that provide laboratory experience in chemistry a practical laboratory manual, *Experiments in General, Organic and Biochemistry*, by two of the authors is available to accompany the text.

Since this book is of necessity rather limited in size it is not possible to present a complete coverage of general, inorganic, organic, and biochemistry. The primary aim of the text is to cover the fundamental aspects of general and inorganic chemistry in the first section, followed by the essential features of organic and biochemistry in the following sections. The first section is planned to indoctrinate the student in the field of chemistry in general, to introduce him to inorganic reactions, and to prepare him for an understanding of organic chemistry. The second section lays the groundwork for the biochemistry of living systems in the final section.

The authors hope that an understanding of the material in the book will result in more students planning careers in science—perhaps in chemistry, nursing, pharmacy, dentistry, medicine, medical technology, or other health-related fields. If the text is used in the only chemistry course taken in his career, the material should prepare the student to better understand the impact of current chemical developments on modern living.

The authors are indebted to the many students that have provided an opportunity to teach chemistry at a level represented by the scope of the present book. They would also like to thank their graduate students and colleagues who shared the responsibility of this teaching for valuable comments and criticisms. The constant assurance and patience of our wives, Dorothy Routh, Joy Eyman, and Marge Burton, and their generous assistance in the preparation and typing of the manuscript have our heartfelt appreciation. Finally, the authors are most grateful to their publisher, W. B. Saunders Company, for pertinent advice and continued assistance in the preparation of the manuscript for publication.

JOSEPH I. ROUTH

DARRELL P. EYMAN

DONALD J. BURTON

CONTENTS

Chapter 14

Chapter 15

Chapter 16

Chapter 17

Chapter 25

Chapter 26

Chapter 27

CHAPTER 1

FUNDAMENTAL CONCEPTS

To set the stage for a study of chemistry some chemical terms must be defined and then some fundamental concepts must be considered. Knowledge of elementary concepts of the science, such as the scientific method, chemical properties and changes, and the metric system of measurement, will serve as valuable tools in all fields of chemistry. In fact, recognition of the fundamental concepts of chemistry will facilitate the task of learning and of applying these concepts to the study of inorganic, organic, and biological chemistry.

THE SCIENTIFIC METHOD

The transition from alchemy to chemistry as it is now known occurred in the late seventeenth century. Of all the factors responsible for the change from mysticism to the science of chemistry, the most important was the development of the scientific method.

The **scientific method** is primarily a means of experimentally testing and verifying proposed laws of natural behavior. All the fundamental laws of chemical behavior were established, altered, and confirmed by experiment. Chemistry is a physical science, and physical science is primarily an experimental science, although many chemists devote all their efforts to developing theories to explain the experimental observations of others.

The fundamental steps in applying the scientific method to a study of chemical phenomena are as follows:

1. A clear statement of the problem to be studied.

2. Collection and organization of the experimental facts regarding the phenomenon.
3. Evaluation of the observed facts and formulation of a hypothesis to explain the phenomenon.
4. Testing and retesting the hypothesis experimentally under varying conditions to establish it as a theory.

The sequence of application of the steps in the scientific method as stated here is idealized. In reality it is often virtually impossible to follow this sequence. For example, a scientist may make many observations of a phenomenon before he realizes that he can make a clear statement of the problem to be studied.

In chemistry, experiments may be designed to study reactions of chemical compounds under controlled conditions of temperature, pressure, and the like, in an attempt to discover a uniformity of behavior that may be stated as a chemical law. Proper design of experiments and rigid control of conditions, as well as unbiased observation and reporting of results, are prerequisites of the scientific method. The fundamental laws of chemistry today were obtained by experiments devised and carried out by many investigators in many different laboratories. A modern investigator is never completely satisfied with his explanation of a process or a reaction until his experimental results have been confirmed in other laboratories.

Preconceived theories based on limited experience and training may prevent an investigator from reaching an unbiased conclusion from his research experiments. Many theories for the explanation of chemical processes have been proposed as the result of a series of experiments, only to be altered as new experimental evidence is obtained. Research in all fields of chemistry is based on the scientific method of testing ideas and theories by means of well designed and controlled experiments.

MATTER AND ENERGY

Chemistry is the science that deals with the composition of substances and with the changes that they may undergo. It is also concerned with the properties of substances and with their energy relationships. All substances are forms of matter, and **matter** is anything that possesses **mass** and occupies **space.** Matter exists in three states, **solid, liquid,** or **gas,** depending on the temperature and pressure. Matter in one physical state may often be changed into another by suitable energy changes.

Energy is defined as the ability to do work. It exists in many different forms, each of which may be converted into any of the other forms. Heat, light, motion, sound, and electricity are all familiar forms of energy. To measure the amount of energy in any of its forms, it is ordinarily converted to heat energy and expressed in calories. A **calorie** is a unit of heat energy that will raise the temperature of one gram of water one degree centigrade (at 15° centigrade).

The most commonly used unit is the large calorie, or **kilocalorie,** which equals 1000 calories.

Chemical energy is the energy that is stored up in chemical substances, and is released or consumed during chemical changes. Chemical processes, in which chemical energy is released as heat, are called **exothermic,** and those in which energy is absorbed are called **endothermic.** Sometimes these energy changes become the most important part of a chemical reaction—for example, the combustion of fuel in the furnace to produce heat, or the "burning" of foods in the body to produce heat and energy. It should be kept in mind that every chemical change is accompanied by a change in energy.

In 1906 Albert Einstein proposed that matter and energy are readily converted into each other and are related by the following expression:

$$E = mc^2$$

In this equation, E represents energy, m is mass, and c is the velocity of light. This equation is often quoted as the basis for the harnessing of atomic energy and the production of the atomic bomb. That matter can be converted into energy was proved in nuclear reactions. This relationship also affects the law of conservation of mass and the law of conservation of energy, which state that matter and energy can be neither created nor destroyed. A more appropriate statement might be: matter and energy can be neither created nor destroyed; they can only be interconverted.

THE CHARACTERIZATION OF MATTER

Substances are usually recognized by their appearance, taste, odor, feel, and other similar characteristics. Such characteristics are called the **properties** of the substances and are divided into two classes, physical and chemical. The **physical properties** of a substance are the characteristics other than those involved in chemical processes. The chemical properties of a substance, however, are made manifest only when the substance undergoes a chemical change. Such characteristics as the physical state (solid, liquid, or gaseous), crystalline form, density, hardness, color, and luster are common physical properties. The **chemical properties** of a substance are its characteristic reactions with other substances, such as oxygen, water, acids, and bases, or its decomposition.

Properties are the signs by which substances are recognized. If all the physical and chemical properties of two substances are studied and it is found that they are identical, then the substances must be the same. If the properties are different, two substances have been characterized.

Substances are constantly undergoing physical and chemical changes. **Physical changes** are changes in the condition or state of a substance. They do not result in the formation of new substances, nor do they involve a change in composition. An example of a physical change would be the breaking of a bottle. Although there has been a marked change, the substance is still glass.

No new substance has resulted, nor has there been a change in the composition of the glass. The boiling of water involves a physical change from the liquid to the gaseous state, but the composition of the matter is unchanged.

If a piece of iron is filed into small pieces, a definite change is observed, yet the particles are readily identified as iron, for they have the same properties as the original piece. If the iron filings are exposed to moisture, however, the iron will soon be changed into rust. A magnet will no longer attract the particles; the metallic luster is gone; the properties are different from those of the original substance; and it is therefore concluded that a new substance has been formed. When a piece of wood is heated in a test tube, it is observed that dense fumes are formed, and a black, charred mass remains behind. The rusting of iron and the destruction of wood by heat are examples of chemical changes. **Chemical changes** are defined as those changes that result in the formation of new substances and involve alterations in the composition of the original substance.

UNITS OF MEASUREMENT

The quantitative characterization of a property or an object requires a set of standard dimensions called **units.** In the United States, the **English system** of measurements is commonly used. It is always rather confusing for those beginning the study of a scientific subject to be introduced to a system of measurement that differs from the system with which they are familiar. The **metric system,** used by the scientist for weights and measures, is convenient because it is a decimal system, in which the divisions and multiples are in ratios of tens. Various units are expressed as fractions or multiples of a standard unit by applying prefixes to the name of a reference unit. These prefixes and their meanings are given in Table 1–1. A great deal of confusion would be avoided if this system could be adopted by everyone. As evidence of a trend in this direction, many pharmaceutical manufacturers have switched to the metric system for their products, and the U.S. Army has adopted the system for arms and ammunition. Recently, at the Eleventh Conference on Weights and Measures in Paris, the world adopted a new international standard of length.

Length. The standard unit of length in the metric system is called the **meter,** which was originally based on one ten-millionth of the distance from the equator to the North Pole. For many years the world relied on a material standard of length, the distance between two engraved lines on the International

TABLE 1–1 PREFIXES IN THE METRIC SYSTEM

Prefix	Value
Kilo-	1000
Hecto-	100
Deka-	10
Deci-	0.1 or 10^{-1}
Centi-	0.01 or 10^{-2}
Milli-	0.001 or 10^{-3}
Micro-	10^{-6}
Nano-	10^{-9}

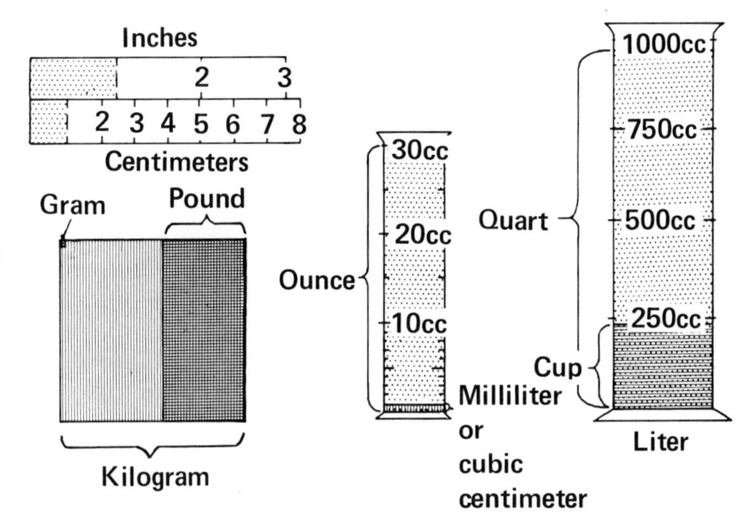

Figure 1-1 A comparison of the metric and English systems of measurement.

Meter Bar kept at Paris. The new definition established by the Eleventh General Conference states that the meter is 1,650,763.73 wavelengths of the orange-red line of krypton 86, which is equivalent to 39.37 inches. The **centimeter** is one-hundredth the length of a meter, and the **millimeter** is one-thousandth of a meter, or one-tenth of a centimeter. An **Angstrom,** Å, is 10^{-10} meters. There are approximately 30 centimeters in a foot, or about 2.54 centimeters in an inch (Fig. 1–1).

These units of length are abbreviated as m (meter), cm (centimeter), and mm (millimeter), and their relation to each other can be stated simply as follows: 1 m = 100 cm = 1000 mm.

Mass. The standard unit of mass is the **kilogram,** which is the mass of a block of platinum-iridium kept by the International Bureau of Weights and Measures. The **gram** is one-thousandth of a kilogram, and the **milligram** is one-thousandth of a gram (Fig. 1–1). These units are abbreviated as follows: kg (kilogram), g (gram), and mg (milligram). A **microgram** is one-thousandth of a milligram and is abbreviated μg, a unit often called a **gamma** (γ). These relationships may be represented as follows: 1 kg = 1000g; 1g = 1000 mg; and 1 mg = 1000 μg. There are 453.6 g per pound.

Volume. The **liter** is a unit of volume occupied by a kilogram of pure water at 4° centigrade (the temperature at which a given volume of water weighs the most). (Fig. 1–1). A milliliter is one-thousandth of a liter and is approximately the same as a **cubic centimeter.** A liter is commonly expressed as 1000 milliliters or 1000 cubic centimeters. A **microliter,** sometimes called a **lambda** (λ), is one-thousandth of a milliliter. Modern micro methods often require micro-liter quantities for analysis. A liter is 1.057 quarts. A fluid ounce is approximately 30 cubic centimeters, and a teaspoon holds about 4 cubic centimeters. The units of volume are abbreviated as follows: 1 (liter), ml (milliliter), cm^3 or cc (cubic centimeter), and μl (microliter).

Density. The **density** of a material is the mass per unit volume. Two objects of equal volume but of different mass are said to have unequal densities. Air and other gases have very low densities, whereas metals have relatively

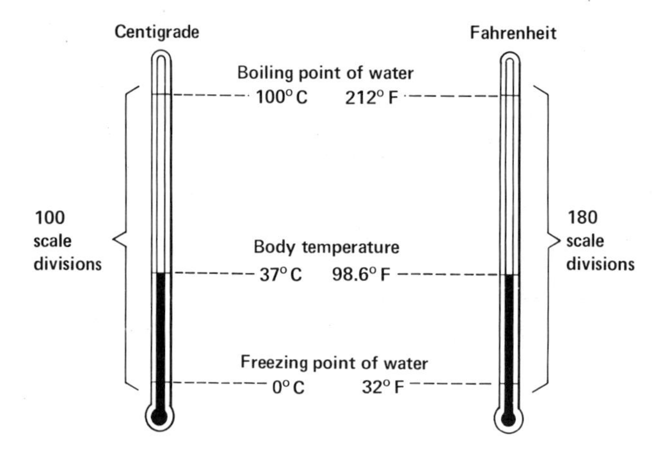

Figure 1-2 Scale divisions of 100 on the centigrade thermometer = 180 on the Fahrenheit. Therefore, 1 scale division centigrade = $\frac{9}{5}$ scale division Fahrenheit, and 1 scale division Fahrenheit = $\frac{5}{9}$ scale division centigrade.

high densities. Common units for expressing density include g/ml, g/l, and lbs/ft^3. The densities of gases are usually expressed in units of g/l, whereas for liquids and solids g/ml is used.

The density of water is 1 g/ml. The weight, or mass, of a given volume of a material divided by the weight of an equal volume of water is the **specific gravity.** This ratio, which is really just the ratio of the density of a material to that of water, is dimensionless.

Temperature. The degree of hotness, or the concentration of heat energy per unit mass, is the **temperature** of a body. In the United States temperature measurements are usually expressed on the familiar **Fahrenheit scale.** On this scale the temperature at which water freezes is 32 degrees (32°F), whereas water boils at 212°F. In scientific work this system has been replaced by the **centigrade,** or **Celsius, scale,** which is based on the freezing and boiling points of water. The freezing point is taken as 0°C, and the boiling point as 100°C. A comparison of the thermometers associated with these two scales is shown in Figure 1-2.

A temperature reading on one scale may readily be converted to the corresponding temperature on the other scale. To convert degrees Fahrenheit to degrees centigrade, add 40, multiply by $\frac{5}{9}$, and subtract 40 from the result. As an example, the following is a conversion of 32°F to degrees centigrade:

$$32°\text{F} + 40 = 72$$
$$72 \times \frac{5}{9} = \frac{360}{9} = 40$$
$$40 - 40 = 0°\text{C}$$

To convert degrees centigrade to degrees Fahrenheit, add 40, multiply by $\frac{9}{5}$, and subtract 40 from the result. For example, the following is a conversion of 100°C to degrees Fahrenheit:

$$100°\text{C} + 40 = 140$$
$$140 \times \frac{9}{5} = \frac{1260}{5} = 252$$
$$252 - 40 = 212°\text{F}$$

In converting negative temperatures it is necessary to make sure that the algebraic sum is used. For example, in the conversion of −15°F to degrees centigrade:

$$-15°F + 40 = +25 \text{ (algebraic sum)}$$
$$+25 \times 5/9 = 125/9 = +14$$
$$+14 - 40 = -26°C \text{ (algebraic sum)}$$

Other methods of conversion from one temperature scale to another may be used, although the one given here seems easiest to remember. In both conversions 40 is added to the original temperature. This sum is multiplied by either $5/9$ or $9/5$, and 40 is subtracted from the result. Using a common point like the boiling point of water, 212°F and 100°C, it can be seen that the Fahrenheit value is higher than the centigrade; therefore, it is necessary to use the larger factor $9/5$ to convert centigrade to Fahrenheit. Conversely, since the centigrade value is lower than the Fahrenheit, the factor $5/9$ can be used to convert Fahrenheit to centigrade.

Another method that is commonly used to convert readings on one scale to readings on the other is as follows:

$$C = 5/9(F - 32°)$$
$$F = 9/5C + 32°$$

ELEMENTS, COMPOUNDS, AND MIXTURES

All substances may be divided into three compositional classes: elements, compounds, and mixtures. **Elements** are considered as basic units of matter that cannot be decomposed by ordinary chemical methods. There are 103 of these elements and they are considered to be the building stones of all matter. Carbon, oxygen, iron, gold, neon, copper, and nitrogen are examples of elements. The relative abundance of elements in the earth's crust, in the ocean, and in the atmosphere is shown in Table 1–2. Eight elements—oxygen, silicon, aluminum, iron, calcium, sodium, potassium, and magnesium—make up 97 per cent of the composition of the earth. The remaining 3 per cent of the earth is composed of relatively small quantities of all the other naturally occurring elements.

By proper combination of elements, millions of more complex substances may be prepared. These substances are called **compounds** and are composed of two or more elements combined chemically and in definite proportions. If a compound is decomposed, or broken down into its constituent elements, it is possible to determine its composition. Water may be decomposed into its constituents, the elements hydrogen and oxygen, by passing an electric current

TABLE 1–2 PERCENTAGE COMPOSITION OF THE EARTH'S CRUST, OCEANS AND ATMOSPHERE

Oxygen	49.5	Magnesium	1.9	Sulfur	0.05
Silicon	25.7	Hydrogen	1.9	Barium	0.05
Aluminum	7.5	Titanium	0.6	Chromium	0.03
Iron	4.7	Chlorine	0.2	Nitrogen	0.03
Calcium	3.4	Phosphorus	0.1	Fluorine	0.03
Sodium	2.6	Carbon	0.09	Nickel	0.02
Potassium	2.4	Manganese	0.08	Strontium	0.02
				All others	0.09

through it. Iron rust, which is iron oxide, may be broken down into its constituent elements, iron and oxygen, by suitable chemical procedure. It will be shown later that every chemical compound contains its constituent elements in definite proportions by weight. A given amount of water, for example, always contains a definite amount of hydrogen combined with a definite amount of oxygen. This prevalence of definite composition helps to identify any chemical compound whose composition is known.

The remaining class of substances is called **mixtures.** A mixture is made up of two or more substances that are not combined chemically and may be present in any proportion. For example, powdered iron and powdered sulfur may be physically mixed in any proportion, and the individual ingredients still retain their characteristic properties. The iron can be separated from the mixture by the use of a magnet, whereas the sulfur can be separated from the mixture by dissolving it in the solvent carbon disulfide. The ingredients in a mixture can usually be separated by physical methods without changing the chemical identity of the individual substances that are present. In contrast to mixtures, if the iron and sulfur are heated in the proper proportion by weight to form the compound iron sulfide, new properties result, and the compound cannot then be separated into its constituents by physical methods. Most natural and prepared foods are mixtures, as are many of the therapeutic agents used in medicine. Air is a very important mixture consisting mainly of the gaseous elements oxygen and nitrogen.

THE LAWS OF CHEMICAL CHANGE AND DALTON'S ATOMIC THEORY

Although the alchemists carried out many chemical reactions in their search for gold, they did not apply the scientific method to a study of these reactions. In the seventeenth and eighteenth centuries a few chemists investigated the changes that occurred in chemical reactions. Several common elements were known early in the nineteenth century, and three laws of chemical change had been stated.

The Law of Conservation of Mass. The law states that *in a chemical change mass is neither created nor destroyed.* Before men became interested in the energy changes that accompany chemical reactions, several chemists had demonstrated the conservation of mass. By conducting reactions in closed containers, they showed that the products of a chemical change had the same mass as the starting materials. Careful experiments involving combustion and precipitation, with the formation of new compounds, resulted in the formulation of the law of conservation of mass.

The Law of Definite Proportion. *When two or more elements combine, they always combine in a fixed, or definite, proportion by weight.* For example, if water is formed from a mixture of hydrogen and oxygen gas, it will always contain 2g of hydrogen for every 16g of oxygen. When hydrogen is present in excess of the amount needed to combine with oxygen to form water, the resulting mixture will consist of water and the excess (uncombined) hydrogen gas.

It follows that each compound has a definite composition. The electrolysis of water always results in the formation of 2 parts of hydrogen gas and 16 parts of oxygen gas by weight.

The Law of Multiple Proportions. *If an element unites with another element in more than one proportion by weight to form two or more compounds, these proportions bear a ratio to one another that may be expressed in small whole numbers.* In the early studies of chemical change it was found that the same two elements would combine to form different compounds under different experimental conditions. For example, hydrogen and oxygen gas usually react to form water, but under conditions of a high-energy electric discharge they may form hydrogen peroxide. In a similar fashion it can be shown that carbon will combine with oxygen to form either carbon monoxide or carbon dioxide, depending on the experimental conditions. The weights of oxygen per unit weight of carbon in the two examples bear a ratio to one another of 1:2.

In an attempt to explain the facts outlined in the laws of chemical change, John Dalton, an English schoolteacher, proposed his atomic theory in 1803. He began by assuming that all elements are composed of minute, invisible particles called **atoms.** The atom may then be considered the smallest unit of an element. Dalton stated further that all atoms of the same element had the same properties and the same weight, but that they differed from the atoms of all other elements in these respects. In chemical changes the atoms could combine to form small particles of compounds, or they could separate or change places in these compounds. When the atoms enter into chemical combination, the weight of each individual atom does not change.

To explain the law of definite composition, he stated that chemical compounds are formed by the union of two or more elementary atoms. When one element combines with another, the combination always takes place between definite numbers of atoms of each kind. To explain the law of multiple proportions, he stated that combination always takes place in the simplest possible ratios, for example, 1 atom of A with 1, 2, or 3 atoms of B. The compound carbon monoxide would represent the combination of 1 atom of carbon with 1 atom of oxygen, whereas carbon dioxide would represent the combination of 1 atom of carbon with 2 atoms of oxygen. His theory also explains the law of conservation of mass, since he postulated that atoms are indivisible and can be neither created nor destroyed.

Dalton's original theory has been modified, and many exceptions have been made to his statements. The importance of his theory should not be underestimated, however, since for the study of the structure of chemical compounds it marks the beginning of the modern era. It was over a century later that a clear conception of the internal structure of atoms was formulated.

ATOMIC SYMBOLS

In studying the elements and the chemical reactions that they undergo, it is often inconvenient to write out the complete name of each element every time it occurs. For this reason the chemist has assigned a symbol to each of the 103 elements as a sort of chemical shorthand. The earlier symbols used

TABLE 1-3 SOME ELEMENTS AND THEIR SYMBOLS

Element	*Symbol*	*Element*	*Symbol*
Aluminum	Al	Mercury (hydrargyrum)	Hg
Barium	Ba	Nitrogen	N
Bromine	Br	Oxygen	O
Calcium	Ca	Phosphorus	P
Carbon	C	Platinum	Pt
Chlorine	Cl	Potassium (kalium)	K
Copper (cuprum)	Cu	Radium	Ra
Hydrogen	H	Silver (argentum)	Ag
Iodine	I	Sodium (natrium)	Na
Iron (ferrum)	Fe	Sulfur	S
Lead (plumbum)	Pb	Uranium	U
Magnesium	Mg	Zinc	Zn

by the alchemists were associated with heavenly bodies and were used to keep their discoveries secret. The modern system of atomic symbols has been kept as simple as possible to achieve common understanding of chemical reactions by all chemists. Some elements are represented by the first letter of their name; thus O stands for oxygen, N for nitrogen, C for carbon, and H for hydrogen. Since the names of several elements have the same first letter, in some instances another identifying letter has been added to distinguish these elements, for example, Ca for calcium, Ba for barium, Cl for chlorine, and Br for bromine. Some of the elements were known in ancient times and were given Latin names, since that language was then in more common usage. The symbols for these elements are derived from the Latin instead of the English name. **Cuprum** is the Latin name for copper and the symbol is Cu, iron is represented by the symbol Fe from the Latin **ferrum,** and **plumbum** is the Latin name for lead and the symbol is Pb. These three elements were used by the ancient Romans.

The symbols for some of the important elements are given in Table 1-3. A complete listing is found inside the book cover. When the symbol is derived from the Latin name of the element, the Latin name is given in parentheses.

MOLECULES

In Dalton's atomic theory it is stated that the atom may be considered the smallest unit of an element that can take part in a chemical change. It is further stated that in chemical changes the atoms can combine to form small particles of compounds, or they can separate or change places in these compounds. Each small particle of a compound contains a definite number of atoms. This means that each unit particle of a compound must have the same number and kinds of atoms as all the other unit particles. These small unit particles of which every compound is composed are called **molecules.** The molecule can be considered the indivisible unit for compounds, much as the atom is the unit particle for elements.

The simplest compound would be one whose molecule contains 1 atom

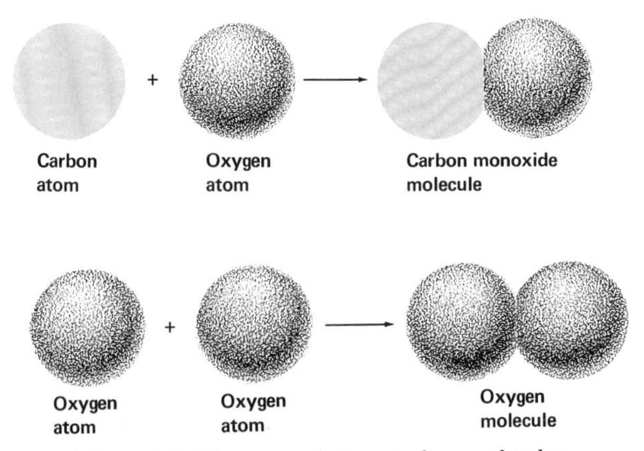

Figure 1-3 The union of atoms to form molecules.

of each of the 2 elements that unite to form the compound. An example of such a simple compound is carbon monoxide, whose molecule is composed of 1 atom of carbon and 1 atom of oxygen (Fig. 1–3). Every molecule of water contains 2 atoms of hydrogen and 1 atom of oxygen.

The majority of molecules are composed of 2 or more different atoms, but some atoms of the same element are capable of uniting with each other to form a molecule of that element. This is particularly true of elements that are gases at room temperature (oxygen, hydrogen, nitrogen, and chlorine). Such gases always exist in molecular form when in the free state, each molecule containing 2 atoms of the element. For example, a molecule of oxygen is represented in Figure 1–3.

ATOMIC AND MOLECULAR WEIGHTS AND THE MOLE CONCEPT

Although it has never been possible to weigh a single atom of an element or a single molecule of a compound, it is possible to obtain the weights of equal numbers of different atoms or molecules and thus the relative weights of individual atoms or molecules. For example, by weighing the same number of hydrogen, oxygen, and sulfur atoms, it was found that the oxygen atoms weigh approximately sixteen times as much as the hydrogen atoms and one-half as much as the sulfur atoms. After several trials and errors, oxygen was assigned an atomic weight of 16 and was established as the reference element for the purpose of assigning atomic weights to all the elements. Atomic weights assigned in this manner have been used satisfactorily by chemists for many years. The physicists have used a slightly different scale based on a particular isotope of oxygen (isotopes will be discussed in Chapter 2). In 1961, however, the International Union of Pure and Applied Chemistry and the International Union of Pure and Applied Physics both agreed to use the same atomic weight values, based on carbon 12 = 12.0000 as the standard.

The actual masses of atoms are very small compared to common mass units and thus is it convenient to introduce the **atomic mass unit, amu,** which is $\frac{1}{12}$ the mass of one carbon-12 atom. The mass of any atom can be expressed simply as the relative atomic weight in units of amu. For practical reasons, the most common mass unit used with the relative atomic weight scale is the gram. When the relative atomic weight of an element is given in units of grams, it is known as the **gram atomic weight.** A gram atomic weight of oxygen is 16 grams, whereas that of carbon is 12 grams. A gram atomic weight of each of these elements contains the same number of atoms. That number, traditionally referred to as **Avogadro's number,** has been experimentally determined in several ways to be 6.023×10^{23} [*]. Avogadro's number of particles is referred to as a **mole.** Thus, a mole of atoms weighs one gram atomic weight. The terms **mole** and **gram atomic weight** are differentiated easily if the former is thought of as a specific number of things (e.g., atoms, molecules, particles), whereas the latter is thought of as a given quantity of matter (i.e., mass). The mass of a carbon-12 atom in units of grams is calculated to be

$$\frac{12.0000\text{g C}}{6.023 \times 10^{23}\text{ atoms C}} = \frac{1.990 \times 10^{-23}\text{g C}}{\text{atom C}}$$

The mass of one million oxygen atoms is

$$10^{6}\ \cancel{\text{atoms O}} \times \frac{16.0\text{g O}}{6.023 \times 10^{23}\ \cancel{\text{atoms O}}} = 2.66 \times 10^{-17}\text{g O}$$

In this calculation one million oxygen atoms is multiplied by a **conversion factor** which relates the mass of a mole of oxygen atoms to the number of atoms in a mole. Using the same calculation technique, it is seen that the number of atoms in 96 grams of nitrogen is

$$96\text{g}\ \cancel{\text{N}} \times \frac{6.023 \times 10^{23}\text{ atoms N}}{14\text{g}\ \cancel{\text{N}}} = 4.13 \times 10^{24}\text{ atoms N}$$

The number of moles of helium, which has a mass of 26 grams, is

$$26\text{g}\ \cancel{\text{He}} \times \frac{1\text{g}\ \cancel{\text{atomic wt He}}}{4\text{g}\ \cancel{\text{He}}} \times \frac{1\text{ mole He}}{1\text{g}\ \cancel{\text{atomic wt He}}} = 6.5\text{ moles He}$$

In this calculation technique, called the **factor dimensional method,** conversion factors are applied to reach the desired units. The conversion factor applied must have a numerator and a denominator which are equivalent by definition. The conversion factor must be applied in such a way that after cancellation of units the desired units remain.

If the composition of a compound and the atomic weights of its constituent atoms are known, the molecular weight can be calculated readily. The **molecular weight** of a compound is the sum of the atomic weights of all the atoms present in 1 molecule of the substance. A molecule of carbon dioxide contains 1 atom of carbon and 2 atoms of oxygen. Since the atomic weight of carbon

---

[*] Extremely small values or large values are often represented as a number times ten to a power. For example, 10,000,000,000 is expressed as ten to the tenth power or 10^{10}. The fraction 1/10,000,000,000 can be expressed as 0.0000000001 or $1/10^{10}$ or 1×10^{-10}.

is 12 and the atomic weight of oxygen is 16, the molecular weight may be calculated as follows:

Element		*Weight*	
Carbon		12 amu	
Oxygen	2 × 16 amu =	32 amu	
Carbon dioxide		44 amu	= Molecular weight

A molecule of carbonic acid contains 2 atoms of hydrogen, 1 atom of carbon and 3 atoms of oxygen. Since the atomic weight of hydrogen is ordinarily taken as 1 in chemical calculations, the molecular weight of carbonic acid is 62 as shown by the calculation:

Element		*Weight*	
Hydrogen	2 × 1 amu =	2 amu	
Carbon		12 amu	
Oxygen	3 × 16 amu =	48 amu	
Carbonic acid		62 amu	= Molecular weight

Each atom of hydrogen, carbon, or oxygen in these compounds has a weight of its own, and this weight must be used in a calculation of the molecular weight each time the atom occurs in the compound.

If the molecular weight of a compound is given in units of grams, it is known as the **gram molecular weight.** Thus carbon dioxide has a gram molecular weight of 44g; carbonic acid, 62g; and water, 18g. A gram molecular weight of a compound contains the same number of molecules as a gram molecular weight of any other compound, namely, a mole, or 6.023×10^{23} molecules. The weight of a mole of molecules is one gram molecular weight.

FORMULAS

A **formula** expresses in symbols the composition of a substance. Since compounds are composed of atoms combined in definite proportions, they can be represented by a combination of the symbols of the atoms. A molecule of hydrogen chloride is composed of 1 atom of hydrogen and 1 atom of chlorine and is represented by the formula HCl. Where there is more than 1 atom of the same kind in the **formula unit** or in the molecule, the symbol is not repeated, but the number of atoms is indicated as a subscript to the symbol for the element. For example, the formula for water is written H_2O, meaning that 1 molecule of water contains 2 atoms of hydrogen and 1 atom of oxygen. The small subscript 2 which follows the H indicates that 2 atoms are present in the molecule. A molecule of sulfuric acid consists of 2 atoms of hydrogen, 1 atom of sulfur, and 4 atoms of oxygen. Its formula may be written as H_2SO_4. This formula represents 1 molecule, 1 molecular weight, 1 gram molecular weight, or 1 mole of sulfuric acid. The gram molecular weight equals the sum of the gram atomic weights of the atoms in the molecule, or 2 gram atomic weights of hydrogen plus 1 gram atomic weight of sulfur plus 4 gram atomic weights of oxygen. One formula unit of sodium carbonate is composed of 2 atoms of sodium, 1 atom of carbon, and 3 atoms of oxygen; the formula is Na_2CO_3. Again it should be observed that the subscript numbers are written after the atomic symbols to which they belong. Since the atoms of most gases do not exist by themselves, but unite to form molecules of the gas, the molecule of hydrogen is written H_2, of nitrogen N_2, and of oxygen O_2.

To designate more than 1 molecule of a substance, the appropriate number is placed in front of the formula. The term $3BCl_3$ represents 3 molecules of boron chloride, and $2O_2$ represents 2 molecules of oxygen gas. To avoid confusion of the subscript numbers and the numbers written in front of the formulas, it would be wise for the beginner to write the full meaning of the different terms. For example, $3H_2SO_4$ could read 3 molecules of sulfuric acid, each molecule consisting of 2 atoms of hydrogen, 1 atom of sulfur, and 4 atoms of oxygen.

The mass of 0.46 gram molecular weights of H_2SO_4 is calculated to be

$$0.46\cancel{\text{g molecular wt } H_2SO_4} \times \frac{98\text{g } H_2SO_4}{1\cancel{\text{g molecular wt } H_2SO_4}} = 45\text{g } H_2SO_4$$

The number of formula units in 28g of Na_2CO_3 is

$$28\cancel{\text{g } Na_2CO_3} \times \frac{1\cancel{\text{g molecular wt } Na_2CO_3}}{106\cancel{\text{g } Na_2CO_3}} \times \frac{1\cancel{\text{ mole } Na_2CO_3}}{1\cancel{\text{g molecular wt } Na_2CO_3}} \times \frac{6.023 \times 10^{23} \text{ formula units } Na_2CO_3}{1\cancel{\text{ mole } Na_2CO_3}}$$

$$= 1.59 \times 10^{23} \text{ formula units } Na_2CO_3$$

In this calculation the first and second conversion factors could have been combined since 1 mole of Na_2CO_3 has a mass of 106g.

The mass of H_2O which contains the same number of oxygen atoms as 2.5 grams of H_2SO_4 is

$$2.5\cancel{\text{g } H_2SO_4} \times \frac{1\cancel{\text{ mole } H_2SO_4}}{98\cancel{\text{g } H_2SO_4}} \times \frac{4\cancel{\text{ mole O}}}{1\cancel{\text{ mole } H_2SO_4}} \times \frac{1\cancel{\text{ mole } H_2O}}{1\cancel{\text{ mole O}}} \times \frac{18\text{g } H_2O}{1\cancel{\text{ mole } H_2O}}$$

$$= 1.84\text{g } H_2O$$

The formula of a chemical entity is correctly interpreted as a qualitative and quantitative statement of composition. For elements the symbol states qualitatively the identity of the atoms, whereas its quantitative interpretation involves numbers of atoms. The symbol of an element is commonly used to represent quantitatively either 1 atom or 1 mole of atoms as determined by the stated convention.

The formula of a chemical compound states qualitatively the identities of the atoms comprising the compound. In addition the formula states the relative number of each type of atom in the compound. The formula may state the actual number of each atom per molecule, or per formula unit, or the actual number of moles of each type of atom per mole of molecules or formula units. In this statement molecules and formula units are differentiated primarily as covalently bonded species and ionic species, respectively. The bonds in these compounds are discussed in Chapter 3. Covalently bonded species, which exist as discrete molecules under normal conditions, include CO_2, H_2 and H_2O, as examples. Ionic compounds and covalent network solid compounds do not exist as discrete "small" molecules under normal conditions, but have structures involving extensive three dimensional lattices as discussed in Chapter 4. For example, sodium chloride exists as a crystalline solid in which each sodium ion is surrounded by six chloride ions, and each chloride ion is surrounded by six sodium ions. It is impossible to identify a discrete sodium chloride "molecule" in this structure, and yet the formula NaCl qualitatively and quantitatively describes the composition. For this reason NaCl is referred to as a formula unit, and 6.023×10^{23} NaCl as a mole of formula units.

The **empirical formula** of a compound states the relative numbers of atoms of each element in the compound, whereas the **molecular formula** states the actual numbers of atoms of each element in a molecule of the compound. The empirical and molecular formulas are the same for some compounds, including HCl, CO_2, H_2O, and BF_3; but for many compounds the empirical formula and molecular formula are different as indicated in the following tabulation:

Empirical formula	*Molecular formula*
BCl_2	B_2Cl_4
P_2O_5	P_4O_{10}
NH_2	N_2H_4
CH_3O	$C_2H_6O_2$
CH	C_2H_2, C_6H_6

The empirical formula or the molecular formula of a compound can be used to calculate the mass per cent composition as follows:

What is the mass per cent of each element in the compound $SOCl_2$?

One mole of $SOCl_2$ contains;

$$\begin{array}{rl} & 32\text{g S} \\ & 16\text{g O} \\ 2 \times 35.5 = & \underline{71\text{g Cl}} \\ & 119\text{g/mole} \end{array}$$

$$\% \text{ S} = \frac{32\text{g S}}{119\text{g SOCl}_2} \times 100 = 13.44\% \text{ S}$$

$$\% \text{ O} = \frac{16\text{g O}}{119\text{g SOCl}_2} \times 100 = 26.88\% \text{ O}$$

$$\% \text{ Cl} = \frac{71\text{g Cl}}{119\text{g SOCl}_2} \times 100 = 59.68\% \text{ Cl}$$

In order to determine the empirical formula of a new compound, it is necessary to determine experimentally the relative masses of each element in the compound. If the relative masses of each component of a compound are known the empirical formula is determined by a technique illustrated in the following calculations:

A 15g sample of a compound contains 7.0g Si and 8.0g O. What is the empirical formula of this compound?

In 15g of this compound the number of moles of each element is

$$7.0\text{g}\ \cancel{\text{Si}} \times \frac{1 \text{ mole Si}}{28.09\text{g}\ \cancel{\text{Si}}} = 0.250 \text{ moles Si}$$

$$8.0\text{g}\ \cancel{\text{O}} \times \frac{1 \text{ mole O}}{16.00\text{g}\ \cancel{\text{O}}} = 0.500 \text{ moles O}$$

The empirical formula expresses the relative numbers of atoms of each element or the relative numbers of moles of atoms of each element. The calculation indicates the relative number of moles of each element is 0.250 moles Si to 0.500 moles O or, expressed in small whole numbers, 1 mole Si to 2 moles O. The empirical formula is thus SiO_2. The **empirical formula weight** is 60 amu (28 + 2 × 16).

A compound is found to contain 75.0 per cent C, 7.5 per cent H and 17.5 per cent N. What is its empirical formula?

A 100g sample of this compound would contain 75.0g C, 7.5g H, and 17.5g N. In 100g of the compound the number of moles of each element would be

$$75.0\text{g}\,\cancel{\text{C}} \times \frac{1 \text{ mole C}}{12.0\text{g}\,\cancel{\text{C}}} = 6.25 \text{ moles C}$$

$$7.5\text{g}\,\cancel{\text{H}} \times \frac{1 \text{ mole H}}{1.0\text{g}\,\cancel{\text{H}}} = 7.5 \text{ moles H}$$

$$17.5\text{g}\,\cancel{\text{N}} \times \frac{1 \text{ mole N}}{14.0\text{g}\,\cancel{\text{N}}} = 1.25 \text{ moles N}$$

This calculation indicates that the relative number of moles of each element is 6.25 moles C to 7.5 moles H to 1.25 moles N, or 5 moles C to 6 moles H to 1 mole N. The empirical formula is thus C_5H_6N and the empirical formula weight is 80 amu (5 × 12 + 6 × 1 + 14).

In order to establish the relationship between the empirical formula and the molecular formula, it is necessary to know that the molecular weight will always be a whole number multiple of the empirical formula weight. The whole number multiple which relates these formulas is the molecular weight divided by the empirical formula weight. The following calculation illustrates this:

A compound having the empirical formula CH has a molecular weight of 78 amu. What is the molecular formula?

The empirical formula weight is 12 + 1 = 13 amu.

$$\frac{\text{molecular weight}}{\text{empirical formula weight}} = \frac{78 \text{ amu}}{13 \text{ amu}} = 6$$

The molecular formula is determined by multiplying each of the subscripts in the empirical formula by 6; thus, the molecular formula is C_6H_6.

Questions

1. Explain the relation to each other of the three laws of chemical change.

2. List in outline form the essential points of Dalton's atomic theory.

3. Name several types of energy. Which form of energy is most easily measured?

4. Approximately how many calories of heat would be required to raise the temperature of a quart of water from 10° to 20°C?

5. The straightway speed of an Indianapolis 500 racing car is 180 miles per hour. What speed would this correspond to in kilometers per hour?

6. A popular professional baseball player can command an annual salary equivalent to his weight in gold. If gold sells at $1000 a kilogram and he weighs 176 pounds, what is his annual salary?

7. A child appeared to be running a high fever. The only thermometer available was calibrated in degrees centigrade and gave a reading of 40°C. What was the child's temperature in degrees Fahrenheit?

8. How would you define a molecule?

9. Why was it necessary to choose a reference element for atomic weights?

10. How would you distinguish between compounds and mixtures? Name a compound that contains oxygen and a mixture that contains oxygen.

11. Indicate below the items totally unrelated to one mole of $CaCO_3$.
(a) 100 grams/mole
(b) one formula-unit
(c) 3 gram-atoms of oxygen
(d) 6.023×10^{23} molecules of $CaCO_3$
(e) one gram molecule
(f) 6.023×10^{23} atoms of Ca

12. What is the molecular weight of SO_2Br_2? of $(NH_2)_2POCl$?

13. One formula-unit of $PbSO_4$ contains which of the following?
(a) one atom of Pb
(b) one gram-atom of Pb
(c) one mole of sulfur
(d) one mole of $PbSO_4$

14. How many moles are contained in 569 grams of sodium? How many atoms?

15. What is the mass of 7.52×10^{25} atoms of carbon?

16. Calculate the mass of 0.400 moles of H_2S.

17. How many grams of nitrogen are there in 8.2 gram atomic weights of nitrogen?

18. Find the percentage composition of each of the following compounds:
(a) H_2O
(b) Na_2SO_4
(c) $Ca(CN)_2$
(d) NO_2

19. An analysis of a compound indicates that the mass % composition is: 22.77% Na; 21.78% B; and 55.45% O. Find the empirical formula.

20. Given the empirical formula and molecular weight, find the molecular formulas of each of the following:
(a) CH_2O, 88 amu
(b) P_2O_5, 284 amu
(c) CH_2, 84 amu
(d) $AlCl_3$, 267 amu

Suggested Reading

Bauman: Can Matter Be Converted to Energy? Journal of Chemical Education, Vol. 43, p. 366, 1966.
Dinga: The Elements and the Derivation of their Names and Symbols. Chemistry, Vol. 41, No. 2, p. 20, 1968.
Guggenheim: The Mole and Related Quantities. Journal of Chemical Education, Vol. 38, p. 86, 1961.
IUPAC Adopts a New Look in Atomic Weights. Chemistry and Engineering News, Vol. 48, No. 4, p. 38, 1970.
Kieffer: The Mole Concept in Chemistry. New York, Reinhold Publishing Co., 1962.
Margolis: Formulation and Stoichiometry. New York, Appleton Century Crofts, 1968.
Socrates: SI Units, Journal of Chemical Education, Vol. 46, p. 710, 1969.

CHAPTER 2

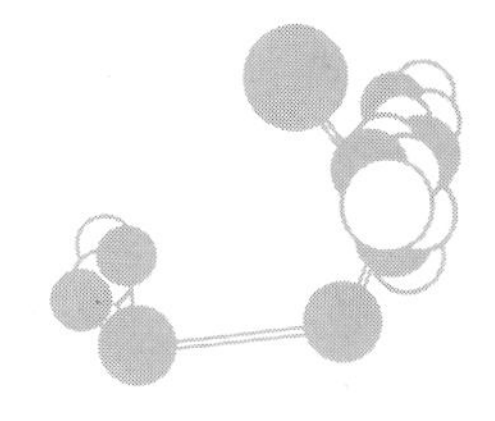

ATOMIC STRUCTURE AND PERIODICITY OF THE ELEMENTS

During the nineteenth century, chemists accepted Dalton's atomic theory and were extremely productive in uncovering weight relationships and establishing the composition of compounds. They accepted the atom as a simple entity incapable of changing its character during a chemical reaction. The atom was thought to be the fundamental particle of matter, and the actual nature of the atom was not of concern until the end of the nineteenth century. By then the results of Michael Faraday's studies of electrical current passed through liquids and solutions were interpreted as an indication that the components of matter were electrical in nature.

SUBATOMIC PARTICLES

As an aid in understanding the complexities of atomic structure, it is helpful to first consider the units, or subatomic particles, involved in the construction of an atom.

Electrons. The first of the subatomic particles to be discovered was the **electron.** In 1859 it was observed that the application of electricity to electrodes in a highly evacuated glass tube, a **cathode-ray tube,** produced peculiar radiations from the negative electrode, or **cathode** (Fig. 2–1). If the tube was placed between electrically charged plates, the rays were deflected toward the positively charged plate. In addition, the rays were deflected when a magnetic field was placed about the tube.

From these observations, cathode rays were described as a stream of

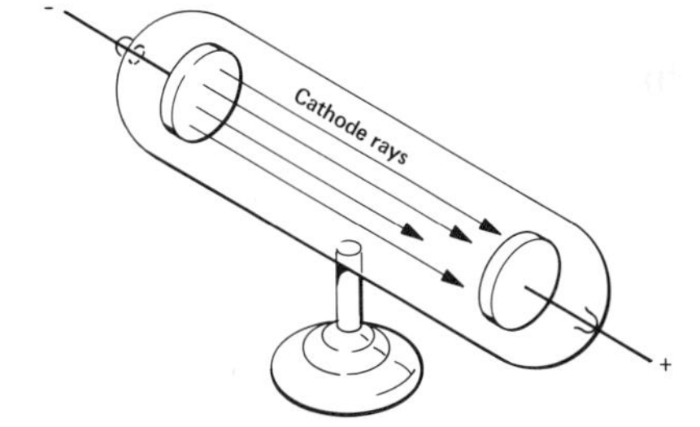

Figure 2-1 Cathode rays emanating from the negative electrode.

negatively charged particles ejected from the cathode by the electric current. Since the same particles were given off by cathodes made of many different elements, and also when different gases were present in the tube, they were considered as constituent parts of the atoms of all elements.

By 1900 these negatively charged particles had been named **electrons,** and their mass and charge had been accurately determined. The mass of an electron is 9.1×10^{-28} grams, or about 1/1837 of the mass of the hydrogen atom, which is the lightest of all atoms. The charge on the electron is -4.8×10^{-10} electrostatic units, and since this is the smallest unit charge of electricity known, it is referred to as -1.

Protons. From common observation it has been assumed that all substances are electrically neutral. For example, a piece of metal, rock, or wood does not convey an electric shock when touched, nor does it indicate a flow of electric current when attached to delicate measuring instruments. For this reason, it probably can also be assumed that all substances or compounds are composed of an equal number of positive and negative electrical units.

More extensive observations with vacuum tubes containing perforated cathodes revealed the presence of positive rays that passed through the holes, or canals, in the cathode. These positive or canal rays were formed from the small number of gas molecules left in the evacuated tube. When the stream of electrons in the cathode ray struck these electrically neutral gas molecules, the impact split off some of the external electrons, leaving a positively charged particle called an **ion.** The cathode immediately attracted the positively charged units, and, if the cathode was perforated, they passed through it, traveling in the direction opposite to that of the cathode rays (Fig. 2-2). The positive or **canal rays** had properties similar to those of the cathode rays—although the mass varied with the nature of the gas in the tube—and the charge on the particles was always $+4.8 \times 10^{-10}$ electrostatic units or a simple multiple of this value. When hydrogen gas was in the tube, the mass of the particles was

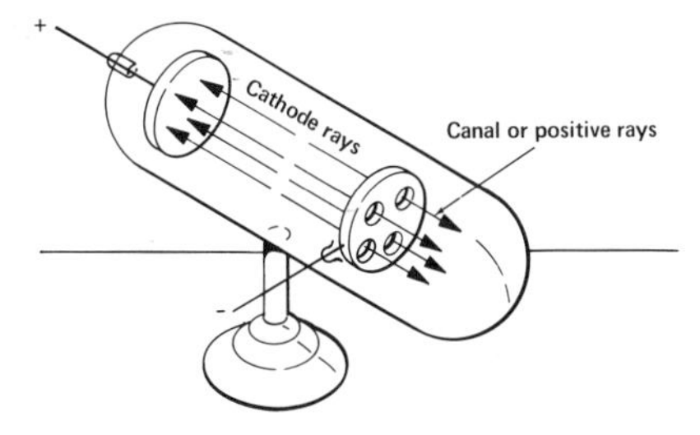

Figure 2-2 An illustration of the canal or positive rays formed from the residual gas in a cathode ray tube.

TABLE 2–1 MASSES AND CHARGES ON SUBATOMIC PARTICLES

Particle	*Mass (amu)*	*Charge*
Electron	0.00055	−1
Proton	1.00759	+1
Neutron	1.00898	0

found to be practically equal to that of the hydrogen atom (1.66×10^{-24} grams). Since one of these particles represented the lightest unit of mass with a positive charge of 4.8×10^{-10} electrostatic units, it was established as a fundamental unit of atomic structure and called a **proton.** An electrical charge of 4.8×10^{-10} electrostatic units has been adopted as the magnitude of a unit charge for the electron and the proton. Common usage refers to a charge of −1 for the electron and +1 for the proton.

Neutrons. For several years the electron and proton were thought to be the only subatomic particles. Until 1932, in order to account for the structure of atoms, a combination of protons and electrons was thought to exist in the nucleus. Such a combination would be electrically neutral and possess a mass approximately equal to that of a proton. In 1932 the discovery of a new structural unit, the **neutron,** increased our understanding of the nucleus. This new particle is not electrically charged, and thus it was not observed in studies on cathode-ray or positive-ray vacuum tubes.

It is convenient to think of the neutron as a combination of a proton and an electron. Recent research, in fact, strongly suggests that this combination exists, and that the mass of the neutron is equal to the sum of the masses of a proton and an electron. The neutron also has been shown to be unstable when in the free state.

Summary of Knowledge of Subatomic Particles. Before proceeding with a discussion of the structure of the atom, it may be well to summarize our knowledge of the three fundamental subatomic particles. These units are: (1) **electrons,** which are negatively charged particles 1/1837 of the mass of the hydrogen atom; (2) **protons,** which are positively charged particles of mass one, and which may be considered hydrogen atoms minus their one electron; (3) **neutrons,** which are uncharged or electrically neutral, particles of mass one, and which may be considered to be composed of a proton plus an electron. More precise values for the masses of these three particles are shown in Table 2–1.

THE STRUCTURE OF ATOMS

Shortly after the discovery of protons in 1911 several investigators were attempting to describe the structure of atoms. One type of experiment involved the bombardment of very thin sheets of gold foil with **alpha particles,** which are helium ions with a mass of 4 amu, and a charge of +2. It was known that alpha particles are given off by radium, and that they are positively charged and possess a small mass. When these particles struck the gold foil, several phenomena were observed. Some of the particles passed through the foil and continued in a straight line; others passed through but had their paths altered; in addition, others were deflected from the surface of the foil. The British

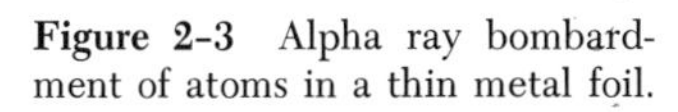
Figure 2-3 Alpha ray bombardment of atoms in a thin metal foil.

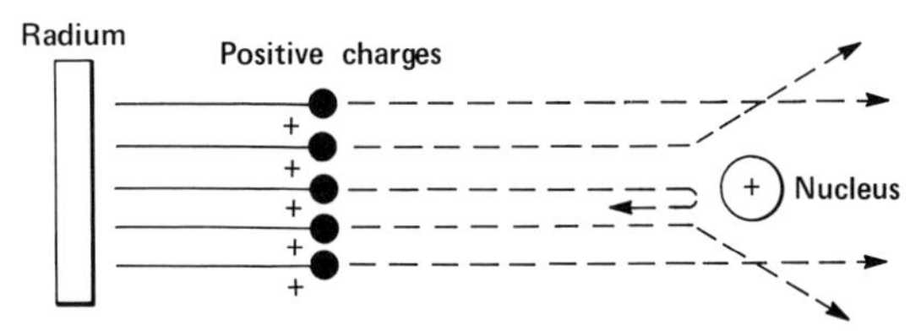

physicist Rutherford explained this scattering of alpha particles by suggesting that the atoms of the metal foil consist mostly of space, with a small, positively charged, heavy nucleus surrounded by electrons at relatively large distances from the nucleus. Most of the alpha particles pass through the empty space of the atoms, but a few are deflected from their original paths. Deflection of the alpha particles occurs when they come too close to the positively charged nucleus and are repelled by it (Fig. 2-3).

The nucleus contains the protons and neutrons, while the electrons are arranged in a series of shells at varying distances from the nucleus. Some conception of the relatively enormous space that exists between electrons and the nucleus of an atom may be gained from the following example. If the electrons and nucleus of each atom from every compound and substance on earth could be stripped of their space and combined in a dense, compact mass, the resultant body would be a sphere ½ mile in diameter. This sphere would weigh as much as the earth and would possess such a high density that a cube of the material one centimeter on a side would weigh about 100 million tons. We may, then, summarize our concept of the atom as a particle of matter in which a cloud of electron density is located about a small, dense mass of protons and neutrons.

The Nucleus. The nucleus, which accounts for practically all the mass of the atom, is composed essentially of protons and neutrons. In size its diameter is approximately 1/10,000 that of the atom. As stated earlier, the proton represents a unit charge of positive electricity, in contrast to the electron, which is the unit charge of negative electricity. The mass of the proton is similar to that of the hydrogen atom and is taken as one unit of atomic mass. The neutron has a charge of zero and a mass approximately the same as that of the proton. The number of positively charged protons in the nucleus is equal to the **atomic number, Z**, of the atom, which is also equal to the number of negatively charged electrons outside the nucleus. Since each proton or neutron contributes one unit to the weight of an atom, its **atomic weight** is equal to the total number of protons and neutrons in the nucleus.

ISOTOPES

For many years it was believed that all atoms of a given element had the same mass. The atoms of a few elements produced only a single spot on the photographic plate in a special type of cathode-ray tube called a **mass spectrograph,** thus indicating that all the atoms of this element had the same mass. But other experiments with the mass spectrograph indicated that some elements contained atoms of a weight differing from that originally determined by other methods. In 1932 Harold Urey discovered some hydrogen with a mass approximately twice that of the mass of ordinary hydrogen. This heavy hydrogen was called **deuterium,** and water containing deuterium in place of hydrogen was called **heavy water.** Atoms of the same element but possessing different masses are called **isotopes.** The existence of many isotopes of different elements was proved by the use of the mass spectrograph. Nearly all elements possess at least two isotopes, while lead, for example, possesses 14 isotopes, whose atomic masses vary from 203 to 216. The atomic weight of an element reported in tables of atomic weights represents the average weight of all of its atoms.

Isotopes of an element can be more readily explained on the basis of the theory of atomic structure. For example, ordinary hydrogen has a nucleus containing 1 proton surrounded by a shell containing 1 electron, whereas deuterium has a nucleus that contains 1 proton and 1 neutron, accounting for its atomic weight of 2. Another example of an element with two naturally occurring isotopes is chlorine. One isotope contains 17 protons and 18 neutrons, and the second has 17 protons and 20 neutrons. The naturally occurring mixture of these two isotopes contains 23 per cent of the heavier isotope, resulting in an observed atomic weight of 35.46 amu.

On the basis of atomic structure, isotopes can be defined as atoms that have the same atomic number but different atomic weights. This means that isotopes have different numbers of neutrons in their nuclei. They have the same number of protons and electrons, however, and behave alike chemically. About 300 different isotopes similar to those of hydrogen and chlorine have been found occurring in natural mixtures.

THE ELECTRONIC STRUCTURE OF ATOMS

Having discussed the gross structure of the atom, it is now necessary to discuss the details of the positions, energies, and properties of the electrons in atoms. Since the electrons of an atom are determining factors of, and are involved in, all chemical reactions, it is necessary to consider the factors which influence their positions and energies. Just as in the case of Rutherford's studies of the structure of the atom, other experiments also can be used to test proposed models for the electronic structure of an atom.

The Bohr Atom. If the light emitted by atoms of thermally or electrically excited elements is passed through a prism in a **prism spectrograph** (Fig. 2–4), it appears as a series of lines referred to as an **emission spectrum** (Fig. 2–5).

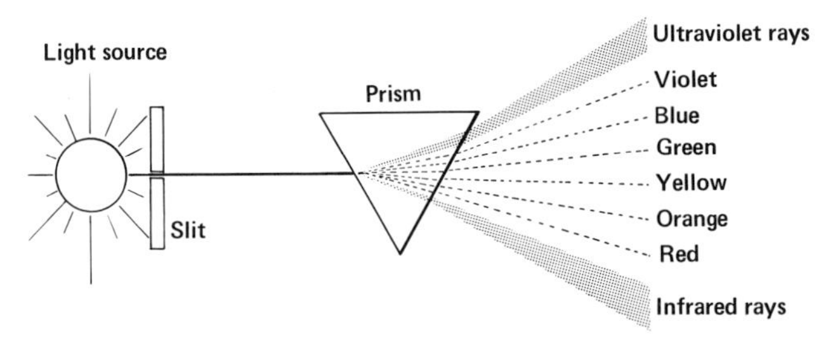

Figure 2–4 Schematic representation of a prism spectrograph.

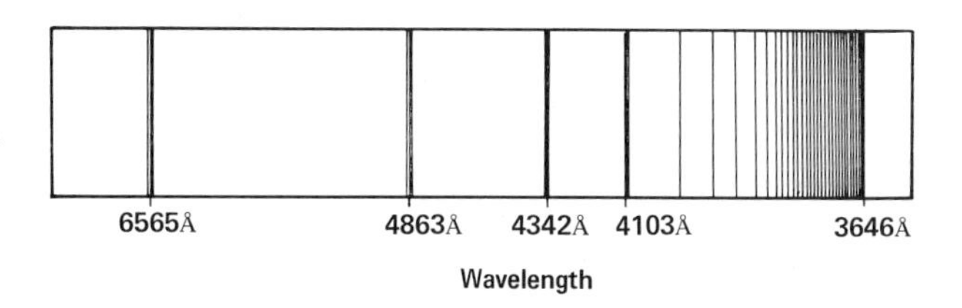

Figure 2–5 A portion of the emission spectrum of the hydrogen atom.

Each of these lines corresponds to light of a given energy, indicating that atoms cannot emit light of all possible energies but only of certain levels. The lines in the spectrum of a given element are found to exhibit a regular pattern.

In 1913 Niels Bohr proposed an explanation for the line spectra, and a model for the electronic structure of the hydrogen atom, consistent with the experimental facts. He assumed a Rutherford nuclear atom in which the electron moves about the nucleus in circular orbits of given radii. This means that the kinetic energy, or energy of motion, of the electron in the atom, which is directly related to the radius of the orbit, can assume only certain discrete values. That is, the possible energies of the electron are **quantized.** Bohr suggested that the single electron in hydrogen during thermal or electrical excitation can add only discrete quantities of energy and thus shift from one energy level to another. When the electron goes from a higher energy level to one of lower energy, it emits light of an energy corresponding to the difference between energy levels. That is, the energy emitted or absorbed, ΔE, is $E_{upper} - E_{lower}$. The energy levels are assigned **quantum numbers,** *n*, ranging from 1 to ∞, the relative energies of which are displayed in the energy level diagram in Figure 2–6. The **quantum levels** associated with the quantum numbers are often referred to as **"shells,"** or **"orbits,"** and are lettered K, L, M, N, and so forth, rather than $n = 1, 2, 3, 4, \ldots$. In the hydrogen atom, the radius of the "Bohr orbit" associated with a given quantum number, *n*, is $0.53n^2$ Å. The size of the "shell," or orbit, increases with increasing *n*.

The Bohr theory of the hydrogen atom was truly a breakthrough in the attempt to find a suitable quantitative model for the atom. It quantitatively explained the emission spectrum of hydrogen by predicting the exact energies of the light emitted. In addition, it successfully forecast many other experimentally measured properties of hydrogen. However, the theory could not be successfully extended to other elements, and consequently it was abandoned for another model capable of incorporating elements of higher atomic number.

The Wave-Mechanics Atom. In 1924 Louis De Broglie proposed that electrons could have wave-like properties as well as particle properties. This

Figure 2–6 Energy levels for the hydrogen atom.

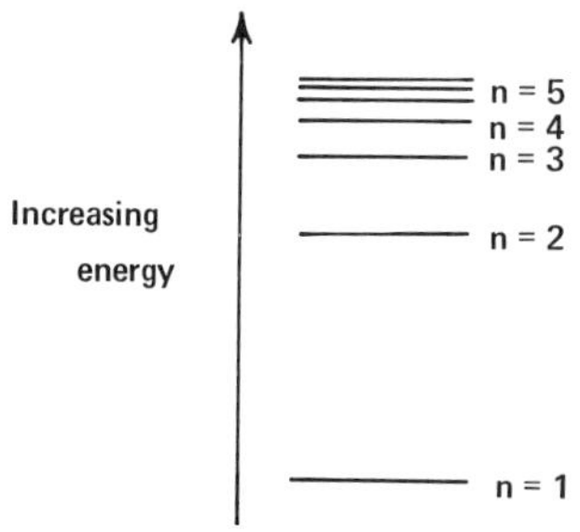

proposal was experimentally verified by Davisson and Germer in 1927. In 1926 Erwin Schrödinger assumed that the motion, position, energy and other properties of an electron in an atom could be explained by using an equation which assumed wave-like properties for the electron. The solutions to the wave equation do not give the exact position of the electron when it has a given energy, but they do predict the **probability** of finding the electron in a given position when it has a given energy. These positions, called **wave functions,** or more commonly **orbitals,** are distinguished from one another by a set of special numbers, the **quantum numbers.** The quantum numbers used to describe the orbitals are as follows:

1. n, the **principal quantum number,** which corresponds to the quantum number n in the Bohr atom and is generally related to the average distance of the electron from the nucleus. The electrons with a given n are said to occupy a given quantum level, the energy of which increases with increasing n. The values of n are 1, 2, 3, 4, . . . ∞.

2. ℓ, the **azimuthal quantum number,** which is related to the shapes of the orbitals. The values of ℓ are 0, 1, 2, 3, . . . $n - 1$; that is, the highest ℓ value for a given n is $n - 1$. For a given n, there are n values of ℓ. The orbitals with ℓ values of 0, 1, 2, and 3 are commonly referred to as s, p, d, and f orbitals, respectively. The orbital with $n = 1$ and $\ell = 0$ is denoted 1*s*, whereas the orbital with $n = 2$ and $\ell = 1$ is 2*p*. The relative energies of the orbitals in a given quantum level increase with increasing ℓ values, as is indicated in the energy level diagram in Figure 2–7.

3. m, the **magnetic quantum number,** is related to the orientation of a given orbital in a magnetic field. In a magnetic field, orbitals of a given ℓ value have $2\ell + 1$ possible orientations, and thus there are $2\ell + 1$ m values for each ℓ value. These values range from $-\ell$, through zero, to $+\ell$.

One additional quantum number is required to completely describe the electron in an atom. This quantum number which does not arise in the solution

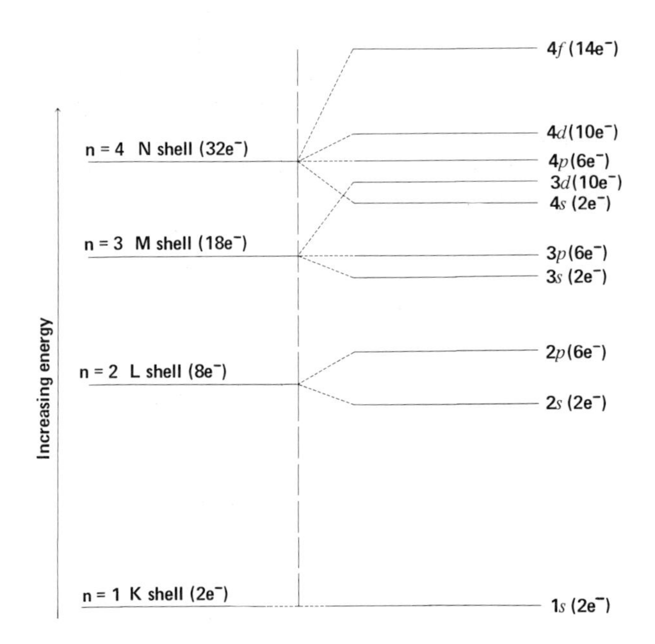

Figure 2–7 Energy levels and sublevels in the first four quantum levels or shells.

TABLE 2-2 QUANTUM NUMBERS FOR THE ELECTRONS IN THE FIRST TWO QUANTUM LEVELS

Quantum Level (n)	*Shell*	ℓ	*Type of Orbital*	m	m_s	*Electrons in Quantum Level*
1	K	0	s	0	$+\frac{1}{2}$	2
		0		0	$-\frac{1}{2}$	
2	L	0	s	0	$+\frac{1}{2}$	6
		0		0	$-\frac{1}{2}$	
		1	p	+1	$+\frac{1}{2}$	
		1		+1	$-\frac{1}{2}$	
		1		0	$+\frac{1}{2}$	
		1		0	$-\frac{1}{2}$	
		1		−1	$+\frac{1}{2}$	
		1		−1	$-\frac{1}{2}$	

of the wave equation is m_s, the **spin quantum number.** The spin quantum number, which is related to the spin of the electron, can have values of $+\frac{1}{2}$ and $-\frac{1}{2}$ for each m value. Each orbital can contain two electrons with opposing spins. That is, the m_s values must have opposite signs. The number of electrons which can be contained in any quantum level, n, is $2n^2$, as the first level can contain 2 electrons, the second level 8 electrons, and the third level 18 electrons.

A set of four quantum numbers, n, ℓ, m, and m_s, completely describes the electron in an atom. The quantum numbers associated with each of the electrons in the first two quantum levels are given in Table 2–2.

The orbital described by the quantum numbers is pictured as an electron cloud in Figure 2–8 which represents the 1*s* orbital. In this cross sectional representation of the electron cloud, the number of dots per unit area is related to the probability of finding the electron in that area, and can be thought of as an **electron density.** In the 1*s*-orbital, the probability is greatest at the center of the atom, the nucleus, and decreases with increasing distance from the nucleus. The most common diagrammatic representation of the orbitals is that defined by a surface which encloses 95 per cent of the probabilities of the electron. This type of diagram is shown in Figure 2–9 for s-, p-, and d-orbitals. Notice that the p-orbitals consist of two **lobes** of electron density on either side of the nucleus, but the electron density at the nucleus is zero. The d-orbitals have even more complex shapes, but they too have zero electron density at the nucleus.

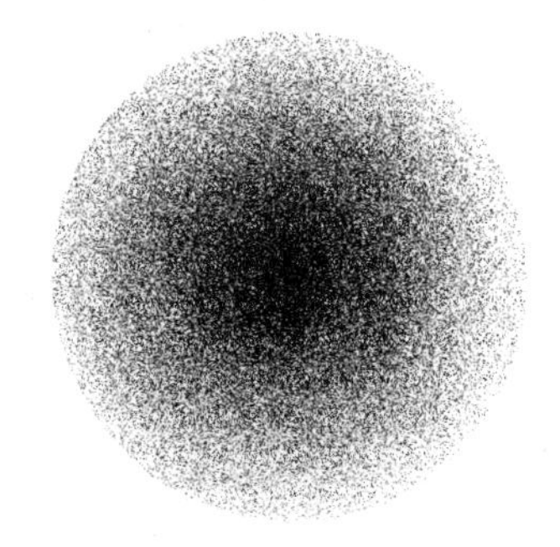

Figure 2–8 The electron cloud representation of an orbital.

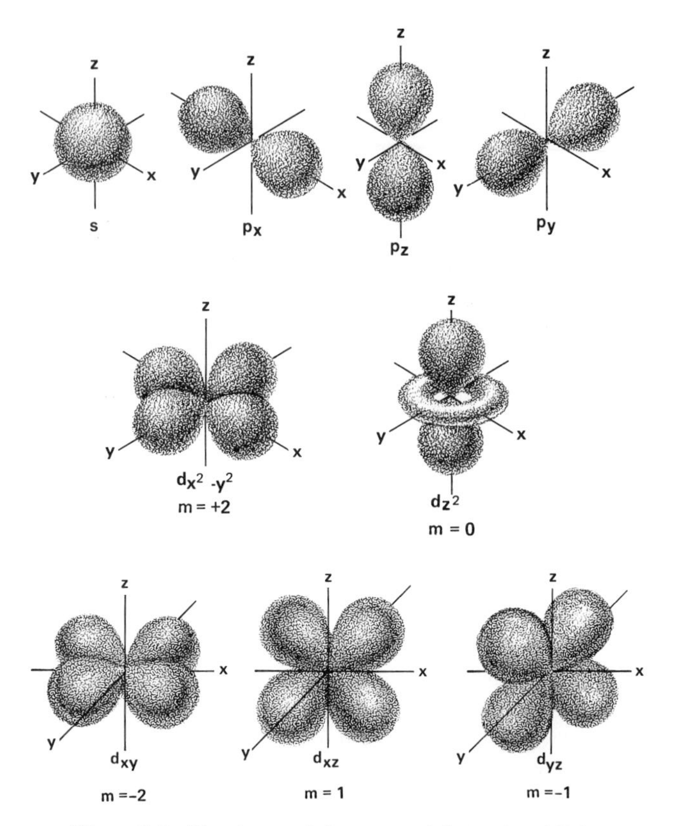

Figure 2–9 The shapes of the s, p, and d atomic orbitals.

The schematic tabulation of the electrons in a given atom is the **electronic configuration.** The electronic configurations of the first ten elements, found in Table 2–3, lead to several fundamental observations. The first of these observations, already mentioned, is that *only two electrons can occupy any orbital.* Also, it is observed that *electrons occupy an orbital only if all orbitals of lower energy are filled.* In addition, it can be seen that *an orbital is not occupied by a pair of electrons until other orbitals of equivalent energy are each occupied by one electron.*

A shorthand method of expressing electronic configuration is by indicating the principal quantum number and the number of electrons in each type of orbital as follows:

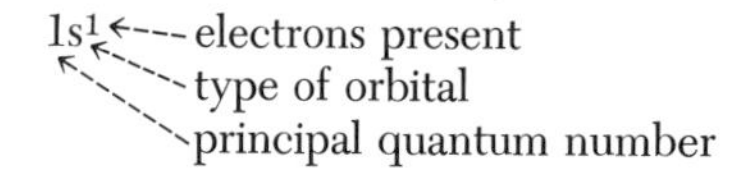

This notation is for the hydrogen atom which has one electron in an s-orbital in the first principal quantum level. The electronic configuration for helium with an atomic number of 2 is $1s^2$. For an atom the sum of the right superscripts always equals the atomic number. For ions the electronic configuration is indicated by simply increasing or decreasing the sum of the right superscripts to indicate the numbers of electrons added or lost.

TABLE 2-3 ELECTRONIC CONFIGURATIONS OF THE FIRST TEN ELEMENTS

	1*s*	2*s*		2*p*		3*s*	
H	①						$1s^1$
He	⑪						$1s^2$
Li	⑪	①					$1s^2\ 2s^1$
Be	⑪	⑪					$1s^2\ 2s^2$
B	⑪	⑪	①				$1s^2\ 2s^2\ 2p^1$
C	⑪	⑪	①	①			$1s^2\ 2s^2\ 2p^2$
N	⑪	⑪	①	①	①		$1s^2\ 2s^2\ 2p^3$
O	⑪	⑪	⑪	①	①		$1s^2\ 2s^2\ 2p^4$
F	⑪	⑪	⑪	⑪	①		$1s^2\ 2s^2\ 2p^5$
Ne	⑪	⑪	⑪	⑪	⑪		$1s^2\ 2s^2\ 2p^6$
Na	⑪	⑪	⑪	⑪	⑪	①	$1s^2\ 2s^2\ 2p^6\ 3s^1$

PERIODIC TABLES

Prior to 1850 about one-half of the elements now known had been discovered. Their chemical and physical properties and their combinations with other elements to form compounds were studied by many chemists. Considerable controversy existed as to the assignment of correct atomic weights for the elements.

As chemists gained more information about chemical and physical properties and atomic weights it was only natural that attempts would be made to discover a possible relationship among the elements. In fact, as early as 1820 an effort was made to separate elements into groups that possessed similar properties. In that year Johann Wolfgang Döbereiner observed that members of a group of three elements such as lithium (at. wt. = 7), sodium (23), and potassium (39) had very similar properties, and that, in addition, the atomic weight of the middle element was about the average of the atomic weights of the other two. Another of these so-called triads consisted of chlorine (35.5), bromine (79.9), and iodine (126.9). The announcement of the **Döbereiner triads** stimulated other investigators in their search for larger groups of elements with similar properties.

In 1866 Newlands in England reported further correlation between the atomic weights of elements and their properties. He prepared a list of the known elements, starting with lithium, atomic weight 7, as follows:

^{7}Li	^{9}Be	^{11}B	^{12}C	^{14}N	^{16}O	^{19}F
^{23}Na	^{24}Mg	^{27}Al	^{28}Si	^{31}P	^{32}S	$^{35.5}Cl$
^{39}K	^{40}Ca					

Newlands pointed out that similar properties occurred in every eighth element, an arrangement very like the octave of the musical scale. Many chemists at that time ridiculed his ideas, but later he was awarded a medal in honor of his discovery.

The Periodic Table as Developed by Mendeleev. The most extensive attempt to classify the elements was carried out by the Russian chemist Mendeleev in 1869. When he started to prepare a chemistry manual for his students at St. Petersburg University, his goal was to discover a logical interconnection between the properties of the chemical elements and their compounds. He arranged the elements in order of increasing atomic weights in such a way that elements with similar properties were placed in the same vertical columns. From his studies he concluded that *both the chemical and physical properties of the elements vary in a periodic fashion with their atomic weights.* His arrangement of elements in vertical columns was called a **periodic table.** It contained horizontal rows called *periods* and vertical columns called *groups* or *families.* This periodic table enabled chemists to classify their knowledge and to concentrate their study on the physical and chemical properties of eight groups of elements rather than on each element individually.

Some of the elements, however, seemed out of place in the table. When placed in the group with similar properties, the atomic weight of some of the elements was less than that of the element immediately preceding it. This resulted in redetermination of many of the atomic weights and correction of some experimental errors. It also led Mendeleev to note other systematic discrepencies in properties and to predict the properties of as yet undiscovered elements. In spite of the more accurate determination of atomic weights, there were still a few elements out of place in the table; for example, argon and potassium, and iodine and tellurium. After other difficulties had been encountered and repeated atomic weight determinations did not change the position of these elements that were out of place, it was finally admitted that the periodic grouping of elements according to their atomic weights might not be entirely correct.

Moseley's Contribution to the Periodic Table. In 1914 Moseley, a young English scientist, was studying the characteristics of x-rays given off by x-ray tubes. A simple x-ray tube is an evacuated electron tube in which cathode rays, or streams of electrons, strike the target, or anode, which sends out penetrating x-rays. Moseley constructed x-ray tube targets out of metallic elements and measured the frequencies of the x-rays emanating from these different metallic elements. His results established that the square root of the x-ray frequency for each element is nearly proportional to the atomic number. By using this nearly constant increase in the square root of the x-ray frequency of elements, he was able not only to place elements in their proper position in the periodic table but also to locate the positions of undiscovered elements. When a **periodic table based on the atomic numbers** was prepared, it was found that various properties agreed perfectly with the arrangement of the elements in the table. This new system places argon and potassium, and iodine and tellurium, in their proper positions in the table even though their atomic weights do not follow in proper progression. More accurately stated, then, **the physical and chemical properties of the elements are a periodic function of their atomic numbers.**

The Long Form of the Periodic Table. With this information in mind, it is possible to construct a periodic table (Table 2–4) that separates and classifies the elements in a more satisfactory manner. In this table, main group elements are designated A and subgroup elements are designated B. Within each group, the atomic number increases proceeding down the column and there are many similarities of properties. In the main groups, the elements become more metallic with increasing atomic number; whereas, all of the subgroup elements are metallic. Within each period the atomic number increases proceeding to the right and there is a pronounced lack of similarity of the properties of adjacent elements. Metallic character decreases proceeding from left to right in each period.

The **nonmetals** are located in the upper right corner of the table. The **light metals** are located in the upper left portion of the table, including the upper elements in Group IIIA, whereas the **heavy metals** are located in the bottom half of the table. The most **active metals** are found at the bottom of the extreme left of the table in Groups IA, IIA, and IIIB. The most active **nonmetals** are found on the extreme right of the table in Groups VA, VIA and VIIA. The **transition elements** are located in the central portion of the bottom half of the table in Group IIIB to Group VIII inclusive.

The **rare earth elements** have very similar properties and are found in Group IIIB. For convenience, the complete list of rare earth elements is shown in a separate section at the bottom of the table. The heaviest elements, starting with actinium (with an atomic number of 89) and including uranium and the elements recently discovered, have been named **actinides** and are also located in Group IIIB, since they have very similar properties to the other elements in this group and to the rare earth elements. The actinides are also shown in a separate section at the bottom of the periodic table.

Although it is convenient to include the rare earth and actinide elements in separate sections in the table, the most accurate portrayal of the periodic table would represent the elements from 57 to 71 in a vertical column in Group IIIB. Elements from 72 to 89 would still proceed in a horizontal row to complete Period 6 and start Period 7. Elements from 89 to 103 would again be shown in a vertical column in Group IIIB. This form of the periodic table also possesses the advantage that properties of an element can be predicted by considering its location within the table.

PERIODIC PROPERTIES

Some elements form acids, whereas others have the property of forming bases. **Acids** are compounds that usually contain hydrogen, oxygen, and another element. The characteristics of an acid are due to the hydrogen ion, which is hydrogen with a unit positive electrical charge. **Bases** are compounds of hydrogen and oxygen with another element, in which the oxygen and hydrogen combine to form a hydroxyl group. The hydroxyl group (OH^-) carries a unit negative charge and is responsible for the characteristics of a base. The property of forming an acid or base depends on the position of the element in the periodic table. For example, in the periodic table the elements of Group I form strong bases, those of Group II, moderately strong bases, and Group III and Group

TABLE 2-4 LONG FORM OF THE PERIODIC TABLE

IA	IIA	IIIB	IVB	VB	VIB	VIIB	VIII			IB	IIB	IIIA	IVA	VA	VIA	VIIA	O Inert Gases
		Transition Elements															
1 **H** 1.008																1 **H** 1.008	2 **He** 4.003
3 **Li** 6.939	4 **Be** 9.012											5 **B** 10.81	6 **C** 12.011	7 **N** 14.007	8 **O** 15.999	9 **F** 19.00	10 **Ne** 20.183
11 **Na** 22.990	12 **Mg** 24.31											13 **Al** 26.98	14 **Si** 28.086	15 **P** 30.974	16 **S** 32.064	17 **Cl** 35.453	18 **Ar** 39.948
19 **K** 39.102	20 **Ca** 40.08	21 **Sc** 44.956	22 **Ti** 47.90	23 **V** 50.94	24 **Cr** 51.996	25 **Mn** 54.94	26 **Fe** 55.85	27 **Co** 58.93	28 **Ni** 58.71	29 **Cu** 63.54	30 **Zn** 65.37	31 **Ga** 69.72	32 **Ge** 72.59	33 **As** 74.92	34 **Se** 78.96	35 **Br** 79.909	36 **Kr** 83.80
37 **Rb** 85.47	38 **Sr** 87.62	39 **Y** 88.91	40 **Zr** 91.22	41 **Nb** 92.91	42 **Mo** 95.94	43 **Tc** 99	44 **Ru** 101.07	45 **Rh** 102.91	46 **Pd** 106.4	47 **Ag** 107.870	48 **Cd** 112.40	49 **In** 114.82	50 **Sn** 118.69	51 **Sb** 121.75	52 **Te** 127.60	53 **I** 126.90	54 **Xe** 131.30
55 **Cs** 132.91	56 **Ba** 137.34	57-71 **La-Lu** **Rare** **Earths**	72 **Hf** 178.49	73 **Ta** 180.95	74 **W** 183.85	75 **Re** 186.2	76 **Os** 190.2	77 **Ir** 192.2	78 **Pt** 195.09	79 **Au** 196.97	80 **Hg** 200.59	81 **Tl** 204.37	82 **Pb** 207.19	83 **Bi** 208.98	84 **Po** 210	85 **At** 210	86 **Rn** 222
87 **Fr** 223	88 **Ra** 226.05	89-103 **Ac-Lw** **Actinides**	104 257														
Rare Earths		57 **La** 138.91	58 **Ce** 140.12	59 **Pr** 140.91	60 **Nd** 144.24	61 **Pm** 145	62 **Sm** 150.35	63 **Eu** 151.96	64 **Gd** 157.25	65 **Tb** 158.92	66 **Dy** 162.50	67 **Ho** 164.93	68 **Er** 167.26	69 **Tm** 168.93	70 **Yb** 173.04	71 **Lu** 174.97	
Actinides		89 **Ac** 227	90 **Th** 232.04	91 **Pa** 231	92 **U** 238.03	93 **Np** 237	94 **Pu** 242	95 **Am** 243	96 **Cm** 247	97 **Bk** 249	98 **Cf** 251	99 **Es** 254	100 **Fm** 253	101 **Md** 256	102 **No** 253	103 **Lw** 257	

IV elements may form weakly basic or weakly acidic compounds. The elements in Groups V, VI, and VII form strong acids, such as phosphoric, sulfuric and hydrochloric acids. In general, the elements on the left form strong bases, those on the right form strong acids, and those in the middle can form either weak acids or weak bases. The **melting point** of elements has also been shown to be a periodic function of their atomic weights. Many other physical and chemical properties pertain to changes that are similar to those in melting point, valence, acidity, and basicity.

Correlation of Atomic Structure and Periodic Properties. The electronic configuration of an element determines its chemical properties as well as many of its physical properties. The electrons most important in determining chemical properties are those found in the outermost quantum level, or shell. These electrons are called the **valence electrons.** The periodically repeated occurrence of elements with similar properties is due to the periodically repeated occurrence of elements with the same number of valence electrons. For example, in all of the elements in Period 2, starting with lithium, the first shell is filled with 2 electrons. Lithium has 1 electron in the second shell; beryllium, 2; boron, 3; carbon, 4; and so on to neon, in which the second shell is completed with 8 electrons. The element with the next highest atomic number would naturally start a new period under lithium, since its first two shells are completely filled with electrons, and it has 1 electron in the third shell. Following sodium, magnesium has its first two shells completely filled with electrons and 2 extra electrons in the third shell; aluminum has 3 electrons in the third shell, and so on to argon, which has the third shell filled with 8 electrons. The element with the next highest atomic weight would then start a new row, or series, being placed under lithium and sodium. This is potassium, with the first two shells completely filled, with 8 electrons in the third shell and 1 electron in the fourth shell. Since the number of electrons necessary to fill a given shell (starting with shell number one) is 2, 8, 18 and 32, it is not surprising that 2, 8, 18 or 32 elements are needed to complete a horizontal row, or period, in the periodic table. The relationships just discussed may be seen in the following tabulation:

Electrons in	*Li*	*Be*	*B*	*C*	*N*	*O*	*F*	*Ne*
First shell	2	2	2	2	2	2	2	2
Second shell	1	2	3	4	5	6	7	8

	Na	*Mg*	*Al*	*Si*	*P*	*S*	*Cl*	*Ar*
First shell	2	2	2	2	2	2	2	2
Second shell	8	8	8	8	8	8	8	8
Third shell	1	2	3	4	5	6	7	8

	K	*etc.*
First shell	2	etc.
Second shell	8	etc.
Third shell	8	etc.
Fourth shell	1	etc.

Another symbolism for indicating the electronic configuration of an atom is very helpful in predicting chemical properties. The **Lewis symbol** is written as the letter(s) denoting the element surrounded by dots symbolizing the number of valence electrons. The letter(s) of the symbol represent the nucleus and all electrons in inner closed quantum levels, or shells, and the dots represent the valence electrons. Examples of Lewis symbols are:

Li· Be: ·B: :Ċ· :Ṅ· :Ö· :F̈· :N̈e:

The Lewis symbols for elements in one vertical column, or group, in the periodic table differ only in letters, since each has the same number of valence electrons. Thus, the Lewis symbols for nitrogen and phosphorus, both in Group VA, are :Ṅ· and :Ṗ·.

Periodic Variation of Atomic Properties. Many properties of atoms are observed to vary periodically with their location in the periodic table. Three very important properties which show periodic variation are atomic size, ionization potential, and electron affinity.

The size of an atom is determined by the radius of a sphere which encloses 95 per cent of its electron cloud. The outermost portion of the electron cloud will be occupied by the valence electrons, and thus the distance of the valence electrons from the nucleus determines the radius of the atom. Several factors can influence atomic radii. However, for an atom outside of the influence of other atoms, electrical charges, and other forces, the major factors are three in number. Recalling that the quantum level, or shell, of the valence electrons increases in size with increasing principal quantum number, n, the sizes of atoms would be expected to increase as progressively higher quantum levels are occupied. This is found to be true. However, within a given principal quantum level, or period, the size of atoms decreases with increasing atomic number or nuclear charge. This occurs because the electron-nuclear force of attraction increases with increasing nuclear charge. The third factor which influences atomic sizes is the repulsion between valence electrons and electrons in inner filled quantum levels. This effect tends to "shield" the valence electrons from the nucleus and decreases the force of attraction toward the nucleus.

To summarize the periodic variation of atomic sizes, it can be stated that there is an increase in size within a group with increasing atomic number, and a decrease in size within a period with increasing atomic number. Thus, the smaller atoms are found in the upper right corner of the periodic table, and the larger atoms are in the lower left corner. These trends are illustrated by the atomic radii listed in Table 2–5 for the elements of Period 2 and the elements of Group IA.

Table 2–5 Trends in the Atomic Radii of Elements. The Numbers Given are Actually Covalent Radii in Å.

Period 2	*Li*	*Be*	*B*	*C*	*N*	*O*	*F*	*Ne*
	1.23	0.89	0.80	0.77	0.74	0.73	0.72	0.71
Group IA	*H*	*Li*	*Na*	*K*	*Rb*	*Cs*		
	0.37	1.23	1.57	2.03	2.16	2.35		

The actual force of interaction between the valence electrons and the nucleus can be measured by determining the energy involved in removing an electron from an atom. The process of removing an electron from an atom generates a positive ion and is called **ionization.** Using Lewis symbols, the removal of an electron from a neutral boron atom is shown as:

$$:\dot{B}\cdot \rightarrow :B^{+1} + e^{-}$$

In this representation, the charge on the boron ion is indicated with a +1 right superscript, and the electron which the boron loses is represented by e^{-}. The *energy which must be added to a gaseous atom to remove a valence electron* is called the **ionization potential.** The ionization potential is largest for electrons which are held very tightly by the atom, and consequently increases with decreasing atomic radius. Ionization potentials are largest for elements in the upper right corner of the periodic table and smallest for those in the lower left corner. The periodic variation of the ionization potentials within the first 65 elements is shown in Figure 2–10.

In many chemical reactions an atom adds an electron to become a negative ion. This process, which might also be called ionization, is represented as follows for fluorine adding an electron to become a negative ion.

$$:\ddot{F}: + e^{-} \rightarrow :\ddot{F}:^{-1}$$

The tendency of an atom to gain an electron is called **electron affinity.** The electron affinity is the *energy released when an electron is added to a gaseous atom.* Since added electrons will be held most tightly if they enter an orbital near the nucleus, the highest electron affinities are associated with the smallest atoms. Electron affinities are largest for elements in the upper right corner of the periodic table and smallest for those in the lower left corner.

The ionization of an atom results in ions which have radii quite different from those of the parent atoms. Positive ions are always smaller than their parent atoms, whereas negative ions are always larger than their parent atoms.

The variation of ionization potentials and electron affinities among the elements serves to differentiate between two classes of elements, **metals** and **nonmetals.** In general, metals which are located in the left portion of the periodic table have low ionization potentials and electron affinities, and, consequently, the highest tendencies to lose electrons and become positive ions.

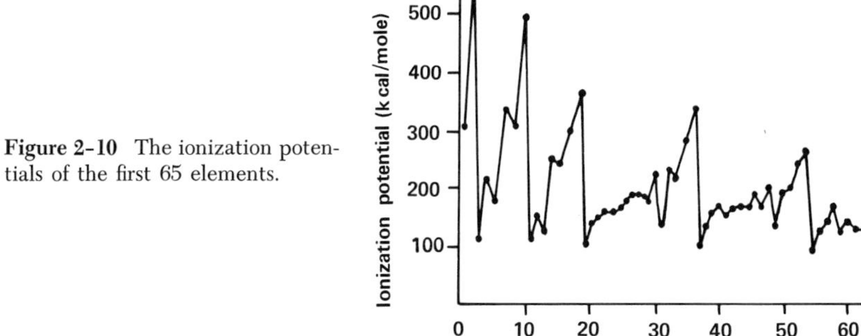

Figure 2–10 The ionization potentials of the first 65 elements.

The number of electrons lost is most commonly that number which will remove all electrons from the valence quantum level. The resulting ion will have an electronic configuration equivalent to that of an inert gas. The gases in Group O of the periodic table are very unreactive or chemically inert, but the heavier members are known to react with the very reactive element fluorine. The electronic configurations of the inert gases are unique in that each involves a set of p-orbitals which is completely filled with electrons. These electronic configurations are more stable than those of any other elements. The elements in Group IA of the periodic table form ions with a positive electrical charge of one, whereas the elements in Group IIA form ions with a positive electrical charge of two. Using atomic symbols, the common sodium and potassium ions in the first group would be represented by Na^+ and K^+. Magnesium and calcium ions would be written Mg^{++} and Ca^{++}. Occasionally, 3 electrons may be removed from the outer shell of an atom, but this is not common. The removal of 4 electrons from the outer shell, which is rare, would result in an ion with four positive charges.

In the nonmetallic elements on the right side of the periodic table there occurs an entirely different behavior. Nonmetals have high ionization potentials and high electron affinities. Instead of losing or giving up an electron, the **nonmetallic elements tend to gain electrons.** Chlorine, in Group VIIA, the same series as sodium and magnesium, is an example. Chlorine has 7 electrons in its outer shell. When the chlorine atom takes on an extra electron from a metallic atom, it is converted into a chloride ion. The **chloride ion,** written Cl^-, has a negative electrical charge of one unit and a stable arrangement of 8 electrons in the third principal quantum level.

Going back to Group VIA in the table, an atom of sulfur can gain 2 electrons from a metallic atom to form a **sulfide ion.** Nonmetallic elements rarely gain 3 electrons, so there are few ions with 3 negative charges.

RADIOACTIVITY

In 1895 Röntgen, a German physicist, discovered that the cathode-ray tube emitted invisible rays capable of penetrating opaque substances. These rays were called x-rays and their properties also aroused considerable interest. In the following year the French physicist Becquerel investigated several fluorescent substances as possible sources of these penetrating rays. Of the substances tested, only a **uranium** compound affected a photographic plate that had been protected by a wrapping of black paper. Becquerel found that all uranium compounds gave off penetrating rays, which he called **Becquerel rays,** and the intensity of the rays was proportional to the amount of uranium contained in the compound. He used the term **radioactivity** to refer to the production of radiation by uranium compounds. Madame Curie and her husband Pierre investigated radioactivity and discovered the radioactive elements *polonium* and *radium* as constituents of pitchblende.

Properties of Radioactive Elements. Chemically, radium acts like other elements of the same family. When freshly prepared it has the appearance of metallic calcium or barium and exhibits similar properties. The properties of radium associated with the radioactivity, however, are striking. Rutherford studied the radiations given off by radium by placing some of the radioactive material in the bottom of a thick lead well. The lead shielded the apparatus from any other radiations and allowed him to focus the rays on a photographic plate. Under the influence of a magnetic field, the rays were deflected in such a way that three types of radiations were observed (Fig. 2-11). Rutherford named the rays alpha, beta, and gamma, and found that they were given off from all radioactive elements.

1. **Alpha rays,** which consist of **alpha particles,** are nuclei of helium atoms and consist of 2 protons and 2 neutrons with a positive charge of two units. When released from the nucleus of a radioactive element, alpha particles have a velocity of over 10,000 miles per second. They immediately collide with thousands of air molecules, which slow them down and finally stop them after they have gone about 8 cm. These collisions not only produce heat, but also produce ionized particles, since some of the air molecules have electrons knocked out of them by the alpha particles.

2. **Beta rays** are streams of electrons. Their initial velocity approaches that of the speed of light, or over 150,000 miles per second. They are more penetrating than alpha rays and are of the opposite electrical charge.

3. **Gamma rays** are similar to x-rays, but have even greater power of penetration. They do not consist of individual particles, but are electromagnetic waves or light. They are not deflected in a magnetic field, as are alpha and beta rays. Although alpha and beta rays cause fogging of photographic plates, gamma rays are more penetrating and will fog films more rapidly.

It is of interest to compare the penetrating power of these high-velocity particles or rays. From Rutherford's original experiments he learned that alpha particles could pass through very thin sheets of metal foil. They are larger particles than beta or gamma rays and can be stopped by a sheet of paper. Beta rays have about 100 times the penetrating power of alpha particles and will pass through thin sheets of metal. Gamma rays are difficult to stop, since they have 10,000 times the penetrating power of alpha particles and about 2 inches of lead or 12 inches of iron are required to screen them out.

Radioactivity arises because of the instability of the nucleus. In radioactive elements the nucleus is apparently not satisfied with its portion of protons and neutrons and is continually giving off alpha particles and beta particles until a stable arrangement is reached. Since radioactive elements are constantly undergoing disintegration, or natural decay, it is reasonable to assume that eventually all the radioactivity of a given element will be dissipated. This period

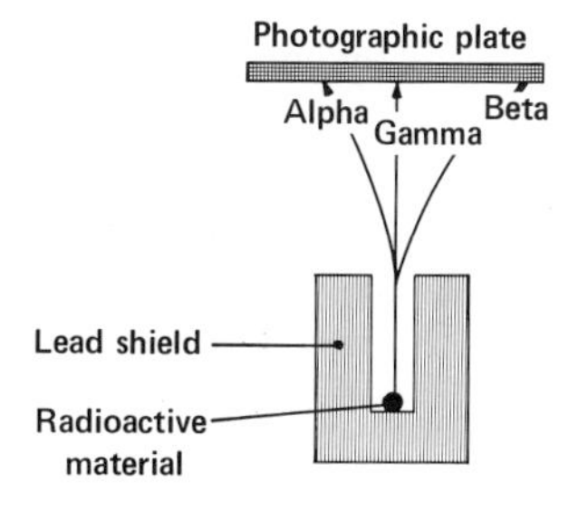

Figure 2-11 The effect of a magnetic field on the three types of radiations given off by radioactive material. In the illustration the magnetic field is perpendicular to the plane of the paper.

of activity can be determined with a fair degree of accuracy. By counting the number of alpha particles emitted per second from a sample of radium containing a known number of radium atoms, it has been calculated that half of the radium atoms will have decayed in about 1590 years. In general, the time necessary for half the weight of a sample of radioactive element to decay or disintegrate is called the **half-life** of that element. Using radium as an example, at the end of 1590 years one half of an original sample would remain; at the end of another 1590 years one fourth of the original sample would be left; and at the end of another 1590 years one eighth of the sample would remain. The half-life of radioactive elements shows considerable variation. Uranium has a half-life of approximately four and a half billion years, whereas a form of radium called radium C_1 has a half-life of 0.0001 second. The half-life of most radioactive elements falls somewhere in between these two extremes. Apparently the rate of disintegration of a radioactive element depends only on its atomic structure, since changes in temperature or pressure or chemical combination or any other measurable conditions have no effect on the length of its half-life.

TRANSFORMATION OF ELEMENTS

Natural Transformation. As elements possessing natural radioactivity disintegrate, there is an accompanying **transformation** into other radioactive elements or isotopes. The elements or isotopes are represented by an appropriate symbol, with the atomic number at the lower left of the symbol and the atomic weight at its upper left as in $^{238}_{92}U$.

Transformation reactions usually involve the loss of an alpha or a beta particle from the nucleus of an isotope, with the resultant formation of a new element or isotope. Since alpha particles have the same structure as the nucleus of the helium atom, the loss of an alpha particle from the nucleus of uranium results in the formation of a new element whose atomic number is two less and atomic weight is four less than that of the original atom. This is illustrated by the following equation:

$$^{238}_{92}U \rightarrow {}^{234}_{90}Th + {}^{4}_{2}He$$

The thorium isotope formed from uranium by the loss of an alpha particle is unstable and produces another new element by the loss of a beta particle. This beta particle, or electron, must be emitted from the nucleus if a new element is to be formed. The loss of an electron from a radioactive isotope results in the formation of another element whose atomic number is increased by one, but whose atomic weight is unchanged, as illustrated by the following:

$$^{234}_{90}Th \rightarrow {}^{0}_{-}e + {}^{234}_{91}Pa$$

In the nuclear reactions so far described, the electrons outside of the nucleus (extranuclear) have not been considered. Obviously the loss of an alpha particle must be accompanied by the loss of two extranuclear electrons to maintain electrical neutrality. Also, the loss of a beta particle involves the gain of an electron. Electrons apparently can be readily shifted back and forth in the surrounding environment to establish neutrality of newly formed elements or isotopes. For example, some of the alpha particles emitted from nuclei pick up 2 electrons and become helium atoms.

Artificial Transformation. The natural transformation of elements suggests the possibility of artificial transformation, or the conversion of one element into another by the chemist. To effect the transformation of elements, it is necessary, however, to alter the charge on the nucleus, or, in other words, to bring about a rearrangement of protons and neutrons in the nucleus of an element. The charge could be changed by the addition of a negative charge or a positive charge to the nucleus. However, the extranuclear electrons of an atom with their negative charge would act as a barrier to the addition of a negative charge, while the positively charged nucleus would repel the addition of a positive charge.

Rutherford, in 1919, conceived the idea of bombarding the nucleus of an element with alpha particles from radium, traveling at high velocities. He succeeded in knocking protons out of the nucleus of nitrogen atoms with the resultant formation of atoms of hydrogen. Later it was shown that this reaction also produced an isotope of oxygen:

$$\underset{\text{Nitrogen atom}}{{}^{14}_{7}\text{N}} + \underset{\alpha\text{ particle}}{{}^{4}_{2}\text{He}} \rightarrow \underset{\text{Proton}}{{}^{1}_{1}\text{H}} + \underset{\text{Oxygen isotope}}{{}^{17}_{8}\text{O}}$$

Rutherford's experiment was the first in which an element was transformed into another element artificially by the bombardment of the nucleus with atomic particles.

Since 1919, other transformations have been studied, and several radioactive isotopes have been produced by nuclear bombardment. The important subatomic particle, the neutron, was discovered in 1932 by the British physicist Chadwick in his study of the nuclear bombardment of beryllium with alpha particles:

$$\underset{\text{Beryllium atom}}{{}^{9}_{4}\text{Be}} + \underset{\alpha\text{ particle}}{{}^{4}_{2}\text{He}} \rightarrow \underset{\text{Carbon isotope}}{{}^{12}_{6}\text{C}} + \underset{\text{Neutron}}{{}^{1}_{0}\text{n}}$$

It was shown in 1934 by Frédéric and Irène Curie Joliot, son-in-law and daughter of Marie Curie, while studying the bombardment of aluminum, boron, and magnesium atoms with streams of alpha particles from polonium, that radioactive isotopes could be artificially produced by transformation reactions. The transformation of boron produced a radioactive isotope of nitrogen:

$$\underset{\text{Boron atom (stable)}}{{}^{10}_{5}\text{B}} + \underset{\alpha\text{ particle}}{{}^{4}_{2}\text{He}} \rightarrow \underset{\text{Nitrogen isotope (radioactive)}}{{}^{13}_{7}\text{N}} + \underset{\text{Neutron}}{{}^{1}_{0}\text{n}}$$

The radioactive nitrogen isotope with an atomic weight of 13, which is commonly known as ^{13}N, has a short half-life and rapidly disintegrates to form a stable isotope of carbon with an atomic weight of 13.

When the results of these experiments were published, investigators immediately began bombarding other elements with various types of subatomic particles in an attempt to produce new radioactive isotopes. Some of the important subatomic particles that were used to bombard elements include alpha particles, neutrons, protons, electrons, and deuterium nuclei.

NUCLEAR FISSION. In 1939 the process of nuclear fission was discovered by the German scientists Hahn and Strassman. When they bombarded the uranium isotope ^{235}U with neutrons, it underwent splitting, or **fission,** into fragments of lower atomic weight, and large amounts of energy were released. The size of the fission products is not always constant, but apparently one atom of ^{235}U splits into two unequal fragments whose nuclei have atomic numbers between 30 and 60. During the fission process several neutrons are released which strike other ^{235}U atoms, causing additional fission products to be formed along with more neutrons and more energy. If this fission process proceeds for any period of time, an explosion results. This **chain reaction** fission process of uranium isotope ^{235}U was the basis for the first **atomic bomb.**

The energy released in nuclear fission reactions, in which elements are changed into other elements, is much greater than that involved in ordinary chemical reactions, where one compound is changed into another compound. Since subatomic particles are often split off in nuclear reactions, there results a small but definite change in mass. This change of mass is related to change in energy by the equation proposed by Einstein in 1906:

$$E = mc^2$$

where c, the velocity of light, equals 3×10^{10} cm per second. In the nuclear fission reactions described here, it was found that the products accounted for about 99.9 per cent of the weight of the reacting nucleus. Thus 0.1 per cent of the total mass was converted into the energy evidenced by the explosion of the atomic bomb.

NUCLEAR FUSION. The source of the solar energy of the sun and the energy of the stars is the result of nuclear fusion, or the union of light nuclei to form heavy nuclei. A series of reactions involving the nuclei of hydrogen atoms with those of carbon, nitrogen, and oxygen isotopes results in the formation of helium nuclei and **positrons.** (A positron is a small particle exactly like an electron, except that it is positively charged.) The essential equation may be represented as:

$$4\,^{1}_{1}H \rightarrow 2\,^{4}_{2}He + 2\,^{0}_{+}e$$

Hydrogen nuclei — Helium nuclei — Positrons

A simple calculation will show that a greater proportion of the total mass of the hydrogen nuclei is converted into energy in this fusion reaction than in the fission reaction. This higher rate of conversion accounts for the great energy of the sun and stars.

Questions

1. Explain the origin of the term "periodic table."

2. On what basis were the vertical columns or groups of elements chosen for the periodic table? the horizontal periods?

3. Briefly describe the properties of cathode rays.

4. Why does one commonly speak of a charge of -1 for the electron and $+1$ for the proton?

5. What is an isotope?

6. Name and explain briefly the four quantum numbers.

7. Using Lewis symbols, illustrate ionization.

8. Do ionization potentials increase with decreasing atomic radius? Why?

9. Are the highest electron affinities associated with the largest or smallest atoms?

10. Use the shorthand method to express the electronic configurations for each of the following elements or ions:
(a) O (b) Na (c) P (d) Cl^- (e) Al^{+3}

11. Why was a periodic table based on the atomic weights of the elements not entirely satisfactory?

12. Briefly explain the contribution of Moseley to the construction of an accurate periodic table.

13. Explain the location of the light metals, the heavy metals, and the nonmetals in the long form of the periodic table.

14. What relationship exists between the electronic configurations of the elements in any one vertical column or group? in any one horizontal row or period?

15. How does the electron structure of metals differ from that of nonmetals?

16. What type of substance is usually formed when a metal transfers electrons to a nonmetal?

17. Indicate the position in the periodic table where the following can be found:
(a) elements with largest electron affinities
(b) elements with largest ionization potential
(c) metals
(d) non-metals
(e) inert gases

18. Name the two most important factors influencing the size of an atom.

19. Which of the following reactions requires very high ionization energy? Why?
(a) $Na - e^- \rightarrow Na^+$
(b) $Na^+ - e^- \rightarrow Na^{++}$

20. Group VII atoms (halogens) tend to take on electrons to form negative ions; explain, giving examples.

21. Arrange the following in order of increasing size: Na^+, Ne, Cl^-, Mg^{++}, S^{--}, and Al^{+++}.

22. What are the properties of alpha particles? How could you detect their presence?

23. How does the nucleus of a radioactive element differ from that of a stable element?

24. In general, what happens to a radioactive element when it loses an alpha particle? a beta particle?

25. Why was Rutherford's bombardment of nitrogen atoms with alpha particles such an important experiment in the development of nuclear chemistry?

26. Name and describe briefly the important subatomic particles used in the bombardment of various elements.

27. Explain the principle involved in the synthesis of new elements possessing atomic weights greater than that of uranium.

28. What is meant by nuclear fission?

Suggested Reading

Berry: Atomic Orbitals. Journal of Chemical Education, Vol. 43, p. 283, 1966.
Campbell: Atomic Size and the Periodic Table. Journal of Chemical Education, Vol. 23, p. 259, 1946.
Redfern and Salmon: Periodic Classification of the Elements. Journal of Chemical Education, Vol. 39, p. 41, 1962.
Sanderson: Chemical Periodicity. New York, Reinhold Publishing Corp., 1960.
Seaborg: Prospect for Further Considerable Extension of the Periodic Table. Journal of Chemical Education, Vol. 46, p. 627, 1969.
Sisler: Electronic Structure, Properties, and the Periodic Law. New York, Reinhold Publishing Corp., 1965.

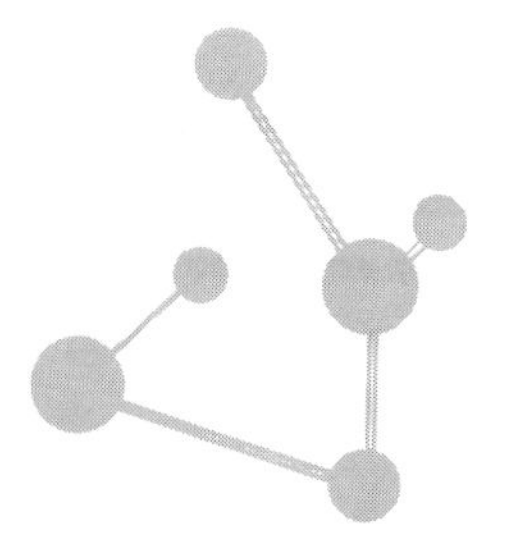

CHAPTER 3

CHEMICAL BONDS

Dalton's atomic theory described the formation of compounds and chemical reaction as the combination of atoms. In this manner he provided an explanation for the laws of definite composition, multiple proportions, and conservation of mass. As more information concerning the nature of the atom became available, chemists sought a solution for the fundamental question of why elements combine with each other. For a time chemists explained compound formation as the chemical affinity of one atom for another. The degree of affinity was expressed as the combining power of the atom. It was soon realized, however, that the terms combining power and chemical affinity of atoms were merely expressions that oversimplified a complex problem.

In 1812 Berzelius, a Swedish chemist, proposed a theory of electrostatic attraction between atoms in a compound. He assumed an electrochemical nature for acids, bases, and salts, with electrical forces between the atoms. Another theory of bonding, based on bonds between the atoms with a definite combining power for each atom in the molecule, was proposed by Kekulé, a German chemist. After several years of argument by advocates of both theories, it was further theorized that in inorganic compounds the bonds are electrostatic in nature, whereas in carbon compounds electrical forces between the atoms are weak or nonexistent. These theories were forerunners of electrovalence and convalence, which will be explained in some detail.

As was pointed out in the last chapter, the electrons in the outer shell of an atom play an important role in the periodic properties of the element. The combining power of an atom is closely related to the number of electrons in the outer shell. In the early 1900's chemists called attention to the stability of helium with its pair of valence electrons. It was also stated that many compounds contained an even number of valence electrons, with the inert gases (except helium) having a stable configuration of eight valence electrons. These observations led to the statement that when atoms combine to form molecules, the valence electrons tend to group in pairs or octets. A more useful statement, however, might be that when elements combine to form compounds, the valence electrons of their atoms show a tendency to assume the stable configuration of an inert gas. It is found that this **rule of octets** can be used to predict the combining capacities of most elements in the first three periods.

IONIC BONDS

The nature of ionic bonding may be illustrated by considering the formation of sodium chloride. In the overall reaction it appears that sodium atoms react with chlorine atoms to form the compound sodium chloride. As seen in Figure 3–1, however, an electron is transferred from the outer shell of the sodium atom to the outer shell of the chlorine atom. This loss of an electron from sodium occurs readily and requires a relatively small amount of energy, which is called the **ionization potential.** When this electron is presented to the chlorine atom, the chlorine atom readily accepts it, with the release of energy, which is called the **electron affinity.** The sodium ions and the chloride ions that are formed in the process are much more stable than the atoms, and resemble the inert gases neon and argon, respectively, in their valence electron configuration. The ions differ from the inert gas atoms in that they are no longer neutral but bear positive (Na) and negative (Cl) charges. These opposite charges attract each other and are responsible for the strong **ionic bond** between sodium ions and chloride ions that forms the compound sodium chloride. Compounds in which the atoms are held together by ionic bonds are called **ionic compounds.**

Another method of illustrating the steps that occur in the formation of an ionic compound is as follows:

1. $Na\cdot_{(g)} \rightarrow Na^+_{(g)} + e^-$
2. $:\ddot{\underset{..}{Cl}}\cdot_{(g)} + e^- \rightarrow :\ddot{\underset{..}{Cl}}:^-_{(g)}$
3. $Na^+_{(g)} + :\ddot{\underset{..}{Cl}}:^-_{(g)} \rightarrow (Na^+:\ddot{\underset{..}{Cl}}:^-)_{(c)}$

In steps (1) and (2), ionization of gaseous Na and Cl occurs. In step (3), gaseous Na^+ and Cl^- combine to form crystalline NaCl. This last step plays a major role in determining the tendency of the reaction to occur. The overall reaction

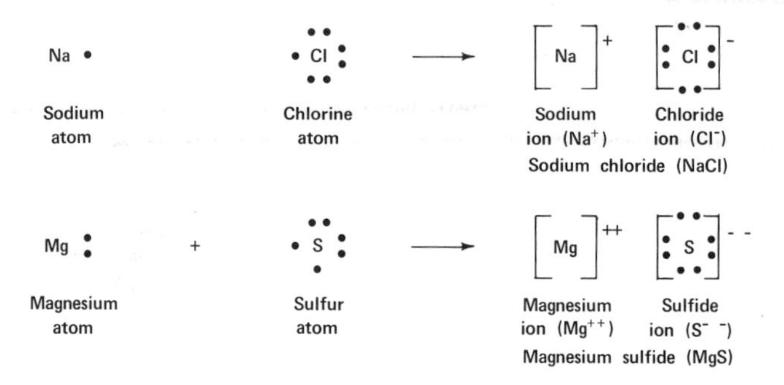

Figure 3–1 The process of electron transfer commonly occurs in the formation of inorganic salts.

may be represented by the following equation:

$$\mathrm{Na}\cdot_{(g)} + :\ddot{\underset{..}{\mathrm{Cl}}}\cdot_{(g)} \rightarrow (\mathrm{Na}^+:\ddot{\underset{..}{\mathrm{Cl}}}:^-)_{(c)}$$

A compound formed by the transfer of 2 electrons to yield an outer shell of 8 valence electrons in each ion is magnesium sulfide. A relatively small amount of energy is required to remove the 2 outer electrons of magnesium. They are accepted by the outer shell of the sulfur atom with the release of energy. This electron transfer is illustrated in Figure 3–1. The ionic bond between magnesium and sulfide ions results from the strong electrostatic attraction between the oppositely charged ions.

In simple chemical reactions, atoms are commonly represented by the symbol alone. However, ions alone, or in compounds, are represented as the symbol bearing either positive or negative charges. For example, note ions such as Na^+ and Mg^{++} and compounds such as Na^+Cl^- and $Mg^{++}S^{--}$.

The electrical charge carried by an ion represents its **combining capacity.** The bonding resulting from electron transfer is called **electrovalence,** or **ionic bonding.** It can readily be seen that the combining capacity is equal to the number of electrons gained or lost by an atom when it is converted into an ion.

Ionic solids are composed of oppositely charged ions occupying sites in the crystal lattice. Each positive ion and each negative ion surrounds itself with the largest possible number of oppositely charged ions at the shortest possible interionic distance. The number of nearest neighbors is called the **coordination number.** In ionic solids such as CaF_2, in which the ions have charges of different magnitude, the coordination numbers of the ions are in the ratio of the reciprocal of their charge magnitudes. For example, in the CaF_2 structure, each calcium ion is surrounded by eight fluoride ions at the corners of a cube, and each fluoride ion is surrounded by four calcium ions at the corners of a tetrahedron. Ionic solids having ions with charges of equal magnitude have crystal lattices which are determined by the relative sizes of the positive and negative ions. Figure 3–2 shows the unit cell of NaCl in which both Na^+ and Cl^- have coordination numbers of six. The interionic forces in ionic crystals are quite strong and result in high melting points, low vapor pressures, and hardness. These properties are typical of salts which are ionic solids.

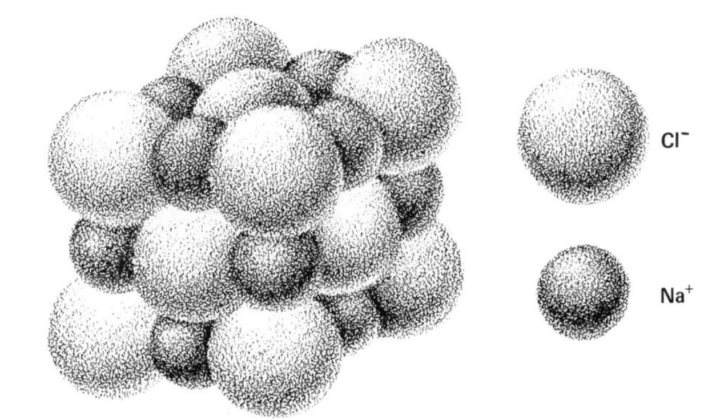

Figure 3–2 A representative segment of crystalline NaCl.

COVALENT BONDS

About 1916 it was suggested that two atoms may combine by sharing valence electrons. The process of joining atoms to form molecules by the sharing of electrons is called **covalence.** As an example, hydrogen gas consists of molecules that contain 2 hydrogen atoms held together by a force resulting from the sharing of a pair of electrons, as shown in Figure 3–3. This figure illustrates the use of Lewis symbols for molecules with covalent bonds. In the symbol for the molecule, a pair of dots placed between the atoms represents a bond.

Hydrogen, chlorine, nitrogen, and other diatomic gaseous elements show similar behavior in sharing electrons to the extent that their valence shells are filled. In the chlorine molecule, each chlorine atom with 7 valence electrons is able to attain a complete shell by sharing one pair of electrons with the other chlorine atom. In the nitrogen molecule, a filled shell is attained only if three pairs of electrons are shared. These examples are represented schematically in Figure 3–3 with Lewis formulas.

It is stated that the hydrogen and chlorine molecules are held together by a **single bond,** but that the nitrogen molecule has a **triple bond. Double bonds** are also known to exist in many molecules.

In some compounds, a single covalent bond exists in which one of the bonded atoms furnishes both of the electrons which are shared. Such a covalent bond is called a **coordinate covalent bond.** For example, the formation of the trimethyl ammonium ion by the interaction of trimethyl amine, $(CH_3)_3N$, and a hydrogen ion, H^+, involves the formation of a coordinate covalent bond. This is represented in the following equation:

$$(CH_3)_3N: + H^+ \rightarrow \left[(CH_3)_3N{-}H\right]^+$$

The rationalization of the existence of a force between atoms which share a pair of electrons is aided by the following hypothetical experiment. In such an experiment, diagrammatically represented in Figure 3–4 a, two hydrogen atoms are allowed to approach one another until their 1s atomic orbitals overlap and mutually occupy the space between the nuclei. The result of the overlap is an electron cloud associated with the molecule. An electron cloud associated with more than one nucleus is referred to as a **molecular orbital.** In the molecular orbital formed here by the combination of atomic orbitals, there is a high probability of finding the electrons in the region between the nuclei.

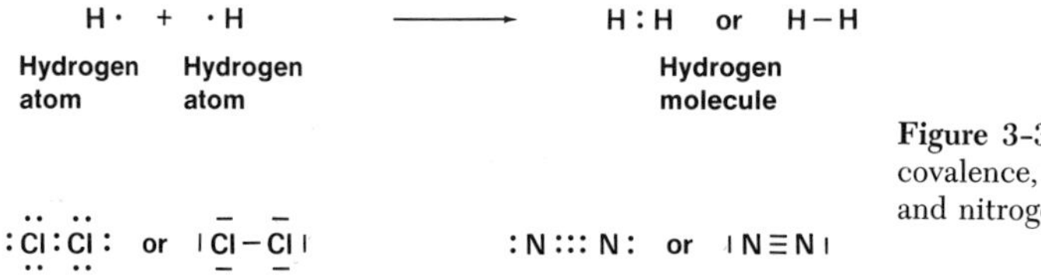

Figure 3–3 The sharing of electrons, or covalence, illustrated by hydrogen, chlorine and nitrogen molecules.

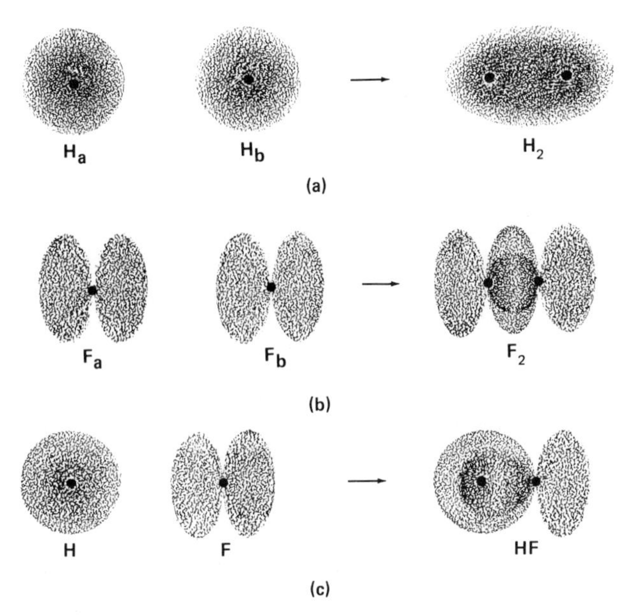

Figure 3-4 (*a*) Hypothetical approach of the electron clouds of two hydrogen atoms and the resulting electron cloud of the hydrogen molecule. (*b*) A similar approach of the p-orbitals of two fluorine atoms. (*c*) The approach of a 1s-orbital of a hydrogen and a 2p-orbital of a fluorine.

Before the atoms approach one another, the only forces which exist are between the electron and the nucleus of a given atom. When the atoms are brought close together, four new forces arise. Two of these tend to destabilize the molecule relative to the separated atoms. These are the repulsions of the nuclei for one another, and the repulsions of the electrons for one another. These repulsions are expected, since it is known that similarly charged objects repel one another. One of the attractive forces arises because the electron originating from atom H_a is attracted to the nucleus of atom H_b. The same force holds for the electron on atom H_b interacting with the nucleus of atom H_a. The second attractive force arises from the interaction of the opposed small magnetic fields associated with the oppositely spinning electrons. The summation of the contributions of these four forces determines the total force of interaction between the two atoms, and consequently, the net stabilization of the molecule relative to the separated atoms.

The results of a similar experiment involving the approach of fluorine atoms is shown in Figure 3–4 b, where it is seen that atomic 2p-orbitals on the approaching atoms overlap to form a region between the nuclei where the probability of finding electrons is quite high. Figure 3–4 c shows a third type of experiment in which a 1s-orbital on hydrogen overlaps a 2p-orbital on fluorine to form a covalent bond.

In each of the three hypothetical experiments shown in Figure 3–4, the overlap of the electron clouds of approaching atoms results in an increased electron density along the internuclear axis. A covalent bond, resulting from increased electron density along the axis connecting the nuclei of adjacent atoms, is called a **sigma-bond** (σ-bond).

A covalent bond between adjacent atoms can also result from the overlap

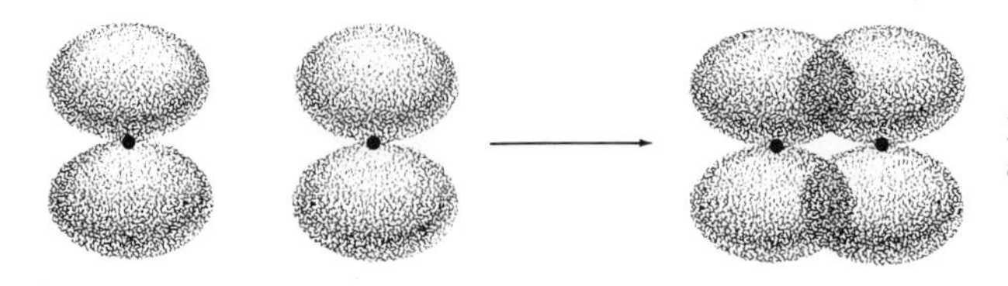

Figure 3-5 Side-by-side overlap of atomic p-orbitals to give a pi-bond.

of p-orbitals in a side-by-side fashion, as shown in Figure 3-5. Overlap of electron clouds in this manner results in increased electron density above and below the internuclear axis. A bond of this type is called a **pi-bond** (π-bond).

The force of interaction, or the energy of interaction, between the atoms in a molecule varies with the internuclear separation of the atoms. This variation, the **stabilization energy** (*PE*) variation, is shown as a function of the **internuclear separation** (*R*) for hydrogen in Figure 3-6. In this figure it is seen that at large distances the stabilization energy approaches zero, representing no interaction of the atoms. As *R* decreases, the stabilization energy increases. However, below $R = 0.74$Å, the stabilization energy decreases with decreasing *R* due to the nuclear repulsion force. At very small values of *R*, the system has a negative *PE*, which means that the system is less stable than the separated atoms. The value of *R* at which *PE* is largest is the **equilibrium internuclear separation** in the hydrogen molecule.

The distance of internuclear separation in all molecules of which the atoms are covalently bonded is determined by the combination of attractive and repulsive forces. The distance of internuclear separation at which the stabilization energy is largest varies from molecule to molecule. It is dependent on the sizes of the atomic orbitals which are being combined, the nuclear charges, the ionization potentials, the electron affinity, and other fundamental atomic properties. One half the internuclear distance of separation in homonuclear diatomic molecules (e.g., H_2, O_2, N_2, F_2, and so forth), as indicated in Figure 3-7, is defined as the **covalent radius.** Some covalent radii are given in Table 2-5.

POLAR BONDS

"Pure" covalent bonds, as are found in homonuclear diatomic molecules, and "pure" ionic bonds, as are found in salts like NaCl, represent the extremes

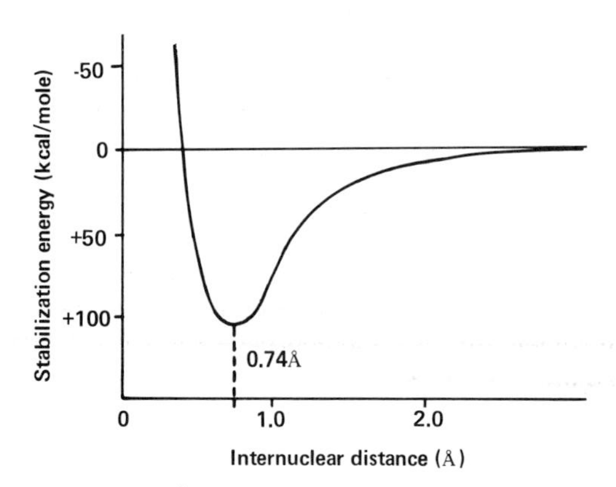

Figure 3-6 Stabilization energy of the hydrogen molecule as a function of the internuclear distance of separation.

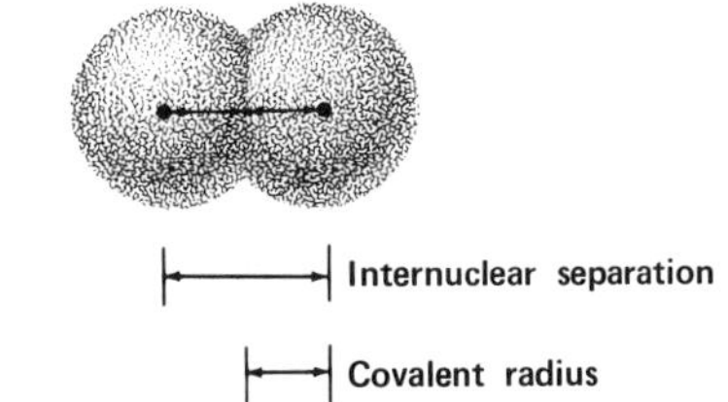

Figure 3-7 Covalent radius of a homonuclear diatomic molecule.

of bonding. In the former, the electron pair is shared equally between atoms, whereas in the latter there is no sharing of electrons, but rather a complete transfer of an electron from one atom to another. Bonding intermediate between "pure" covalent and "pure" ionic is very common. Since electrons may not always be shared equally between atoms, but yet are not necessarily completely transferred from one atom to another, there exists the possibility of unequal charge distribution in a covalent bond.

For example, in the compound HCl, or hydrogen chloride, the shared pair of electrons is attracted more by the chlorine end of the molecule than by the hydrogen end. The result is an unequal charge distribution, with the chlorine end more negative and the hydrogen end of the molecule more positive. Hydrogen chloride may be represented as follows:

$$\text{HCl} \quad \text{or} \quad \text{H}:\ddot{\underset{..}{\text{Cl}}}: \quad \text{or} \quad \text{H}^+:\ddot{\underset{..}{\text{Cl}}}:^- \quad \text{or} \quad (+\ -)$$

Even though the molecule is electrically neutral, the center of the positive charge does not coincide with the center of the negative charge. The molecule is called a **dipole** and is said to possess a **dipole moment.** Dipole moments are often designated by the symbol $\leftrightarrow$. In this symbol, the pointed end corresponds to the negative end of the dipole; the length is related to the magnitude; and the orientation of the dipole is indicated by the orientation of the symbol. When placed in an electrical field, such molecules will line up with their negative ends facing the positive plate (electrode) and their positive ends facing the negative plate (electrode). The dipole character of these compounds gives rise to the name **polar bonds** and **polar covalent compounds.**

Compounds whose atoms equally share a pair of electrons will have the center of their positive charge coinciding with the center of their negative charge. These compounds do not exhibit dipole characteristics and are called **nonpolar molecules** with **nonpolar bonds.** Hydrogen molecules (H_2) and chlorine molecules (Cl_2) are examples of nonpolar covalent compounds. In general, if a molecule is composed of two of the same kind of atoms, the bond between them will be nonpolar and the molecule will be nonpolar. If two different atoms make up the molecule, the bond is polar and the molecule is polar.

In molecules with more than two atoms, there exists the possibility of having more than one polar bond. In such molecules, the dipole moment will be related to the summation of all polar bond contributions. In the water molecule, which is non-linear, each of the two bonds has an associated dipole. The sum of the bond dipoles is a dipole which is not aligned with either bond, but which bisects the HOH angle, as shown in Figure 3–8. The water molecule has a rather large dipole moment.

In the carbon dioxide molecule, CO_2, which is linear, each bond has a

Figure 3-8 Bond dipoles and the molecular dipole of water.

dipole; but the summation of these dipoles is zero, since they point in opposite directions. The dipoles are oriented as follows:

$$\overleftarrow{+} \quad \overrightarrow{+}$$
$$:\ddot{O}=C=\ddot{O}:$$

The carbon dioxide molecule has no net dipole moment.

ELECTRONEGATIVITY

It has already been stated that in the compound HCl the chlorine atom has a greater attraction for the electron pair than does the hydrogen atom. The attraction of an atom for shared electrons depends on the amount of energy required in the transfer of electrons from or to its outer electron shell. The force of attraction for valence electrons varies from element to element and is called its **electronegativity.** The relative electronegativity values that have been determined for some of the common elements are represented in a partial periodic table as shown in Table 3–1. These values are related to the ability of the atoms to attract shared electrons and thus increase the negative charge of their end of a molecule.

The electronegativity of an element is related to its ionization potential and its electron affinity. Elements which have high ionization potentials and high electron affinities exhibit high electronegativities. Inspection of Table 3–1 will reveal that elements with the highest electronegativities are found in the upper right corner of the periodic table. There is also a tendency toward increasing values across a period to the right and toward decreasing values moving down the elements of a group. If two elements with greatly different electronegativities combine, the bond will be highly polar or ionic in nature. Metals in Group IA and IIA combining with the nonmetals in group VIA and VIIA almost always form ionic compounds with ionic bonds. When the combination involves two elements with similar values, the bonds are usually covalent. If the compound is covalent, the atoms with the greatest difference in electronegativity will form the more polar bonds. Two like atoms with no difference in their electronegativity values will obviously form nonpolar covalent bonds.

COMBINING CAPACITY OR VALENCE

In the theoretical consideration of bonding, based on the electronic structure of the atoms, and discussed in the foregoing paragraphs, it was learned that atoms could combine with each other to form molecules by the loss or gain of electrons (electrovalence or ionic bonding), or by the sharing of pairs

Table 3-1 Relative Electronegativity Values of the Common Elements

IA	IIA	IIIB	IVB	VB	VIB	VIIB	VIII			IB	IIB	IIIA	IVA	VA	VIA	VIIA	O
1 **H** 2.1																	2 **He** 0
3 **Li** 1.0	4 **Be** 1.5											5 **B** 2.0	6 **C** 2.5	7 **N** 3.0	8 **O** 3.5	9 **F** 4.0	10 **Ne** 0
11 **Na** 0.9	12 **Mg** 1.2											13 **Al** 1.5	14 **Si** 1.8	15 **P** 2.1	16 **S** 2.5	17 **Cl** 3.0	18 **Ar** 0
19 **K** 0.8	20 **Ca** 1.0	21 **Sc** 1.3	22 **Ti** 1.5	23 **V** 1.6	24 **Cr** 1.6	25 **Mn** 1.5	26 **Fe** 1.8	27 **Co** 1.8	28 **Ni** 1.8	29 **Cu** 1.9	30 **Zn** 1.6	31 **Ga** 1.6	32 **Ge** 1.8	33 **As** 2.0	34 **Se** 2.4	35 **Br** 2.8	36 **Kr** 0
37 **Rb** 0.8	38 **Sr** 1.0	39 **Y** 1.2	40 **Zr** 1.4	41 **Nb** 1.6	42 **Mo** 1.8	43 **Tc** 1.9	44 **Ru** 2.2	45 **Rh** 2.2	46 **Pd** 2.2	47 **Ag** 1.9	48 **Cd** 1.7	49 **In** 1.7	50 **Sn** 1.8	51 **Sb** 1.9	52 **Te** 2.1	53 **I** 2.5	54 **Xe** 0
55 **Cs** 0.7	56 **Ba** 0.9	57-71 — 1.1-1.2	72 **Hf** 1.3	73 **Ta** 1.5	74 **W** 1.7	75 **Re** 1.9	76 **Os** 2.2	77 **Ir** 2.2	78 **Pt** 2.2	79 **Au** 2.4	80 **Hg** 1.9	81 **Tl** 1.8	82 **Pb** 1.8	83 **Bi** 1.9	84 **Po** 2.0	85 **At** 2.2	86 **Rn** 0
87 **Fr** 0.7	88 **Ra** 0.9	89- 1.1-															

of electrons (covalence). Since compounds are often formed by a combination of bonding factors, it may aid in the understanding of bonding to consider the combining capacity from a simple practical standpoint.

When atoms combine to form molecules they unite in different proportions in different compounds. Since hydrogen is the lightest element known, the other elements are often compared with it. The holding power of hydrogen for other atoms is 1, or, more simply stated, the combining capacity of hydrogen is 1. One atom of hydrogen combines with 1 atom of chlorine. The chlorine atom, and any other atom that unites with a hydrogen atom in a 1:1 ratio, also has a combining capacity of 1. Aluminum unites with 3 chlorine atoms, giving it a holding power three times that of hydrogen; therefore, its combining capacity is 3. Carbon tetrachloride is a compound in which 4 chlorine atoms are held in combination with 1 carbon atom; hence, the combining capacity of carbon is 4. The formulas of these compounds may be used to illustrate combining capacity as follows:

```
                    Cl              Cl
                    |               |
H—Cl            Cl—Al—Cl        Cl—C—Cl
                                    |
                                    Cl

(HCl)            (AlCl3)          (CCl4)
```

Each bond represents a holding power, a combining capacity, or a valence, of 1. In the ordinary empirical formulas given in parentheses, the number of chlorine atoms that combine with H, Al, and C is given by the subscript following the Cl.

THE SHAPES OF MOLECULES

The shape of a molecule is determined by the distances between bonded atoms in the molecule, called **bond lengths,** and the angles between bonds, called **bond angles.** In simple diatomic molecules, such as H_2 and CO, the bond length completely describes the geometry of the molecule. In triatomic and higher polyatomic molecules, more than bond lengths must be stated to completely describe the geometry of the molecule. In such molecules the geometry is determined by the nature of the central atom, such as oxygen in H_2O and carbon in CCl_4.

In triatomic molecules, two molecular geometries are found, **linear** and **angular.** The use of Lewis formulas allows one to predict which of these two geometries a given molecule will exhibit. The molecule BeH_2, which has the Lewis formula H:Be:H, is linear, just as is CO_2, which has the Lewis formula :Ö=C=Ö:. In these Lewis formulas, the symbols (—) and (:) have been used interchangeably to represent pairs of electrons. In the BeH_2 molecule, all electrons are involved in bonding, whereas, in CO_2 there are two pairs of electrons on each oxygen which are not involved in bonding. These electrons are called **non-bonding electrons,** as opposed to those involved in bonding, which are called **bonding electrons.** The molecule H_2O with the Lewis formula

H—Ō—H (angular Lewis formula, with non-bonding pairs on O), and the molecule NO_2 with the Lewis formula ⟨O=Ṅ—O⟩ or ⟨O—Ṅ=O|, are both angular. Notice that each of these molecules, in contrast to BeH_2 and CO_2, has non-bonding electrons on the central element. Since electrons repel one another, it is expected in these molecules that the bonding electrons and the non-bonding electrons associated with the central atom will orient themselves so that the repulsion interactions will be minimized. That is, they will seek the maximum distance of separation.

These principles, coupled with the nature of the Lewis formulas, would lead to the prediction that BeH_2 and CO_2, each having only bonding electrons on the central element, would be linear. This geometry would give the largest separation of the bonding pairs of electrons and would minimize the repulsions. In H_2O and in NO_2 the interaction of bonding pairs with non-bonding electrons on the central atom would lead to an angular molecule.

Typical examples of tetratomic molecules are BF_3 and NH_3, which have the Lewis formulas:

|F̄| bonded to B, which is bonded to two further ⟨F⟩ atoms; and N| bonded to three H atoms.

These formulas indicate that the electronic environment of the central atoms in these molecules is quite different. The BF_3 molecule has a central atom with three bonding pairs of electrons, whereas the NH_3 molecule has a central element with three bonding pairs and a non-bonding pair. The geometries of the two molecules are predicted by the application of the electron repulsion principle to be **planar triangular** (**trigonal**) for BF_3 and **triangular based pyramid** (**pyramidal**) for NH_3 (Fig. 3–9). In BF_3 electron repulsions are minimized by placing the bonding pairs of electrons at the corners of an equilateral triangle.

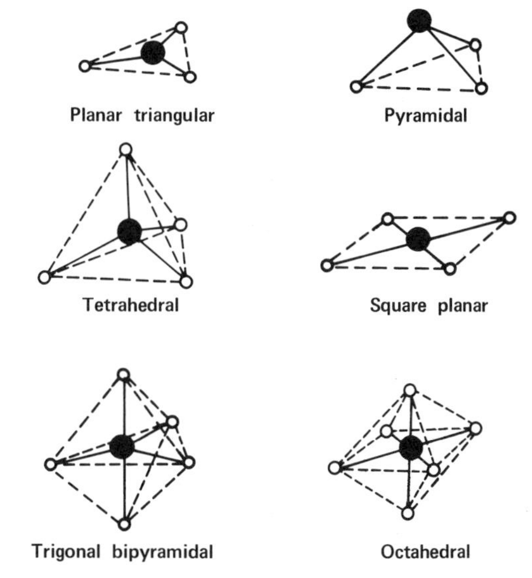

Figure 3–9 The most common geometries of simple molecules.

The interaction of a fourth pair of electrons, the non-bonding pair, in NH_3 results in the non-planar geometry.

The predominant geometry found for molecules with four atoms bonded to a central atom is **tetrahedral,** as shown in Figure 3–9. In this type of molecule the central atom is located at the center of the tetrahedron and the bonded atoms are found at the apices. This geometry occurs when four bonding pairs of electrons seek the maximum distance of separation in order to minimize the repulsions. The only other geometry observed for such pentatomic molecules is **square planar,** in which the bonded atoms are located at the corners of a square with the central element located in the center of the square. This geometry is limited to transition metal compounds.

The most common geometry for molecules with five atoms around the central atom is **trigonal bipyramidal,** and that for six atoms around a central atom is **octahedral** (Fig. 3–9).

The shapes of complex molecules can usually be predicted by considering the principles discussed here for simple molecules. Shapes of specific simple and complex molecules will be discussed in later chapters.

For simple molecules containing second period elements (i.e., B, C, and so forth) as the central atom, the approximate bond angles can be predicted by considering the sum of the nonbonding electron pairs and the bonding electron pairs associated with the central atom. For example, it was indicated that pentatomic molecules, such as CH_4, exhibit tetrahedral geometry. In this molecule the H—C—H bond angle is 109.5°, as the result of repulsions of the bonding electron pairs. However, the ammonia molecule, which has one non-bonding pair of electrons and three atoms bonded to nitrogen, and which has pyramidal geometry, has an H—N—H bond angle of 107°, nearly the same as that in CH_4. The water molecule, with two pairs of non-bonding electrons on the central atom and two bonded hydrogen atoms, displays an H—O—H bond angle of 105°. In general for second period elements, a total of four non-bonding electron pairs and atoms bonded to the central element will result in bond angles which equal or approximate those found in CH_4.

ATOMIC ORBITAL HYBRIDIZATION

The shape of a molecule can be enlightening as to the bonding interactions among the constituent atoms. In the cases of molecules having all single bonds, the relative orientations of the atoms establish the relative orientations of the sigma-bonds. For example, in methane, CH_4, previously stated to have tetrahedral geometry, four sigma-bonds, bonding a carbon atom to each of four hydrogen atoms, are pointed toward the apices of a tetrahedron. The assumption of inter-bond angles of 109.5° is consistent with the experimental observation that each of the four hydrogen atoms bonded to carbon has the same chemical properties, and thus must be involved in identical bonding. These experimental observations are inconsistent with a bonding model which involves the use of pure atomic orbitals by carbon.

If the carbon atom in methane uses four pure atomic orbitals to bond to each of four hydrogen atoms, the molecule would not be tetrahedral. This is

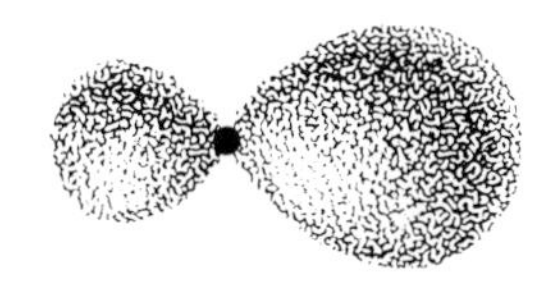

Figure 3-10 An sp^3 hybrid orbital.

because the four orbitals of the valence shell of carbon, one s-orbital and three p-orbitals, are not equivalent. The use of pure atomic orbitals would result in three sigma-bonds which are mutually perpendicular, due to the overlap of the three 2p-orbitals of carbon with three 1s-orbitals of hydrogen. The fourth sigma-bond would arise by overlap of the carbon 2s-orbital with a hydrogen 1s-orbital. Clearly this bonding model would not produce a tetrahedral arrangement of four equivalent bonds, as is the case in methane.

A set of four equivalent sigma-bonds in methane is explained by considering the combination of the carbon 2s-orbital and the three carbon 2p-orbitals as generating a set of four equivalent **hybrid orbitals.** These four hybrid orbitals, **designated sp^3-orbitals,** are not like s-orbitals or p-orbitals. Each appears similar to a p-orbital with the two lobes being of different sizes (Fig. 3–10). The relative orientations of the four hybrid sp^3-orbitals is such that the larger lobe of each points towards the apex of a tetrahedron (Fig. 3–11).

The H—N—H bond angle in ammonia, 107°, is best explained by considering the hybridization of a 2s-orbital and three 2p-orbitals on nitrogen to give a set of four sp^3-hybrid orbitals. This model also explains the H—O—H bond angle of 105° in water.

For simple molecules having a total of three bonded atoms and non-bonding electron pairs around the central atom, the concept of atomic orbital hybridization can be used to explain the observed geometry. The bonded atoms or non-bonded electron pairs found at the corners of an equilateral triangle are bonded to a set of three equivalent **sp^2-hybrid orbitals** on the central atom. This set of hybrid orbitals, arising from the combination of an s-orbital and two p-orbitals, results in bond angles of 120° (Fig. 3–11). In this bonding scheme, one p-orbital on the central element which is not included in the hybridization assumes an orientation perpendicular to the plane of the three hybrid orbitals.

Molecules having a total of two bonded atoms and nonbonding electron pairs on the central atom are linear. This geometry is explained by considering a set of two equivalent **sp-hybrid orbitals** on the central element and two unhybridized p-orbitals oriented perpendicular to the sp-hybrid orbitals and perpendicular to one another. The use of sp-hybrid orbitals by a central element results in a bond angle of 180° (Fig. 3–11).

In many simple molecules containing central elements from the third and higher periods, hybrid orbitals are not prevalent, and the bonding is best

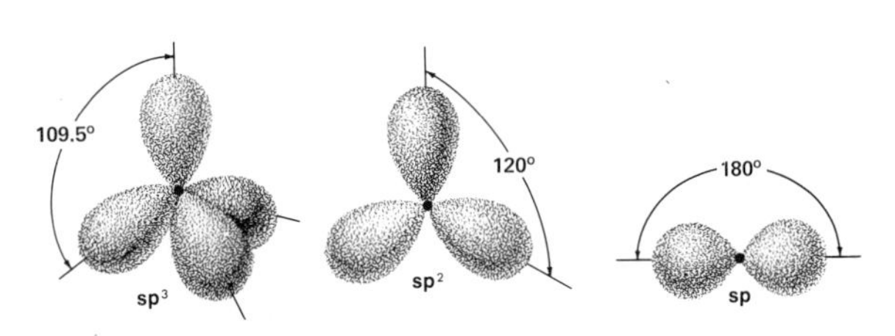

Figure 3-11 The relative orientations of equivalent hybrid orbitals.

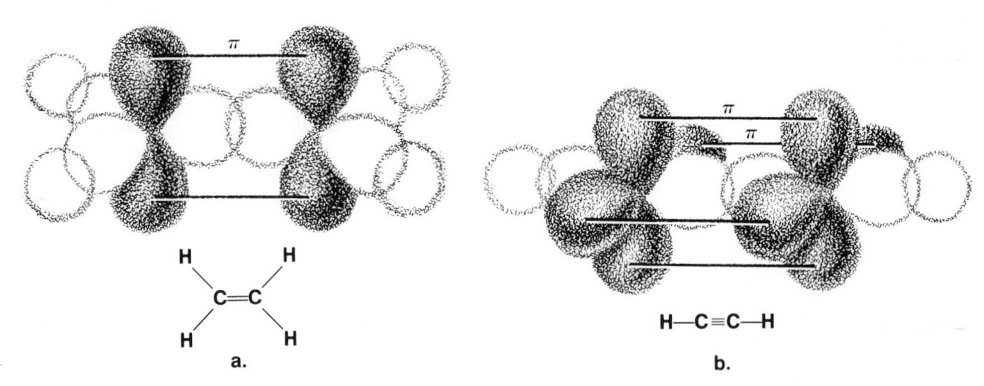

Figure 3-12 The approach of hybridized (unshaded) orbitals to form σ-bonds and unhybridized p-orbitals (shaded) to form π-bonds in ethylene and acetylene.

described by the use of pure atomic orbitals. For example, the H—P—H bond angle in phosphine, PH_3, is 92°, which is nearly that expected if each hydrogen is bonded to one of the three mutually perpendicular 3p-orbitals on phosphorus. The same is true for hydrogen sulfide, H_2S, in which the H—S—H bond angle is found to be 92°.

DOUBLE AND TRIPLE BONDS

As indicated earlier, in many molecules more than one pair of electrons is shared between two atoms. In such multi-bonded molecules, there must be π-bonds involved since only one σ-bond can form between any two atoms.

The molecule ethylene, C_2H_4, involves one σ-bond and one π-bond between the carbon atoms as shown in Figure 3–12 a. The sp^2-hybrid orbitals on each carbon form two σ-bonds with hydrogen atoms in addition to the carbon-carbon σ-bond. The remaining unhybridized p-orbitals of the two carbon atoms form a π-bond.

The molecule acetylene, C_2H_2, has one σ-bond and two π-bonds between the carbon atoms as indicated in Figure 3–12 b. The carbon atoms use sp-hybrid orbitals in the formation of σ-bonds between themselves and with the hydrogen atoms. The remaining p-orbitals on the carbon atoms combine to form two π-bonds.

Questions

1. Which of the electrons in an atom's structure are termed valence electrons? What relation do valence electrons have to atomic orbitals?

2. Explain and give an example of an ionic bond.

3. What is a covalent bond? Does the formation of this bond involve the transfer of valence electrons? Explain.

4. Describe a polar bond. What is meant by a dipole?

5. What is meant by the electronegativity of an atom?

6. Employing the concept of electronegativity, how is it predicted whether the combination of two elements will produce (a) an ionic bond (b) a covalent bond?

7. Using Lewis formulas, illustrate the bonding in each of the following: CO_2, NH_4^+, H_2O, CH_4, OH^-, BCl_3, HCl, O_2, SO_2.

8. Indicate the more electronegative element in each of the following pairs:
(a) S($Z = 16$) or Cl($Z = 17$)
(b) S($Z = 16$) or Se($Z = 34$)
(c) N($Z = 7$) or P($Z = 15$)

9. Predict which of the following molecules have dipole moments:

(a)

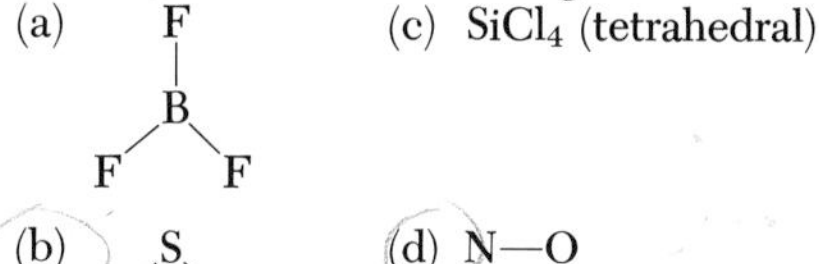

(c) $SiCl_4$ (tetrahedral)

(b) S with two O (bent: O—S—O)

(d) N—O

10. Assuming that all bonds are single bonds, predict the shapes of each of the following molecules:
(a) NO_2 (c) SCl_4
(b) PF_3 (d) $AsCl_5$

11. For molecules (a) and (b) in question 10, identify the type of hybrid orbitals used by the central atom.

12. Which of the following molecules must have at least one multiple (double or triple) bond? In each case determine whether the bond is double or triple.
(a) HCCH (b) HNNH (c) HOOH

13. What are the major differences between a sigma-bond and a pi-bond?

14. Using the rule of octets, predict the combining capacities of each of the following elements:
(a) Be (b) Si (c) P (d) Al (e) Na (f) Cl (g) O (h) Ne

15. Use the results of question 14 to predict the formulas of compounds containing:
(a) Be and P (b) Si and O (c) Al and Cl

16. In the TiO_2 crystal, each Ti^{+4} is surrounded by six O^{-2}. What is the coordination number of O^{-2}?

Suggested Reading

Ferreira: Molecular Orbital Theory, An Introduction. Chemistry, Vol. 41, No. 6, p. 8, 1968.
Gillespie: The Electron-Pair Repulsion Model for Molecular Geometry. Journal of Chemical Education, Vol. 47, p. 18, 1970.
Griswold: Chemical Bonding and Structure. Lexington, Mass., D. C. Heath and Co., 1968.
House: Ionic Bonding in Solids. Chemistry, Vol. 43, No. 2, p. 18, 1970.
Howald: Bond Energies in the Interpretation of Descriptive Chemistry. Journal of Chemical Education, Vol. 45, p. 163, 1968.
Margolis: Bonding and Structure. New York, Appleton-Century-Crofts, 1968.

CHAPTER 4

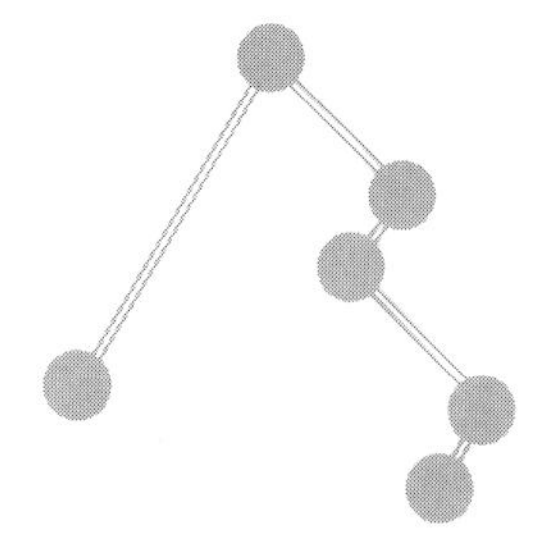

THE STATES OF MATTER

In the preceding chapter, the material presented included the nature of matter, its units of composition, atoms and molecules, and the forces of interaction between atoms in molecules. Because of the sizes and masses of atoms and molecules, man's study of matter always involves samples which contain large numbers of these units. The **macroscopic** properties of matter are determined by the properties of collections of large numbers of atoms and molecules. However, the properties of the individual atoms and molecules of a substance are significant in determining the properties of a collection. One of the most important considerations in predicting macroscopic properties is the forces between molecules, between ions, or between atoms which are not bonded. The nature and magnitudes of these interparticle interactions determine the **states of matter.** The states of matter are three in number: **solid, liquid,** and **gas.** These states are sufficiently different that only a few gross distinguishing features need be indicated to establish a clear differentiation.

Solids have a fixed shape and a fixed volume. They are incompressible; that is, they do not decrease their volumes when large forces are applied. The density of a solid generally is higher than that of a liquid or a gas.

Liquids have no fixed shape but assume the shape of their container. They do have a definite volume. Liquids are nearly incompressible and have densities which are usually lower than those of solids.

Gases have no fixed shape or volume but rather occupy the entire space afforded within the walls of their container. The densities of gases are much lower than those of liquids and solids. The ratio of the density of a gas to the density of a liquid or solid is usually less than 1/100.

THE PROPERTIES OF GASES

Several properties are common to the gaseous state. For example, gases expand when heated and contract when cooled or when external pressure is applied. Gases diffuse rapidly, filling any vessel and mixing with any other gas that is present in the vessel. Also, gases exert a pressure on the walls of any vessel in which they are stored. All the properties of gases can be explained on the basis of the kinetic molecular theory of matter to be presented later in this chapter.

The discovery of gases and a study of their properties occurred rather late in the early history and development of chemistry. Van Helmont was responsible for the name gas, and around 1630 he described carbon dioxide, which he called "gas sylvestre," whereas other investigators called it "fixed air." He was of the opinion that gas could not be contained in a vessel, but Robert Boyle later demonstrated the collection of a gas. In 1754 Joseph Black reported extensively on the preparation and properties of fixed air (carbon dioxide). Cavendish, who was given credit for the discovery of hydrogen in 1766, devised many pieces of apparatus for the preparation, collection, and handling of gases.

Some gases possess characteristic colors or odors that aid in their detection. The majority of gases are invisible, however, and must be detected by one or more of their other characteristic properties. In general, all gases possess properties that distinguish them from solids and liquids.

Pressure and its Measurement. **Pressure** is the property of matter which determines the direction of bulk flow, in that matter flows from regions of high pressure to regions of lower pressure. Pressure is expressed as force per unit area, often in units of pounds per square inches, lb/in.^{2} As stated previously, gases exert a pressure on the walls of the vessel in which they are stored. This pressure is most easily measured with a **barometer.**

A **Torricelli barometer** consists of a glass tube more than 760 mm in length, sealed at one end, completely filled with mercury, and inverted in a beaker also filled with mercury (Fig. 4–1). When inverted, the mercury drops somewhat

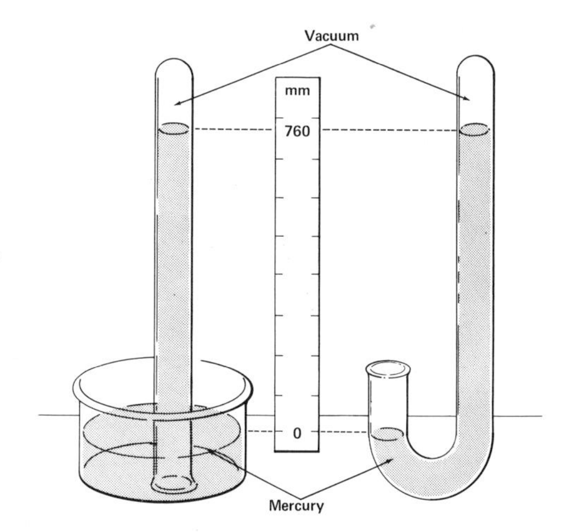

Figure 4–1 Simple barometers at sea level, or 1 atmosphere of pressure.

in the tube, creating a vacuum in the upper portion of the tube. The extent to which the mercury drops is determined by the force per unit area which the earth's atmosphere exerts on the mercury in the beaker. The mercury drops until the pressure exerted by the atmosphere equals the pressure exerted by the column of mercury at the level of the mercury surface in the beaker. Under these conditions the pressure of the atmosphere is said to be "supporting" a column of mercury.

The standard unit of pressure used in studies of gases is **one atmosphere.** This is the average pressure which the earth's atmosphere exerts at sea level, and is equivalent to 14.7 lbs/in^2, or a pressure which will support a column of mercury 760 mm in height. This pressure is commonly referred to as 760 torr. The pressure unit **torr,** which is 1/760 atmosphere, is named in honor of the inventor of the barometer.

Almost any liquid can be used in a tube to make a barometer. One atmosphere of pressure will support a column of water 34 feet high, but this length is impractical to measure. Consequently the denser liquid mercury is used in most barometers.

If a barometer is carried to the top of a mountain, where the distance to the outer limit of the atmosphere is less than at sea level, it is observed that the atmosphere does not support a column of mercury 760 mm in length. At higher elevations the atmosphere supports even shorter columns of mercury.

Boyle's Law. As early as 1660, Boyle observed that *if the temperature of a sample of gas is kept constant, the volume varies inversely with the pressure applied to the gas.* In proportionality form, **Boyle's law** is given as $V_T \alpha 1/P$. In this expression, V_T is the volume of the gas at a given temperature and P is the pressure. This relationship is shown graphically in Figure 4–2.

Since gases are not always kept at atmospheric pressure, Boyle's law can be applied to calculate the volume of a gas at any known pressure. As a problem, assume that a sample of gas has a volume of 240 ml at a pressure of 500 torr. If the pressure is reduced to 300 torr, keeping the temperature constant, what will be the new volume of the gas? First it is seen by simple reasoning that the new volume will be greater than the original, since the pressure is reduced, thereby increasing the volume of the gas. The new volume V is equal to the old volume, which is 240 ml multiplied by a ratio of the two pressures, or stated numerically,

$$V = 240 \text{ ml} \times \frac{500 \not{\text{torr}}}{300 \not{\text{torr}}} = 400 \text{ ml}$$

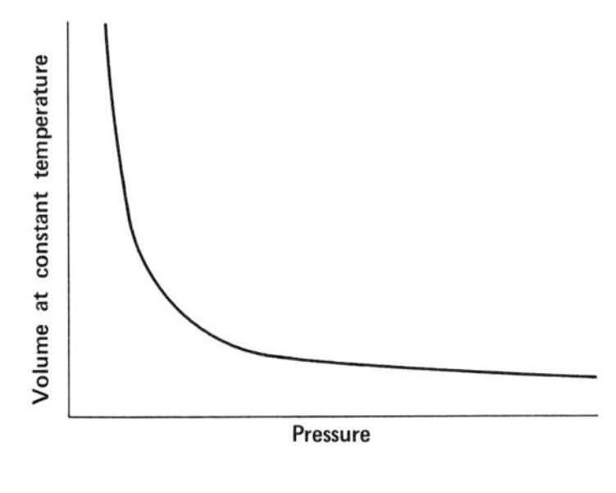

Figure 4–2 The volume-pressure relationship of a gas at constant temperature.

As a check on the calculation, it is seen that the pressure has been approximately halved, so that the volume should be approximately doubled. Often it is necessary to know what volume a sample of gas would occupy at standard pressure, or 760 torr. If a sample of oxygen has a volume of 380 ml at a pressure of 700 torr, what would be the volume at standard pressure? In this problem the pressure must be increased to 760 torr, and therefore the volume will decrease. The calculation is

$$V = 380 \text{ ml} \times \frac{700 \cancel{\text{torr}}}{760 \cancel{\text{torr}}} = 350 \text{ ml}$$

It can be seen that the most important point to be considered in problems of this type is whether the resulting volume will be larger or smaller than the original volume. If the final pressure is greater than the original pressure, then the resulting volume will be less than the original; however, if the resulting pressure is less than the original pressure, the resulting volume will be larger than the original.

Charles's Law and the Absolute Temperature Scale. In 1787 the French physicist Charles observed that cooling a gas caused it to contract $\frac{1}{273}$ of its volume per degree centigrade fall in temperature (Fig. 4–3). This immediately suggested that if it were possible to cool a gas to $-273°C$, the gas would have a zero volume. In practice, however, gases become liquefied before they reach a temperature of $-273°C$ and therefore no longer follow the gas laws. Nevertheless, the value of $-273°C$ was established as absolute zero, and defined the zero point of the **absolute,** or **Kelvin,** scale. Each absolute or Kelvin degree equals a centigrade degree and consequently centigrade temperatures are converted to absolute or Kelvin temperatures simply by adding 273.

Charles's Law states that the volume of a *gas varies directly as the absolute temperature, if the pressure is kept constant.* In proportionality form, Charles's law is given as $V_P \propto T$.

By the application of Charles's law, it is possible to calculate the volume a sample of gas would occupy at a temperature different from the original, if the pressure is kept constant. As a problem, assume that a sample of gas has a volume of 100 ml at 30°C and calculate the volume at 100°C. First, add 273° to each of the centigrade temperatures to convert them to absolute temperatures. This gives temperatures of 303°K and 373°K, respectively. Since the volume changes directly as the temperature, the new volume is equal to the old volume multiplied by the ratio 373/303, or the calculations may

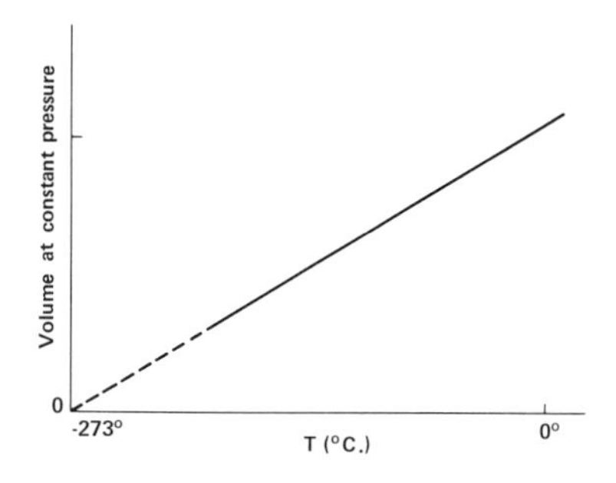

Figure 4-3 The volume-temperature relationship of a gas at constant pressure.

be expressed as:

$$V = 100 \text{ ml} \times \frac{373\,°\text{K}}{303\,°\text{K}} = 123 \text{ ml}$$

In general, if the temperature of a gas increases, the volume increases, and in calculations, the ratio of the two temperatures is written with the higher one over the lower. In problems in which the temperature is decreased, the volume decreases, and the ratio is written with the lower temperature over the higher.

Combining the knowledge obtained from the application of Boyle's law and Charles's law, it is possible to calculate the effect of changes in both pressure and temperature on the volume of a sample of gas. As a problem, assume that a sample of gas has a volume of 600 ml at 800 torr and a temperature of 80°C. Calculate the volume at 600 torr and a temperature of 10°C. The simplest way to solve this problem is to consider the pressure change and temperature change separately. Since the volume varies inversely with the pressure, write

$$V = 600 \text{ ml} \times \frac{800 \text{ torr}}{600 \text{ torr}}$$

This would provide a correction for the pressure change, and the resulting volume would be larger than that of the original sample. Since the volume varies directly as the absolute temperature, write

$$V = 600 \text{ ml} \times \frac{(10 + 273)°\text{K}}{(80 + 273)°\text{K}}$$

Combining these two operations to correct for both temperature and the pressure, write

$$V = 600 \text{ ml} \times \frac{800 \text{ torr}}{600 \text{ torr}} \times \frac{283\,°\text{K}}{353\,°\text{K}} = 641 \text{ ml}$$

Recall that the pressure of 760 torr, which is the average pressure at sea level, is established as the **standard pressure.** The **standard temperature** is 0°C or 273°K. **Standard conditions** refer to standard temperature and standard pressure, **STP.** It is often convenient to calculate the volume a certain sample of gas would occupy under standard conditions. For example, if a sample of gas has a volume of 1000 ml at a pressure of 900 torr and a temperature of 40°C what is the volume under standard conditions? Correcting first for pressure, write

$$V = 1000 \text{ ml} \times \frac{900 \text{ torr}}{760 \text{ torr}}$$

The resulting volume is larger than the original since the pressure is decreased under standard conditions. Correcting for the temperature, write

$$V = 1000 \text{ ml} \times \frac{(0 + 273)°\text{K}}{(40 + 273)°\text{K}}$$

Again combining these two operations, write

$$V = 1000 \text{ ml} \times \frac{900 \text{ torr}}{760 \text{ torr}} \times \frac{273\,°\text{K}}{313\,°\text{K}} = 1033 \text{ ml}$$

In problems of this type, it is important to remember that the volume of a gas varies inversely with the pressure and directly with the absolute temperature, and corrections must be made accordingly.

Avogadro's Hypothesis: the Molar Volume. In 1808 the French chemist Gay-Lussac investigated the chemical reactions that occur between gaseous substances. He observed that *under similar conditions of temperature and pressure the volumes of gases that react with each other are always in a ratio of small whole numbers.*

In 1811, Avogadro, an Italian physicist, proposed a fundamental explanation to account for Gay-Lussac's observations. He stated that *equal volumes of all gases under the same conditions of temperature and pressure contain the same number of molecules.* In proportionality form this expression is

$$V_{T,P} \, \alpha \, n$$

where n is the number of moles of molecules.

Knowing the gram molecular weight of oxygen, 32g/mole, and its density at 0°C and 760 torr, 1.429g/l, it is possible to calculate the volume occupied by one mole of oxygen at standard conditions. Thus,

$$\frac{32\text{g}\,\cancel{O_2}}{\text{mole}} \times \frac{1\,\text{l}\,O_2}{1.429\text{g}\,\cancel{O_2}} = \frac{22.4\,\text{l}}{\text{mole}}$$

One mole of oxygen gas occupies a volume of 22.4 liters at standard conditions. In fact a volume of 22.4 liters of any gas at standard conditions contains a gram molecular weight of the gas. For example, this volume of hydrogen weighs 2.016g. Therefore, the gram molecular weight of hydrogen is 2.016g, whereas 22.4 liters of nitrogen weighs 28g, giving nitrogen gas a gram molecular weight of 28g. The volume occupied by one mole of a gas, or 22.4 liters, is called the **molar volume** of a gas. A mole, or molar volume, of any gas under standard conditions contains the same number of molecules. This number of molecules equals 6×10^{23}, and is called **Avogadro's number.** In other words, 1 mole, or molar volume, or 22.4 liters, of a gas contains 6×10^{23} molecules of that gas.

The gram molecular weight of a gas can be determined by merely weighing 22.4 liters of the gas at standard temperature and pressure. Since it is much more convenient to work with smaller volumes of gas, a definite volume of gas is usually weighed under prevailing temperature and pressure conditions, and then the weight of 22.4 l at standard conditions is calculated. For example, assume that 2.0 liters of a gas at a temperature of 20°C and a pressure of 750 torr weighs 2.30g. What is the gram molecular weight? Realizing that the density of the gas, 2.30g/2.0 liters, has the same units as the gram molecular weight, it is possible to convert the density to gram molecular weight as follows:

$$\frac{2.30\text{g}}{2\,\cancel{\text{liters}}} \times \frac{760\,\cancel{\text{torr}}}{750\,\cancel{\text{torr}}} \times \frac{293\,\cancel{°\text{K}}}{273\,\cancel{°\text{K}}} \times \frac{22.4\,\cancel{\text{liters}}}{1\text{ mole}} = \frac{28\text{g}}{\text{mole}}$$

In this calculation, the conversion factor involving pressure is greater than one, because the density increases as the pressure increases. Likewise, the temperature factor is greater than one, since decreasing the temperature increases the density. The calculated gram molecular weight of the gas is 28g/mole.

If the formula of a gas is known, it can be used to calculate the molecular weight and the density. For example, the formula for ammonia is NH_3, and the molecular weight is 14 amu plus 3 amu, or 17 amu. To calculate the density of ammonia gas in grams per liter at standard conditions, it is necessary to divide the gram molecular weight by the molar volume. Thus,

$$\frac{17\text{g}}{\cancel{\text{mole}}} \times \frac{1\ \cancel{\text{mole}}}{22.4\ \text{liters}} = 0.771\text{g/l}$$

Dalton's Law of Partial Pressures. Dalton observed that *when two or more gases are present in a mixture, each gas exerts a pressure that is apparently not affected by the presence of the other gases.* Stated in another way, the total pressure of a mixture of gases is the sum of the pressures of each individual gas. This is known as **Dalton's law of partial pressures.** In equation form, it is stated that $P_T = p_a + p_b + p_c + \cdot \cdot \cdot \ p_n$. To illustrate, consider two vessels of equal size and volume, one containing oxygen under a pressure of three atmospheres and the other containing helium under a pressure of two atmospheres. If both gases are placed in one vessel, they exert a total pressure of five atmospheres, since each gas exerts the same pressure that it would exert if it occupied the entire space by itself. In this mixture of oxygen and helium gas, the portion of the total pressure exerted by oxygen is called its **partial pressure,** p_{O_2}, which is three atmospheres, whereas the partial pressure of helium is two atmospheres.

In the laboratory gases are commonly collected by displacement of water from a bottle that is inverted in a pneumatic trough. Gases collected in this fashion always contain **water vapor** mixed with the gas. The total pressure of a given volume of oxygen collected over water is therefore the sum of the pressure of the oxygen and the pressure of the water vapor. For example, if the gas were collected under atmospheric pressure (barometer reading, 760 torr), and the pressure exerted by the water vapor at 25°C is equal to 25 torr, the partial pressure of oxygen is 760 − 25, or 735 torr. In this mixture of gases the partial pressure of water vapor is 25 torr.

As an illustration of these principles, calculate the volume at standard conditions of a 240 ml sample of gas collected over water at 23°C if the barometric pressure is 731 torr and the partial pressure of water is 21 torr. The actual pressure exerted by the gas is 731 − 21, or 710 torr. The volume at standard conditions, smaller than 240 ml since pressure on the gas increases, is calculated to be

$$V = 240\ \text{ml} \times \frac{710\ \cancel{\text{torr}}}{760\ \cancel{\text{torr}}} \times \frac{273°\cancel{\text{K}}}{(23 + 273)°\cancel{\text{K}}} = 207\ \text{ml}$$

Graham's Law of Effusion. At the beginning of this chapter it was stated that one of the important properties of a gas is its diffusibility. The diffusibility of a gas is its ability to permeate any space in which it may be placed. This property is often demonstrated by releasing a small quantity of a gas, such as ammonia or hydrogen sulfide, in the laboratory. In a short time the odor fills the entire room, indicating the rapid diffusion of the gas molecules. When a gas is introduced into an evacuated vessel, it immediately fills all the space in that vessel. If another gas is present in the vessel, the second gas mixes with the first by the process of diffusion and is soon found in equal concentration in all regions of that vessel.

Thomas Graham, a British physicist, observed that the rate of **effusion,** which is the rate of diffusion of a gas through a small orifice into an evacuated chamber, depends on the density of the gas. He stated that *the speed of effusion of two gases varies inversely as the square roots of the densities or molecular weights of the gases.* In proportionality form this is written as $v \propto 1/\sqrt{d}$, or $v \propto 1/\sqrt{MW}$. If the effusion of the inert gas neon is compared with that of hydrogen gas, it is found that the rate of effusion of hydrogen is four times that of neon. A similar comparison of the effusion rates of oxygen and of helium reveals that helium effuses two and one-half times as fast as oxygen.

Observation of effusion rates can be used to determine molecular weights of gases as illustrated by the following calculation: a sample of an unknown gas, X, equal in size to a sample of oxygen, takes 80 seconds to diffuse through a hole, whereas the oxygen sample takes 40 seconds. Graham's law states that

$$\frac{\text{rate of effusion of } O_2}{\text{rate of effusion of X}} = \frac{\sqrt{\text{MW of X}}}{\sqrt{\text{MW of } O_2}}$$

Realizing that the rate of effusion is inversely proportional to the time, and using the data given and a molecular weight of 32 amu for O_2, the following result is obtained:

$$\frac{80 \text{ sec}}{40 \text{ sec}} = \frac{\sqrt{\text{MW of X}}}{\sqrt{32 \text{ amu}}}$$

$$\text{MW of X} = 32 \text{ amu} \times \frac{(80)^2}{(40)^2} = 128 \text{ amu}$$

The molecular weight of the unknown gas is 128 amu.

IDEAL GASES AND THE EQUATION OF STATE

Restating Boyle's law, Charles's law, and Avogadro's hypothesis as

$$V_T \propto 1/P, \quad V_P \propto T, \quad \text{and} \quad V_{T,P} \propto n,$$

it is possible to write the following combined relationship:

$$V \propto nT/P$$

If this proportionality is changed to an equality using the proportionality constant R, and both sides of the equation are multiplied by P, the following relationship is obtained:

$$PV = nRT$$

This equation shows the relationship of all of the variables required to completely describe the behavior of a gas. A gas which rigorously obeys this equation is called an **ideal gas.** The equation is called the **equation of state for an ideal gas.** The value of the **gas constant,** R, can be determined by using known values of P, V, n, and T. For example, it is known that one mole of a gas ($n = 1$) occupies the molar volume (22.4 liters) at standard temperature (273°K) and pressure (1 atmosphere). The constant R is thus found to be:

$$R = \frac{PV}{nT} = \frac{(1\ \text{atm})(22.4\ \text{l})}{(1\ \text{mole})(273°\text{K})} = \frac{0.082\ \text{l atm}}{\text{mole}°\text{K}}$$

The ideal gas equation can be used for many calculations already discussed. For example, the volume occupied by three moles of a gas at 4 atmospheres and 500°K is

$$V = \frac{RnT}{P} = \frac{0.082\ \text{l}\ \cancel{\text{atm}} \times 3\ \cancel{\text{mole}} \times 500\cancel{°\text{K}}}{\cancel{\text{mole}°\text{K}} \times 4\ \cancel{\text{atm}}} = 30.7\ \text{l}$$

In this calculation, the procedure is to solve for the variable which is not given. As a second illustration, calculate the number of moles of gas which will exert a pressure of 16.6 atmospheres when contained in a 100 liter tank at 25°C.

$$n = \frac{PV}{RT} = \frac{16.6\ \cancel{\text{atm}} \times 100\ \text{l}}{0.082\frac{\text{l}\ \cancel{\text{atm}}}{\text{mole}\ \cancel{°\text{K}}} \times 298\cancel{°\text{K}}} = 68.0\ \text{moles}$$

THE KINETIC MOLECULAR THEORY

The observation of many facets of the behavior of gases under many conditions has led to a theory which is consistent with most experimental facts. This theory, called the **kinetic molecular theory,** is based on a set of fundamental assumptions which serve as a model for predicting properties. These assumptions are:

1. Gases are composed of molecules or atoms which have negligible volume, and which are at relatively large distances of separation.
2. The molecules move at high velocities in straight paths, but in random directions.
3. The molecules collide with each other and with the container walls, but the collisions are totally elastic so that energy is conserved. As the result of these collisions, some molecules move faster than others, and a distribution of molecular velocities results.
4. The average kinetic energy of the molecules of all gases is the same at a given temperature, and is directly proportional to the absolute temperature.
5. There are no attractive or repulsive forces between molecules.

The consistency of this theory and experimental observation can be checked by considering a gaseous sample in a cylinder with a movable piston. External pressure is varied by way of weights on the piston.

Experimentally, it has been found that the volume of the gas under two units of external pressure is equal to one-half the volume under one unit of pressure. Since, at constant temperature, the velocities of the gas molecules do not change, and the masses of the particles remain the same, there is no change in the kinetic energy of the gas. The only way the force on the under surface of the piston can be increased to balance the two units of external pressure is for the molecules to strike the piston twice as often. This can be accomplished by moving the piston downward into the cylinder until the volume of the gas is half of the original volume.

The observations leading to Charles's law are also consistent with the kinetic molecular theory. An increase in temperature will increase the velocity and kinetic energy of the gas molecules. In effect this causes an increase in the internal forces, since the gas molecules strike the under surface of the piston more often and with greater force. The piston moves upward to increase the space between the molecules, thus reducing the number of impacts and the force exerted by the gas. The upward movement also increases the volume of the gas. A decrease in temperature causes a decrease in the volume of the gas, since the kinetic energy of the molecules is less and the internal forces are decreased.

The kinetic molecular theory is seen to be consistent with Avogadro's hypothesis if it is realized that increasing the number of molecules in a sample increases the frequency of collisions per unit area of the walls of the container. The frequency of collisions per unit area of the walls, and thus the pressure, can remain constant at a given temperature only if the surface area is increased along with increasing the sample size.

Assumption 5 of the kinetic molecular theory indicates that molecules act independently of one another regardless of their identities. Consequently, at constant volume and temperature, the total pressure exerted by a gas mixture should be dependent only on the total number of molecules present, as stated in Dalton's law of partial pressures.

The fundamental assumptions of the kinetic molecular theory are consistent with Graham's law of effusion. The fourth assumption states that the average kinetic energy for all gases is the same at a given temperature. Thus, one half the product of the mass times the velocity squared must be the same for all gases. It is seen that if

$$m_A v_A^2 = m_B v_B^2 \quad \text{then} \quad m_A/m_B = v_B^2/v_A^2$$

Taking the square root of each side of this expression gives

$$v_B/v_A = \sqrt{m_A}/\sqrt{m_B}$$

which indicates that the theory is in agreement with experimentally observed fact.

There are no known gases which obey the ideal equation of state at all temperatures and pressures. As pointed out earlier, the forces of interaction between real gas molecules, although small, do exist. These forces become very significant under conditions when the average kinetic energy is small. Thus, at low temperatures or at high pressures, when these conditions exist, it is found that the ideal gas equation of state is not obeyed. That is, the ideal gas equation does not accurately describe the relationships among pressure, temperature, volume, and the number of moles for real gases.

Another property which distinguishes real and ideal gases is the real volumes of molecules. The kinetic molecular theory assumes that gaseous molecules occupy no space. Consequently, it is impossible for a real gas to adhere to Charles's law at very low temperatures, or to adhere to Boyle's law at very high pressures.

At high pressures or at low temperatures, attractive forces between gas molecules tend to pull the molecules together. If this effect is sufficiently large, the gas molecules coalesce to form a liquid. This process is called **liquefaction.**

Liquefaction can be induced more readily at high pressures, since the average distance of molecular separation is smaller. At very high temperatures, gases often cannot be liquefied by increasing the pressure. The **critical temperature,** T_c, is the temperature above which a gas cannot be liquefied by increasing the pressure. The **critical pressure,** P_c, is the pressure required to liquefy a gas at the critical temperature.

In addition to serving as a consistent explanation for the behavior of gases, the kinetic molecular theory can be used to qualitatively differentiate the properties of materials in the three states of matter. Not all matter can exist in each of the three states, but most materials (e.g., elements or compounds) can. The physical conditions under which each state exists define a set of unique characteristics of the material.

The kinetic molecular theory has assumptions, some of which are obviously inconsistent with the observed macroscopic properties of liquids and solids, and yet it is useful in explaining many of the characteristics of these states. In liquids it is found that rates of diffusion are considerably slower than in gases, but much faster than in solids, where rates of diffusion are often virtually immeasurable. These facts suggest that the molecules in liquids are moving at random, but in solids the rate of motion is very small or zero. The relative densities of the three states suggest a rather significant variation in the number of particles occupying a given volume, or in the average interparticle distance. The particles in a gas apparently have greater distances of separation than those of a liquid, and the particle separation in solids must be the smallest. The distance of separation of the particles in a material is a function of the magnitudes of the intermolecular forces of attraction. For compounds with very strong, attractive, intermolecular interactions, the distance of separation of the particles tends to be small, and this tendency is diminished as the attractive intermolecular interactions become smaller. In solids and liquids, for which the intermolecular forces are much larger than in gases, the average distance of separation of the molecules is quite small. This condition is in contrast to ideal gases, where it is assumed that no molecular interactions prevail.

In liquids, just as in gases, random motion of molecules and their resulting collisions lead to a distribution of molecular velocities and kinetic energies. The average kinetic energy of the molecules of a liquid is dependent on the temperature, but the nature of this dependence is not the same for all liquids.

In solids, molecules or ions oscillate about an average position in the crystal. An increase in the temperature of the solid results in increased violence of the oscillation. Through collisions with molecules or ions occupying neighboring sites in the crystal, a distribution of oscillation violence arises.

INTERMOLECULAR FORCES

Several types of interactions give rise to the forces which cause gases to deviate from ideal behavior. These same interactions cause liquids and solids, which are composed of atoms or covalent molecules, to remain in the condensed state or **phase.** The most important of these interactions are **dipole-dipole** and those which give rise to **London forces.** These belong to a general class of forces called **van der Waals' forces.**

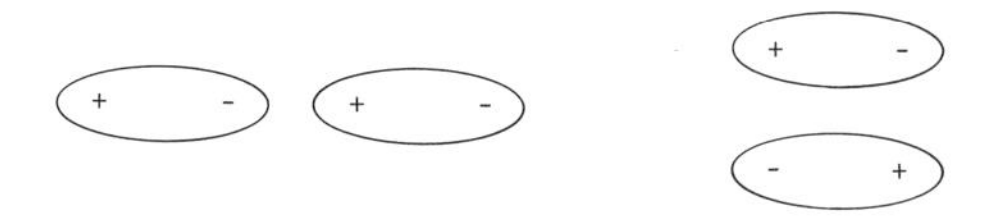

Figure 4–4 Two possible orientations of dipoles which lead to net attraction.

Dipole-dipole interactions develop from the attraction of oppositely charged portions of the dipoles of molecules. Figure 4–4 shows a representation of two orientations of dipoles which lead to a net attraction. The magnitude of the attraction increases with increasing size of the dipoles of the molecules. In addition, it increases with decreasing distance of separation of the molecules. However, at very small distances of separation, the electron clouds on adjacent molecules repel one another.

London forces explain the fact that even atoms of the inert gases have intermolecular attractions and can be liquefied. If the electron cloud of an atom at some instant does not have spherical symmetry about the nucleus, but is distorted or **polarized,** the atom will have a net dipole moment. Such a dipole moment could influence a nearby atom by inducing a distortion in its electron cloud. Figure 4–5 represents the instantaneous shapes of the electron clouds of two adjacent atoms at three different times. In the center the atoms have spherical electron clouds, no dipole moments, and therefore no net force of attraction. In both the other configurations, the polarized atoms have aligned dipole moments, which is to be expected since these dipole moments are mutually influential. For any configurations in which the polarized atoms have aligned dipoles, there will be a net force of attraction. It can be concluded that all configurations, other than that of spherical electron clouds, will result in forces of attraction.

The magnitudes of London forces between atoms increase with increasing ease of electron cloud distortion or **polarizability.** Because of the distance of the valence electrons from the nucleus, and because of the screening effect of the inner filled shells of electrons, large atoms of the heavy elements have the largest polarizabilities, and hence the largest London forces. This accounts for the increase in boiling points of the inert gases in going from helium to xenon.

In different liquids, the variation of intermolecular forces leads to distinguishing characteristics. For example, the **viscosity,** or resistance to flow, of a liquid is largely due to attractions between adjacent molecules. The movement of molecules will be fastest when they have low masses, symmetrical shapes, and weak intermolecular interactions. As the temperature is increased, the increased kinetic energies of the molecules overcome the intermolecular forces and the viscosity decreases.

Another distinguishing characteristic of liquids which is determined by molecular forces is **surface tension.** A molecule in the center of a liquid sample experiences forces of attraction in all directions, whereas a molecule at or near

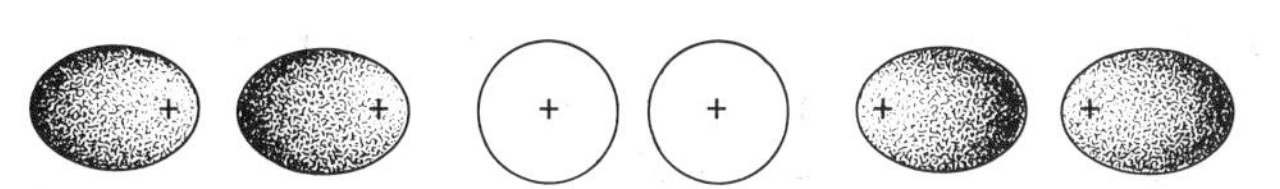

Figure 4–5 Instantaneous electron cloud shapes on neighboring atoms at three different times.

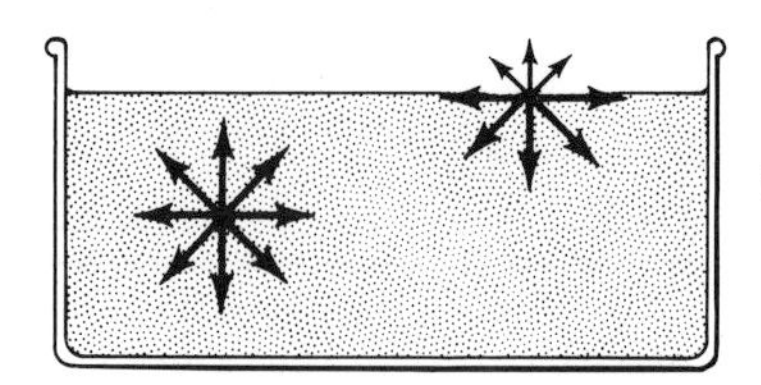

Figure 4-6 The forces acting on molecules in a liquid.

the surface of the liquid experiences forces which are greatest in the direction toward the center of the sample (Fig. 4-6). This inward pull on the surface molecules results in surface tension, which is the resistance to expansion of the surface area. A small liquid sample in the absence of any external influence will assume the shape of a sphere, since this shape has the smallest surface area for a given volume. As expected, increasing the temperature of a liquid decreases the surface tension because of decreased intermolecular interaction.

CHANGES OF STATE

Vaporization and Sublimation. In order to bring about liquefaction of a gas, it is necessary to decrease the average kinetic energy of the molecules sufficiently to allow intermolecular forces of attraction to pull the molecules together. In order to bring about the reverse process, **evaporation** or **vaporization,** it is necessary to increase the average kinetic energy of the molecules of a liquid until it is sufficiently large to overcome the intermolecular forces of attraction. Since there is a distribution of kinetic energies, some molecules will have kinetic energies higher than the average and some lower. The fraction of molecules with the highest kinetic energies will have the greatest tendency to overcome intermolecular forces and leave the liquid. If the highest energy molecules leave the liquid, the average kinetic energy of the molecules remaining behind will be decreased, and the temperature of the liquid will drop. Consequently, evaporation of a liquid always results in cooling the liquid, unless heat is supplied by the surroundings to maintain constant temperature.

The amount of heat, or **thermal energy,** which must be supplied to a given quantity of a liquid to bring about vaporization is a characteristic property of the liquid. This property is called the **enthalpy of vaporization.** Refrigerators and air conditioners use a cyclic process of **liquefaction,** or **condensation,** and evaporation of a suitable refrigerant to bring about cooling. It is the evaporation part of the process which is used to cool. The compressor increases the pressure on the gaseous refrigerant in order to bring about liquefaction.

The process of molecules leaving the solid state and entering the gaseous state without going through the liquid state is called **sublimation.** In solids the intermolecular attractive forces are usually considerably larger than those found in liquids. Consequently, the tendencies of liquids to evaporate at a given temperature are usually greater than the tendencies of solids to sublime. The amount of thermal energy, or heat, required to sublime a given quantity of solid is called the **enthalpy of sublimation.** The enthalpy of sublimation of a material is always greater than the enthalpy of vaporization.

Vapor Pressure. Liquids and solids placed in closed containers do not completely evaporate or sublime. This is because the gaseous molecules are confined to the container and cannot diffuse away, and consequently eventually reenter the condensed state. The gaseous molecules moving about at random will reenter the condensed state at a rate which is dependent on the number of molecules in the gaseous state. Eventually the rate at which molecules are reentering the condensed state will equal the rate at which they are entering the gaseous state. Under these conditions, the number of gaseous molecules occupying the volume above the condensed state will be invariant with time, even though a dynamic process prevails. This situation is referred to as a **dynamic equilibrium.** Under such equilibrium conditions, the pressure exerted by the gas over the condensed state is called the **equilibrium vapor pressure.** The vapor pressure of a material does not depend upon the relative amounts of condensed state or gaseous state present, but it does depend upon the nature of the material.

Boiling. The vapor pressures of liquids and solids increase with increasing temperature, as is shown in Figure 4–7 for water, iodine, carbon tetrachloride, and bromine. The temperature at which the vapor pressure of a material equals the pressure exerted by the surrounding atmosphere is called the **boiling point.** The temperature at which the vapor pressure of a substance is one atmosphere is called the **normal boiling point.** A liquid boils at lower temperatures if the pressure of its surrounding atmosphere is decreased. At higher altitudes, where the atmospheric pressure is relatively low, water boils at lower temperatures. Cooking food by boiling at high altitudes may be difficult, because water boils at a temperature which is not high enough for efficient cooking. This difficulty is overcome by the use of a pressure cooker, in which the steam that is formed in boiling is confined to increase the pressure. As the result of the increased pressure, the boiling point of the water is increased. For example, a steam pressure of 760 torr added to the atmospheric pressure will result in a boiling point of about 120°C.

Chemists often purify liquids by heating to convert the liquid to a gas and then cooling the gas to condense it back to a liquid. This process, called **distillation,** is performed using a distillation apparatus such as the one shown in Figure 4–8. Distillation results in separation of impurities from a sample because of differences in the boiling points. Occasionally it is necessary to distill a material at a relatively low temperature in order to avoid thermally induced

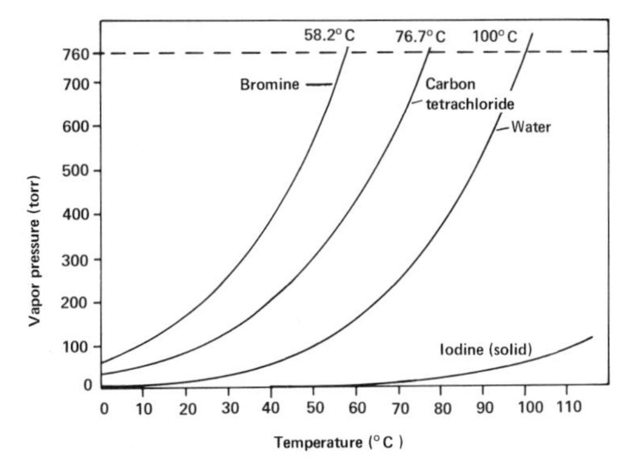

Figure 4-7 The variation of vapor pressure with temperature.

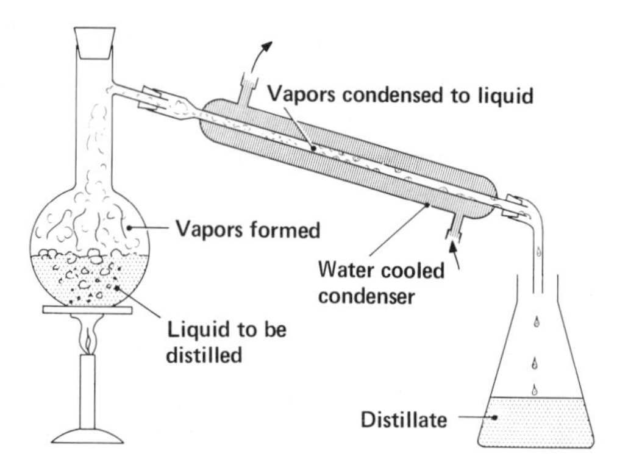

Figure 4-8 A simple distillation apparatus.

decomposition. This can be accomplished through decreasing the pressure of the atmosphere surrounding the liquid by creating a partial vacuum in the distillation apparatus. The pressure in the system can be lowered until the desired boiling temperature is attained.

The change of a liquid to a gas at the boiling point requires the addition of the enthalpy of vaporization. At 760 torr and 100°C all the heat added (540 cal) to boil one gram of water is consumed in overcoming intermolecular attractions and increasing the average distance of separation between the molecules. The temperature and the average kinetic energy of the molecules do not change during the process. This is schematically represented in the heating curve in Figure 4–9. It is seen that as a liquid is boiled, its temperature remains constant. Before boiling starts and after boiling is completed, all heat added serves to increase the kinetic energy, which increases the temperature of the system. When a gas condenses to a liquid, it releases the enthalpy of vaporization to its surroundings. This is why steam is used so efficiently in many processes to transfer thermal energy.

The boiling points of liquids increase with increasing magnitudes of intermolecular or interatomic forces. The lowest boiling points are observed for non-polar species such as H_2, O_2 and He. Somewhat higher boiling points are observed for liquids of non-polar molecules involving larger atoms whose electron clouds are more polarizable (e.g., Br_2, I_2, Xe). Liquids of polar mole-

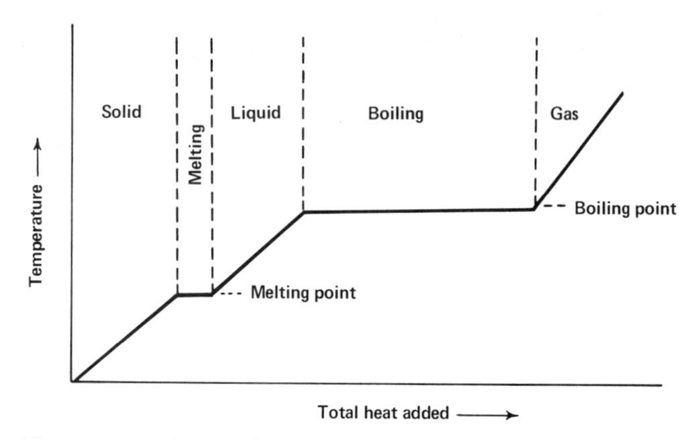

Figure 4-9 A heating curve showing the temperature of a system as it is influenced by the addition of heat at a constant rate.

cules show higher boiling points than liquids of non-polar molecules containing similar atoms. The highest boiling points are observed for the metallic elements and ionic compounds.

MELTING AND FREEZING. Most solids, when heated sufficiently, undergo a solid to liquid transition. This transition is called **melting** or **fusion,** whereas the reverse process is called **freezing** or **crystallization.** The former of these processes consumes heat from the environment, whereas the latter releases heat to the environment. The temperature at which this transition takes place is called the **melting point** or the **freezing point.** At this temperature, the vapor pressures of the solid and liquid are equal and the system is at equilibrium. Pure materials are thus predicted to remain at one temperature throughout a melting or freezing process. Impure materials have a gradual change in temperature during the solid-liquid transition. During the melting process, the temperature of an impure sample rises. This difference in the characteristics of pure and impure solids can be used as a simple test for purity. The melting points of solids are rather insensitive to pressure changes, although they do have a small dependence on them.

The heat added to bring about melting is called the **enthalpy of fusion.** The addition of 80 calories to 1 gram of ice at 0°C completely converts it to water at 0°C. All the heat added during the melting process is consumed in liberating molecules from their rather static positions in the solid. Consequently during the melting process the average kinetic energy of the molecules, and therefore the temperature of the system, remains constant. This is represented in the heating curve in Figure 4–9 where it is seen that heat added is consumed in increasing the temperature of the system only before melting starts and after it is completed. The increase in average distance of separation of the molecules in going from solid to liquid is very small compared to that involved in vaporizing a liquid. Therefore, it is expected, as is indicated for water, that the enthalpies of fusion of solids are usually smaller than the enthalpies of vaporization of corresponding liquids.

The melting points of solids increase with increasing magnitudes of intermolecular and interatomic forces. In general they vary in the same way as the boiling points.

THE NATURE OF THE CONDENSED STATES

THE LIQUID STATE

Although various macroscopic physical properties (e.g., viscosity, vapor pressure, surface tension) of liquids are well-characterized, the liquid state is not well understood on the microscopic level. It is known that long-range structure does not exist in liquids because of the random motion of constituent molecules; however, the rather strong short-range forces do give rise to structuring over short distances. That is, each liquid molecule can be pictured as surrounded by other molecules at a very short distance with the number of nearest neighbors determined by spatial considerations as well as intermolecular forces. The fluidity prevails because these clusters are continually interchanging members and each cluster can move relative to others.

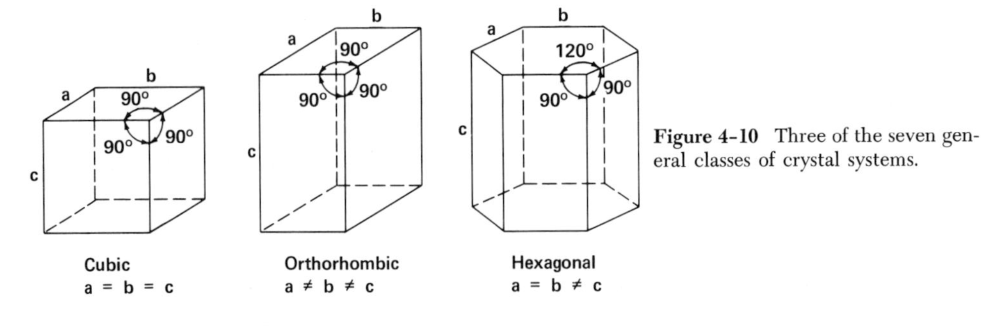

Figure 4-10 Three of the seven general classes of crystal systems.

THE CRYSTALLINE STATE

In many solids the atoms, molecules, or ions are arranged in a systematic, highly ordered pattern which repeats itself in three dimensions. Such solids are called **crystalline.** The three dimensional arrangement of the units is called the **space lattice** or **crystal lattice.** The smallest portion of a crystal which can be used to describe the crystal lattice is called the **unit cell.** The three dimensional extension of the unit cell results in highly regular macroscopic properties of crystals, such as angles between faces.

There are seven general classes of space lattices into which all crystal systems may be classified. Simple space lattices may be visualized as geometric entities in which the corners are occupied by the units of the solid. These space lattices vary in angles between faces and in the relative lengths of the edges. The simplest space lattice is the **cubic** lattice, in which all edges are the same length and all angles between faces are 90°. The **orthorhombic** space lattice has all edges unequal, but the interfacial angles are all 90°. The **hexagonal** space lattice has two edges equal and two angles equal. These three space lattices are represented in Figure 4–10.

Within each of the seven classes of space lattices there are several subclasses which are differentiated by units placed at the center or on the faces. Figure 4–11 shows the three subclasses within the cubic system: **simple cubic, body-centered cubic,** and **face-centered cubic.**

TYPES OF SOLIDS

Crystalline solids may be divided into four classes on the basis of the predominant forces holding the lattice together. These classes are **molecular, covalent network, ionic,** and **metallic.**

Molecular Solids. Molecular solids are composed of discrete covalent molecules held together in the lattice by either dipole-dipole interactions or

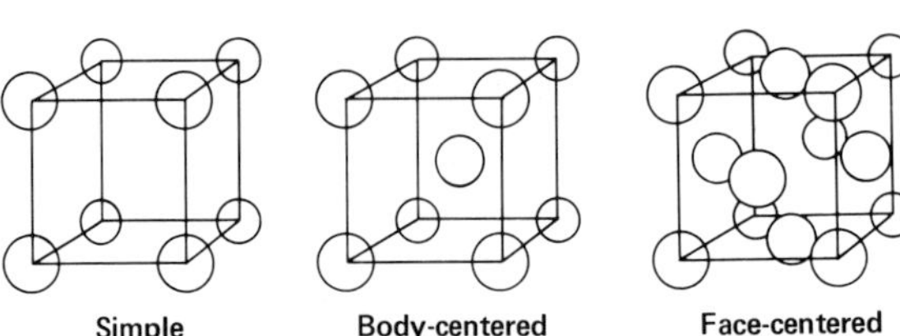

Figure 4-11 The subclasses of the cubic crystal system.

London forces. These are the weakest interparticle forces found in solids, and consequently, solids in this class have the smallest enthalpies of fusion, the lowest melting points, and the highest vapor pressures. They tend to be soft, are not good thermal conductors, and act as electrical insulators. Examples include ice, dry ice, naphthalene (moth balls), sulfur, and white phosphorus. The exceptionally strong dipole-dipole interactions found in water give ice a set of physical properties which deviate from those of most molecular solids.

Covalent Network Solids. Covalent network solids are composed of atoms which are covalently bonded with their neighbors to give a three dimensional network. A crystal of such a solid may be visualized as a gigantic molecule. The forces between atoms in these solids are due to very strong covalent bonds and result in very high melting points, low vapor pressures, and hardness. Covalent network solids are poor conductors of electricity and heat. Examples include diamond and quartz.

Ionic Solids. Ionic solids, discussed previously in Chapter 3, are composed of oppositely charged ions which occupy lattice sites so that each ion is surrounded by nearest neighbors of opposite charge. The lattice forces are quite strong, resulting in hardness, high melting points, and low vapor pressure.

Metallic Solids. Metallic solids are composed of metal atoms occupying the lattice sites. Over 60 per cent of all metallic solids have the **closest packed structures,** in which each atom touches its twelve nearest neighbors. Unlike the majority of other solids, metals are very good electrical and thermal conductors. The electrons of the metal atoms are thought to be responsible for carrying an electrical current. For this reason metallic solids are often thought of as a lattice of metal ions which is occupied by electrons which are free to move about in the lattice. This freedom of electron motion would account for the observed electrical conductivity. The properties of metallic solids show a wide variation. For example, the hardness ranges from very hard in the case of tungsten to very soft for the alkali metals. The melting points also show a wide variation, with mercury melting at $-38.9°C$ and tungsten melting at $3370°C$.

Questions

1. What factors determine whether a certain substance exists as a solid, liquid, or gas? Explain.

2. How does the kinetic molecular theory account for the pressure that a gas exerts on the walls of a container?

3. How is the volume of a sample of gas affected by changes in pressure when the temperature is kept constant?

4. What is meant by "atmospheric pressure"? Name and describe the instrument that is used to measure atmospheric pressure.

5. How does the absolute, or Kelvin, temperature scale differ from the centigrade scale? What is meant by "absolute zero"?

6. If a sample of gas has a volume of 700 ml at 600 torr and a temperature of 60°C, calculate the volume it would occupy at 400 torr and a temperature of 30°C.

7. What is meant by the term "partial pressure"? Why is it necessary to consider the partial pressure of water vapor when gas samples are collected over water in the laboratory?

8. Nitrogen gas has a molecular weight of 28. Calculate the weight of 1 liter of the gas under standard conditions of temperature and pressure. Express the answer as grams per liter.

9. Calculate the pressure exerted by the following:
   (a) 2.6 moles of a gas occupying 8.3 l at 15°C.
   (b) 4 moles of a gas occupying 22.4 l at 0°C.
   (c) 0.08 moles of a gas occupying 20 l at 10°C.

10. Find the molecular weight of a gas if 500 ml was found to weigh 0.5 g at 750 torr and 30°C.

11. Given two gases (X and Y) which occupy the same volume (4 l) at standard conditions; gas X weighs twice the weight of gas Y, which has a molecular weight of 8; find the molecular weight of gas X.

12. Calculate the volume of the dry gas at standard conditions which occupied 12.2 l when collected over water at 16°C and a total pressure of 740 torr.

13. What is the molecular weight of a gas X which effuses at a rate which is 0.25 times the rate of effusion of carbon monoxide?

14. Match the following:

(a) dynamic equilibrium	(v) low intermolecular forces
(b) volatile liquid	(w) intermolecular forces of attraction
(c) high viscosity	(x) heat content change
(d) London forces	(y) equal rates for opposing processes
(e) enthalpy	(z) high intermolecular forces

15. The enthalpy of vaporization of water is 540 cal/g. How much heat is required to vaporize one mole of water at its boiling point?

16. When a liquid freezes, does the mobility of its particles increase or decrease?

17. How is the enthalpy of fusion related to intermolecular forces?

18. The unit cell of topaz has interfacial angles of 90°, and its cell dimensions are all different. What space lattice does its structure represent?

19. Which compresses more easily and why: liquids or gases?

20. From among the following, choose one from each pair which is appropriate for molecules having as their only intermolecular forces large London forces:
   (a) large polarizability: small polarizability
   (b) large atoms: small atoms
   (c) high boiling points: low boiling points
   (d) very viscous: free-flowing

21. Explain the relationship between intermolecular forces and surface tension.

22. When an impure liquid boils, does the temperature tend to increase, decrease, or remain constant?

23. The boiling point of water is always 100°C. True or false? Explain.

24. An increase in temperature has what effect on intermolecular forces? on surface tension? on vapor pressure?

25. What simple, convenient method would you recommend for purifying a liquid contaminated with table salt?

26. In solids, are the intermolecular attractive forces larger or smaller than those in corresponding liquids?

27. Is the tendency for liquids to evaporate at a given temperature greater or less than that of a solid to sublime?

Suggested Reading

Bernal: The Structure of Liquids. Scientific American, Vol. 203, No. 2, p. 124, 1960.
Feifer: The Relationship Between Avogadro's Principle and the Law of Gay-Lussac. Journal of Chemical Education, Vol. 43, p. 411, 1966.
Gehman: Standard Ionic Crystal Structures. Journal of Chemical Education, Vol. 40, p. 54, 1963.
House: Ionic Bonding in Solids. Chemistry, Vol. 43, No. 2, p. 18, 1970.
Neville: The Discovery of Boyle's Law, 1661–62. Journal of Chemical Education, Vol. 39, p. 356, 1962.
Sanderson: The Nature of Ionic Solids. Journal of Chemical Education, Vol. 44, p. 516, 1967.
Slabaugh: The Kinetic Structure of Gases. Journal of Chemical Education, Vol. 30, p. 68, 1953.

CHAPTER 5

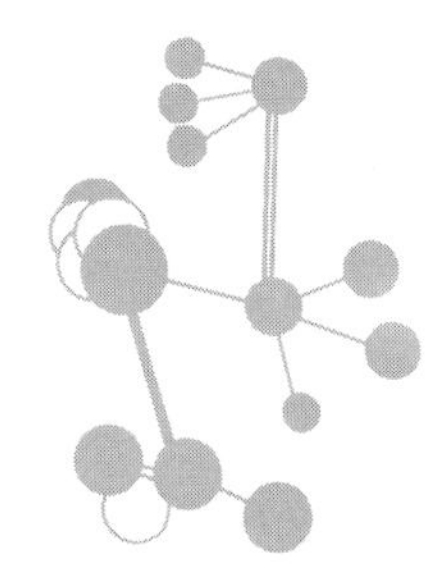

THE NATURE OF CHEMICAL REACTIONS

A total characterization of a chemical process or reaction includes the composition of the matter consumed and generated and the energetics of the reaction process. The energetics of chemical reactions influence the rates and the spontaneities, with the latter manifested in the final composition of chemical systems. Many chemical reactions have several common features which make classification possible.

Oxidation Numbers. In addition to being a statement of composition, a formula yields information which can be used to predict properties. The systematic assignment of **oxidation numbers** to the elements within a formula facilitates this prediction of chemical properties. Oxidation numbers are arbitrarily assigned to elements within a formula to indicate whether that element in the formation of the compound has lost or gained electron density. Elements which gain electron density are assigned negative oxidation numbers, whereas those which lose electron density are assigned positive values. The following set of rules is arbitrarily followed in assigning oxidation numbers:

1. Uncombined elements have an oxidation number of 0.
2. In a molecule the sum of the oxidation numbers of all the atoms must equal zero.
3. Simple ions, which are charged monatomic species, have oxidation numbers equal to the charge.
4. For complex ions, which are charged polyatomic species, the sum of the oxidation numbers of all the atoms must equal the charge on the complex ion.
5. Chemically combined hydrogen has an oxidation number of +1, except when combined with active metals such as Li, Na, Be and Mg.

6. Chemically combined oxygen has an oxidation number of -2, except in peroxides where it is -1.

The commonly observed oxidation numbers of the elements show periodic variation as determined by that number of electrons which must be lost or gained to attain the electronic configuration of an inert gas. Consequently, the elements in Group I of the periodic table show only the $+1$ oxidation number when chemically combined. The elements of Group VII, the halogens, show only -1 oxidation numbers when combined with elements of lower electronegativity. Many elements show a variation of oxidation numbers in different compounds.

The following examples illustrate the assignment of oxidation numbers in some representative formulas:

1. K_2O Oxygen is -2 and potassium $+1$. The sum of the oxidation numbers is 0. The assignment is represented as follows:

$$\begin{array}{c} 2(+1)\,(-2) \\ K_2O \end{array}$$

2. C_4H_8O

$$\begin{array}{lr} 1 \text{ (oxygen at } -2) & = -2 \\ 8 \text{ (hydrogen at } +1) & = \underline{+8} \\ \text{total oxidation number} & = +6 \end{array}$$

So that the total oxidation number of four carbon atoms must be -6. Therefore, the oxidation number of each is $-6/4 = -3/2$.

$$\begin{array}{c} (-3/2)4\,(+1)8\,(-2) \\ C_4H_8O \end{array}$$

The last example of oxidation number assignment points out the necessity to realize that the assignment of oxidation numbers is only a "bookkeeping" technique and should not be used as an indication of the actual charge on an atom in a molecule or ion.

Classification of Chemical Reactions. A chemical reaction is said to have taken place when two or more substances combine to give new substances, or when one substance decomposes to give different substances. In the broadest sense, these two types of reactions might be called **combination** and **decomposition** reactions, respectively.

Another classification of reactions, more useful than combination and decomposition, is based on the loss or gain of electrons by atoms. **Oxidation-reduction** reactions are those reactions in which an element has a change in oxidation number. The element or elements which undergo an oxidation number change may be in a compound. All other chemical reactions are called **metathesis** reactions. These reactions always involve exchange of groups between two interacting substances without any changes in oxidation numbers.

The simplest example of an oxidation-reduction reaction is the combination of any two elements. Before combination the elements have oxidation numbers of zero, but after the reaction takes place, the oxidation numbers of the elements in the reaction product are non-zero. The reaction of hydrogen and oxygen to produce water is an oxidation-reduction reaction.

The reaction of HCl and NaOH to produce NaCl and H_2O is an example of a metathesis reaction. In this reaction there is an exchange of groups, or partners, in that Cl^-, originally combined with H^+, is combined with Na^+ in the product, and H^+, originally combined with Cl^-, combines with OH^- to produce water.

The chemist makes use of symbols and formulas to state the facts of chemical reactions. To represent the formation of water, a **reaction equation** is written as follows:

$$2H_2 + O_2 \rightarrow 2H_2O$$

This equation states that two molecules (or two moles) of hydrogen and one molecule (or one mole) of oxygen combine, or react, to give two molecules (or two moles) of water. Hydrogen and oxygen are called **reactants** and water is called the **product.** Although, as indicated, the equation can be interpreted either on a molecular basis or on a mole basis, the latter of these interpretations has the advantage that the coefficients can immediately be interpreted in useable mass or volume units. One mole of oxygen, O_2, is 32 grams or 22.4 liters at standard temperature and pressure.

Reaction Stoichiometry. A chemical equation is a quantitative and qualitative statement of what happens in a chemical reaction. Any chemical reaction can be written in the form of a chemical equation if the correct formulas for all the reactants and products are known, and if the mass relationships of reactants and products are known. The chemical equation must obey the law of conservation of mass in that the reactants and products must contain the same total mass of each type of atom in the system. When this requirement is met, the equation is said to be **balanced.**

The simplest method of balancing chemical reactions is by inspection. This method always works for metathesis reactions, but it is often very difficult for oxidation-reduction reactions. The following equation, representing the interaction of phosphoric acid and calcium hydroxide, is readily balanced by inspection:

$$H_3PO_4 + Ca(OH)_2 \rightarrow Ca_3(PO_4)_2 + H_2O$$

Inspection of the products indicates that there are $3Ca^{+2}$ present, and therefore there must be $3Ca(OH)_2$ on the reactant side of the equation. This inspection also indicates that there must be $2H_3PO_4$ on the reactant side. These two adjustments lead to the following partially balanced equation:

$$2H_3PO_4 + 3Ca(OH)_2 \rightarrow Ca_3(PO_4)_2 + H_2O$$

A simple count of atoms indicates that in this equation there are far too many hydrogens and oxygens on the reactant side, $6H^+$ and $6OH^-$. To give complete balancing, water should have the coefficient 6:

$$2H_3PO_4 + 3Ca(OH)_2 \rightarrow Ca_3(PO_4)_2 + 6H_2O$$

The coefficients assigned to reactants and products in a balanced chemical equation are referred to as the **reaction stoichiometry.** These coefficients indicate the relative numbers of atoms, molecules, formula units, and ions involved in the reaction, and therefore can be interpreted as an indication of the mass relationships. The following equation and calculation illustrate this:

How much oxygen is required to convert 93.7g phosphorus to its oxide, P_2O_5?

The first step in any problem of this type, which relates masses of reactants or products, is writing and balancing the chemical equation:

$$2P + \tfrac{5}{2}O_2 \rightarrow P_2O_5$$

This equation has been balanced using a fractional coefficient for O_2. Two moles of phosphorus atoms react with 2.5 moles of oxygen molecules. This relationship allows a conversion from grams of phosphorus to grams of oxygen as follows:

$$93.7\text{g}\,\cancel{P} \times \frac{1\,\cancel{\text{mole P}}}{31.0\text{g}\,\cancel{P}} \times \frac{2.5\,\cancel{\text{moles }O_2}}{2\,\cancel{\text{moles P}}} \times \frac{32.0\text{g } O_2}{1\,\cancel{\text{mole }O_2}} = 121\text{g } O_2$$

Metathesis Reactions. Metathesis reactions which involve no change in oxidation numbers most commonly occur between ions in solutions. These reactions are most favorable and have the highest tendency to proceed when one of the products does not exist as discrete ions in solution. This happens when the product is a **gas,** a **precipitate** (insoluble solid species), or a **covalent undissociated molecule.** The following three metathesis reactions are examples of these:

$$2HCl + Na_2CO_3 \rightarrow 2NaCl + H_2O + CO_2\uparrow \text{ (gas)}$$
$$HCl + AgNO_3 \rightarrow HNO_3 + AgCl\downarrow \text{ (solid)}$$
$$HCl + NaOH \rightarrow NaCl + H_2O \text{ (covalent molecule)}$$

For reactions which occur in solution, reaction equations are often abbreviated to include only those reactants which are directly involved in chemical change. This requires a knowledge of the nature of the reactants in solution, that is, whether or not they are dissociated into ions. The following equations are the **net equations** corresponding to those given above:

$$2H^+ + CO_3^{-2} \rightarrow H_2O + CO_2\uparrow \text{ (gas)}$$
$$Ag^+ + Cl^- \rightarrow AgCl\downarrow \text{ (solid)}$$
$$H^+ + OH^- \rightarrow H_2O \text{ (covalent molecule)}$$

Oxidation-Reduction Reactions. In an oxidation-reduction reaction some species lose electrons and some species gain electrons. The process of removing electrons from an atom, thereby giving a more positive oxidation number, is called **oxidation.** After oxidation has occurred, the atom is said to have been **oxidized.** The process of adding electrons to an atom, and so giving a more negative oxidation number, is called **reduction.** After reduction has occurred, the atom is said to have been **reduced.**

The element which gains electrons in an oxidation-reduction reaction is called the **oxidizing agent,** whereas the element which loses electrons is the **reducing agent.** In addition to the conservation of mass in a chemical reaction, there must also be conservation of electrons, electrical charges, and oxidation numbers. If electrons are lost by an element being oxidized, they must be gained by the oxidizing agent. For this reason reference is made less often to separate oxidation or reduction reactions than to combined oxidation-reduction reactions. In any reaction in which an oxidation occurs, it is always accompanied by a reduction, and the oxidation and reduction always take place to an equal degree.

The assignment of oxidation numbers to reactants and products in an equation enables immediate detection of the oxidizing agent and the reducing agent. The oxidation number of the oxidizing agent becomes more negative while that of the reducing agent becomes more positive. This is illustrated with the following unbalanced equation in which oxidation numbers have been assigned:

$$\overset{(+4)\,(-2)2}{MnO_2} + \overset{(+4)\,(-2)2}{PbO_2} \rightarrow \overset{+2}{Pb^{+2}} + \overset{(+7)\,(-2)4}{(MnO_4)^{-1}}$$

The oxidation number of manganese increases from +4 to +7 during the reaction, indicating that it is the reducing agent. Lead has its oxidation number decreased from +4 to +2, and it is the oxidizing agent. Manganese is oxidized while lead is reduced.

Oxidation-reduction equations which are difficult to balance by inspection can usually be balanced using a simple electron bookkeeping technique. The principal feature of this technique involves accounting for all electrons transferred between reducing agent and oxidizing agent. The sequence followed is:

1. Assign oxidation numbers and determine which species undergo a change in oxidation number.
2. Assign coefficients to the reactants and products so that the total increase in oxidation number equals the total decrease.
3. Balance the other species in the equation.

The utility of this method is demonstrated in balancing the following equation:

$$Cr_2O_7^{-2} + H^+ + Cl^- \rightarrow Cr^{+3} + Cl_2 + H_2O$$

In this reaction, Cr is reduced and Cl is oxidized. It is necessary to determine the oxidation-number change per formula unit of $Cr_2O_7^{-2}$ and to give Cr^{+3} a coefficient of 2 before oxidation-number balance is performed.

$$\overset{(+6)2\ (-2)7}{(Cr_2O_7)^{-2}} + H^+ + 6Cl^{-1} \rightarrow 2Cr^{+3} + 3Cl_2 + H_2O$$

Balancing oxygen, hydrogen, and water gives:

$$(Cr_2O_7)^{-2} + 14H^+ + 6Cl^{-1} \rightarrow 2Cr^{+3} + 3Cl_2 + 7H_2O$$

Mass and electrical balance have been established. The final equation is an expression of the net process occurring in water.

The previously balanced equation indicates that the ratios of the coefficients of the reducing agent and the oxidizing agent are in inverse ratio to their oxidation number changes. This is expected, since all electrons lost by the reducing agent must be gained by the oxidizing agent.

The mass of a reducing agent or oxidizing agent which is involved in the transfer of one mole of electrons is a **gram equivalent weight.** A gram equivalent weight is the weight of one mole of the material divided by the total oxidation number change. In the second example above, the equivalent weight of $(Cr_2O_7)^{-2}$ is the weight of one mole of $(Cr_2O_7)^{-2}$, 216g, divided by the oxidation-number change, 6, or 36g/g.equiv.wt. Since one gram equivalent weight is involved in the transfer of one mole of electrons, one gram equivalent weight of reducing agent will react completely with one gram equivalent weight of oxidizing agent.

ELECTROCHEMICAL PROCESSES

Oxidation-reduction reactions can be induced by passing an electric current, and it is also possible to use an oxidation-reduction reaction to produce an electric current. The former phenomenon, **electrolysis,** reported by Faraday in 1834, is performed in an **electrolytic cell,** as shown in Figure 5–1. In this apparatus the source of direct current furnishes electrons to the *cathode,* which is consequently negatively charged. Reduction always takes place at the *anode,* which is positively charged in electrolytic cells. Electric current can be generated by chemical reactions under the special conditions which exist in a **voltaic cell,** which is quite similar to an electrolytic cell. In order for an oxidation-reduction reaction to produce an electric current, it must occur spontaneously. Electrolysis proceeds with the consumption of energy, electrical energy in this case. The study of chemical reactions involving either the consumption or generation of electrical current is called **electrochemistry.**

Electrolysis. The reactions taking place in an electrolytic cell are not like the oxidation-reduction reactions previously discussed in that the reactants do not react by direct exchange of electrons between one another. The reducing agent reacts by transferring an electron to the anode. The electrons given to the anode travel through the external circuit to the cathode, where they are added to the oxidizing agent. The overall oxidation-reduction reaction can be thought of as the sum of two independent **half-reactions,** the anode half-reaction and the cathode half-reaction. In the electrolysis of molten NaCl these would be represented as follows:

anode	$2Cl^- \rightarrow Cl_2\ (g) + 2e^-$
cathode	$2Na^+ + 2e^- \rightarrow 2Na$
overall	$2Na^+ + 2Cl^- \rightarrow 2Na + Cl_2$

In an electrolytic cell the source of direct current in the external circuit serves as an "electron pump" which pulls electrons off the reducing agent at the anode and pushes them into the oxidizing agent at the cathode. The energy with which this "pump" must operate is dependent upon the energy change accompanying the overall reaction. Reactions which occur spontaneously require no "electron pumping," and in fact spontaneously generate a current in the external circuit. Reactions which do not occur spontaneously require "electron pumping."

Faraday's Law. The extent to which a chemical reaction proceeds in an electrochemical apparatus is directly proportional to how much electricity

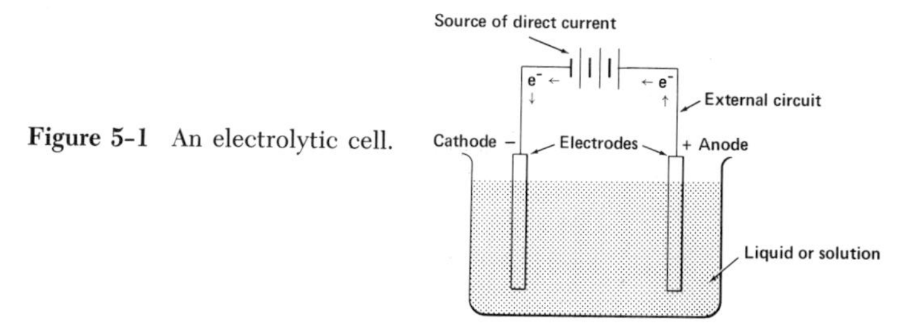

Figure 5–1 An electrolytic cell.

is passed. If one mole of electrons is passed through the electrochemical cell, then one gram equivalent weight of reducing agent will be oxidized at the anode, and one gram equivalent weight of oxidizing agent will be reduced at the cathode. This is a statement of **Faraday's law.** One mole of electrons, called a **faraday,** has a total negative charge of **96,500 coulombs.** If a current is passed at a known rate for a known length of time, it is possible to calculate the number of coulombs, and therefore the number of faradays, involved in an electrolysis. The rate at which a current is passed is expressed using the unit **ampere,** which is *one coulomb per second.* The following sample calculation illustrates the treatment of data obtained in an electrolysis experiment:

A current of 3.00 amperes is passed through an electrolytic apparatus for 10.00 minutes. How many coulombs have passed? How many faradays? What weight of Cu^{+2} from an aqueous solution would be reduced to Cu and deposited at the cathode?

The current, having units of amperes or coulombs per second, when multiplied by the time gives the number of coulombs passed:

$$10.00 \, \cancel{\text{min}} \times 3.00 \, \cancel{\text{amperes}} \times \frac{1 \text{ coul}/\cancel{\text{sec}}}{\cancel{\text{ampere}}} \times \frac{60 \, \cancel{\text{sec}}}{\cancel{\text{min}}} = 1800 \text{ coul}$$

Since 96,500 coul is 1 faraday,

$$1800 \, \cancel{\text{coul}} \times \frac{1 \text{ faraday}}{96{,}500 \, \cancel{\text{coul}}} = 0.01865 \text{ faraday}$$

Since 1 faraday will deposit 1 gram equivalent weight of Cu at the cathode,

$$0.01865 \, \cancel{\text{faraday}} \times \frac{1 \, \cancel{\text{g.equiv.wt. Cu}}}{1 \, \cancel{\text{faraday}}} \times \frac{1 \, \cancel{\text{g.atom.wt. Cu}}}{2 \, \cancel{\text{g.equiv.wt. Cu}}} \times \frac{63.5\text{g Cu}}{1 \, \cancel{\text{g.atom.wt. Cu}}} = 0.592\text{g Cu}$$

VOLTAIC CELLS. Spontaneous oxidation-reduction reactions, when physically separated into half-reactions in an electrochemical cell, can produce an electric current in an external circuit which connects the half-reactions. An electrochemical cell in which an oxidation-reduction reaction spontaneously produces a current in an external circuit is called a **voltaic cell.**

The reaction between metallic zinc and Cu^{+2} in aqueous solution proceeds spontaneously and can be used in a voltaic cell as seen in Figure 5–2. This cell is called the **Daniell cell.** It is seen that in this cell the reaction at the anode is the oxidation of Zn to Zn^{+2}. The Zn^{+2} produced migrates toward the cathode, leaving electrons on the anode which is consequently negatively charged. At the cathode Cu^{+2} ions are reduced to Cu by obtaining electrons

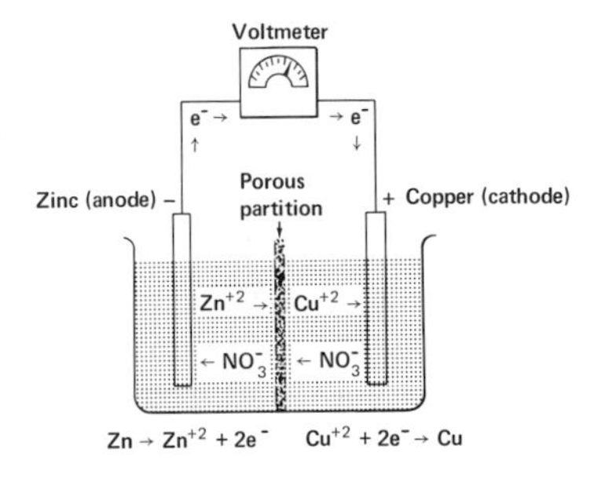

Figure 5-2 The Daniell cell.

from the electrode, leaving it positively charged. The electron current flows from anode to cathode via the external circuit, as in electrolytic cells, but the signs on the electrodes are reversed. In order to eliminate the possibility of Cu^{+2} reacting directly with the Zn of the anode, the cell is divided into two portions. These portions are separated by a **porous partition** which allows only slow migration of the ions.

Cell Potentials. The number of electrons which flow through the external circuit in a voltaic cell is related by Faraday's law to the quantity of material which reacts. The potential energy difference between electrons on the anode and electrons on the cathode is the **cell potential,** or the **electromotive force,** of the voltaic cell, and is expressed in units of **volts.** The cell potential of a given cell is dependent on the spontaneity, or driving force, with which the oxidation-reduction reaction takes place. Reactions having a high degree of spontaneity, that is, a large tendency to proceed, have large cell potentials.

Commercial Voltaic Cells: Batteries. The **lead storage battery** used in automobiles acts as a voltaic cell when furnishing current, or discharging, and as an electrolytic cell when charging. When discharging, the anode consists of lead and the cathode is lead dioxide. The solution in which these electrodes are immersed is sulfuric acid. The reactions associated with discharging are

$$
\begin{array}{ll}
\text{anode} & Pb\,(s) + SO_4^{-2} \rightarrow PbSO_4\,(s) + 2e^- \\
\text{cathode} & 2e^- + PbO_2\,(s) + 4H^+ + SO_4^{-2} \rightarrow PbSO_4\,(s) + 2H_2O \\
\hline
\text{overall} & Pb\,(s) + PbO_2\,(s) + 4H^+ + 2SO_4^{-2} \rightleftharpoons 2PbSO_4\,(s) + 2H_2O
\end{array}
$$

As this reaction proceeds and reaches equilibrium, the battery discharges and the cell potential decreases. Lead sulfate is deposited on both the cathode and anode, and the concentration of H_2SO_4 decreases. As the H_2SO_4 concentration decreases, the density of the battery liquid decreases.

A lead storage battery acts as an electrolytic cell and recharges when a direct current is passed. During recharging the reactions given above are reversed. The $PbSO_4$ on the cathode and anode are consumed, and the concentration of H_2SO_4 increases, as does the density of the battery liquid. When fully charged, a lèad storage battery can produce about two volts per cell. Automobile batteries use three or six cells in series to generate six or twelve volts, respectively.

Another very common voltaic cell is the **dry cell** (Fig. 5–3) or flashlight

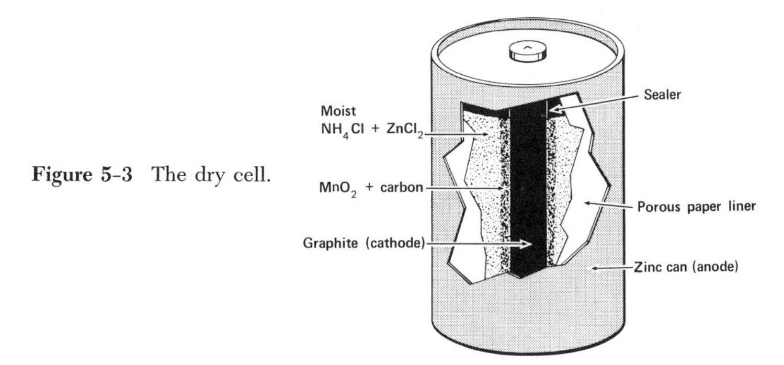

Figure 5-3 The dry cell.

battery. In this cell a zinc metal cylindrical container, the anode, is filled with a moist paste of ammonium chloride and zinc chloride. A graphite rod inserted into this paste is surrounded by manganese dioxide. The container is sealed to prevent the escape of the moisture. This cell, which produces about 1.3 to 1.5 volts, has complex electrode reactions which are not well understood. It is known that zinc is oxidized to Zn^{+2} and that MnO_2 is reduced.

RATES OF REACTIONS

The study of rates of chemical reactions and the actual steps involved in converting reactants to products is called *chemical kinetics*. The rate of a reaction refers to the amount of reactant consumed or product generated in a given amount of time. The easiest way to express the rate is to state the change in concentration per unit time (i.e., concentration change/time). The rate of a reaction is determined experimentally by periodic measurements of the concentration of the reacting species or the products.

It has also been established experimentally that the primary factors influencing rates include **concentration of reactants, temperature, nature of the reactants,** and **catalysts.** In order to understand these factors, the actual step by step procedure of a reaction will be considered. The step by step description of a reaction is called the **reaction mechanism.**

The Rate Law. For *any reaction involving a single step,* as opposed to an overall reaction involving a number of steps, the rate is directly proportional to the concentrations of the reactants. Consider the general reaction

$$A + B \rightleftharpoons C + D$$

The rate of the forward reaction, $\underset{\rightarrow}{R}$, is related to the concentration of A and B as follows:

$$\underset{\rightarrow}{R} \; \alpha \; [A][B]$$

where [A] and [B] are the concentrations of A and B. The proportionality constant which must be applied to convert this proportion to an equality is called the **rate constant.** The statement of the relationship between rate and concentration is called the **rate law.**

$$\underset{\rightarrow}{R} = k[A][B]$$

For many reactions which proceed by a single step, the form of the rate law can be predicted by considering the coefficients on the reactants in the balanced equation. This is seen in the following reaction and its associated rate law:

$$2NO_2 \rightarrow 2NO + O_2$$
$$\underset{\rightarrow}{R} = k[NO_2][NO_2] = k[NO_2]^2$$

Many chemical reactions do not occur in a single step, but may proceed through two or more steps, as illustrated by the following general reaction:

$$\begin{array}{ll} A + B \rightarrow C & \text{step 1} \\ C + A \rightarrow D & \text{step 2} \\ \hline A_2 + B \rightarrow D & \text{overall reaction} \end{array}$$

For such reactions there is often no simple relationship between the coefficients on the reactants in the balanced equation and the form of the rate law. The following balanced reaction equation and its experimentally determined rate law illustrate this:

$$H_2O_2 + 2HI \rightarrow 2H_2O + I_2$$
$$\underset{\rightarrow}{R} + k[H_2O_2][HI]$$

It is for this reason that rate laws are always established using experimental data. The experimentally determined rate law for a chemical reaction often is used to help establish the reaction mechanism.

The summation of the superscripts in the rate law is called the **order** of the reaction. Thus, in the last rate law it is seen that the order of the reaction is $1 + 1 = 2$. The reaction is referred to as a second order reaction.

Theory of Reaction Rates. In the transition from reactants to products in a chemical reaction, an intermediate stage, referred to as the **transition state,** is often postulated. For a simple reaction this might be represented as follows:

$$A + B \rightarrow \text{(transition state)} \rightarrow C + D$$

This transition state can be formed only when A and B collide. Therefore, reaction rates are influenced by how often reactant atoms or molecules collide. As the concentration of reactant increases, the number of collisions per unit time increases, and, as a result, the reaction rate increases. This simple model is consistent with experimental observations which indicate that reaction rates increase with increasing concentration of the reactants.

Rates of chemical reactions increase with increasing temperature, suggesting that molecules must have a certain amount of energy before chemical reactions will take place. Apparently the energy is furnished in the form of thermal energy as the temperature of the reacting system is increased. Consequently, it is assumed that the transition state of a given reaction is higher in energy than the average energy of the reactants or the products. This is indicated in Figure 5–4. The transition state is apparently formed by the collision of A and B only when they have sufficiently high energy. The difference in the energy of the transition state and that of the reactants is the energy which must be added to the internal energy of A and B to form the transition state. This energy difference is called the **activation energy, E_a.** When the reaction energy, ΔE, is negative, the energy content of the products is lower than that of the reactants, resulting in energy loss to the environment during the reaction. Such a reaction is called *exothermic*. In this situation the reverse

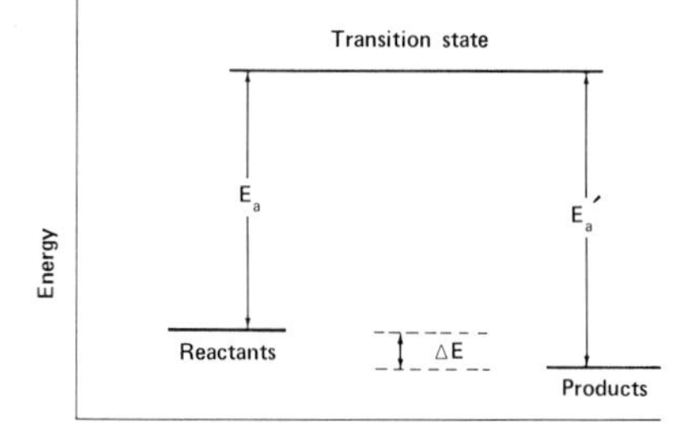

Figure 5–4 The relative energies of the reactants, the transition state, and the products for an exothermic reaction.

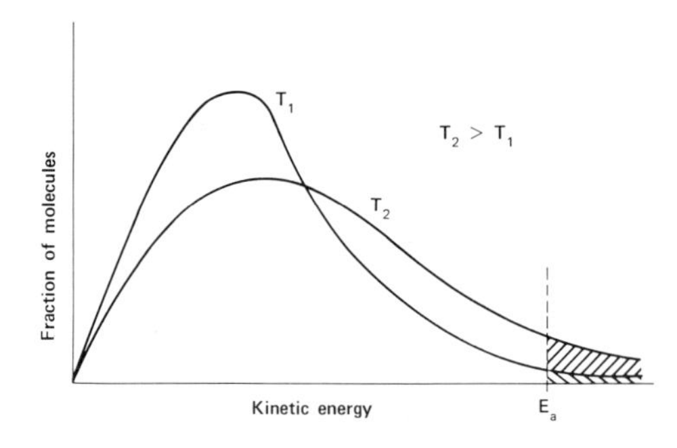

Figure 5-5 The distribution of molecular kinetic energies at two temperatures.

reaction, C + D → A + B, has an activation energy, E_a', which is greater than E_a. Reactions which give products with higher energy content than the reactants proceed with consumption of energy from the environment, $+\Delta E$, and are called *endothermic*.

Increasing the temperature increases both the number of collisions per unit time and the fraction of collisions which results in the formation of the transition state. At a given temperature there is a distribution of the kinetic energies of A and B, so that there will be some A and some B which possess the activation energy. As shown in Figure 5–5, an increase in temperature increases the fraction of A and B having energy in excess of the activation energy. With increasing temperature, the number of A and B collisions resulting in formation of the transition state increases, and consequently, the rate of reaction increases. This explanation is consistent with experimental observation.

The transition state might be thought of as a special intermediate molecule in which new bonds have been formed. If the transition state decomposes by the exact reverse of the way it is formed, reactants are re-formed, but if it decomposes in any other way, products are formed. Since the transition state can and will decompose to form both reactants and products, the reaction rate is always less than the rate of formation of the transition state.

The number of molecules, ions, or atoms which combine to form the transition state is defined as the **molecularity** of the reaction. Reactions involving only one species in forming the transition state are called **unimolecular,** whereas those involving two and three species are called **bimolecular** and **trimolecular.** The molecularity of a given step in a reaction can sometimes be deduced with high certainty from the rate law, but it must be emphasized that it is a theoretical concept which cannot be determined directly.

The actual energy involved in forming the transition state is related to the nature of A and B. If A and B have a high attraction for one another, the activation energy is expected to be relatively small. If A and B repel one another, the activation energy is large. These considerations indicate that every chemical reaction will have a unique activation energy dependent upon the nature of A and B. In general the rates of chemical reactions are inversely related to the activation energy, so that reactions with high activation energies have relatively low rates.

CATALYSTS. Occasionally an extra component added to a reaction system will take part in the reaction and result in an increased rate, but it will not be incorporated in the products. This component increases the rate of the

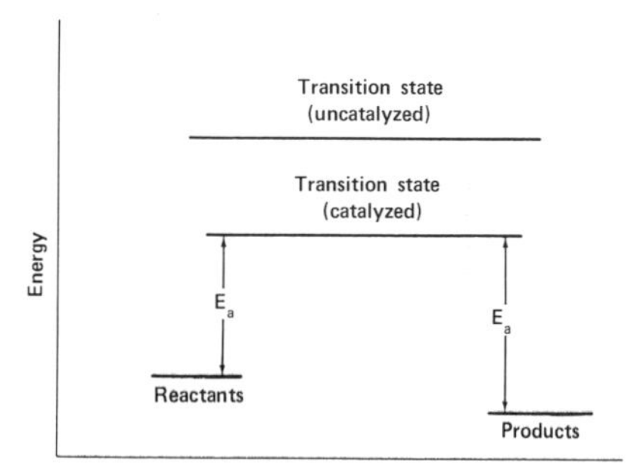

Figure 5-6 The relative energies of the reactants, the transition state, and the products for a catalyzed exothermic reaction.

reaction by furnishing an alternate reaction mechanism with lower activation energy. Components which accomplish this are called **catalysts.**

The effect of a catalyst upon the activation energy of a reaction is seen in Figure 5–6. This figure indicates that the catalyst does not alter the energy of the reactants or the products but only lowers the energy of the transition state.

The phenomenon of catalysis is basic to life. In animals, catalysts called **enzymes** play special roles in metabolism. These compounds and their catalytic role will be discussed in the biochemistry section of this book.

CHEMICAL EQUILIBRIUM

The nature of dynamic equilibrium in systems undergoing physical changes was discussed in Chapter 4. In such equilibria, it was pointed out that two opposing processes were proceeding at the same rate, so that there was no net change in the system. Many chemical reactions are also reversible processes, in that the products formed can react to regenerate the reactants. When the rates of the opposing reactions are the same, no net changes in chemical composition take place, and the chemical system is in a state of **dynamic chemical equilibrium.**

A reversible chemical reaction has a characteristic rate for the forward reaction and another for the reverse. For the reaction

$$A + B \rightleftharpoons C + D$$

the rate of the forward reaction will depend in some way on the concentrations of A and B, whereas the rate of the reverse reaction will depend in some way on the concentrations of C and D. If one performs an experiment in which A and B are mixed, and the rate of reaction is observed, it is found that the fastest rate prevails at the instant of mixing. The rate of disappearance of A and B decreases with time until eventually the concentrations of these reactants are constant. If, on mixing A and B, the concentrations of C and D are observed, it is found that they increase with the greatest rate at the moment of mixing. The rate of concentration increase slows with time, and eventually the concentration remains constant. The concentrations of A, B, C, and D become invariant with time simultaneously. When the concentrations of all species are invariant with time, chemical equilibrium prevails. The results of this hypothetical experiment are plotted in Figure 5–7.

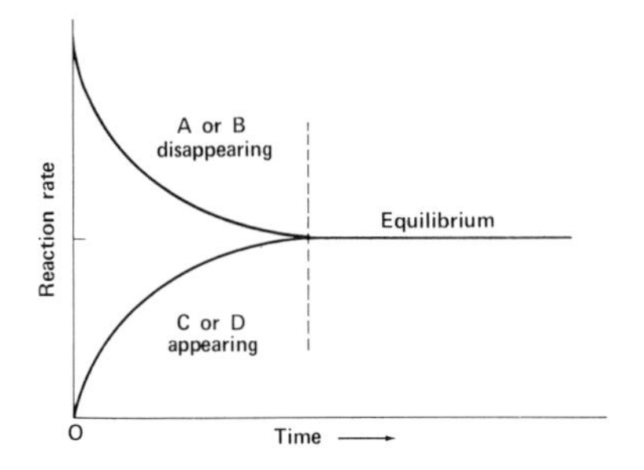

Figure 5-7 A chemical reaction attaining chemical equilibrium.

If a reversible chemical reaction is at chemical equilibrium, the rates of the opposing reactions are equal. For the simple case where both the forward and reverse reactions have a one-step mechanism, the rate laws are

$$\underset{\rightarrow}{R} = K[A][B] \qquad \underset{\leftarrow}{R} = k'[C][D]$$

At chemical equilibrium these rates are equal, so that

$$k[A][B] = k'[C][D]$$

$$\frac{k}{k'} = \frac{[C][D]}{[A][B]}$$

At a given temperature k and k′ do not change, and k/k′ does not change. The ratio of the rate constants is called the **equilibrium constant,** K.

$$K = \frac{[C][D]}{[A][B]}$$

This treatment claims that, at a given temperature for a given reaction, the product of the concentrations of C and D divided by the product of the concentrations of A and B is a constant.

It can be shown that the results of the preceding treatment are valid regardless of the mechanism of the reaction. For the general reversible reaction

$$aA + bB + cC + \cdots \rightleftharpoons zZ + yY + xX + \cdots$$

the composition of the system at equilibrium must obey the following:

$$K = \frac{[Z]^z[Y]^y[X]^x \cdots}{[A]^a[B]^b[C]^c \cdots}$$

This expression is the **law of chemical equilibrium** in equation form. For a given reversible reaction system, the composition of the system at equilibrium must satisfy the conditions imposed by the equation. By convention, the equilibrium constant is written with the components on the right hand side of the reaction equation in the numerator. The equilibrium constant for the reaction written in the reverse direction is 1/K.

Equilibrium Constants. Equilibrium constants are fundamental characteristics of chemical reactions, in that they are dependent on the nature of the products and the reactants. They are also dependent on the temperature, but are independent of the concentration and the size of the system. The magnitude of the equilibrium constant indicates the extent to which a reaction proceeds. Larger equilibrium constants are associated with those reactions which proceed more nearly to completion. The following sample calculation illustrates the variation of K with extent of reaction.

At 50°C a mole of N_2O_4 gas in a 5 liter container is 15% dissociated, according to the equation

$$N_2O_4\ (g) \rightleftharpoons 2NO_2\ (g)$$

What is the equilibrium constant?

The equilibrium constant expression is $K = \frac{[NO_2]^2}{[N_2O_4]}$.

The first step in this problem is to determine the concentration of reactant and product at equilibrium. Before any N_2O_4 dissociates, its concentration is 1 mole/5 liters. If 15% is dissociated at equilibrium, then the concentration is

$$1\ \text{mole}/5\ \text{liters} - (1 - 0.15)\ \text{mole}/5\ \text{liters} = \frac{0.85\ \text{mole}}{5\ \text{liters}}$$

The coefficients in the reaction equation indicate that 2 moles of NO_2 are formed from each mole of dissociated N_2O_4. If 0.15 mole of N_2O_4 dissociates, 0.30 mole of NO_2 is formed, and the concentration of NO_2 is 0.30 mole/5 liters. Substituting into the equilibrium expression

$$K = \frac{(0.30\ \text{mole}/5\ \text{liters})^2}{(0.85\ \text{mole}/5\ \text{liters})} = 0.021\ \text{mole/liter}$$

For this reaction which proceeds only 15%, the equilibrium constant is much less than one. It is seen in solving this problem that equilibrium constants occasionally have units. In fact, K will have units whenever the sum of the concentration exponents in the numerator does not equal that in the denominator. This situation prevails whenever there is a change in the number of moles going from reactants to products.

Le Chatelier's Principle. If a system is at chemical equilibrium, the composition of the system can be altered by certain perturbing influences. These include varying the temperature, changing the concentration of some component in the system, or, for systems of gaseous reactants, changing the total pressure of the system. In the case of gaseous systems, the total pressure may be changed by changing the volume. The general effects of perturbing influences is summarized in **Le Chatelier's principle,** which states that *a system at equilibrium will react to an applied stress so as to remove that stress.*

As an illustration of Le Chatelier's principle, consider the reaction used in the Haber process to make ammonia,

$$N_2\ (g) + 3H_2\ (g) \rightleftharpoons 2NH_3\ (g)$$

which proceeds with a decrease in the total number of moles and is exothermic. If N_2 is added to the system at equilibrium, $[N_2]$ is instantaneously increased, so that the reaction proceeding to the right progresses at a greater rate. At the moment N_2 is added, the rate of the reaction proceeding to the left is unaltered, since $[NH_3]$ has not been changed. If the reaction consuming N_2 proceeds for a short while at a rate greater than the reaction generating N_2, then the equilibrium "shifts." The concentration of NH_3 increases, and the concentration of H_2 decreases. A new equilibrium composition arises within the limits imposed by K, the equilibrium constant. In general if a component is added to a system at equilibrium, a reaction will take place to consume that component.

Increasing the temperature of the above system imposes a stress which can also be removed by shifting the equilibrium. The reaction producing NH_3 proceeds with the evolution of heat; it is exothermic. The reverse reaction is endothermic and consumes heat. Increasing the temperature of the system at equilibrium favors the endothermic reaction, which consumes the added heat, and the equilibrium is shifted so that the concentration of NH_3 decreases. In general for exothermic reactions, temperature increases decrease K, whereas for endothermic reactions, K increases with increasing temperature.

An increase in the total pressure of the above system imposes a stress which can be relieved by a reaction which decreases the total number of moles of gaseous material present, and thereby decreases the pressure. The reaction proceeding to the right to form NH_3 would relieve this stress. In general, for gaseous systems at equilibrium, the application of pressure will induce a net reaction which will decrease the pressure. Pressure has no effect on gaseous systems such as the following, in which the reaction involves no change in the number of moles of gaseous material.

$$H_2\,(g) + Cl_2\,(g) \rightleftharpoons 2HCl\,(g)$$

Questions

1. Balance the following oxidation-reduction reaction:
$Fe + CuSO_4 \rightarrow Fe_2(SO_4)_3 + Cu$
What element is oxidized? reduced? What is the oxidizing agent? the reducing agent?

2. Indicate whether each of the following reactions are oxidation-reduction or metathesis:
(a) $2HNO_3 + 3H_2S \rightarrow 2NO + 4H_2O + 3S$
(b) $HCl + H_2O \rightarrow H_3O^+ + Cl^-$
(c) $4FeS_2 + 11O_2 \rightarrow 2Fe_2O_3 + 8SO_2$
(d) $PCl_3 + 3H_2O \rightarrow P(OH)_3 + 3HCl$

3. Balance the following equations:
(a) $Al_2(SO_4)_3 + Pb(NO_3)_2 \rightarrow PbSO_4 + Al(NO_3)_3$
(b) $Mg_3N_2 + H_2O \rightarrow NH_3 + Mg(OH)_2$
(c) $H_2 + N_2 \rightarrow NH_3$
(d) $MgO + PCl_5 \rightarrow MgCl_2 + P_4O_{10}$
(e) $Sn + HNO_3 \rightarrow SnO_2 + NO_2 + H_2O$
(f) $KI + H_2SO_4 \rightarrow K_2SO_4 + I_2 + H_2S + H_2O$
(g) $Cr_2O_3 + Na_2CO_3 + KNO_3 \rightarrow Na_2CrO_4 + CO_2 + KNO_2$

4. Indicate the reducing agent and oxidizing agent in each of the following reactions:
(a) $Zn + H_2SO_4 \rightarrow ZnSO_4 + H_2$
(b) $2Fe^{+++} + 2I^- \rightarrow 2Fe^{++} + I_2$
(c) $2HNO_3 + 3H_2S \rightarrow 2NO + 4H_2O + 3S$
(d) $Ca + 2H_2O \rightarrow Ca(OH)_2 + H_2$

5. Calculate the gram equivalent weights of the underlined compound in each of the following reactions:

(a) $\underline{3K_2SO_3} + 2KMnO_4 + H_2O \rightarrow 2MnO_2 + 3K_2SO_4 + 2KOH$

(b) $2MnO_4^- + \underline{5C_2O_4^{--}} + 16H^+ \rightarrow 10CO_2 + 2Mn + 8H_2O$

6. Indicate the oxidation numbers of the following:

(a) N in NO_2
(b) Fe in Fe_2O_3
(c) N in HNO_3
(d) Mn in $(MnO_4)^-$
(e) S in $(SO_4)^{--}$
(f) C in $(C_2O_4)^{--}$

7. In an electrolytic cell, the reducing agent does what?

(a) accepts electrons at the cathode
(b) transfers electrons at the anode
(c) transfers electrons directly to the oxidizing agent

8. A current of 5.00 amperes is passed through an electrolytic apparatus for 20.00 minutes. Calculate the number of coulombs passed, the number of faradays, and the weight of Zn^{+2} which would be reduced to Zn and deposited at the cathode.

9. What is a voltaic cell? Give an example.

10. How is the spontaneity of the overall oxidation-reduction reaction related to the half-cell potentials?

11. An automobile battery exemplifies what kind of electrochemical cell when charging? When discharging?

12. Match item on the left with the more closely related item on the right.

(a) E^0	(t) electrons added at the cathode
(b) volt	(u) reactions induced by passing electric current
(c) one ampere	(v) one mole of electrons
(d) oxidizing agent	(w) standard electrode potential
(e) reducing agent	(x) one coulomb per second
(f) electrolysis	(y) unit used in expressing cell potential
(g) one faraday	(z) transfers electrons to anode

13. Is the reaction rate always greater or less than the rate of formation of the transition state? Why?

14. Do low rates of chemical reactions indicate high or low activation energy?

15. Does increasing the temperature increase or decrease the frequency of collisions between reacting particles?

16. The addition of a catalyst increases the rate of a reaction by

(a) increasing the energy of reactants
(b) increasing the energy of products
(c) increasing the energy of the transition state
(d) decreasing the energy of the transition state

17. With increasing temperature the rates of chemical reactions increase by more than that predicted when considering only the increased rate of molecular collisions. Explain.

18. Why is chemical equilibrium referred to as a dynamic equilibrium?

19. For the reaction $H_2O\ (g) + CO\ (g) \rightleftharpoons H_2\ (g) + CO_2\ (g)$, which of the following would result in a change in the equilibrium composition, that is, a "shift" in the equilibrium?

(a) increase in total pressure
(b) addition of a catalyst
(c) increasing the concentration of CO
(d) decreasing volume of reaction container
(e) increasing temperature

20. A reaction will go furthest toward completion when K is large. Explain.
21. Write the equilibrium constant expression for each of the following reactions.
    (a) $Al_2Cl_6\ (g) \rightleftharpoons 2AlCl_3\ (g)$
    (b) $CO\ (g) + H_2O\ (g) \rightleftharpoons CO_2\ (g) + H_2\ (g)$
    (c) $4NO\ (g) + 6H_2O\ (g) \rightleftharpoons 4NH_3\ (g) + 5O_2\ (g)$
    (d) $2NO_2\ (g) \rightleftharpoons 2NO\ (g) + O_2\ (g)$
    (e) $2Cl_2\ (g) + 2H_2O\ (g) \rightleftharpoons 4HCl\ (g) + O_2\ (g)$

Suggested Reading

Campbell: Why Do Chemical Reactions Occur? Englewood Cliffs, N. J., Prentice-Hall, Inc., 1965.
Kieffer: The Mole Concept in Chemistry. New York, Reinhold Publishing Corp., 1963.
Mahan: Temperature Dependence of Chemical Equilibrium. Journal of Chemical Education, Vol. 40, p. 293, 1963.
Margolis: Formulation and Stoichiometry. New York, Appleton Century Crofts, 1968.
Nyman and Kamm: Chemical Equilibrium. Lexington, Mass., D. C. Heath and Co., 1968.
Yalman: Writing Oxidation-Reduction Equations. Journal of Chemical Education, Vol. 36, p. 215, 1959.

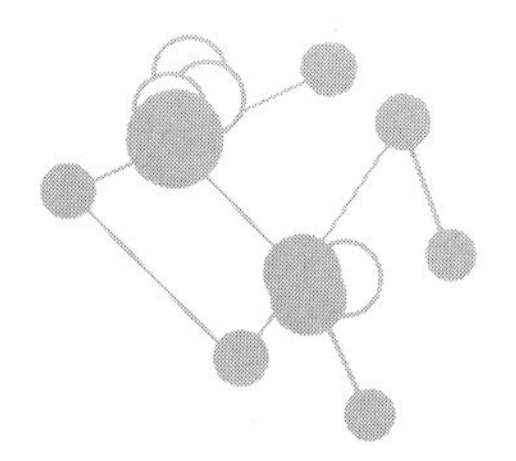

CHAPTER 6

SOLUTIONS

It is not necessary to be a chemist to recognize solutions in the home, the supermarket, and the drug store. Water softeners, detergents and cleaning compounds, perfumes, lotions, and many medications are readily available in solution.

The common examples of sugar or salt dissolved in water illustrate a mixture that is called a solution. The sugar molecules or salt particles become uniformly distributed among the water molecules. A **solution,** then, may be defined as a homogeneous mixture of two or more substances in which the particles are of atomic or molecular size. The particles dissolved in a salt solution are essentially sodium ions and chloride ions, whereas the particles dissolved in a sugar solution are sugar molecules (Fig. 6–1). The composition of solutions can vary only within the limits of the ability of one substance to dissolve in the other. Sugar and salt can be mixed in any proportion in the solid state, but there is a limit to the amount of sugar or salt that can be dissolved in a given quantity of water.

In solution, the substance that is dissolved is called the **solute,** whereas the substance in which the solute is dissolved is called the **solvent.** If there is any doubt as to which substance dissolves in the other, it is common practice to call the solvent the substance present in the greatest amount. Since matter

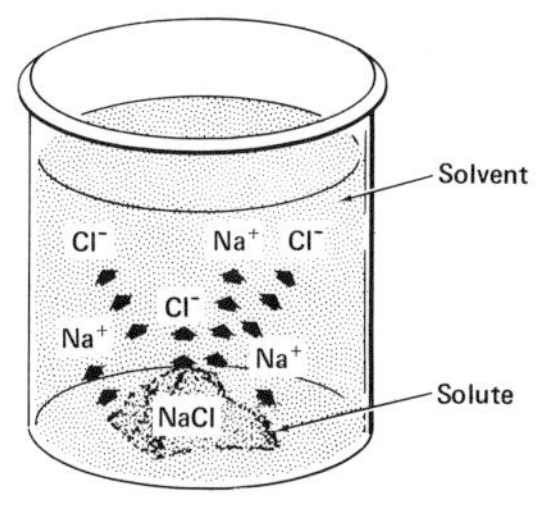

Figure 6–1 A representation of common salt (the solute) dissolving in water (the solvent).

TABLE 6–1 TYPES OF SOLUTIONS

Types of Solutions	*Examples*
Solid in liquid	Salt in water
Liquid in liquid	Gasoline
Gas in liquid	Ammonia in water
Solid in solid	Carbon in iron (steel)
Liquid in solid	Mercury in silver (amalgam)
Gas in solid	Hydrogen in palladium
Solid in gas	Iodine vapor in air
Liquid in gas	Water vapor in air
Gas in gas	Carbon dioxide in oxygen

exists as a solid, a liquid, or a gas, there are nine types of two-component solutions theoretically possible. These are listed in Table 6–1.

The most common types of solutions are those in which a liquid is the solvent. In liquid solutions and gaseous solutions, the solute molecules diffuse, or move about freely, in the solvent. For example, if one drops a crystal of a colored solute, such as potassium permanganate ($KMnO_4$), into a vessel of water, a purple color is soon observed in the water immediately surrounding the crystal. In a few hours the purple color is evenly scattered throughout the entire solution.

By far the most common and most important solvent is water. Its solutions are called **aqueous solutions.** The oceans are gigantic aqueous solutions containing many dissolved solutes. Aqueous solutions play an important role in many naturally occurring processes by serving as a transport medium for solutes. When food is digested in the body, it is dissolved and carried into the circulatory system in aqueous solution. The growth of plants depends on the transport of food and wastes in aqueous solutions.

WATER

Water is the most abundant of all chemical compounds. About three fourths of the surface of the earth is covered with water, either as a liquid or in the arctic regions as ice. Virtually all liquid water on the earth contains dissolved materials. The soil contains large quantities of water, which are essential for the growth of plants. Its presence in the atmosphere is readily recognized, because it often condenses into dew, fog, rain, or snow. As a substance essential to our existence, water ranks next to oxygen in importance. The body can survive several weeks without food, but only a few days without water. The digestion of food, the circulation, the elimination of waste materials, the regulation of acid-base balance and body temperature, as well as other vital functions, depend on an adequate supply of water. Approximately two thirds of the body weight is water, and most of the foods we eat have a water content of from 10 to 90 per cent. Bread, for example, is about 35 per cent water, meat about 70 per cent, and most vegetables are over 75 per cent.

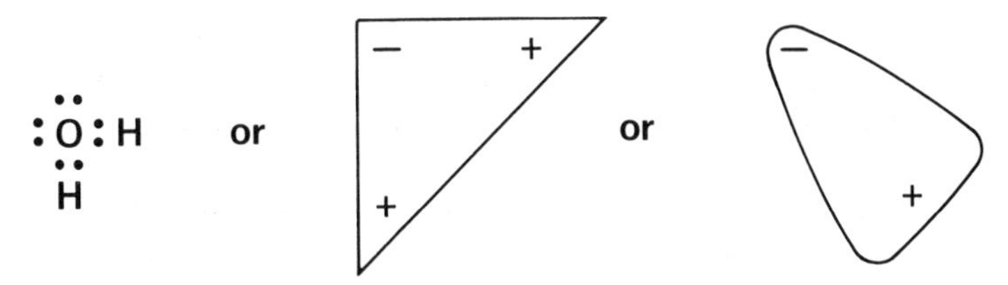

Figure 6-2 A representation of water as a polar molecule.

Physical Properties. Pure water has no odor, taste, or color. Water freezes at 0°C (32°F) and boils at 100°C (212°F). These values are abnormally high for a compound with such a low molecular weight.

Other physical properties of water are also exceptional. Its maximum density occurs at 4°C, with the density decreasing above and below this temperature. When water freezes, it becomes less dense, allowing ice to float. Also, the volume increases by nearly one-tenth in changing from water to ice, which explains the cracked automobile blocks and broken water pipes that occur in freezing weather. Since ice is less dense than water, the surface of a river or lake freezes first and the layer of ice protects the aquatic plants and animals from the cold. These abnormal properties of water are attributed to extensive hydrogen bonding which is a rather strong intermolecular interaction. It occurs in compounds having hydrogen atoms bonded to highly electronegative atoms such as oxygen, nitrogen, and fluorine.

Chemical Properties. Water is a polar molecule; i.e., a covalent compound in which the electronic charge is not uniformly distributed. Apparently the electrons, including the shared pairs, are closer to the oxygen, with the result that the oxygen end of the molecule has a negative charge and the hydrogen end has a positive charge. This result may be represented in several ways (Fig. 6–2). Since water is a polar compound, it has the property of attracting ions of electrovalent compounds and disrupting their crystal structure. When solid crystals of sodium chloride are placed in water, the forces holding the sodium ions and chloride ions together are overcome by the attraction of the water molecules for the ions. This process results in the solution of the sodium chloride as represented in Figure 6–3. In a similar fashion, water will also attract other polar molecules. Since so many substances are either electrovalent or polar in nature, water has the property of dissolving a large majority of common inorganic compounds. In the laboratory many chemical reactions are studied in water solution. In the living organism the constituents of the cells are kept in solution by water, and most of the reactions that take place in the tissues will not take place in its absence.

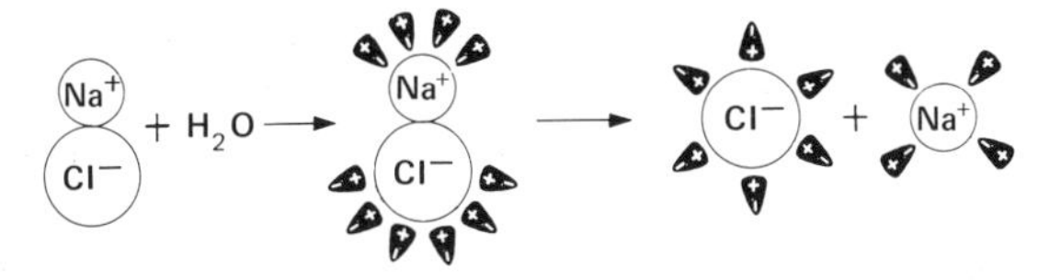

Figure 6-3 Sodium chloride going into solution illustrates the effect of water on ionic compounds.

Water is one of the most stable compounds known, and for many years it was thought to be an element. It may be heated to very high temperatures (2000°C) without appreciable decomposition. However, if an electric current is passed through it, decomposition occurs; two volumes of hydrogen are produced for each volume of oxygen. The oxygen atom is approximately sixteen times as heavy as the hydrogen atom, so that water, by weight, is eight-ninths oxygen and one-ninth hydrogen.

Another interesting chemical property of water is its action with certain metals. If a small piece of metallic potassium is placed in water, a violent reaction takes place with the formation of a hydroxide (any compound that contains the OH^- ion) and hydrogen gas:

$$2K + 2H_2O \rightarrow 2KOH + H_2\uparrow$$

The compound that is formed is called potassium hydroxide. Hydrogen gas, like oxygen, exists as molecules, each containing 2 hydrogen atoms.

Water will combine with the oxides of some metals to form a metallic hydroxide, which is also called a **base.**

$$MgO + H_2O \rightarrow Mg(OH)_2$$

Magnesium oxide — Magnesium hydroxide

Certain oxides of nonmetals react with water to form **acids.**

$$SO_3 + H_2O \rightarrow H_2SO_4$$

Sulfur trioxide — Sulfuric acid

Acids and bases combine to form water and salts in a chemical reaction called **neutralization.**

One of the most important chemical properties of water is concerned with the process of hydrolysis. In **hydrolysis,** or breaking apart with water, the compounds are split into two parts, the hydrogen of the water uniting with one part to make an acid and the hydroxyl uniting with one part to make a hydroxide. An example of hydrolysis is the reaction of water and a compound such as ammonium sulfate.

$$(NH_4)_2SO_4 + 2HOH \rightleftharpoons 2NH_4OH + H_2SO_4$$

Ammonium sulfate — Ammonium hydroxide — Sulfuric acid

The double arrow indicates that the reaction can go in both directions. Actually, only a small amount of the ammonium sulfate reacts with water in this way.

The process of digestion in the body is mainly one of hydrolysis. For example, a complex molecule such as fat is hydrolyzed as follows:

$$\text{fat} + HOH \rightarrow \text{fatty acid} + \text{glycerol}$$

The H of water goes into the fatty acid, while the OH is an integral part of the glycerol molecule. Many examples of hydrolysis will be discussed later in the study of the chemistry of the reactions that occur in the body.

Hydrates. Water molecules combine with the molecules of certain substances, forming loose chemical combinations called **hydrates.** These hydrates form well defined crystals when their solutions are allowed to evaporate slowly. For example, copper sulfate forms blue crystals when a solution of this substance evaporates slowly. The formula for crystalline copper sulfate is $CuSO_4 \cdot 5H_2O$. The water held in combination is called **water of crystallization** and is written separately to indicate its loose chemical attachment. When this hydrate is heated, it loses its water of crystallization and changes into a white powder whose formula is $CuSO_4$. Examples of other common hydrates are washing soda, $Na_2CO_3 \cdot 10H_2O$; alum, $K_2Al_2(SO_4)_4 \cdot 24H_2O$; gypsum, $CaSO_4 \cdot 2H_2O$; and crystalline sodium sulfate, $Na_2SO_4 \cdot 10H_2O$.

When the water of crystallization has been removed from a hydrate, the resulting compound is said to be **anhydrous.** Substances which give up water of crystallization on exposure to air at ordinary temperatures are called **efflorescent.** Other substances take up water on exposure to atmospheric conditions and are said to be **hygroscopic.** If they take up so much water from the air that they are finally dissolved in it, they are called **deliquescent** substances. Compounds such as sodium hydroxide and calcium chloride are so hygroscopic that they take up water from other materials. Calcium chloride is commonly used as a drying agent or desiccating agent by the chemist.

A hydrate of special interest is gypsum, or calcium sulfate ($CaSO_4 \cdot 2H_2O$), which on heating gives up part of its water to form plaster of Paris, $(CaSO_4)_2 \cdot H_2O$. When the plaster of Paris is mixed with water, it "sets" in a few minutes to reform hard crystalline gypsum. In setting, it expands slightly to form a tight cast or mold. Plaster of Paris is used extensively in making surgical casts. Chemically, the reaction may be represented as follows:

$$\underset{\text{Plaster of Paris}}{(CaSO_4)_2 \cdot H_2O} + 3H_2O \rightarrow \underset{\text{Gypsum}}{2CaSO_4 \cdot 2H_2O}$$

When water is removed from a substance the process is called **dehydration.** Food may be preserved by drying, because the bacteria and other microorganisms that spoil food must have water in order to live.

Purification of Water. Naturally occurring water contains impurities dissolved from the rocks and soil. Even rain water contains particles of dust and dissolved gases from the air. The impurities present in water may be classified as either mineral or organic matter.

The mineral matter found in natural water usually consists of common salt and various compounds of calcium, magnesium, and iron. Water that contains these dissolved salts does not readily form a lather with soap and is called **hard water,** whereas water with little or no mineral matter lathers easily and is called **soft water.**

The organic matter in water is derived from decaying animal and vegetable material. Bacteria utilize this type of material for food, and may cause diseases unless they are removed before the water is used for drinking purposes.

A source of pure drinking water is extremely important to the health of a community. For this reason it would be appropriate to consider several of the methods for the purification of water.

Distillation. In the process of distillation, water is boiled and the resulting steam is cooled and condensed in a different container. The condensed steam is called the distillate, or distilled water. The chemist uses distillation to produce water free from bacteria and dissolved mineral matter. Distilled water is used widely in the preparation of solutions in the laboratory and in the hospital. Distillation is the most effective method for the purification of water, but is too expensive to be employed by large towns or cities.

Boiling. Water from natural sources may be made safe for drinking by boiling for 10 to 15 minutes. This process does not remove the impurities but does kill any pathogenic bacteria that might be present. This method of water purification is reliable in emergencies, but is not generally employed for civilian water supplies.

Filtration. In the laboratory, suspended material is separated from water by passing the liquid through a porous material that has holes that are smaller than the suspended matter, thereby holding it back. The dissolved impurities are not removed by this method. The bacteria are on the suspended organic matter and are largely removed by filtration. To destroy bacteria completely, it has become general practice to treat the water with chlorine, a substance that kills the remaining microorganisms.

Treatment with Ozone. A special form of oxygen, O_3, called **ozone,** is often used in water purification. When used in place of chlorine, it eliminates many undesirable tastes and odors by oxidation. It is also a more effective germicidal agent than chlorine. The expense of the electricity required to produce ozone from oxygen has prohibited its use in the purification of city water supplies in this country, although several installations are in operation in Europe.

Aeration. Water may be purified by exposure to air for long periods. The oxygen of the air dissolves in the water and oxidizes organic material, thus depriving bacteria of their source of food. It also kills bacteria by direct chemical reaction. Most cities do not depend on this process alone for water purification, but use it to remove objectionable tastes and odors from the water. Aeration of water supplies is usually accomplished by spraying the water into the air from fountains, or by allowing it to flow in thin sheets over tiles.

HARD AND SOFT WATER. Since the common methods of water purification do not remove the dissolved inorganic matter, many cities have hard water. The inorganic matter present in hard water usually consists of bicarbonates, sulfates, or chlorides of calcium, magnesium, and iron. Water that contains only calcium or magnesium bicarbonate is called **temporary hard water,** because these salts can be removed by heating. When heated, they are converted into the insoluble carbonates that form most of the scale on boilers and teakettles. Temporary hard water can therefore be softened by boiling.

$$\underset{\substack{\text{Calcium}\\ \text{bicarbonate}\\ \text{(soluble)}}}{Ca(HCO_3)_2} \xrightarrow{\text{heat}} \underset{\substack{\text{Calcium}\\ \text{carbonate}\\ \text{(insoluble)}}}{CaCO_3\downarrow} + H_2O + CO_2\uparrow$$

Water that contains sulfates or chlorides of calcium, magnesium, or iron is called **permanent hard water,** because it does not lose these salts on heating. Permanent hard water can be softened by adding a chemical compound that will convert the soluble calcium, magnesium, or iron salts into insoluble precipitates, which may be removed by filtration.

Water softeners used in homes, hospitals, laundries, and small industries often employ synthetic **ion exchange resins** or a material called **Zeolite,** which is a natural sodium aluminum silicate. As the water passes through a bed or column of the resin or Zeolite, the calcium, magnesium, or iron ions are retained and exchanged for less objectionable ions that pass through the softener with the water. For example, the sodium ions in the Zeolite are exchanged for calcium ions, as represented in the following equation:

$$Ca^{++} + 2Na^{+}\ Zeolite \rightarrow 2Na^{+} + Ca^{++}\ Zeolite$$

Since about 1940 many types of ion exchange resins have been produced and studied. The two major types are the cation and anion exchangers, which are made with a great variety of resins and active groups. The most common cation exchangers contain sulfonic acid groups, whereas an important group of anion exchangers contain quaternary ammonium groups. The resins are commonly polymers, such as polystyrenedivinylbenzene resins. Water that is used in the laboratory and in many industries must not only be softened but must be free from inorganic ions. To produce this de-ionized or **demineralized water,** a bed of mixed resins is used. For example, a combination of a strongly acid cation exchanger and a strongly basic anion exchanger would remove sodium and chloride ions, as represented in the following reactions:

$$Na^{+} + Cl^{-} + H^{+}SO_3^{-}\ resin \rightarrow Na^{+}SO_3^{-}\ resin + H^{+} + Cl^{-}$$
$$H^{+} + Cl^{-} + NH_2\ resin \rightarrow Cl^{-}NH_3^{+}\ resin$$

Since ion exchange resins will not remove nonelectrolytes or organic matter from water, for strict analytical purposes the water is first distilled and then passed through a mixed-bed, ion-exchange resin.

FACTORS AFFECTING SOLUBILITY

Except for gases as solutes or solvents, the dissolving process, **dissolution,** involves increasing the average distance of separation between solute molecules and between solvent molecules. Each of these processes necessitates overcoming intermolecular forces of attraction and consequently consumes energy. If the dissolving process is to be at all favored, this consumption of energy must be compensated for in some way. If the separated solute molecules can have significant attractive intermolecular interactions with the solvent molecules, then the dissolving process may be energetically favorable. This solute-solvent interaction is called **solvation.** Solutes consisting of polar molecules are not highly soluble in solvents consisting of non-polar molecules, because the solute-solvent interaction is too weak to overcome the energy required to separate the solute molecules. Non-polar solute molecules, although easily separated, are not highly soluble in polar solvents, because the energy required to separate solvent molecules is not compensated for by the weak solute-solvent interaction. In general, it can be stated that ionic and polar solutes are most soluble in polar solvents, and non-polar solutes are most soluble in non-polar solvents.

Water is a common solvent for polar solutes, such as sugar and salts, but it is unsatisfactory for non-polar solutes, such as fat or paint. Iodine is only slightly soluble in water, but will dissolve readily in alcohol. Ether, carbon tetrachloride, and gasoline are good solvents for fatty material, whereas turpentine is used to dissolve paint.

The summation of the energy terms discussed in the previous paragraphs

TABLE 6–2 SOLUBILITY AT DIFFERENT TEMPERATURES

Substance	*Grams Dissolved by 100 ml of Water at*		
	0°C	*20°C*	*100°C*
Potassium nitrate	13.3	31.6	246.0
Copper sulfate	14.3	20.7	75.4
Sodium chloride	35.7	36.0	39.8
Calcium hydroxide	0.185	0.165	0.077

determines whether the **enthalpy of solution** of a solute is endothermic or exothermic. For those solutes which undergo an endothermic dissolution, the solubility increases with increasing temperature. Solutes with exothermic dissolution have decreasing solubility with increasing temperature, but the solubility of most solutes increases with a rise in temperature. Table 6–2 illustrates the change in solubility of certain solutes at different temperatures.

The first two substances listed in the table show a definite increase in solubility as the temperature is raised, but the solubility of sodium chloride is only slightly affected by the change in temperature. The solubility of calcium hydroxide, on the other hand, decreases with a rise in temperature.

The solubility of gases in liquids and solids is decreased by a rise in temperature and is increased by an increase in pressure. In the preparation of carbonated drinks, large amounts of carbon dioxide are forced into solution by pressure at a low temperature. If a cold bottle of soda water is opened, the pressure is released and the gas escapes slowly from the solution, forming bubbles in the water. If a warm bottle is opened, the carbon dioxide escapes rapidly, causing foam to spurt out of the bottle.

CONCENTRATIONS OF SOLUTIONS

The **concentration** of a solution states the quantity of solute contained in a given amount of solution or solvent. When a solution contains a small amount of the solute, it is said to be **dilute;** when it contains a large amount, it is said to be **concentrated.**

If a solution is prepared by stirring in an excess of solute until no more will dissolve, it is said to be a **saturated solution.** At any given temperature a saturated solution will contain a definite quantity of solute in a given volume of solvent. For example, the addition of 40g of sodium chloride to 100g of water at 20°C would result in a solution that contains 36g of sodium chloride and 4g of undissolved crystals. A state of equilibrium would exist between the dissolved and undissolved solute. Particles of the crystalline solute would be continually going into solution at a rate that is exactly equal to the rate at which solute particles are crystallizing out of the solution (Fig. 6–4). A **saturated solution** may be defined as a solution in which the dissolved solute exists in a state of equilibrium with the undissolved solute.

If a saturated solution is prepared at a higher temperature and is then allowed to cool, the extra solute that was dissolved at the higher temperature usually settles out of the solution. If the hot solution is cooled slowly and is not disturbed, the excess solute may not settle out. In this case the solution

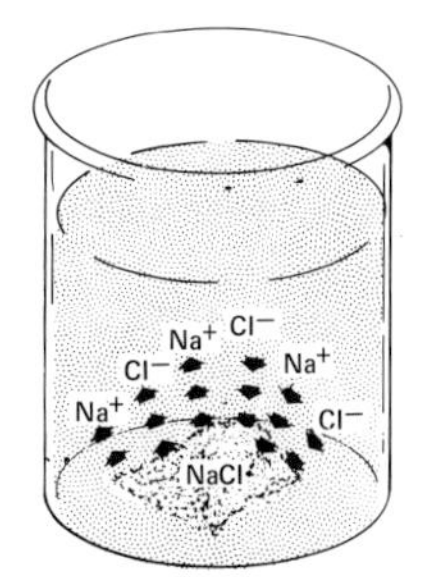

Figure 6-4 A state of equilibrium existing in a saturated solution of sodium chloride.

will contain more solute than it can ordinarily dissolve at room temperature. Such a solution is called a **supersaturated** solution. If this solution is disturbed by the addition of a crystal of the solute, the material in excess of that required to saturate the solution at that temperature will immediately crystallize out.

METHODS OF EXPRESSING CONCENTRATIONS. It is often necessary to have a more quantitative expression of the concentration of a solution than that represented by the terms dilute, concentrated, saturated, or supersaturated. A common method of designating the concentration of an aqueous solution is based on **percentage by weight or by volume.** Unfortunately, there are so many ways to express percentage concentration that the term is often confusing. The three most commonly used are: (a) as grams of solute in 100 ml of solution, (b) as grams of solute in 100g of solution, and (c) as milliliters of solute in 100 ml of solution. The weight-per-volume units, (a), are preferred by laboratories and industries where the solute is most important and the solvent is merely used as a vehicle for the solute. In analytical procedures the accurate measurement of a volume of the solution provides a definite weight of solute for a chemical reaction. The weight-per-weight units, (b), are preferred in many studies, because the concentration does not change with changes in temperature.

The volume-per-volume units, (c), are the least accurate and are often used when the solute is a liquid. For example, a 5 per cent solution of alcohol is prepared by diluting 5 ml of alcohol to a volume of 100 ml with water. This type of solution is often expressed as per cent by volume to distinguish it from (a) or (b).

To determine the amount of any substance required to prepare a definite volume of a solution of a given weight-volume per cent, one should start with the number of grams present in 100 ml. For example, to find the amount of glucose that must be used to prepare 300 ml of a 5 per cent solution, the following calculations would be made:

$$300\ \cancel{\text{ml}} \times \frac{5\text{g}}{100\ \cancel{\text{ml}}} = 15\text{g}$$

A concentration expression which states the mass ratios of solutes to the total mass of the solution, but in units of moles, is called the **mole fraction, (X).** The mole fraction of a component of a solution is the number of moles of that component divided by the total number of moles of all components. The following calculation illustrates the use of this concentration expression:

What is the mole fraction of H_2SO_4 in an aqueous solution which is 30% H_2SO_4 by weight? In 100g of this solution there would be 30g H_2SO_4 and 70g

H_2O. The number of moles of H_2SO_4 and H_2O respectively would be

$$30\text{g } \cancel{H_2SO_4} \times \frac{1 \text{ mole } H_2SO_4}{98\text{g } \cancel{H_2SO_4}} = 0.306 \text{ moles } H_2SO_4$$

$$70\text{g } \cancel{H_2O} \times \frac{1 \text{ mole } H_2O}{18\text{g } \cancel{H_2O}} = 3.89 \text{ moles } H_2O$$

The mole fraction equals

$$\frac{\text{moles } H_2SO_4}{\text{moles } H_2SO_4 + \text{moles } H_2O} = \frac{0.306}{0.306 + 3.89} = 0.073$$

A commonly used mass per volume concentration expression is **molarity, (M),** which is the number of moles of solute per liter of solution. This method of expressing the concentration is convenient because the concentration bears a definite relationship to the molecular weight of the solute. A solution that contains 1 mole of the solute in 1 liter is called a **molar** solution. In calculating the amount of a compound that is used to prepare solutions of a given molarity, the following scheme may be used:

What is the molarity of a solution which is prepared by adding enough water to 36g NaOH to prepare 150 ml of solution?

The units sought are those of molarity, mole/l, and thus the problem is worked as follows:

$$\frac{36\text{g } \cancel{\text{NaOH}}}{150 \text{ } \cancel{\text{ml sol'n}}} \times \frac{1 \text{ mole NaOH}}{40\text{g } \cancel{\text{NaOH}}} \times \frac{1000 \text{ } \cancel{\text{ml sol'n}}}{1 \text{ l sol'n}} = \frac{6 \text{ moles NaOH}}{\text{liter}} = 6\text{M}$$

This calculation involves the calculation of the gram molecular weight of NaOH and the application of conversion factors to get the units of moles of NaOH and liters of solution.

By the use of molarity, it is possible to obtain any desired fraction of a mole of a substance by merely measuring a volume of the solution.

Another mass per mass concentration expression is **molality, (m),** which is the number of moles of solute per kilogram of solvent. For example, a 1 **molal** solution contains 1 mole of the solute in 1 kg of solvent. For aqueous solutions, a comparison of a molal with a molar solution reveals two important differences. The molar solution has a final volume of 1000 ml, whereas the molal solution has a volume that exceeds 1000 ml because the solute has been added to 1000g of solvent. Temperature changes affect the volume of a solution and will therefore slightly alter the molarity of a solution, but will not change the molality of a solution.

Chemists often have occasion to express the concentration of a given solution in both molarity and molality. Because of the units involved, the conversion of molarity to molality, or vice versa, requires a knowledge of the density of a solution. The following calculation illustrates this conversion:

A 0.50**m** aqueous solution of $BaSO_4$ has a density of 1.095 g/ml at 20°C. What is its molarity?

A 0.50**m** solution contains 0.50 moles of $BaSO_4$ (233.3 g/mole) in 1000g of water. Therefore the total weight of a solution containing 0.50 moles of $BaSO_4$ would be

$$\frac{233.3}{2}\text{g} + 1000\text{g, or } 1116.7\text{g}$$

Knowing this, the density is used to calculate the molarity.

$$\frac{0.50 \text{ mole } BaSO_4}{1116.7\text{g} \cancel{\text{sol'n}}} \times \frac{1.095\text{g} \cancel{\text{sol'n}}}{1 \cancel{\text{ml sol'n}}} \times \frac{1000 \cancel{\text{ml sol'n}}}{1 \text{ sol'n}} = \frac{0.49 \text{ moles } BaSO_4}{1 \text{ sol'n}}$$

This illustrative calculation indicates that for aqueous solutions the molarity and the molality are approximately the same. This approximation is closest for dilute aqueous solutions. This approximation does not hold for solutions which do not have water as the solvent, that is, **nonaqueous solutions,** unless the density of the solvent is near that of water.

Another mass volume concentration expression is **normality, (N),** which is the number of gram equivalent weights of solute per liter of solution. The gram equivalent weight of an acid yields one mole of protons, or has one gram of replaceable hydrogen, whereas one gram equivalent weight of a base will neutralize one gram equivalent weight of acid. The gram equivalent weight of a species involved in an oxidation-reduction reaction is the weight of one mole divided by the oxidation number change. A 1 **normal** solution of an acid, base, reducing agent, or oxidizing agent contains one gram equivalent weight of solute in one liter of solution.

The following sample problems illustrate the utility of this concentration expression:

Calculate the weight of H_2SO_4 in 150 ml of a 0.3N solution of H_2SO_4.

The quantity of solution is 150 ml, and 0.3N allows a conversion of H_2SO_4 to gram equivalent weights. The gram equivalent weight of H_2SO_4 is one half the gram molecular weight (98g), since this acid can yield two moles of protons per mole of acid. The answer is obtained as follows:

$$150 \cancel{\text{ml sol'n}} \times \frac{0.3 \cancel{\text{g.eq.wt.}}}{1000 \cancel{\text{ml sol'n}}} \times \frac{1 \cancel{\text{g.mol.wt.}}}{2 \cancel{\text{g.eq.wt.}}} \times \frac{98\text{g } H_2SO_4}{\cancel{\text{g.mol.wt.}}} = 2.2\text{g } H_2SO_4$$

Calculate the normality of a liter of solution containing 27g of $KMnO_4$ which is used in a reaction producing MnO_2.

The final units, gram equivalent weights per liter, are mass over volume, and so the calculation is started with these units. The oxidation number change of manganese is +7 to +4, and therefore there are three gram equivalent weights per mole of $KMnO_4$.

$$\frac{27\text{g} \cancel{KMnO_4}}{1 \text{ l sol'n}} \times \frac{1 \cancel{\text{mole } KMnO_4}}{158.0\text{g} \cancel{KMnO_4}} \times \frac{3 \text{ g.eq.wt. } KMnO_4}{1 \cancel{\text{mole } KMnO_4}}$$
$$= \frac{0.51 \text{ g.eq.wt. } KMnO_4}{1 \text{ sol'n}}, \text{ or } 0.51\text{N}$$

The expression of concentration as normality is especially convenient when dealing with the combination of solutions whose solutes react with one another.

Titration is the process of adding a solution of one reactant to another reacting sample until the number of gram equivalent weights of the reactants are the same. If the concentration of the solution added is known, it is called a **standard solution,** and the quantity added can be used to determine the amount of the second reactant. The **equivalence point** of the titration is commonly detected using indicators which change color upon addition of excess standard solution. The most common indicator used in titrations of acids and bases is **red phenolphthalein,** which is red in basic solution and colorless in acidic solution.

In calculations involving acid-base or oxidation-reduction titrations, the use of normality concentration units is most convenient. At the equivalence point in a titration, the number of gram equivalent weights of each of the reactants is the same; that is, for the reaction $A + B \rightarrow C$, the number of g.eq.wts. of A = number of g.eq.wts. of B. The product of normality and volume is the number of gram equivalent weights (i.e., $N \times V$ = number of g.eq.wts.). Thus, at the equivalence point of a titration,

$$N_A \times V_A = N_B \times V_B$$

The following examples illustrate calculations involving several types of titrations with different concentration expressions.

In an acid-base titration, 100 ml of 2.0M HCl would be required to neutralize what weight of NaOH?

The quantity of HCl is expressed by the volume of HCl solution used, since this can be converted to number of moles using the concentration. A knowledge of the stoichiometry of the reaction and the molecular weight of NaOH results in the following:

$$V_{(acid)} \times M_{(acid)} \times \text{Stoichiometry} \times \text{g.mol.wt.base}$$

$$100\ \cancel{\text{ml HCl}} \times \frac{2\ \cancel{\text{moles HCl}}}{1000\ \cancel{\text{ml HCl}}} \times \frac{1\ \cancel{\text{mole NaOH}}}{1\ \cancel{\text{mole HCl}}} \times \frac{40\text{g NaOH}}{1\ \cancel{\text{mole NaOH}}} = 8.0\text{g NaOH}$$

What volume of 0.020M $Ca(OH)_2$ will react with 50.00 ml of 0.100M H_2SO_4?

The conversion sequence used in this calculation is volume of acid to moles of acid, moles of acid to moles of base, and moles of base to volume of base.

$$V_{(acid)} \times M_{(acid)} \times \text{Stoichiometry} \times 1/M(\text{base})$$

$$50.00\ \cancel{\text{ml } H_2SO_4} \times \frac{0.100\ \cancel{\text{moles } H_2SO_4}}{1000\ \cancel{\text{ml } H_2SO_4}} \times \frac{1\ \cancel{\text{mole } Ca(OH)_2}}{1\ \cancel{\text{mole } H_2SO_4}} \times \frac{1000\text{ ml } Ca(OH)_2}{0.020\ \cancel{\text{moles } Ca(OH)_2}} = 250\text{ ml } Ca(OH)_2$$

What is the molarity of a $KMnO_4$ solution if 50.00 ml is required for the titration of 50.00 ml of 0.050M $K_2C_2O_4$ solution? The balanced reaction equation is:

$$2MnO_4^- + 5C_2O_4^{-2} + 16H^+ \rightarrow 2Mn^{+2} + 10CO_2 + 8H_2O$$

The starting point is the quantity of $K_2C_2O_4$ solution which is converted to moles. The ratio of the moles of $KMnO_4$ and $K_2C_2O_4$ at the equivalence point is obtained from the reaction stoichiometry. To obtain units of molarity, the number of moles of $KMnO_4$ must be divided by the volume involved.

$$V_{(red.)}/V_{(ox.)} \times M_{(red.)} \times \text{Stoichiometry} \times \text{Volume conversion}$$

$$\frac{50.00\ \cancel{\text{ml } K_2C_2O_4}}{50.00\ \cancel{\text{ml } KMnO_4}} \times \frac{0.050\ \cancel{\text{moles } K_2C_2O_4}}{1000\ \cancel{\text{ml } K_2C_2O_4}} \times \frac{2\text{ moles } KMnO_4}{5\ \cancel{\text{moles } K_2C_2O_4}} \times \frac{1000\ \cancel{\text{ml } KMnO_4}}{1\text{ l } KMnO_4}$$

$$= \frac{0.02\text{ moles } KMnO_4}{1\text{ sol'n}} = 0.02M$$

What volume of 0.67N H_3PO_4 solution is required to neutralize 47 ml of 0.43N NH_4OH solution?

Recalling the use of $V_A \times N_A = V_B \times N_B$, simply substitute and solve for V_A.

$$V_A \times \frac{0.67 \text{ g.eq.wt. } H_3PO_4}{1000 \text{ ml } H_3PO_4} = 47 \cancel{\text{ml } NH_4OH} \times \frac{0.43 \text{ g.eq.wt. } (NH_4OH)}{1000 \cancel{\text{ ml } NH_4OH}}$$

$$V_A = 47 \times \frac{0.43}{0.67} \text{ ml } H_3PO_4 = 30 \text{ ml } H_3PO_4$$

This calculation does not necessitate a consideration of the stoichiometry of the reaction, since normalities are used. The problem could have also been worked by applying conversion factors to the volume of NH_4OH solution.

What volume of 7N HNO_3 is required to completely dissolve 46 grams of Ag metal which is oxidized to Ag^{+1}?

The equivalent weight of Ag is the atomic weight (107.9), since the oxidation number change is 1.

$$47\cancel{\text{g Ag}} \times \frac{1 \cancel{\text{ g.eq.wt. Ag}}}{107.9\cancel{\text{g Ag}}} \times \frac{1 \cancel{\text{ g.eq.wt. } HNO_3}}{1 \cancel{\text{ g.eq.wt. Ag}}} \times \frac{1000 \text{ ml } HNO_3}{1 \cancel{\text{ g.eq.wt. } HNO_3}} = 435 \text{ ml } HNO_3$$

PHYSICAL PROPERTIES OF SOLUTIONS

Many of the physical properties of liquids are changed when solutes are introduced. For example, vapor pressure and melting points are lowered and the boiling point is elevated. In colder climates aqueous solutions of ethylene glycol are used in the cooling systems of automobiles because they lower the freezing point.

Many properties of solutions are quantitatively related to the mole fraction of the solute or number of solute particles. These properties are called *colligative properties*. Another property of solutions quite important in life processes is osmotic pressure.

Many plant or animal membranes are semipermeable in that they allow one component of a solution to pass through, while they hold back another component. The roots of a plant are covered with a semipermeable membrane that allows the passage of water into the plant but will not allow the substances in the sap to pass out into the ground. If the solutions on either side of a semipermeable membrane are unequal in concentration, there is a tendency to equalize the concentration.

The selective flow of a diffusible component through a membrane is called **osmosis.** The diffusible component, usually water, will tend to flow from the more dilute solution into the concentrated solution resulting in an increased pressure in the more concentrated solutions. This pressure tends to force the solvent back out of the region of higher concentration, so that after a while an equilibrium is established by the balancing of the rate of outward and inward flow of solvent. The pressure at which this equilibrium is reached is called the **osmotic pressure.** The osmotic pressure of a dilute solution is proportional to the molal concentration of the solution, and increases with the temperature. For example, the molal osmotic pressure for water at 0°C is about 22.4 atmospheres, compared to 24.4 atmospheres at 25°C. When the solutions on each side of a semipermeable membrane have established equilibrium and have an equal concentration of components, they are said to be **isotonic.**

COLLOIDS

All the solutions described in previous chapters have consisted of soluble solutes dissolved in a solvent. Aqueous solutions have been stressed, in which the solutes have broken up into either molecules or ions that formed homogeneous mixtures with the solvent molecules. These solutions, whose particles are of molecular or ionic dimensions and will not settle out on standing, are called **true solutions.** If finely divided clay is shaken with water and allowed to stand, the particles will slowly settle to the bottom. The clay particles are insoluble in the solvent and can be seen by the naked eye. Such a mixture is called a **suspension.**

Some mixtures of insoluble solutes have particles so small that they will not settle out of the solvent on standing, and cannot be seen by the naked eye. These substances, whose particles are intermediate in size between those in true solutions and those in suspensions, are called **colloids.** The term colloid was first applied to certain biological materials by Thomas Graham, a Scottish chemist, in 1861. He classified several substances into two categories, depending on their ability to pass through the pores of animal or vegetable membranes. Substances that readily diffused through the membranes and were easy to crystallize from solution Graham classified as **crystalloids.** Examples were sodium chloride and copper sulfate. Those that did not pass through the membranes and were not readily crystallized were named colloids, from the Greek word meaning "glue-like." Examples of colloids were gelatin and glue.

The particles in the three types of solutions just discussed vary from atomic or molecular size in the true solution to large, visible particles in the suspension. The size of the particles in a colloidal solution are usually considered to vary from 1 to 100 nanometers (nm). A nanometer is 10^{-9} meters or one millionth of a millimeter. The relations of these three types of solutions and some of their properties are shown in Table 6–3.

In colloid chemistry the particles are called the dispersed phase, and the fluid in which they are dispersed is called the dispersion medium. The correct term for a colloidal solution is therefore a colloidal dispersion, although the term "solution" is commonly used.

Types of Colloids. Colloidal solutions may be classified by the same system employed for true solutions. Instead of solute and solvent, the particles

TABLE 6–3 TYPES OF SOLUTIONS AND THEIR PROPERTIES

True Solutions	*Colloidal Solutions*	*Suspensions*
1. Particle size less than 1 nm	Particle size 1 nm to 100 nm	Particle size 100 nm or more
2. Invisible	Visible only in ultra- or electron microscope	Visible to naked eye
3. Will pass through filters and membranes	Will pass through filters but not membranes	Will not pass through filters
4. Possesses molecular movement	Exhibits Brownian movement	Moves only by force of gravity

TABLE 6–4 TYPES OF COLLOIDAL SYSTEMS

Dispersed, or Discontinuous, Phase	*Dispersion Medium, or Continuous, Phase*	*Examples*
Solid	Solid	Gems, ruby glass, alloys
Solid	Liquid	Plasma, inks, gold sols
Solid	Gas	Smoke, dust clouds
Liquid	Solid	Pearls, opals
Liquid	Liquid	Milk, mayonnaise
Liquid	Gas	Fog, sprays, mists
Gas	Solid	Pumice, meerschaum
Gas	Liquid	Foams, meringue, whipped cream

are called the dispersed or discontinuous phase, and the solvent in which they are dispersed is called the dispersion medium or continuous phase. Theoretically, there are nine possible types of colloidal solutions, but since a gas in a gas system would not produce a colloid, we shall consider only eight, as shown in Table 6–4. A colloidal solution is called a **suspensoid** if the dispersed phase is a solid, an **emulsoid** if the dispersed phase is a liquid.

A general classification of colloidal systems depends on the physical state of the dispersed phase relative to the dispersion medium. The most common systems are known as sols, gels, aerosols, and emulsions. Both sols and gels are formed by the dispersion of solids, such as proteins, starch, and soaps, in an aqueous medium. A **sol** is a colloidal solution that is liquid at room temperature, with water as the continuous phase and the solid as the discontinuous phase. A **gel** is a similar colloidal dispersion that is a solid at ordinary temperatures.

Gels. When hot water is added to certain substances, such as gelatin and agar, they readily form colloidal solutions. On cooling, a semisolid system called a **gel** is formed. Apparently the colloidal particles adsorb water molecules on their surface as the solution cools, causing fibers or filaments to form in a "brush heap" structure. The water in the colloidal system is trapped between the filaments, and a semisolid gel-like structure results. This "brush heap" structure of a gel may be readily destroyed by cutting or whipping the gel, with the resultant production of a syrupy solution.

Aerosols. In recent years the number of products prepared in the form of aerosols has increased remarkably. Solids and liquids of all types are dispersed in air or gas and dispensed from a pressure can with a push-button nozzle. Aerosol spray dispensers have become so popular that any industry involved in the manufacture of cosmetics, insecticides, paints, furniture polish, shoe polish, detergents, lubricating oils, and chemicals in general is behind the times if its line does not include several products in aerosol form. The advantages of instant whipped cream, penetrating oil, and hair spray in aerosol dispensers do not have to be stressed.

When colloids are dispersed in air in the form of smoke or chemical dust-laden gases, they may be precipitated by the application of a high-voltage electric current. This principle is applied industrially in the **Cottrell precipitator.**

Emulsions. If two liquids that do not mix, such as oil and water, are shaken together, the result is a milky-appearing solution called an **emulsion.** Small globules of oil remain suspended in the water for a short time, but the two liquids soon separate. For this reason, a mixture of oil and water is called a **temporary emulsion.** Milk and cream are examples of **permanent emulsions.** If a certain type of colloid is added to a temporary emulsion, it coats the globules of the fat or oil and prevents them from running together, thereby making a more permanent emulsion of oil and another liquid, such as water. Colloids that act in this fashion are called **emulsifying agents.** Milk is an emulsion of butter fat in water, with casein acting as an emulsifying agent. Mayonnaise is an emulsion of oils and vinegar to which the colloids of egg yolks are added as emulsifying agents. An **emulsifying agent** is a type of **protective colloid,** and a protective colloid is any colloid that stabilizes another colloidal dispersion. A soap or detergent serves as a protective colloid to prepare a permanent emulsion of oil and water. The use of soap or detergents to clean an oily or greasy surface with water is a practical example of a protective colloid. The soap or detergent serves as a link between the oil and water phase, allowing the water to wet the oily surface and wash away the dirt clinging to the oil.

PROPERTIES OF COLLOIDS. ***Size.*** As has been stated, any substance that is subdivided into particles ranging from 1 to 100 nm is a colloid. Particles of this size have a relatively large surface compared with their small weight. For example, if a tennis ball were subdivided into pieces the size of colloidal particles and spread into a layer, they would cover a surface equal to 20 tennis courts. Powdered charcoal is an example of a colloid whose particles have an extremely large surface in comparison to their weight.

Movement. The molecules in a true solution are in a state of constant rapid motion. Both solute and solvent molecules exhibit this **molecular motion.** Since colloidal particles are composed of an aggregate of many molecules, the movement of the particles is very slow compared with that of an individual molecule. Apparently the major motion in a colloidal dispersion is caused by the bombardment of the particles by the molecules of the dispersion medium. This erratic movement of colloidal particles was first observed under an ultramicroscope by Robert Brown and is known as **Brownian movement** (Fig. 6–5).

If a strong beam of light is passed through a colloidal solution, the path of the beam is clearly outlined because of the reflection of the light from the surfaces of the moving colloidal particles. This phenomenon is called the **Tyndall effect;** it may be used as a simple test to distinguish between true solutions and colloidal solutions. A similar effect is observed when a bright ray of sunlight enters a darkened dusty room. The path of light is clearly outlined by the reflection from the surfaces of the dust particles. The scattering of light by dust and water particles in the earth's atmosphere accounts for the red color of the sun and sky at sunrise and sunset.

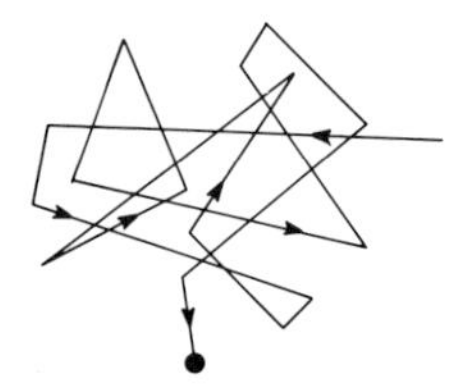

Figure 6–5 Erratic random movement of colloidal particles known as Brownian movement.

Questions

1. Place after the solution on the left the number of each statement that correctly applies to it.

Colloidal Solution	1. Will pass through a membrane.
	2. Particles are invisible.
True Solution	3. Particles are less than 100 mμ in diameter.
	4. Will not pass through filter paper.
Suspension	5. Exhibits Tyndall effect in strong beam of light.
	6. Usually consists of inorganic molecules.
	7. Particles will settle out on standing.

2. How many grams of sodium chloride are there in 500 ml of a 0.9 weight per weight per cent solution having a density of 1.1 g/ml?

3. How many grams of K_3PO_4 would be used to prepare 700 ml of a 2M solution?

4. How many grams of H_2SO_4 would be used to prepare 400 ml of a 1N solution?

5. A 20.0 ml portion of 1.0N HCl required exactly 10.0 ml of a sodium hydroxide solution to reach the chemical equivalent point. What is the normality of the NaOH solution?

6. What are the colligative properties of solutions?

7. Explain the process of osmosis. What is meant by the term "isotonic"?

8. What is a solution? a solute? a solvent?

9. Give an example of a solution of: (1) a gas dissolved in a liquid, (2) a solid dissolved in a liquid, and (3) a liquid dissolved in a liquid.

10. What are the main factors that affect the solubility of a solid solute?

11. For solutes which undergo an endothermic dissolution, does the solubility increase or decrease with increasing temperature?

12. What is the mole fraction of $KClO_3$ in an aqueous solution which is 25% $KClO_3$ by weight?

13. What is the molarity of a solution prepared by adding enough water to 194g of H_3PO_4 to prepare 600 ml of solution?

14. Will temperature changes alter the molarity or the molality of a solution?

15. An aqueous solution has a density of 1.83g/ml in 95.0% H_2SO_4. Calculate the molality and the molarity of this solution?

16. A volume of 46.00 ml of 0.200M H_2SO_4 solution is required for the titration of 32.00 ml of a $Ca(OH)_2$ solution. What is the molarity of the $Ca(OH)_2$ solution?

17. What is the normality of an NH_4OH solution if 25.0 ml are required to react with 47.00 ml of 0.4N H_3PO_4?

18. What is the molarity of a $Na_2C_2O_4$ solution if 11.3 ml of this solution reacts with 47.4 ml of 0.05M $KMnO_4$ solution? (See reaction equation in text)

19. (a) How can water be purified in an emergency?
    (b) How is water treated to remove bacteria and dissolved mineral matter?

20. How is water ordinarily purified for use by the residents of a city?

21. Describe a method for softening (a) temporary hardness in water, and (b) permanent hardness in water.

22. What is meant by the statement that water is a polar molecule?

23. Define and give an example of hydrolysis.

24. Write the formulas and explain the difference between anhydrous and crystalline copper sulfate.

Suggested Reading

Chinard: Colligative Properties. Journal of Chemical Education, Vol. 32, p. 377, 1955.
Choppin: Water—H_2O or $H_{180}O_{90}$? Chemistry, Vol. 38, No. 3, p. 7, 1965.
Jenkins: Fresh Water From Salt. Scientific American, Vol. 196, No. 3, P. 37, 1957.
Margolis: Formulation and Stoichiometry. New York, Appleton Century Crofts, 1968.

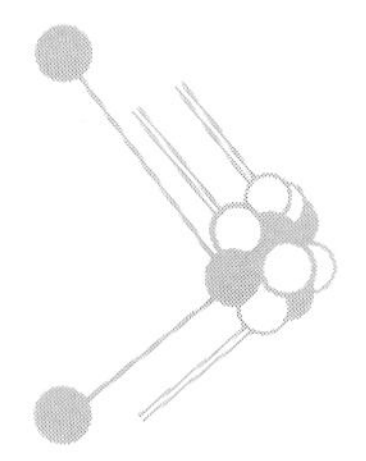

CHAPTER 7

ELECTROLYTE SOLUTIONS

Early in the seventeenth century Volta, Davy, and Berzelius independently proposed electrochemical theories to explain the chemical effects of electricity. Michael Faraday, who started his career as an assistant to Davy, extended the investigations of the three men, especially with respect to the quantitative nature of the electrolysis of solutions. He introduced the terms anode for the positive electrode and cathode for the negative electrode, and the term **electrolyte** for a solution that conducted an electric current between the electrodes. He also proposed the term **ion** for the charged particles in the solution; more specifically, he called the negatively charged ion that moves to the anode an **anion** and the positively charged ion that moves to the cathode a **cation.**

Solutions of electrolytes conduct an electric current because of the presence of ions in solution. A simple apparatus for demonstrating the conductivity of a solution is shown in Figure 7–1. It consists of a source of current connected to a light bulb in such a way that the bulb will not burn until a conductor is placed between the two electrodes. If the solution being tested will conduct the electric current, the open circuit between the two electrodes is closed, and

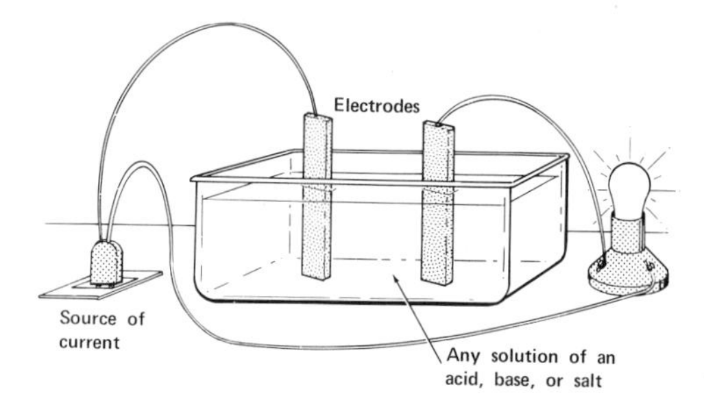

Figure 7-1 Diagram of a simple conductivity apparatus.

the light bulb will glow. Solutions of acids, bases, and salts are found to conduct an electric current readily and to cause the bulb to burn brilliantly. The bulb will not light when the electrodes are placed in distilled water or solutions of sugar, alcohol, or glycerol. As mentioned previously, substances whose solutions will conduct an electric current are called **electrolytes.** Those whose solutions will not conduct the current are **nonelectrolytes.** In general, compounds other than acids, bases, and salts are classified as nonelectrolytes.

The numerical constants obtained for the colligative properties of nonvolatile nonelectrolytes in solution are based on the number of solute particles per unit volume of solution. The fact that the value of these constants increases in solutions of electrolytes suggests the presence of a greater number of solute particles per unit volume. The Swedish chemist Arrhenius was able to show that the degree of dissociation of an electrolyte in solution is nearly the same whether it is calculated from the electrical conductivity or from quantitative treatment of colligative properties. In 1887 Arrhenius proposed a theory to explain more completely the properties of electrolytes in solution, particularly the ability of electrolytes to conduct an electric current. The following points outline the main assumptions of his theory:

1. When an electrolyte is dissolved in water, some of its molecules split or dissociate into positively charged particles and negatively charged particles called ions.
2. The sum of the positive charges that result from the dissociation of the electrolyte is equal to the sum of the negative charges.
3. Nonelectrolytes that fail to conduct an electric current when in solution do not dissociate to form ions.
4. Ions possess properties different from the corresponding uncharged atoms or molecules, and are responsible not only for the electrical properties but also for the chemical properties of a solution.
5. In an extremely dilute solution electrolytes are completely dissociated into ions, but in ordinary concentrations an equilibrium exists between the ions and the undissociated molecules.

Since chemists were not aware of the ionic nature of salts when Arrhenius proposed his theory, he represented the ionization of the electrolyte sodium chloride as an equilibrium between molecules and ions in solution as follows:

$$\underset{\text{sodium chloride molecule}}{NaCl} \rightleftharpoons \underset{\text{sodium ion}}{Na^+} + \underset{\text{chloride ion}}{Cl^-}$$

The theory postulated that a dynamic equilibrium exists between the sodium chloride molecules and the sodium and chloride ions in solution. The electrical charge on the ions is numerically equal to their oxidation number.

The major modification of the Arrhenius theory resulting from later investigations involves the fifth point. It is now known that this assumption does not hold for strong electrolytes but the theory is still valid for weak electrolytes. In 1923 Peter Debye and Erich Hückel proposed the **interionic attraction theory** to account for the abnormal activity of ions of strong electrolytes. They stated that strong electrolytes are 100 per cent dissociated, but in solutions of concentration greater than about 0.1M the attraction between the positive and negative ions hinders their movement in electrolysis and decreases the conductivity of the solution.

In general the dissociation of ionic compounds and dilute solutions of covalent strong electrolytes, such as hydrochloric acid, is nearly complete, whereas weak electrolytes and more concentrated electrolytes are only partially dissociated.

The modern theory of ionization is built on a more complete foundation of knowledge than that proposed by Arrhenius. Many compounds, especially salts, are formed by electron transfer, and consist of ions even when they exist in the crystalline state. When they are dissolved in water, the forces holding them in the tightly packed crystalline state are overcome, and they gain freedom of movement. They then behave like ions in solution; for example, they will migrate under the influence of an electrical current. Covalent polar compounds, though not consisting of ions in the liquid or solid state, will often form ions by hydrolysis when dissolved in water, and are also classed as electrolytes.

TYPES OF ELECTROLYTES

Strong electrolytes are ionic compounds and hydrolyzable covalent polar compounds that dissociate completely into ions when in dilute solutions. **Weak electrolytes** are substances that dissociate only slightly into ions when in solution and exist essentially as undissociated molecules. All electrolytes are either acids, bases, or salts.

The concept of acids and bases was originally severely limited to the special conditions prevailing in aqueous solutions. It has, however, evolved over many years to a concept which not only explains many reactions, but facilitates the prediction of many others. In the evolution to more sophisticated theory, each new idea incorporated earlier ideas and extended the acid-base concept to new cases.

Arrhenius included in his ionization theory definitions for acid and base. An **acid** was defined as a substance containing hydrogen that dissociates to yield hydrogen ions in a water solution. A **base** was defined as a substance that dissociates to give hydroxide ions in a water solution. The Arrhenius definition was readily accepted by chemists, since it not only provided an explanation for many of the earlier puzzling aspects of acid-base behavior, but also increased the understanding of the catalytic effect of acids. This effect had long been considered one of the fundamental properties of an acid. These classic definitions of an acid and of a base are still widely used by chemists when dealing with aqueous solutions. The major limitation of the theory is that it does not apply in nonaqueous solvents such as liquid ammonia.

Brönsted in Denmark and Lowry in England proposed that an acid be defined as a substance whose molecule or ion can give up a proton (hydrogen ion) and a base be defined as a substance whose molecule or ion can combine with a proton. The **Brönsted-Lowry concept** of an acid as a **proton donor** and a base as a **proton acceptor** is useful for work with either aqueous or nonaqueous solutions. It incorporates the Arrhenius concept as a special case where the base is always the hydroxide ion.

The most sophisticated concept proposed to date which attempts to explain metathesis reactions is the **Lewis concept.** According to Lewis, a base is any species capable of donating a pair of electrons for the formation of a coordinate covalent bond, and an acid is any species which is capable of accepting a pair of electrons to form a coordinate covalent bond. This concept incorporates the proton of the Arrhenius and Brönsted-Lowry concepts as an acid. It also considers metal ions and many molecules such as BF_3 as acids.

Acids. Acids are most commonly used in aqueous solution, although when pure they may exist as solids, liquids, or gases. For example, hydrogen chloride, HCl, is a gas; boric acid, H_3BO_3, is a solid; and sulfuric acid, H_2SO_4, is a liquid. When hydrogen chloride is present in a water solution it dissociates as follows:

$$\underset{}{HCl} \rightarrow \underset{\text{hydrogen ion}}{H^+} + \underset{\text{chloride ion}}{Cl^-}$$

This equation illustrates the common definition of an acid as a substance that yields hydrogen ions in aqueous solution. According to the Brönsted and Lowry definition, the hydrogen chloride gas would be called an acid since it is capable of donating a proton to a base. The general concept of their definition would be represented by the equation:

$$\underset{\text{acid}}{HA} + \underset{\text{base}}{B} \rightleftharpoons \underset{\text{acid}}{HB^+} + \underset{\text{base}}{A^-}$$

Therefore, in H_2O the reversible dissociation of hydrochloric acid would be represented as:

$$\underset{\text{acid}}{HCl} + \underset{\text{base}}{H_2O} \rightleftharpoons \underset{\text{acid}}{H_3O^+} + \underset{\text{base}}{Cl^-}$$

The properties of acids in aqueous solutions include the following:

1. All acids in solution have a sour taste (use caution and taste only dilute solutions). Citrus fruits, for example, taste sour because of the presence of citric acid, an organic acid. The sour taste of vinegar is due to acetic acid.

2. Acids change the blue color of litmus dye to red. This is one of the simplest tests for the presence of an acid. A substance such as litmus that has one color in an acid solution and another color in a basic solution is called an **indicator.**

3. Acids react with many metals to form hydrogen gas. The metal reduces the hydrogen of the acid, liberating hydrogen gas. For example, in the following reaction, zinc reduces the hydrogen of sulfuric acid to form zinc sulfate and gaseous hydrogen:

$$Zn + H_2SO_4 \rightarrow ZnSO_4 + H_2\uparrow$$

In this reaction Zn is oxidized by H^+. The hydrogen of an acid can be replaced by most metals indicating that it is **replaceable hydrogen.** The quantity of an acid which will yield one mole of protons, or which has one gram of replaceable hydrogen, is one **gram equivalent weight of the acid.**

4. Acids react with oxides and hydroxides to form water and a **salt.** A salt may be considered as a compound containing neither the initial hydrogen

of an acid nor the hydroxyl group of a base. The action of an acid on an oxide or hydroxide of a metal is illustrated by the following:

$$MgO + 2HNO_3 \rightarrow Mg(NO_3)_2 + H_2O$$
$$NaOH + HCl \rightarrow NaCl + H_2O$$

In the reaction between an acid and a metallic hydroxide, or base, both the acid and the base are neutralized. A reaction of an acid and a base to form water and a salt is therefore called a **neutralization reaction.**

5. Acids react with carbonates and bicarbonates to form carbon dioxide gas, as illustrated by the following equations:

$$Na_2CO_3 + H_2SO_4 \rightarrow CO_2\uparrow + H_2O + Na_2SO_4$$
$$NaHCO_3 + HCl \rightarrow CO_2\uparrow + H_2O + NaCl$$

Baking soda, $NaHCO_3$, is widely used for the neutralization of acid and for the production of carbon dioxide gas.

Baking powders contain a bicarbonate and some acid-forming substance, which, when moisture is added, release gaseous carbon dioxide throughout the cake batter, making it light. The lactic acid of sour milk produces the same effect when mixed with baking soda. The action of a common type of fire extinguisher depends on the reaction between sulfuric acid and sodium bicarbonate.

Acids are usually classified according to the number of hydrogen ions they yield per molecule, and by the number of elements they contain. Those that yield one hydrogen ion per molecule, such as HCl, are called **monoprotic;** those yielding two, such as H_2SO_4, are called **diprotic;** and those yielding three, such as H_3PO_4, are called **triprotic. Binary,** or hydro-acids, are composed of hydrogen and one other element, and **ternary,** or oxy-acids, contain oxygen and another element in addition to hydrogen. The binary acids are named from the element that is combined with hydrogen. They begin with the prefix **hydro-** and end with the suffix **-ic.** Examples:

HCl	HBr	HF
Hydrochloric	Hydrobromic	Hydrofluoric

The most common ternary acids are named after the element other than hydrogen or oxygen, and they end with the suffix **-ic.** Examples:

H_2SO_4	HNO_3	H_3PO_4	$HClO_3$
Sulfuric	Nitric	Phosphoric	Chloric

If the same elements unite to form more than one ternary acid, the acid with one less oxygen than the most common form ends in **-ous.** Examples:

H_2SO_3	HNO_2	H_3PO_3	$HClO_2$
Sulfurous	Nitrous	Phosphorous	Chlorous

Acids of the halogens (chlorine, bromine, iodine) containing oxygen can exist as -ic or -ous acids; chloric and chlorous acids are examples. When these acids contain one more oxygen atom than the -ic acid they are named **per-**

-ic acids, and if they contain one less oxygen atom than the -ous acid they are named **hypo- -ous** acids. Examples:

$HClO_3$	$HClO_4$	$HClO_2$	$HClO$
Chloric	Perchloric	Chlorous	Hypochlorous

Bases. Bases are defined as compounds that contain the hydroxyl group and ionize to form hydroxide ions in a water solution. This definition is by far the most practical and useful when dealing with aqueous solutions. In the Brönsted-Lowry concept, all negative ions, being proton acceptors to some degree, are considered to be bases, with the hydroxide ion being the strongest base that can exist in a water solution.

In the discussion of the properties and reactions of common bases in aqueous solutions, use of the term *base* will be confined to the metallic hydroxides and ammonia. Examples of common bases are: sodium hydroxide, NaOH; potassium hydroxide, KOH; calcium hydroxide, $Ca(OH)_2$; and ammonia.

All of the metals in the bases listed above are located in Groups I and II on the left of the periodic table. The hydroxyl radical is formed from hydrogen and oxygen by a covalent bond, whereas the metal is combined with the hydroxide in an ionic bond.

Ammonia is very soluble in water and gives a basic solution because of the reaction:

$$NH_3 + H_2O \rightleftharpoons NH_4^+ + OH^-$$

The properties of bases in aqueous solutions are as follows:

1. When a base is dissolved in water, the solution has a slippery feeling.
2. Solutions of bases have a bitter, metallic taste. (Caution: they should be tasted only in dilute solutions.)
3. Bases change the red color of litmus to blue. In general, bases reverse the color change that was produced by an acid in an indicator.
4. Bases react with acids to form water and a salt:

$$2KOH + H_2SO_4 \rightarrow 2H_2O + K_2SO_4$$

One **gram equivalent weight of a base** is that quantity which will react with one mole of H^+, or will neutralize one gram equivalent weight of acid.

Bases which consist of a metal combined with the hydroxyl radical are named by starting with the name of the metal and ending with the word **hydroxide.** For example, NaOH is sodium hydroxide, KOH is potassium hydroxide, and $Ca(OH)_2$ is calcium hydroxide.

Salts. As stated before, when an acid and a base react with each other, acidic and basic properties of the solution disappear with the formation of water and a salt. Salts may also be formed by the displacement of hydrogen from an acid with a metal.

$$Zn + 2HNO_3 \rightarrow Zn(NO_3)_2 + H_2\uparrow$$

From this reaction a salt may be defined as a compound formed by replacing the hydrogen of an acid with a metal. A more general definition of a salt would be the combination of any negative ion, except hydroxide, with any positive ion other than hydrogen.

Inorganic salts consist of ions combined by ionic bonds in the solid or crystalline state. The atoms of a complex or polyatomic ion, such as NO_3^-, are usually held together by covalent bonds.

Salts react with themselves, with acids, with bases, and with water in ways which are usually predictable by considering the nature of the possible products. As indicated earlier, these metathesis reactions proceed if they result in the formation of a gas, a solid, or covalent undissociated molecules.

The properties of salts may be summarized as follows:

1. Salts react with each other to form new salts.

$$NaCl + AgNO_3 \rightarrow NaNO_3 + AgCl\downarrow$$

This is a common reaction in chemistry and is used as a test for the presence of a chloride. When silver nitrate is added to a solution that contains chlorides, a positive reaction is indicated by the formation of a white precipitate (insoluble AgCl).

2. Salts react with acids to form other salts and other acids.

$$BaCl_2 + H_2SO_4 \rightarrow BaSO_4\downarrow + 2HCl$$

3. Salts react with bases to form other salts and other bases.

$$MgSO_4 + 2KOH \rightarrow K_2SO_4 + Mg(OH)_2\downarrow$$

Magnesium hydroxide, the base formed in this reaction, is only slightly soluble in water.

4. Salts react with water by hydrolysis to give acidic and basic solutions. These reactions are discussed later in this chapter.

A **normal salt** is one in which all the hydrogen of an acid has been replaced by a metal. For example, Na_2CO_3, $BaSO_4$, and KNO_3 are normal salts. Salts formed from acids that contain more than one replaceable hydrogen atom may retain one or more hydrogen atoms in their molecule, and are called **acid salts.** For example, sulfuric acid may react with sodium hydroxide to form sodium acid sulfate, which is commonly called sodium bisulfate or sodium hydrogen sulfate.

$$H_2SO_4 + NaOH \rightarrow NaHSO_4 + H_2O$$

If both hydrogen atoms are replaced by sodium, the normal salt sodium sulfate is formed.

$$H_2SO_4 + 2NaOH \rightarrow Na_2SO_4 + 2H_2O$$

Other examples of acid salts are sodium bicarbonate ($NaHCO_3$), sodium dihydrogen phosphate (NaH_2PO_4), and disodium hydrogen phosphate (Na_2HPO_4). The latter two compounds are made from phosphoric acid, which has three replaceable hydrogen atoms; therefore, two acid salts are possible.

Salts are named in two parts, the name of the positive ion and the name of the negative ion. Metallic positive ions are referred to by the name of the metal (i.e., sodium, potassium) unless the metal displays more than one oxidation number. For metals of variable oxidation number, the lower oxidation number is referred to by adding the suffix **-ous,** whereas the upper uses **-ic.** Often the Latin names of the elements are used as stems in this system (i.e., fer- for iron, plumb- for lead). It is also common to name the element and indicate the oxidation number in parentheses with Roman numerals. The NH_4^+ ion is called the ammonium ion.

The negative ions of salts are named with reference to the acid from which they come. Simple negative ions are named by adding the suffix **-ide** to the stem of the name of the element. Negative ions derived from ternary acids ending in -ic are given the suffix **-ate,** and those derived from ternary acids ending in -ous are given the suffix **-ite.** The negative ions of acid salts are named by adding as a prefix **bi-,** the word hydrogen, or the word acid.

The following list includes the formulas and names of several common salts:

$NaCl$	sodium chloride	KNO_2	potassium nitrite
$FeBr_2$	ferrous bromide [iron(II) bromide]	$(NH_4)_2SO_3$	ammonium sulfite
		$Ca(ClO)_2$	calcium hypochlorite
$Fe(NO_3)_3$	ferric nitrate [iron(III) nitrate]	Na_2CO_3	sodium carbonate
		$NaHCO_3$	sodium bicarbonate
$Ba_2(PO_4)_3$	barium phosphate	$KHSO_3$	potassium bisulfite
$Al_2(SO_4)_3$	aluminum sulfate		

IONIZATION OF WATER

The Arrhenius theory of electrolytes qualitatively described the properties of certain solutes in aqueous solutions. The behavior of strong and weak electrolytes was explained by considering the extent of the dissociation of solute molecules into ions. For weak electrolytes, the partial dissociation can be treated as a chemical equilibrium of the following type:

$$MX \rightleftharpoons M^+ + X^-$$

For this equilibrium between the molecule MX and the ions M^+ and X^-, the **ionization or dissociation equilibrium constant, K_i,** is

$$K_i = \frac{[M^+][X^-]}{[MX]}$$

Almost all solutes in aqueous solution are either directly involved in, or at least influence, ionic equilibria. Even in the absence of solutes, an ionic equilibrium prevails in water. This equilibrium involves the slight dissociation of water, which is represented in the following net equation:

$$H_2O \rightleftharpoons H^+ + OH^-$$

It must be emphasized that this reaction equation is a net ionic equation, and the H^+ and the OH^- are each associated with a number of water molecules by virtue of extensive hydrogen bonding. The H^+ is known to have at least one water molecule associated, so that it exists as H_3O^+, **the hydronium ion.** The ionization equilibrium constant for this dissociation reaction is

$$K_i = \frac{[H^+][OH^-]}{[H_2O]}$$

Since the concentration of water molecules is constant at a given temperature (i.e., 55.55M at 4°C), the denominator of this expression is incorporated along with K_i into a new constant, K_w, **the ion product constant for water,** as follows:

$$K_i \cdot [H_2O] = K_w = [H^+][OH^-]$$

For water at 25°C, it is found experimentally that $[H^+] = [OH^-] = 10^{-7}$ moles/liter, so that

$$K_w = [H^+][OH^-] = [10^{-7}][10^{-7}] = 10^{-14}$$

The magnitude of K_w indicates the very small extent of the dissociation of water.

Even though the value of K_w was determined by studying pure water at 25°C, this product of $[H^+]$ and $[OH^-]$ is constant in the presence of solutes. For example, if the $[H^+]$ is increased by the addition of a solute, the $[OH^-]$ must decrease until $[H^+][OH^-] = 10^{-14}$. The decrease in $[OH^-]$ on the addition of H^+ is predicted by considering Le Chatelier's principle. The addition of H^+ places a stress upon the equilibrium system, and the stress is removed by a shift which consumes OH^-. The following calculations illustrate the quantitative treatment of this phenomenon:

It is observed that on the addition of 1×10^{-5} moles of gaseous HCl to 1 liter of pure water, the $[H^+]$ increases from 10^{-7}M to 1×10^{-5}M. What is the $[OH^-]$?

Recalling that $K_w = 10^{-14} = [H^+][OH^-]$, and solving for $[OH^-]$ gives

$$[1 \times 10^{-5}][OH^-] = 10^{-14}$$
$$[OH^-] = 10^{-14}/10^{-5} = 10^{-9}\text{M}$$

The answer indicates that while the $[H^+]$ increased, $[OH^-]$ decreased.

What is the $[H^+]$ in an aqueous solution which is 0.1M in NaOH?

Realizing that NaOH is a strong electrolyte, and that $[OH^-]$ is thus 0.1M, $[H^+]$ is obtained in the following way:

$$K_w = [H^+][OH^-] = 10^{-14}$$
$$[H^+][0.1\text{M}] = 10^{-14}$$
$$[H^+] = 10^{-14}/0.1 = 10^{-13}\text{M}$$

In this problem it is seen that increasing the base concentration results in a decrease in the concentration of the acid.

pH, pOH, pK

Chemists commonly use an alternate way of expressing $[H^+]$, $[OH^-]$, and K for various equilibria. This alternate expression involves the use of the negative logarithm of the quantity in question. The **pH** of an aqueous solution is defined as **the negative logarithm of $[H^+]$.** Thus,

$$\text{pH} = -\log [H^+] = \log 1/[H^+]$$
$$\text{and} \quad [H^+] = 10^{-\text{pH}} = 1/10^{\text{pH}} = \text{antilog}(-\text{pH})$$

The **pOH** of an aqueous solution is defined as **the negative logarithm of $[OH^-]$.** Thus,

$$\text{pOH} = -\log [OH^-] = \log 1/[OH^-]$$
$$\text{and} \quad [OH^-] = 10^{-\text{pOH}} = 1/10^{\text{pOH}} = \text{antilog}(-\text{pOH})$$

A similarly defined quantity **pK** is related to equilibrium constants. Thus, the ion product constant for water, K_w, has an associated **pK_w** defined as

$$pK_w = -\log K_w = \log 1/K_w$$

$$\text{and} \quad K_w = 10^{-pK_w} = 1/10^{pK_w} = \text{antilog}\,(-pK_w)$$

For the expression $K_w = [H^+][OH^-] = 10^{-14}$, taking the logarithm of each term and multiplying through by -1 gives

$$-\log K_w = -\log [H^+] - \log [OH^-] = -\log 10^{-14}$$

Substituting the definitions of pK, pH, and pOH into this expression gives

$$pK_w = pH + pOH = -\log 10^{-14}$$

The logarithm of 10^{-14} is -14, and thus,

$$pK_w = pH + pOH = 14$$

This expression leads to the conclusion that for an aqueous solution at 25°C the sum of pH and pOH must always be 14. The following sample calculations illustrate the utility of pH and pOH.

What is the pH and the pOH of an aqueous solution containing 10^{-4}M HCl?

$$pH = -\log [H^+] = -\log 10^{-4}$$
$$= -(-4) = 4$$

Since pH + pOH = 14, pOH = 14 − 4 = 10.

In this solution the $[H^+]$ is greater than that in pure water, 10^{-4}M as compared to 10^{-7}M, but the pH of the solution is smaller than that in pure water, 4 as compared to 7. Aqueous solutions in which the $[H^+]$ is greater than 10^{-7}M, and in which the pH is less than 7, are called **acidic solutions.** Aqueous solutions, in which the $[H^+]$ is less than 10^{-7}M, and thus the $[OH^-]$ is greater than 10^{-7}M, have a pH which is greater than 7 and a pOH less than 7. Such solutions contain a preponderance of OH^- and are called **basic solutions. Neutral solutions** have a $[H^+] = [OH^-] = 10^{-7}$M, and a pH = 7.

TABLE 7-1 pH AND pOH OF AQUEOUS SOLUTIONS AT 25°C

[H⁺] moles/l	*pH*		*[OH⁻] moles/l*	*pOH*
1	0	↑	10^{-14}	14
10^{-1}	1		10^{-13}	13
10^{-2}	2		10^{-12}	12
10^{-3}	3	acidic	10^{-11}	11
10^{-4}	4		10^{-10}	10
10^{-5}	5		10^{-9}	9
10^{-6}	6		10^{-8}	8
10^{-7}	7	neutral	10^{-7}	7
10^{-8}	8		10^{-6}	6
10^{-9}	9		10^{-5}	5
10^{-10}	10		10^{-4}	4
10^{-11}	11	basic	10^{-3}	3
10^{-12}	12		10^{-2}	2
10^{-13}	13		10^{-1}	1
10^{-14}	14	↓	1	0

What is the $[H^+]$ and the $[OH^-]$ for a solution with a pH of 9? From the definition previously stated

$$[H^+] = 10^{-pH} = \text{antilog}(-pH)$$

thus, $[H^+] = 10^{-9}M$. Since pH + pOH = 14, pOH is 5, and $[OH^-] = 10^{-pOH} = 10^{-5}M$. The values of pH and pOH at various H^+ and OH^- are shown in Table 7–1.

DISSOCIATION OF ACIDS AND BASES

It was indicated previously that acids and bases are electrolytes, but that many of these solutes behave as weak electrolytes and are therefore only partially dissociated into ions. The extent of the dissociation, which is dependent on the nature of the acid or base, can be treated as an ionic equilibrium.

Acetic acid, the acidic component of vinegar, is only partially dissociated in aqueous solution and is referred to as a weak acid. The following equation represents the dissociation of acetic acid in water to give hydronium ions and acetate ions.

$$HC_2H_3O_2 + H_2O \rightleftharpoons H_3O^+ + C_2H_3O_2^-$$

The equilibrium constant for this reaction is

$$K = \frac{[H_3O^+][C_2H_3O_2^-]}{[H_2O][HC_2H_3O_2]}$$

For dilute solutions where the concentration of water is a constant, $[H_2O]$ is incorporated along with K into a new constant, **K_a, the acid dissociation constant,** as follows:

$$K \cdot [H_2O] = K_a = \frac{[H^+][C_2H_3O_2^-]}{[HC_2H_3O_2]}$$

In this expression $[H^+]$ has been substituted for $[H_3O^+]$, since the net reaction is

$$HC_2H_3O_2 \rightleftharpoons H^+ + C_2H_3O_2^-$$

At 25°C, K_a for acetic acid is 1.8×10^{-5} and pK_a is 4.75.

Polyprotic acids can yield more than one mole of H^+ per mole of acid. Examples include sulfuric acid, H_2SO_4, a diprotic acid, and phosphoric acid, H_3PO_4, a triprotic acid. Polyprotic acids dissociate in a stepwise manner such that each successive step occurs to a smaller extent than the preceding one. Acid dissociation constants for each step can be written as illustrated in the following equation for sulfuric acid:

$$H_2SO_4 \rightleftharpoons H^+ + HSO_4^- \qquad K_{a1} = \frac{[H^+][HSO_4^-]}{[H_2SO_4]}$$

$$HSO_4 \rightleftharpoons H^+ + SO_4^{-2} \qquad K_{a2} = \frac{[H^+][SO_4^{-2}]}{[HSO_4^-]}$$

The stepwise dissociation of sulfuric acid is especially interesting because the first step is complete, but the second step is not. Consequently, H_2SO_4 is called

TABLE 7-2 ACID AND BASE DISSOCIATION CONSTANTS AT 25°C

Acid	K_a	pK_a
HCl	∞	—
HF	6.7×10^{-4}	3.2
HCN	4×10^{-10}	9.4
$HC_2H_3O_2$	1.8×10^{-5}	4.7
HNO_2	4.5×10^{-4}	3.3
H_2SO_4	$(K_{a1})\ \infty$	—
	$(K_{a2})\ 1.3 \times 10^{-2}$	1.9
H_2CO_3	$(K_{a1})\ 4.2 \times 10^{-7}$	6.4
	$(K_{a2})\ 4.7 \times 10^{-11}$	10.3
H_2S	$(K_{a1})\ 1 \times 10^{-7}$	7.0
	$(K_{a2})\ 1 \times 10^{-15}$	15.0
H_3PO_4	$(K_{a1})\ 7.1 \times 10^{-3}$	2.2
	$(K_{a2})\ 6.3 \times 10^{-8}$	7.2
	$(K_{a3})\ 4 \times 10^{-13}$	12.4
Base	K_b	pK_b
NaOH	∞	—
NH_3	1.8×10^{-5}	4.7

a strong acid, but HSO_4^-, the bisulfate ion, is a weak acid. The acid dissociation constants of several common acids are given in Table 7–2.

In theory the dissociation of weak hydroxy bases can be treated in the same way as weak acids. For the general weak base MOH, the net equation for the dissociation and the expression for K_b are:

$$MOH \rightleftharpoons M^+ + OH^- \qquad K_b = \frac{[M^+][OH^-]}{[MOH]}$$

In reality there are few weak bases of the general formula MOH which have been well characterized in aqueous solutions. There are some with more than one hydroxide per mole of base, but the dissociation equilibria for these are complicated. There are, however, a large number of well characterized molecules which, when placed in water, undergo a hydrolysis reaction to generate OH^-. Among these compounds are ammonia, NH_3, and organic amines.

Gaseous ammonia, when dissolved in water, undergoes a hydrolysis reaction to give a basic solution. The hydrolysis reaction may be represented as follows:

$$NH_3 + H_2O \rightleftharpoons NH_4^+ + OH^-$$

The equilibrium constant for this reaction is

$$K = \frac{[NH_4^+][OH^-]}{[NH_3][H_2O]}$$

If the term $[H_2O]$, a constant in dilute solutions, is incorporated along with the equilibrium constant into a new constant $\mathbf{K_b}$, the following expression is obtained:

$$K \cdot [H_2O] = K_b = \frac{[NH_4^+][OH^-]}{[NH_3]}$$

The constant K_b is the **base dissociation constant.** At 25°C, K_b for ammonia is 1.8×10^{-5}, and $pK_b = 4.75$.

The extent to which an acid or a base dissociates in aqueous solutions is reflected in K_a or K_b, respectively. With increasing strength of acids or bases, the extent of dissociation increases, and K_a and K_b are larger. Correspondingly, the smallest pK_a and pK_b will be associated with the strongest acids and bases.

HYDROLYSIS OF SALTS

Many salts, when placed in water, react with the solvent molecules to give acidic or basic solutions. The net reaction for the hydrolysis of a positive ion, or cation, is

$$M^+ + H_2O \rightleftharpoons MOH + H^+$$

This reaction proceeds furthest to the right when M^+ has a very high affinity for hydroxide, and therefore when MOH is a weak base. The net reaction for the hydrolysis of a negative ion, or anion, is

$$A^- + H_2O \rightleftharpoons HA + OH^-$$

This reaction proceeds furthest to the right when HA is a weak acid. From the reactions above, it can be stated that cations of weak bases hydrolyze to give acidic solutions, and anions of weak acids hydrolyze to give basic solutions.

The cations of strong bases (e.g., Na^+, K^+) and the anions of strong acids (e.g., Cl^-, NO_3^-) do not hydrolyze, and therefore do not influence $[H^+]$ or $[OH^-]$. For this reason salts composed of cations of strong bases and anions of strong acids dissolve in water without any net hydrolysis. Consequently, aqueous solutions of salts, such as NaCl, KCl, $NaNO_3$, KNO_3 and so forth, are neutral, that is, pH = 7.

Any salt which contains either the cation of a weak base, or the anion of a weak acid, will, when dissolving in water, undergo a hydrolysis reaction. This reaction will result in a solution which is not neutral. The salts of strong bases and weak acids, such as $NaC_2H_3O_2$ and KCN, always give basic solutions. Salts of weak bases and strong acids, such as NH_4Cl, give acidic solutions. A salt composed of the cation of a weak base and the anion of a weak acid, such as $NH_4C_2H_3O_2$ or NH_4CN, are acidic or basic depending on the relative extents of hydrolysis of the cation and the anion.

THE COMMON ION EFFECT

The extent of dissociation of a weak acid or a weak base in aqueous solution can be altered by the addition of appropriate salts to the solution. The extent of dissociation of acetic acid can be decreased by adding sodium acetate, $NaC_2H_3O_2$. This phenomenon is predicted by applying Le Chatelier's principle to the dissociation equilibrium:

$$HC_2H_3O_2 \rightleftharpoons H^+ + C_2H_3O_2^-$$

The addition of sodium acetate to an acetic acid solution at equilibrium instantaneously increases the acetate ion concentration, resulting in a stress on the equilibrium. This stress is removed by a shift of the equilibrium to the left, that is, H^+ and $C_2H_3O_2^-$ combine to form undissociated $HC_2H_3O_2$. This effect, called **the common ion effect,** is general for all ionic equilibria, and occurs whenever the concentration of any ionic component of the equilibrium is increased. In acid dissociation equilibria, the equilibrium is also shifted to the left by addition of a strong acid such as HCl, which instantaneously increases the concentration of H^+. Thus, the pH of an acetic acid solution can be increased ($[H^+]$ decreased) by the addition of acetate ion, whereas it can be decreased ($[H^+]$ increased) by adding a strong acid.

The extent of the hydrolysis reaction which takes place when ammonia is added to water is depressed by the addition of ammonium salts, such as ammonium chloride, NH_4Cl. Strong bases, such as sodium hydroxide, NaOH, also shift the equilibrium and depress the extent of the hydrolysis reaction.

BUFFERED SOLUTIONS

It is possible to prepare aqueous solutions which maintain a nearly constant pH, even when relatively large quantities of acid or base are added. Such solutions are called **buffered solutions.** Solutions containing weak acids and their salts, or weak bases and their salts, act as buffered solutions.

The properties of a buffered solution are illustrated by considering an acetic acid-sodium acetate solution and the associated equilibrium:

$$HC_2H_3O_2 \rightleftharpoons H^+ + C_2H_3O_2^-$$

The addition of a small quantity of sodium hydroxide to this equilibrium system will result in the consumption of H^+, but the excess $HC_2H_3O_2$ present will dissociate to replace most of the H^+ consumed. The addition of a small quantity of hydrochloric acid will induce a shift in the equilibrium to consume the added H^+ and some of the excess $C_2H_3O_2^-$, but will result in only a small change in $[H^+]$. The following quantitative treatment illustrates the extent of each of these effects:

A solution which is 0.1M in $HC_2H_3O_2$ and 0.1M in $NaC_2H_3O_2$ has a $[H^+]$ of 1.8×10^{-5} and a pH of 4.75. The addition of 0.01 moles of HCl to a liter of this buffered solution changes the $[H^+]$ to 2.2×10^{-5}, and the pH to 4.66. Thus the pH is changed by only 0.09 units. The addition of the same quantity of HCl to a liter of water decreases the pH from 7 to 2, a change of 5 units. The addition of 0.01 moles of NaOH to a liter of this buffered solution changes the $[H^+]$ to 1.5×10^{-5}, and the pH to 4.83. The same quantity of NaOH added to a liter of water changes the $[H^+]$ from 10^{-7}M to 10^{-12}M, and the pH from 7 to 12.

Buffered solutions play an important role in many naturally occurring processes. All body fluids have definite pH values that must be maintained within fairly narrow ranges for proper physiological functions. The pH of the blood is normally between 7.35 and 7.45. If the pH of the blood falls below 7.0 or goes above 7.8, death occurs. Since many of the reactions that take place in our tissues form acid substances, the blood must have a mechanism to prevent such changes in pH. The equilibrium system involved in buffering the blood includes bicarbonates and carbonates, phosphates and complex salts of proteins.

PRECIPITATION-DISSOLUTION EQUILIBRIA

In a saturated solution containing excess undissolved solute, a precipitation-dissolution dynamic equilibrium prevails. In this dynamic equilibrium solute particles are leaving and being redeposited on the surface of the undissolved crystalline solute at the same rate. If the solute is an ionic material, dissolution is accompanied by a dissociation into discrete ions. The equilibrium can be represented by the following general equation:

$$MX \text{ (solid)} \rightleftharpoons M^+ \text{ (sol'n)} + X^- \text{ (sol'n)}$$

For this equilibrium the equilibrium constant expression is

$$K = \frac{[M^+][X^-]}{[MX \text{ (solid)}]}$$

Since the concentration of a given ionic solid is a constant at a given temperature, [MX (solid)] is incorporated with K into a new constant **K_{sp}, the solubility product constant:**

$$K \cdot [MX \text{ (solid)}] = K_{sp} = [M^+][X^-]$$

The solubility product constants for different types of salts are written in accordance with the law of chemical equilibrium as follows:

$$MX_2 \quad K_{sp} = [M^{+2}][X^-]^2$$
$$M_2X \quad K_{sp} = [M^+]^2[X^{-2}]$$
$$M_2X_3 \quad K_{sp} = [M^{+3}]^2[X^{-2}]^3$$

The K_{sp} values for various compounds are evaluated from solubility data. Whenever the product of the concentrations of the ions raised to the appropriate powers is in excess of the K_{sp}, a precipitation dissolution equilibrium prevails.

For many electrolytes K_{sp} values cannot be determined from solubility data because of reactions which the dissolved ions undergo. For example, the K_{sp} value for HgS is 1×10^{-52}, but the solubility of HgS is considerably greater than 1×10^{-26}. The majority of Hg^{+2} ions and S^{-2} ions which enter solution undergo hydrolysis reactions to form other species. The K_{sp} value is based not on the total quantity of HgS which has dissolved, but rather indicates the true concentrations of Hg^{+2} and S^{-2} which have not undergone reactions upon entering solution.

Questions

1. What is an electrolyte? A non-electrolyte? Give examples of each.

2. Which of the definitions of an acid and a base are most frequently applied to aqueous solutions?

3. Name the following acids and bases:

(a) HCl (f) $HClO_4$
(b) H_3PO_4 (g) H_2SO_3
(c) HNO_2 (h) HNO_3
(d) H_2SO_4 (i) $Ca(OH)_2$
(e) NH_4OH (j) H_2CO_3

4. Write the formula for a normal salt and for an acid salt. Explain the difference between the two compounds.

5. Write the equation for the dissociation of (a) a strong acid, and (b) a weak acid. Explain the difference between them.

6. How did the Debye-Hückel theory alter Arrhenius's assumptions?

7. Write an equation for the ionization of (a) an acid, (b) a base, and (c) a salt.

8. Will the addition of H^+ to an aqueous equilibrium system cause an increase or decrease in $[OH^-]$? What law or principle allows this prediction?

9. To a 1 liter solution of pure water, 1×10^{-4} moles of gaseous HCl was added, increasing the $[H^+]$ from 10^{-7}M to 10^{-4}. What is the $[OH^-]$?

10. Calculate the pH and pOH for aqueous solutions containing the following H^+ concentrations:

(a) 10^{-6}M (c) 0.5M (e) 10M
(b) 2.5×10^{-3}M (d) 4.2×10^{-8}M (f) 0.097M

11. Which of the solutions in question 10 are acidic and which are basic?

12. As acids and bases decrease in strength, does the extent of dissociation increase or decrease? Do K_a and K_b become larger or smaller?

13. Which of the following ions will hydrolyze upon dissolving in water?

(a) anions of strong acids (f) F^-
(b) anions of weak acids (g) NH_4^+
(c) cations of strong bases (h) CN^-
(d) cations of weak bases (i) $H_2PO_4^-$
(e) Cl^- (j) HSO_4^-

14. What is meant by the common ion effect? What basic chemical principle predicts the common ion effect?

15. What is a buffered solution? Describe how it works.

16. For solute MX_2, when $[M^+][X^-]^2$ is greater in value than the K_{sp}, what phenomenon is observed?

17. What is the pH of a neutral solution at 25°C? Of a basic solution? Of an acidic solution?

18. Does diluting a solution of a weak base increase or decrease the number of moles of OH^- per liter?

19. Is the K_{sp} of a salt altered as the salt becomes more soluble with increasing temperature?

20. In dilute solutions, what can be said about the concentrations of water molecules?

21. The addition of Na_2SO_4 to a H_2SO_4 solution will

(a) increase SO_4^{-2} (c) increase pH
(b) decrease H_2SO_4 (d) decrease pH

22. What can be said about the pH of NaCl and KNO_3 solutions?

Suggested Reading

Banks: Chemical Equilibrium and Solutions. New York, McGraw-Hill, 1967.
Butler: Ionic Equilibrium. Reading, Mass., Addison-Wesley Publishing Co., 1964.
Drago and Matwyioff: Acids and Bases. Lexington, Mass., D. C. Heath and Co., 1968.
Margolis: Principles in the Calculation of Ionic Equilibria. New York, The Macmillan Company, 1966.
Robbins: Ionic Reactions and Equilibria. New York, The Macmillan Company, 1967.
Vanderwerf: Acids, Bases and Chemistry of the Covalent Bond. New York, Reinhold Publishing Corp., 1961.

CHAPTER 8

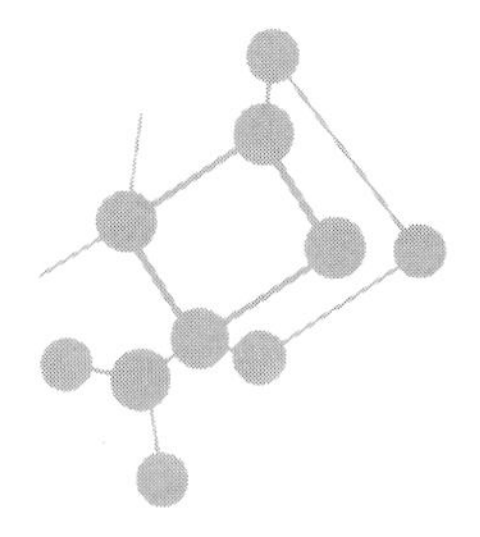

THE NONMETALS

Elements in the upper right side of the periodic table are characterized by their tendency to gain electrons to complete their outer electron shell. Like the halogens, which they include, their electronegativities are relatively high, and they form many important compounds with other elements. The elements shown in Figure 8–1 are classed as **nonmetals.**

Although the nonmetals represent a relatively small group of elements, they possess a wide range of properties. About half of the group, including the inert gases, fluorine, chlorine, nitrogen, and oxygen, exist as gases, although bromine is a liquid and the remainder exist as solids.

The tendency of nonmetallic elements to gain electrons in order to complete their outer electron shell has already been discussed. Atoms of the active nonmetals thereby gain electrons and exist as negatively charged ions, or anions, as opposed to metals, which tend to lose electrons, forming positively charged ions, or cations. The oxides of most active nonmetals react with water to form acids.

				H	He
B	C	N	O	F	Ne
	Si	P	S	Cl	Ar
		As	Se	Br	Kr
		Sb	Te	I	Xe
			Po	At	Rn

Figure 8–1 The nonmetals in the periodic table.

HYDROGEN

The element hydrogen will be treated separately rather than as a member of a family because of its ambivalent chemical nature. Atoms of hydrogen do not occur in the free form but are combined in pairs to form diatomic hydrogen molecules. Very little elemental hydrogen is found in the atmosphere. Hydrogen is found to some extent in natural gas wells, although, in general, the quantity of hydrogen produced from natural sources is extremely limited. In comparison, the hot gases surrounding the sun are apparently rich in hydrogen, and flames of incandescent hydrogen sometimes reach thousands of miles out from the sun.

Many important compounds contain hydrogen as one of their constituents. For example, all acids and bases contain hydrogen, and water is composed of hydrogen and oxygen. Almost all organic compounds present in plant and animal tissue contain hydrogen in combined form. Like oxygen, it is an important constituent in carbohydrates, fats, and proteins, which are used as foods. It occurs combined with carbon in the important products of the petroleum industry, such as gasoline, lubricating oils, and natural gas.

Preparation. An important method for the preparation of both hydrogen and oxygen is the electrolysis of water. When an electric current is passed through water containing a small quantity of acid to increase its conductivity, hydrogen is evolved at the negative electrode while oxygen is evolved at the positive electrode. Pure hydrogen is obtained by this method and is usually stored in steel cylinders under pressure for future use. Hydrogen can also be evolved from water by the addition of **active metals.** It may be recalled that the most active metals are located in Group IA of the periodic table. When, for example, a small piece of sodium is added to cold water, a violent reaction takes place in which hydrogen gas is produced and the metal melts. Heat is given off, and often the hydrogen catches fire with a violent explosion. The equation for this reaction may be written as follows:

$$2Na + 2H_2O \rightarrow 2NaOH + H_2\uparrow$$

A very common method for the preparation of hydrogen in the laboratory involves the addition of dilute acids to certain metals. For example, dilute hydrochloric acid may be added to zinc granules in a bottle connected with a delivery tube to a water trough containing an inverted bottle for the collection of hydrogen. In this reaction zinc displaces hydrogen from the acid forming hydrogen gas and a salt.

$$Zn + 2HCl \rightarrow H_2\uparrow + ZnCl_2$$

Commercially, hydrogen is produced from **water gas.** First, the water gas is prepared by passing steam over burning coke (temperature, about 1000°C) Water gas and steam are then passed over a mixture of iron, thorium, and chromium oxide at a temperature of 500°C. The metallic oxides serve as a catalyst to help in the oxidation of carbon monoxide to form carbon dioxide. These two reactions may be represented as follows:

$$C + H_2O \rightarrow \underbrace{CO + H_2}_{\text{water gas}}$$

$$CO + H_2O \xrightarrow{\text{catalyst}} 2H_2 + CO_2$$

The mixture of hydrogen and carbon dioxide produced in the final equation is passed through water to remove the carbon dioxide, and the hydrogen may then be used immediately or stored in steel cylinders.

Physical Properties. Hydrogen, like oxygen, is a colorless, odorless, and tasteless gas. Approximately 2 ml of hydrogen will dissolve in 100 ml of water, which makes this element less soluble than oxygen. It is the lightest gas known, one liter weighing approximately 0.09g at STP. Hydrogen can be liquefied, but only at extremely low temperatures. The liquid looks like water, but has such a low density that a cork will sink in it. Certain metals can adsorb large quantities of hydrogen gas (adsorption is a surface phenomenon; in this case the hydrogen gas would adhere to the atoms on the surface of the metal). Powdered palladium, for example, can adsorb approximately one thousand times its own volume of hydrogen, and this adsorbed hydrogen is chemically very reactive. This property of hydrogen facilitates the hydrogenation of alkenes with palladium catalysis as discussed in Chapter 12. In the early forms of the periodic table, hydrogen was placed at the top of Group I, followed by Li, Na, K, and other elements of the group. This would classify it as a metal that would readily lose one electron to form ionic compounds with nonmetallic elements, or combine with them by sharing a pair of electrons. It is well known that hydrogen may also combine with metals to form ionic compounds called **hydrides,** in which the hydrogen acts as a negative ion similar to elements in Group VII. This property of hydrogen to act as a metal or a nonmetal is shown in modern periodic tables in different ways. The two most common methods are: (1) to place it at the top of Group IA and Group VIIA, and (2) to center it at the top of the table with lines connecting it to the top of Groups IA and VIIA.

Hydrogen gas does not react readily with other elements at ordinary temperatures; but if the temperature is increased it will react with some elements, including oxygen, sulfur, and the halogens. Under proper conditions and in the presence of a catalyst, hydrogen will combine with other elements and other compounds. For example, hydrogen will combine directly with nitrogen to form the compound ammonia as shown in the following equation:

$$N_2 + 3H_2 \rightarrow 2NH_3$$

Burning hydrogen produces a very hot flame, which has many important applications. By the proper mixing of oxygen and hydrogen, a flame with a temperature over 2000°C is produced. The apparatus that produces this flame is called the **oxyhydrogen torch.** More recently a torch with a flame whose temperature is twice as hot as that of the oxyhydrogen torch has been devised. This is called the **atomic hydrogen torch** and consists of a stream of hydrogen gas passing through an electric arc. As the hydrogen molecules pass through the arc, they absorb energy, and the energy splits the hydrogen molecules into atoms. Immediately outside the arc, the atoms of hydrogen unite again, giving off large quantities of heat. This flame is hot enough to melt tungsten, which has a melting point of 3370°C and may reach temperatures as high as 4000°C.

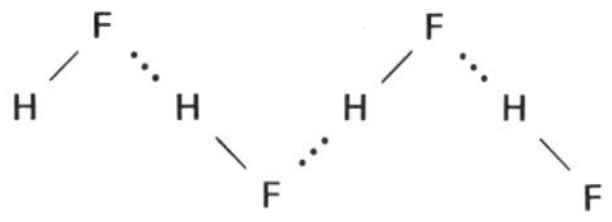

Figure 8-2 Hydrogen bonding in hydrogen fluoride.

Recently temperatures as high as 4500°C have been attained by burning fluorine in hydrogen.

Hydrogen also has the ability to remove oxygen from oxides. A common illustration of this ability is the reaction of hydrogen gas on hot copper oxide to produce copper metal and water.

$$CuO + H_2 \rightarrow Cu + H_2O$$

In recent years several industrial applications have resulted from the characteristic properties of metal hydrides. For example, calcium and lithium hydrides can be used in portable hydrogen generators, since they release large quantities of the gas when added to water. One gram of LiH produces almost three liters of hydrogen gas by this method. The hydrides are good reducing agents and serve as excellent drying agents. The complex hydride, lithium aluminum hydride ($LiAlH_4$), is soluble in organic solvents and is used as a powerful reducing agent by organic chemists. Titanium and zirconium hydrides are used in the manufacture of vacuum tubes. They are sprayed on a part of the tube as a finely divided suspension and flash dried, producing a thin film of the metal. Other hydrides, such as tin and silicon hydrides, have a very high heat of combustion and are being studied as possible rocket fuels.

The Hydrogen Bond. Another important property of hydrogen is that of forming a bridge or bond between two small, strongly electronegative atoms, such as O, N, or F. This bridge, which consists of a proton shared between two atoms, is called a **hydrogen bond.** The compound HF has an abnormally high boiling point compared to that of the other hydrogen halides (HCl, HBr, and HI). This increase in boiling point is thought to result from hydrogen bonding in HF, which could be represented as shown in Figure 8-2.

The fluorine atoms held together by hydrogen bonds form aggregates that change the properties of the simple HF molecule. Later it will be seen that hydrogen bonds formed between O and N atoms in the important biological compounds, the proteins, alter the structure and properties of these compounds. Hydrogen bonding also plays an important role in determining the chemical and physical properties of organic molecules as will be seen in Chapters 16, 18, and 19.

THE INERT GASES

The **inert gases** were so named because of their very nonreactive character. This inert tendency is predictable from the electronic configuration of the elements, which involves a completely filled shell, or quantum level, for each.

Prior to 1962 inert gas atoms were thought to participate in chemical bonding only in ions under high energy conditions. It is now known that xenon will combine directly with fluorine as follows to give three xenon fluorides:

$$Xe + F_2 \rightleftharpoons XeF_2$$
$$Xe + 2F_2 \rightleftharpoons XeF_4$$
$$Xe + 3F_2 \rightleftharpoons XeF_6$$

These compounds, all solids at room temperature, can be stored in glass containers, but they react with water by hydrolysis to give a number of different oxygen and hydrogen containing compounds.

HALOGENS

The group of elements including **fluorine, chlorine, bromine, iodine,** and **astatine** is called the **halogens,** which is derived from a Greek word meaning "salt formers." The last member of the group, astatine, is a short-lived radioactive element that was discovered in 1940. Since only minute quantities of this element have ever been prepared, nothing definite is known concerning its properties. For this reason it will not be considered with the four common halogens in this chapter.

All four of the halogens are too reactive chemically to be found free in nature. They are probably the most typical nonmetals of all the elements in the periodic classification. A comparison of their atomic structures reveals that each has seven electrons in its outer shell. They all possess a high electronegativity and readily form negative halide ions, commonly found in ionic salts. As the atomic weight increases from fluorine to iodine, the physical state of the elements changes. The first two members of the group exist as gases, whereas bromine is a liquid and iodine a solid. The melting point and boiling point temperatures and the size of the individual atoms all increase gradually from fluorine to iodine. The tendencies of the atoms to attract electrons decreases with increasing size, as reflected in the electron affinities and the electronegativity. Fluorine, which is the most electronegative halogen, exists only as the element and with an oxidation number of -1, whereas the other halogens show positive oxidation numbers in addition to 0 and -1.

Fluorine and chlorine are the most abundant of the halogens. Fluorine is found in mineral fluorites or **fluorspar,** CaF_2, and as a constituent of the aluminum ore **cryolite,** Na_3AlF_6. The most common source of chlorine in nature is NaCl, common table salt. Bromine occurs in nature as salts of sodium and magnesium along with chloride salts, but in much smaller concentrations than the chlorides. Iodine is found in nature as $NaIO_3$, an impurity of the $NaNO_3$ found in Chile.

All of the halogens occur as diatomic molecules in the elemental state. Fluorine and chlorine are green-yellow gases under normal conditions. Bromine, the only liquid nonmetallic element under normal conditions, is a dense, reddish-brown liquid which readily vaporizes at room temperature to red vapors that have a strong pungent odor. It is moderately soluble in water, but much more soluble in alcohol, chloroform, or carbon tetrachloride. Iodine exists as a gray-black shiny solid which has violet vapors. It dissolves only slightly in water, but is very soluble in an aqueous solution of KI, as well as in organic solvents such as chloroform, ether, and alcohol.

PREPARATION OF THE HALOGENS. Free fluorine, difficult to prepare because of its extreme chemical reactivity, was first prepared by passing an electric current through a solution of potassium hydrogen fluoride in a platinum U-tube. Fluorine gas (F_2) was given off at the anode, and hydrogen gas was liberated at the cathode. The modern method employs a V-shaped electrolysis vessel made of copper and containing graphite electrodes. The electrolyte is fused potassium hydrogen fluoride, which is kept molten by heating it to a temperature of 70° to 100°C. The net reaction is

$$2KHF_2\ (l) \xrightarrow[\text{current}]{\text{electric}} H_2\ (g) + F_2\ (g) + 2KF\ (l)$$

In the laboratory chlorine is usually generated by oxidizing hydrochloric acid with manganese dioxide as shown in the following reaction:

$$4HCl\ (aq) + MnO_2\ (s) \rightleftharpoons 2H_2O + MnCl_2\ (aq) + Cl_2\ (g)$$

Commercially, chlorine is prepared by the electrolysis of brine, aqueous NaCl, in cells that are designed to keep the products separate. The chlorine gas is given off at the anode, and hydrogen gas and sodium hydroxide are formed at the cathode. A reaction representing this process may be written as follows:

$$2NaCl\ (aq) + 2H_2O \xrightarrow[\text{current}]{\text{electric}} 2NaOH\ (aq) + H_2\ (g) + Cl_2\ (g)$$

Fortunately for industry, this process also produces hydrogen gas and concentrated sodium hydroxide in commercial quantities.

Bromine is prepared by allowing sodium bromide and sulfuric acid to react with each other to liberate hydrobromic acid, which then reacts with manganese dioxide to form bromine gas. The sodium bromide, sulfuric acid, and manganese dioxide are usually heated together as shown in the following reaction:

$$2NaBr\ (aq) + 3H_2SO_4\ (aq) + MnO_2\ (s) \rightleftharpoons 2NaHSO_4\ (aq) + MnSO_4\ (aq) + 2H_2O + Br_2\ (g)$$

The best commercial source of bromine for many years has been the salt brines pumped out of the earth in Michigan. The brine is concentrated to crystallize most of the sodium chloride, and the residue containing bromides is electrolyzed to give bromine gas.

A more recent method for the preparation of bromine employs large volumes of sea water which is acidified and treated with chlorine to liberate the bromine. In this reaction chlorine acts as an oxidizing agent as follows:

$$Cl_2\ (g) + 2Br^-\ (aq) \rightleftharpoons 2Cl^-\ (aq) + Br_2\ (l)$$

Iodine may be prepared in a manner similar to that described for bromine, in which a salt of iodine is heated with sulfuric acid and manganese oxide. A more direct method for the preparation of iodine from its salts is that of passing chlorine gas into an aqueous solution of the salt.

HALIDES. The halogens combine with metals, nonmetals, and among themselves. In combination with metals the halogens always have −1 as an oxidation number and are called **halides.** The halides of metals have high ionic character as indicated by their high melting points. As the metallic character of the combining element decreases, the ionic character of the binary halides gradually decreases, and the bonds become more covalent.

Most binary fluorides of the metals have low solubility in water whereas the binary chlorides, bromides, and iodides of most metals and nonmetals are soluble in water. Most covalent halides hydrolyze on entering water to give an acidic solution. This is illustrated by the hydrolysis of PCl_3 as follows:

$$PCl_3 + 3H_2O \rightleftharpoons H_3PO_3 + 3H^+ + 3Cl^-$$

A common test for the presence of the chloride, bromide, or iodide ion is carried out by adding a solution of silver nitrate to a solution of the halogen salts. When a halogen is present, an insoluble silver salt is precipitated out of solution. The color of the precipitate helps identify the halogen, since silver chloride is white, silver bromide, pale yellow, and silver iodide, a lemon yellow. The chemical reaction involved in this test may be represented by the reaction of silver nitrate and sodium chloride as follows:

$$AgNO_3\,(aq) + NaCl\,(aq) \rightleftharpoons NaNO_3\,(aq) + AgCl\,(s)\,(white)$$

The silver chloride and silver bromide precipitates will dissolve in ammonium hydroxide solution, but silver iodide is insoluble in this reagent. Silver fluoride is soluble in water, and consequently, Ca^{+2} is used to test for F^- since CaF_2 is insoluble in water.

A very common test for the presence of iodine is the blue color that it yields in the presence of starch. A solution of potassium iodide and starch may also be used to detect the presence of small quantities of free chlorine or bromine, since these two elements will liberate iodine from the potassium iodide and produce a blue color in the presence of starch.

Hydrogen Halides. The hydrogen halides can each be prepared by the direct interaction of hydrogen gas and the halogen. In the laboratory where this method is impractical, HF and HCl are made by placing metal halides in warm concentrated sulfuric acid. The reaction proceeds as follows:

$$2MX\,(s) + H_2SO_4\,(l) \rightleftharpoons M_2SO_4\,(s) + 2HX\,(g)$$

This procedure cannot be used for bromides and iodides because H_2SO_4 oxidizes the halides to the free halogen:

$$2MX\,(s) + 2H_2SO_4\,(l) \rightleftharpoons X_2\,(g) + M_2SO_4\,(s) + SO_2\,(g) + 2H_2O$$

Hydrogen bromide and hydrogen iodide can be made using phosphoric acid in a process similar to the aforementioned:

$$3MX\,(s) + H_3PO_4\,(l) \rightleftharpoons 3HX\,(g) + M_3PO_4\,(s)$$

Some properties of the hydrogen halides are given in Table 8–1.

Except for HF, the melting points and boiling points of the hydrogen halides show a decreasing trend on going from the heaviest to the lightest. The abnormally high melting point, boiling point, and enthalpy of vaporization of HF are attributed to strong intermolecular hydrogen bonding.

The hydrogen halides, all very soluble in water, form aqueous solutions in which all except HF behave as strong acids. These acids are called **hydrohalic** (i.e., hydrofluoric, and so forth) **acids.**

TABLE 8-1 SOME PROPERTIES OF THE HYDROGEN HALIDES

	HF	*HCl*	*HBr*	*HI*
Melting Point (°C)	−83.1	−114.6	−86.9	−50.9
Boiling Point (°C)	19.4	−84.8	−66.9	−35.8
ΔH vaporization (kcal/mole)	7.24	3.85	4.21	4.72
Color	none	none	none	none
K_a in water	6.7×10^{-4}	very large	very large	very large

Hydrofluoric acid cannot be kept in glass bottles because it dissolves glass by the reaction:

$$4HF\,(aq) + SiO_2\,(s) \rightleftharpoons 2H_2O + SiF_4\,(g)$$

The SiF_4 produced is volatile and leaves the solution, so that equilibrium is never attained. Hydrofluoric acid, because of this property, is used to etch glass, and consequently solutions of HF are usually stored in special plastic or paraffin-lined bottles.

OXYACIDS. The ternary acids or oxyacids of the halogens, excluding fluorine, have the general formulas HOX, HOXO, $HOXO_2$, and $HOXO_3$, and are called **hypohalous, halous, halic,** and **perhalic acids** in that order. The halogen oxidation numbers in this series are +1, +3, +5, and +7. In this series the compounds $HBrO_2$ and HIO_2 are not stable and have not been characterized.

The hypohalous acids are weak acids, all having a K_a of less than 10^{-7}. Chlorine, bromine, and iodine, when added to water, undergo disproportionation to give HX and HXO as follows:

$$X_2\,(aq) + H_2O \rightleftharpoons H^+\,(aq) + X^-\,(aq) + HOX\,(aq)$$

This reaction proceeds to the greatest extent for chlorine. If the halogens are passed into basic solution, this reaction is shifted to the right by consumption of H^+. The hypohalous acids and their anions are good oxidizing agents, and as such are used commonly as disinfectants and color bleaching agents. Common liquid bleaches are prepared by passing chlorine into mildly alkaline aqueous solutions. Dry bleaching powder is prepared by the reaction:

$$Ca(OH)_2\,(s) + Cl_2\,(g) \rightleftharpoons H_2O + CaCl(OCl)\,(s)$$

Chlorine is the only halogen forming a ternary acid with the formula HOXO. This acid which has a K_a of 10^{-2} can be obtained by adding ClO_2 to aqueous sodium peroxide, followed by acidification as follows:

$$Na_2O_2\,(aq) + 2ClO_2\,(aq) \rightleftharpoons 2Na^+\,(aq) + 2ClO_2^-\,(aq) + O_2\,(g)$$
$$ClO_2^-\,(aq) + H^+\,(aq) \rightleftharpoons HClO_2\,(aq)$$

This acid and its anion have oxidizing abilities similar to HOCl and OCl^-.

The **halates,** XO_3^-, are formed by disproportionation of **hypohalites,** XO^-, in warmed concentrated solutions as follows:

$$3OX^-\,(aq) \xrightleftharpoons{heat} XO_3^-\,(aq) + 2X^-\,(aq)$$

Chlorates can also be prepared by the electrolysis of warmed solution of the chloride as follows:

$$Cl^- (aq) + 3H_2O \xrightarrow[\text{current}]{\text{electric}} ClO_3^- (aq) + 3H_2 (g)$$

The halic acids are all strong acids in aqueous solutions and are relatively unstable in the pure form. Halates, when heated, decompose to the halide and oxygen. The thermal decomposition of $KClO_3$, catalyzed by MnO_2, is commonly used in the laboratory as a preparation for O_2:

$$2KClO_3 (s) \xrightarrow[MnO_2]{\text{heat}} 2KCl (s) + 3O_2 (g)$$

Perchloric acid ($HClO_4$), which can be obtained in the pure state as a liquid, is prepared by sulfuric acid acidification of perchlorate salts. This is a strong acid in aqueous solution and behaves as an oxidizing agent. Perchlorate salts are prepared by electrolytic oxidation of chlorates, or by controlled heating of chlorates.

Additional Uses of the Halogens and Their Compounds. In recent years the use of fluorine compounds by industry has increased considerably. Hydrogen fluoride is used as a catalyst in the petroleum industry, and hydrofluoric acid and its acid salts are used to etch glass in the manufacture of "frosted" electric light bulbs, chemical apparatus, and decorative glassware. The refrigerant Freon (CF_2Cl_2) is a fluorine compound that is commonly used in household refrigerators.

Recent research in rocket and guided-missile propellants has suggested another important use for fluorine. Compounds such as oxygen difluoride and nitrogen trifluoride can serve as high-energy oxidizers, but liquid fluorine apparently has superior properties. Since liquid fluorine boils at $-306°F$, it must be kept cold with liquid nitrogen and transported in nickel alloy containers. This use alone could result in increased production and decreased cost of fluorine.

Large quantities of chlorine are used industrially for the bleaching of paper and cotton textiles. Considerable quantities are used in the manufacturing of bleaching powder, of chloride of lime, and in the preparation of solutions of sodium hypochlorite for household bleaching agents. The water supply of almost all large cities is treated with chlorine to kill bacteria. It is an essential constituent in chlorates, which are used to manufacture matches, fireworks, and percussion caps for rifle bullets.

Large quantities of bromine are used in the manufacture of ethyl gasoline. The second largest use of bromine is in the production of soil and seed fumigants. Organic compounds of bromine are effective fire-extinguishing and flameproofing agents. Potassium bromide is used as a mild sedative in medicine, while the compound benzyl bromide is an effective tear gas. Compounds of bromine are used in the manufacture of dyes and photographic emulsions.

Tincture of iodine, which is used as an antiseptic in medicine, is prepared by dissolving iodine in a solution of potassium iodide and alcohol. Compounds of iodine are used as drugs and medications and in the preparation of certain dyestuffs. For example, potassium iodide is used as an expectorant, and the organic compound iodoform (CHI_3) is an active disinfectant. Iodized salt contains small amounts of iodine compounds, used to prevent the occurrence of common goiter.

OXYGEN

Of all the elements in the earth's surface, oxygen is the most abundant. One fifth of the volume of the air, eight ninths (by weight) of water, and approximately one half of the earth's crust is oxygen. In the air it exists as free molecular oxygen (O_2); elsewhere it is found combined with many other elements in the form of oxides or in constituents of living matter.

Preparation. Oxygen is usually isolated commercially from air, which is essentially a mixture of this element and nitrogen. Air is liquefied by subjecting it to a high pressure at a low temperature. When the liquid air is allowed to evaporate, the more volatile nitrogen escapes first, leaving behind the fairly pure oxygen. Both the oxygen and nitrogen gases are then forced into steel cylinders under high pressure and stored for future use. Fortunately, large quantities of nitrogen gas are used in the manufacture of ammonia, and the process discussed is economically sound.

Pure oxygen is also obtained for commercial purposes by the electrolysis of water. When an electric current is passed through water, oxygen forms at the positive pole, or anode, and hydrogen forms at the negative pole, or cathode. The oxygen and hydrogen gases thus formed are drawn off and stored under pressure. In the laboratory, oxygen is usually prepared by heating potassium chlorate with manganese dioxide which acts as a catalyst.

Properties. Oxygen is a colorless, odorless, tasteless gas that is slightly heavier than air. When subjected to a high pressure at a low temperature, the gas is converted to liquid oxygen. At a temperature of $-118°C$, a pressure of 50 atmospheres is required for liquefaction. This temperature ($-118°C$) is the **critical temperature.** Liquid oxygen is pale blue in color, slightly heavier than water, and has a boiling temperature of $-183°C$.

Approximately 3 ml of gaseous oxygen will dissolve in 100 ml of water at ordinary temperatures. This slight solubility insures a supply of the gas to aquatic plants and animals, and also for the oxidation of sewage and other contaminating substances in natural water into harmless material.

When oxygen takes up 2 electrons to complete its outer shell, it assumes the stable configuration of an atom of inert gas. Its tendency to accept 2 electrons from both nonmetals and metals is responsible for its ability to form compounds with nearly all known elements. These compounds of oxygen with metals or nonmetals are called **oxides.** The formation of oxides can be readily demonstrated by burning such elements as sulfur, phosphorus, iron, and carbon in pure oxygen. The union of a substance with oxygen is called **oxidation.** The burning of wood, the rusting of iron, and the decay of plant and animal matter are examples of oxidation. Oxygen is able to unite with food and tissue substances in the body at relatively low temperatures because these reactions are hastened by catalysts called enzymes.

Ozone. The air in the neighborhood of an electrical discharge, such as in an electrical storm, has a peculiar odor. The odor is caused by the formation of ozone, which is a form of oxygen represented as O_3, in contrast to ordinary molecular oxygen, O_2.

When an element exists in two or more different forms possessing different physical and chemical properties, the forms are known as **allotropic modifications.** Common elements that exhibit allotropic forms are oxygen, phosphorus, and sulfur.

Ozone may be prepared by passing air or oxygen between two plates that are charged with several thousand volts of alternating current. This so-called silent electrical discharge produces a low concentration of ozone, which emerges from the apparatus greatly diluted with air or oxygen. Since ordinary oxygen is in equilibrium with ozone, the reaction for the formation is reversible and prevents the accumulation of high concentrations of ozone.

Ozone is a colorless gas possessing a garlic-like odor and is more soluble and more dense than oxygen. Liquid ozone is dark blue in color and is capable of decomposing with explosive force, producing ordinary oxygen. The most striking difference between ozone and oxygen is the increased chemical activity of ozone. For example, a noble metal such as silver will not combine with oxygen under ordinary conditions, but will form a film of brown silver oxide when exposed to a low concentration of ozone.

Ozone, in the form of ozonized air produced by silent electrical discharge, has been used to destroy objectionable odors, to sterilize bandages, to bleach delicate fabrics, and to disinfect water for drinking purposes. Because of its oxidizing properties, it is sometimes used to age tobacco artificially and to cause rapid drying of the oils used in the manufacture of linoleum. The industrial use of ozone as an oxidizing agent has increased in recent years.

Combustion. When a substance unites with oxygen so rapidly that heat and light are produced in the reaction, it is said that the substance is burning. One commonly speaks of rapid oxidation, or burning, as **combustion,** although combustion may occur in the absence of oxygen.

Substances that will burn readily, such as paper, wood, illuminating gas, and gasoline, are called **combustible;** those that will not burn, such as asbestos, stone, and clay, are called **incombustible.** The terms **flammable** and **inflammable** are both synonymous with combustible and cause confusion when incorrectly compared to combustible and incombustible. Since air is only one-fifth oxygen, substances burn less vigorously in it than in the pure gas, and many substances that are incombustible in air burn in oxygen. For example, powdered sulfur burns feebly in air but blazes up vigorously when thrust into oxygen. Although iron wire is incombustible in air, it burns brightly in pure oxygen.

It is often said that a substance is "burned" in the body to give energy. This merely means that the substance unites with oxygen in the tissues. Although burning commonly occurs with oxygen as one of the reacting substances, it can occur in the absence of this element. Hydrogen will "burn" in the presence of chlorine to form hydrogen chloride. If powdered antimony is sprinkled into chlorine gas, "burning" will start at ordinary temperatures.

Spontaneous Combustion. When a substance unites slowly with oxygen, heat is produced in the reaction. If the heat cannot escape, the temperature of the substance gradually rises, increasing the rate of oxidation. If the process is allowed to continue, the temperature is raised to a point at which the material bursts into flame. This temperature is called the **kindling temperature** and the process which causes the material to burst into flame is called **spontaneous combustion.** Substances with a low kindling temperature, such as illuminating gas, ether, and gasoline, must be handled with care to prevent them from becoming fire hazards. Damp hay in a farmer's barn and oily rags left by painters often catch fire by spontaneous combustion. Coal piles, especially those in underground storage bins, are susceptible to this type of combustion. Temperature-recording devices are often inserted into the center of the pile as safety measures. Peat bogs often smoulder and burn for years, giving rise to the so-called ghost fires.

Extinguishing Fires. There are two common ways to put out a fire: (1) by removing the source of oxygen from the burning material; and (2) by lowering the temperature of the burning substance to below its kindling temperature. Chemical fire extinguishers contain a liquid that cools the burning substance below its kindling temperature and surrounds it with an incombustible gas. A common type of extinguisher contains water and chemicals that react to form carbon dioxide (CO_2) gas. A second type consists of a cylinder of carbon dioxide gas under pressure, connected by a hose to a long, funnel-shaped nozzle for directing the gas at the base of the flame. The cylinders of gas have a greater capacity, cause less damage in use, and are safe to use on electrical fires and on burning gasoline or oil.

Oil or gasoline fires cannot be extinguished with water, because these liquids are lighter than water and will float on the surface and spread the fire. The water layer underneath the burning oil is unable to lower the temperature below that of the kindling point. An effective fire extinguisher for gasoline or oil fires contains a mixture of chemicals and licorice, which forms a thick foam of carbon dioxide bubbles on the surface of the burning material and thus removes the oxygen. These fires may also be extinguished by use of the carbon dioxide gas extinguishers.

In another type of extinguisher, a heavy, incombustible liquid called carbon tetrachloride is used. When the liquid is sprayed on a fire, it forms a blanket of incombustible gas over the burning substances, thus removing the oxygen supply. **Carbon tetrachloride extinguishers** should not be used in closed quarters because of the toxicity of the vapors. There is also the possibility that poison gases such as phosgene may be formed during the use of these extinguishers.

USES OF OXYGEN. When oxygen is breathed into the lungs, it diffuses into the blood and loosely combines with **hemoglobin** (red pigment of the blood). This unstable compound of oxygen and hemoglobin is called **oxyhemoglobin** and is responsible for the bright-red color of arterial blood. The oxyhemoglobin is carried to the tissues, where it releases its oxygen to react with food and waste products in the tissue cells. One of the waste products formed is carbon dioxide gas, which is carried back to the lungs and exhaled. When the oxyhemoglobin loses its oxygen, it changes from a bright red to a deep purple; this accounts for the color of venous blood.

Oxygen can be used to operate a fuel cell in which it reacts with metallic zinc electrodes in a potassium hydroxide solution to produce electrical power.

The Electric Storage Battery Company is carrying out research to develop similar fuel cells as a source of quiet, smog-free power for vehicles in the future. The manned space probes of the United States have used oxygen-hydrogen fuel cells as the major power supply.

The chemical industry is rapidly increasing its use of oxygen in the production of synthetic gas for ammonia and methanol, the production of acetylene, ethylene oxide, carbon black, phenol, and hydrogen peroxide. Another major demand for oxygen is brought about by its use in the fuel for rockets and missiles.

SULFUR, SELENIUM, AND TELLURIUM

The remaining elements of Group VIA of the periodic table are **sulfur, selenium, tellurium,** and **polonium.** The element polonium, a radioactive decay product of uranium and thorium, is quite radioactive and difficult to study because of its rather intense α-emission.

Sulfur occurs in nature in the elemental state, as sulfide ores, such as ZnS, FeS_2, PbS, as sulfates, such as $CaSO_4 \cdot 2H_2O$, and as H_2S and SO_2. Elemental sulfur in large underground deposits is found in Louisiana and Texas. Selenium and tellurium are rare compared to sulfur, and occur as selenide and telluride impurities in metal sulfide ores. They are recovered from combustion chambers of sulfide ores and from the sludge obtained in the electrolytic refining process for copper.

Sulfur normally exists as a yellow solid that is tasteless and odorless, insoluble in water, but soluble in carbon disulfide. It is similar to carbon and oxygen in that it exists in several allotropic forms. For example, below 96°C sulfur crystallizes from solution as **rhombic** crystals, whereas if it is heated above 114°C and allowed to cool, **monoclinic** crystals are formed. An **amorphous** form, plastic sulfur (so-called), results when hot melted sulfur is poured into cold water. This allotrope is thought to consist of long chains of sulfur atoms. The plastic form is metastable and slowly transforms to the rhombic form.

Both the rhombic and monoclinic allotropes of sulfur are composed of molecules containing eight sulfur atoms in a puckered ring, as shown in Figure 8–3. The vapors of sulfur consist of an equilibrium mixture of S_8, S_4, and S_2 molecules. In CS_2 solution and in naphthalene solution, rhombic sulfur has a molecular weight of 256 amu, corresponding to S_8.

Selenium also exists in a number of allotropic forms, including rhombic and monoclinic crystalline forms, which, when dissolved in CS_2, show molecular weights corresponding to Se_8. A dark gray allotrope, the most stable form, contains long chains of selenium atoms.

Tellurium is known to exist in only one form. This form, similar to the

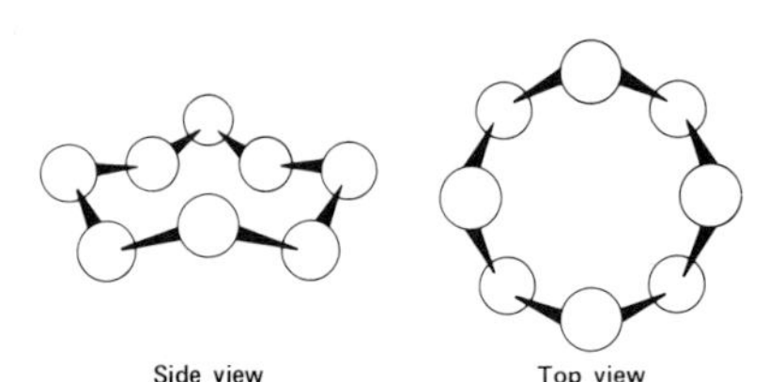

Figure 8–3 The cyclic structure of S_8.

gray form of selenium, is silvery white, insoluble in CS_2, and undoubtedly contains very long chains of tellurium atoms.

The tendency toward greater stability of the non-cyclic allotropes increases in going from sulfur to tellurium. Tellurium, although not a metal, shows more metallic properties than selenium or sulfur.

Sulfides, Selenides, and Tellurides (Chalconides). Sulfur, selenium, and tellurium combine directly with most metals to form metal sulfides, selenides, and tellurides, in which the **chalconides** have an oxidation number of −2.

Metallic sulfides, except for those of all alkali metals and the alkaline earth metals, are quite insoluble in water, probably because of a rather large degree of covalency in the M—S bonds. When alkali metal and alkaline earth metal sulfides are dissolved in water, they dissociate, and the sulfide ion undergoes extensive hydrolysis as follows:

$$S^{-2} + H_2O \rightleftharpoons HS^- + OH^-$$
$$HS^- + H_2O \rightleftharpoons H_2S + OH^-$$

The first hydrolysis step proceeds to a much greater extent than the second. Aqueous solutions containing sulfide ions will dissolve elemental sulfur to form polysulfide ions, $S_x{}^{-2}$:

$$S^{-2}{}_{(aq)} + nS \rightleftharpoons S_{n+1}{}^{-2}{}_{(aq)}$$

Hydrogen Compounds. The binary compounds of hydrogen with sulfur, selenium, and tellurium are usually prepared by acidifying aqueous solutions of the alkali metal chalconides:

$$S^{-2}\,(aq) + 2HCl\,(conc) \rightleftharpoons 2Cl^-\,(aq) + H_2S\,(g)$$

All of these hydrogen compounds are gases at room temperature, have intolerable odors, and are very toxic. In aqueous solutions the hydrogen chalconides behave as weak diprotic acids:

H_2S	$K_{a1} = 1 \times 10^{-7}$	$K_{a2} = 10^{-15}$
H_2Se	$K_{a1} = 1.9 \times 10^{-4}$	$K_{a2} = 1 \times 10^{-10}$
H_2Te	$K_{a1} = 2.3 \times 10^{-3}$	$K_{a2} = 10^{-7}$

Of these three acids, hydrogen sulfide is the most important. It plays a significant role in classical metal ion qualitative analysis schemes, which differentiate many metal ions on the basis of the solubility properties of their various salts.

Oxides and Oxyacids. Sulfur, selenium, and tellurium each form oxides of the type XO_2 and XO_3 in which X assumes oxidation numbers of +4 and +6, respectively.

Each of the dioxides of sulfur, selenium, and tellurium can be obtained by burning the element in air:

$$S\,(s) + O_2\,(g) \rightleftharpoons SO_2\,(g)$$

Sulfur dioxide, SO_2, is also formed as a by-product when metal sulfide ores are roasted in air:

$$2MS\ (s) + 3O_2\ (g) \xrightleftharpoons{heat} 2MO\ (s) + 2SO_2\ (g)$$

The melting points and freezing points of the dioxides increase markedly in going down the group. Sulfur dioxide is the only member of the series which exists as a gas at room temperature (b.p. $-108°C$). The dioxides are all bent molecules as is predictable by considering the total number of bonding and non-bonding pairs of electrons.

Sulfur dioxide, when dissolved in water, hydrolyzes to form the weak diprotic **sulfurous acid,** H_2SO_3($K_{a1} = 1.3 \times 10^{-2}$, $K_{a2} = 5.6 \times 10^{-8}$):

$$SO_2\ (aq) + H_2O \rightleftharpoons H_2SO_3\ (aq)$$

The salts of sulfurous acid, **sulfites** (M_2SO_3), and **bisulfites** ($MHSO_3$), when placed in strongly acidic aqueous solutions, are extensively hydrolyzed, and result in the evolution of SO_2 gas from the solution. These salts also act as reducing agents being oxidized to sulfates:

$$2OH^- + SO_3^{-2} \rightleftharpoons SO_4^{-2} + H_2O + 2e^-$$

Sulfur dioxide finds many industrial uses, including use as a bleach and as a refrigerant; but by far its most important use is in the production of sulfuric acid.

Selenium dioxide and **tellurium dioxide** also hydrolyze in water to give **selenous acid** and **tellurous acid.** These acids are also very weak.

Sulfur trioxide, SO_3, the anhydride of **sulfuric acid,** H_2SO_4, is prepared by the molecular oxygen oxidation of sulfur dioxide in the presence of catalysts:

$$2SO_2\ (g) + O_2\ (g) \xrightarrow{catalyst} 2SO_3\ (g)$$

This process is usually performed at 400 to 500°C using V_2O_5 or spongy platinum as catalysts. The product, which boils at 45°C, consists of planar triangular molecules in the gas phase.

When sulfur trioxide is placed in water, it hydrolyzes to sulfuric acid:

$$SO_3\ (g) + H_2O \rightleftharpoons H^+\ (aq) + HSO_4^-\ (aq)$$

Sulfuric acid is undoubtedly one of the most important compounds of sulfur. It is manufactured most commonly by the **contact process.** In this process, sulfur dioxide is oxidized to sulfur trioxide at increased temperatures in the presence of a catalyst. The sulfur trioxide is then absorbed in 98 per cent sulfuric acid, producing **fuming sulfuric acid,** or **oleum.** By the controlled addition of water, the concentration of the end product is maintained at 98 per cent and is the concentrated acid of commerce. The reactions for this process are as follows:

$$S\ (s) + O_2\ (g) \rightleftharpoons SO_2\ (g)$$

$$SO_2\ (g) + O_2\ (g) \xrightarrow[400°\text{–}500°C]{catalyst} SO_3\ (g)$$

$$SO_3\ (g) + H_2SO_4\ (aq)\ (98\%) \rightleftharpoons H_2S_2O_7\ (l)$$

$$H_2S_2O_7\ (l) + H_2O \rightleftharpoons H_2SO_4\ (aq)$$

Sulfuric acid and its salts, the **sulfates,** M_2SO_4, and **bisulfates,** $MHSO_4$, will not act as reducing agents, but do have weak oxidizing powers. The acid has the best oxidizing ability and, as already mentioned, can oxidize bromides and iodides to the elements.

Alkali metal sulfates, when placed in water, undergo slight hydrolysis to produce basic solutions, whereas alkali metal bisulfates give slightly acidic solutions. The sulfates of calcium, barium, and strontium are only slightly soluble in water and are often used in qualitative testing for these cations.

Selenic acid, H_2SeO_4, in aqueous solutions is very similar to sulfuric acid, except that it is a much better oxidizing agent. **Telluric acid** is quite different from sulfuric acid and selenic acid in that it exists as $Te(OH)_6$. This acid, much weaker than sulfuric or selenic, involves a tellurium atom with a coordination number of six, apparently because of the large size of the tellurium atom.

Additional Uses of the Elements of Group VIA. The most important compound of sulfur is sulfuric acid. The major use of sulfuric acid is in the preparation of fertilizers, although large quantities are also used in chemical industries and in the refining of petroleum products. Other applications include the manufacture of explosives, dyes, textiles, and the pickling of steel. In view of the fact that sulfuric acid is used in so many industrial processes in peace and war, its annual production is often used as a rough measure of the prosperity of a country.

The most important uses of selenium involve its peculiar electric properties. A thin film of selenium over another metal such as iron, copper, or various alloys produces a system that will allow electricity to pass from the selenium to the other metal but not in the reverse direction. This phenomenon makes possible the manufacture of **selenium rectifiers,** which change alternating current into direct current and are used extensively in charging storage batteries. Another property of the selenium-iron or selenium-copper systems is their ability to generate an electric current whose intensity is directly proportional to the intensity of light striking the surface. These are called **photocells** and are used in spectrophotometers, light meters, and in the control of electric circuits. In the glass and ceramic industries selenium is used to impart a red color to the product and to remove undesirable colors due to traces of iron.

THE ELEMENTS OF GROUP VA

The elements of Group VA of the periodic table include **nitrogen, phosphorus, arsenic, antimony,** and **bismuth.** Within this group the trend from nonmetallic character to metallic on going down the group is far more obvious than in the elements of Group VIIA or Group VIA. This gradual transition is evident in physical and chemical properties. Because of the pronounced variation in properties, nitrogen and phosphorus are considered to be nonmetals; arsenic and antimony are considered intermediate between nonmetallic and metallic and are called **metalloids;** and bismuth is metallic.

Nitrogen, the only member of Group VA which occurs in nature as the element, constitutes 78 per cent by volume of the earth's atmosphere. It is colorless, odorless, tasteless, and only slightly soluble in water. In the elemental state, nitrogen exists as diatomic molecules which have a very high bond dissociation energy of 225 kcal/mole due to a triple bond. Because of this extremely high bond dissociation energy, nitrogen is rather inert, but can be rendered very reactive by an electric discharge such as a lightning bolt. The active nitrogen thus generated reacts with oxygen and other components of the atmosphere to generate nitrogen compounds.

Inorganic nitrogen compounds are found in deposits in Chile and South Africa. A bed of sodium and potassium nitrates more than 200 miles long, about 2 miles wide, and 5 feet thick is located in Chile. Organic nitrogen compounds are found in all living matter in the form of plant or animal protein, protoplasm, and more simple organic compounds involved in the synthesis or the breakdown of proteins. Plants possess the ability to convert inorganic nitrogen compounds, such as nitrates, into organic nitrogen compounds, but animals must eat the plants to obtain a source of nitrogen compounds. The extraction of elementary nitrogen from the atmosphere and its incorporation into essential organic constituents of living matter is referred to as nitrogen fixation. This will be discussed in Chapter 22.

Nitrogen is often prepared by removing oxygen from the air by simply passing dry air over hot copper turnings. The nitrogen so produced is not pure; it contains carbon dioxide, argon, and other rare gases. Pure nitrogen may be prepared in the laboratory by heating **ammonium nitrite:**

$$NH_4NO_2\ (s) \xrightarrow{\text{heat}} N_2\ (g) + 2H_2O\ (g)$$

Ammonium nitrite, which is a very unstable compound, is prepared just before use from a mixture of ammonium chloride and sodium nitrite. Large quantities of nitrogen as well as oxygen are produced commercially from the distillation of liquid air.

Phosphorus, like nitrogen, occurs in both inorganic and organic forms in nature. Minerals containing phosphorus, such as **phosphorite,** $Ca_3(PO_4)_2$, and **apatite,** $3Ca_3(PO_4)_2 \cdot CaF_2$, are found in natural deposits in several parts of the world. The bones and teeth of animals contain a complex calcium phosphate salt. Brain and nervous tissue are especially rich in organic phosphorus compounds, and every living cell in plants or animals contains some phosphorus.

Elementary phosphorus is prepared by heating phosphate rock, which is a crude form of phosphorite, with sand and coke in an electric furnace:

$$2Ca_3(PO_4)_2\ (s) + 6SiO_2\ (s) + 10C\ (s) \xrightarrow{\text{heat}} 6CaSiO_3\ (s) + 10CO\ (g) + P_4\ (g)$$

The phosphorus vapor is removed from an opening in the upper part of the furnace and is condensed in cold water. The solid phosphorus is then molded into small sticks and kept under water.

Elementary phosphorus exists in several allotropic forms, the most important of which are white and red phosphorus.

When freshly prepared, phosphorus is a yellowish white, waxy solid which melts at 44°C and ignites in air at about 35°C. It is very poisonous and can be handled with safety only under water. It is insoluble in water, but readily dissolves in carbon disulfide and other organic solvents. This allotrope, **white phosphorus,** is composed in the solid phase, in the gas phase, and in solutions of discrete P_4 tetratomic molecules.

The main use of elemental phosphorus is in the manufacture of matches. The first friction matches were made of white phosphorus, which readily ignited but produced chronic poisoning of the workmen in the match industry. Red phosphorus and a sulfide, P_4S_3, are used in modern matches. For example, the head of a "strike anywhere" match is made of a mixture of P_4S_3 and a combustible substance such as sulfur and an oxidizing agent such as potassium chlorate, with glue added to bind these compounds to the matchstick. The stick is usually impregnated with ammonium phosphate to prevent after-glow when the match is extinguished. The heat of friction produced by drawing the match rapidly over most surfaces ignites the phosphorus sulfide, which sets fire to the remaining material in the head, and finally the wood in the matchstick.

Arsenic, antimony, and bismuth occur most commonly as sulfide impurities in sulfide ores of copper, lead, and silver. The elemental forms are prepared by roasting the sulfide to convert to the oxide, and then reduction with carbon or hydrogen.

Elemental arsenic and antimony can exist in two allotropic forms. A yellow form, which is rather unstable, contains As_4 and Sb_4 molecules. In each case the yellow form is spontaneously converted to a bright gray lustrous allotrope which is metallic in appearance.

Negative Oxidation States. Nitrogen, in combining with less electronegative elements, forms **nitrides** in which it assumes a negative oxidation number. When combined with active metals such as lithium, magnesium, or zinc, these nitrides are **ionic,** and discrete N^{-3} ions are formed. Ionic nitrides have physical properties typical of ionic compounds, and when added to water they hydrolyze to give ammonia and metal hydroxides:

$$Li_3N\,(s) + 3H_2O \rightleftharpoons NH_3\,(g) + 3Li^+\,(aq) + 3OH^-\,(aq)$$

When combined with less electronegative metals and nonmetals, nitrogen forms **covalent nitrides,** including BN, AlN, C_2N_2, Si_3N_4, Sn_3N_4, P_3N_5, and S_4N_4. The properties of these materials vary widely, with some, such as C_2N_2 and S_4N_4, existing as molecular species, whereas others, such as BN and AlN, exist only as covalent network solids. In general the covalent network solids are hard, insoluble, nonvolatile crystals.

Nitrogen forms **interstitial nitrides** with the transition metals in which nitrogen atoms occupy the interstices of the closely packed metal lattice. These materials are metallic in appearance and properties.

Phosphorus combines with active metals to form ionic phosphides, such as Na_3P and Ca_3P_2, which have discrete P^{-3} ions. When added to water these compounds hydrolyze to give **phosphine,** PH_3, as follows:

$$Ca_3P_2\,(s) + 3H_2O \rightleftharpoons 3Ca^{+2}\,(aq) + 3OH^-\,(aq) + 2PH_3\,(g)$$

When combined with the Group IIIA elements, phosphorus forms covalent network solids, such as BP and AlP. The phosphides of transition metals are similar to analogous nitrides and behave as hard, water insoluble, metallic conductors.

Arsenic and antimony form compounds with active metals, which hydrolyze in water to give **arsine,** AsH_3, and **stibine,** SbH_3, respectively.

The stability of the -3 oxidation state in Group V decreases markedly in going from nitrogen to antimony.

Hydrogen Compounds. All of the Group VA elements form hydrogen compounds of the general formula MH_3. As indicated in the preceding section, these can be prepared by the hydrolysis of ionic nitrides, phosphides, and so forth:

$$Li_3M\ (s) + 3H_2O \rightleftharpoons MH_3\ (g) + 3Li^+\ (aq) + 3OH^-\ (aq)$$

Ammonia, NH_3, is also prepared in the **Haber process** by direct combination with hydrogen at high temperatures (1000°C) and pressures (up to 1000 atm) in the presence of catalysts such as iron and iron oxide:

$$N_2\ (g) + 3H_2\ (g) \rightleftharpoons 2NH_3\ (g)$$

Ammonia is a colorless, but very pungent, toxic gas which, when dissolved in water, hydrolyzes to produce basic solutions ($K_b = 1.8 \times 10^{-5}$). The reactivity of this pyramidal molecule is due in large part to a relatively large dipole moment and a lone pair of electrons on the nitrogen atom. Ammonia behaves as a Lewis base in combining with metal ions to form **ammine complexes** such as $[Ag(NH_3)_2]^{+1}$, $[Cu(NH_3)_4]^{+2}$, and $[Ni(NH_3)_6]^{+2}$.

Liquid ammonia (b.p. −33.4°C) has properties which resemble those of water, including a relatively high enthalpy of vaporization (327 cal/g) and a self-dissociation equilibrium:

$$2NH_3 \rightleftharpoons NH_4^+ + NH_2^-$$
$$K_{-50°C} = [NH_4^+][NH_2^-] = 10^{-50}$$

For this reason active metals react with liquid ammonia as follows:

$$2Na\ (s) + 2NH_3\ (l) \rightleftharpoons H_2\ (g) + 2NaNH_2$$

Compounds like $NaNH_2$ are called **metal amides.**

Ammonia will burn in pure oxygen to give elemental nitrogen or **nitric oxide,** NO. The latter product is obtained only when the reaction is allowed to proceed at 900°C in the presence of a platinum catalyst. Industrially this reaction is used in the **Ostwald process** for the production of nitric acid. The hydrides *phosphine* (PH_3), *arsine* (AsH_3), and *stibine* (SbH_3) are all very toxic gases which are less stable than ammonia, in that they tend to decompose to the elements when heated.

Oxides and Oxyacids. Nitrogen forms oxides in which it assumes all possible oxidation numbers from +1 to +5. The important oxyacids of nitrogen are **nitrous acid,** HNO_2, and **nitric acid,** HNO_3. Nitrous acid, in which nitrogen has an oxidation number of +3, is unstable even in aqueous solution and decomposes to nitric acid and nitric oxide:

$$3HNO_2\ (aq) \rightleftharpoons HNO_3\ (aq) + H_2O + 2NO\ (g)$$

Nitrous acid is a weak acid ($K_a = 6.0 \times 10^{-6}$) and is usually prepared by acidifying nitrite salts.

Nitric acid involves nitrogen in the +5 oxidation state. This acid is made commercially by the **Ostwald process.** This process involves catalytic oxidation of ammonia to nitric oxide, which is oxidized by air to nitrogen dioxide and then added to water. In water nitrogen dioxide disproportionates to nitric acid and nitric oxide. The reactions involved in the preparation of nitric acid, starting with nitrogen, are as follows:

$$N_2\ (g) + 3H_2\ (g) \rightleftharpoons 3NH_3\ (g)$$

$$4NH_3\ (g) + 5O_2\ (g) \xrightarrow[Pt]{900^\circ C} 4NO\ (g) + 6H_2O\ (g)$$

$$2NO\ (g) + O_2\ (g) \rightleftharpoons 2NO_2\ (g)$$

$$3NO_2\ (g) + H_2O\ (l) \rightleftharpoons 2H^+\ (aq) + 2NO_3^-\ (aq) + NO\ (g)$$

The nitric oxide generated in the last step is recycled.

Pure nitric acid is a colorless liquid (b.p. 84.1°C) which behaves as a powerful oxidizing agent, with both H^+ and the nitrogen as potential oxidizing sites. However, the reduction products of nitric acid seldom contain hydrogen, but are comprised of lower oxidation states of nitrogen, including NO_2, NO, N_2O, and N_2. The oxidizing power decreases with decreasing concentration. As shown in Figure 8–4, gaseous nitric acid is a planar molecule and the nitrate anion has a triangular planar structure.

Phosphorus, arsenic, and antimony all form the oxides of formula M_4O_6. These elements also form an oxide of empirical formula M_2O_5, but the molecular formula is established only in the case of phosphorus P_4O_{10}. The structures of P_4O_6 and P_4O_{10} involve tetrahedra of phosphorus atoms with oxygen atoms on the edges and apices.

The oxides P_4O_6 and P_4O_{10} are obtained by burning white phosphorus, P_4, in a limited supply of air and in an excess of air, respectively. The oxide P_4O_6 is the anhydride of **phosphorous acid,** H_3PO_3, whereas P_4O_{10} is the anhydride of **phosphoric acid,** H_3PO_4. Probably the most important property of P_4O_{10} is its affinity for water. Because of this, it is used extensively as a thorough drying agent.

The reaction of water with P_4O_{10} can lead to the formation of several phosphoric acids. **Metaphosphoric acid,** $(HPO_3)_n$, the product of limited hydrolysis of P_4O_{10}, is a polymeric acid, which, on addition of more water, gives first **pyrophosphoric acid,** $H_4P_2O_7$, and then **orthophosphoric acid,** H_3PO_4:

$$P_4O_{10}\ (s) + 2H_2O \rightleftharpoons 4(HPO_3)\ (l)$$
$$2HPO_3\ (l) + H_2O \rightleftharpoons H_4P_2O_7\ (l)$$
$$H_4P_2O_7\ (l) + H_2O \rightleftharpoons 2H_3PO_4\ (l)$$

Orthophosphoric acid is a weak triprotic acid ($K_{a1} = 7.1 \times 10^{-3}$; $K_{a2} = 6.3 \times 10^{-8}$; and $K_{a3} = 4 \times 10^{-13}$).

The phosphoric acids and their salts play important roles in many technical processes. Orthophosphoric acid is used in the rust-proofing and preparation of sheet steel for painting. This is the "bonderizing" process employed by the automobile industry. Sodium salts of the phosphoric acids find many important applications in the home and in industry. Trisodium phosphate, often called washing powder, is used to soften water and in boiler-water treatment. Monosodium phosphate is used in the manufacture of baking powders and pharma-

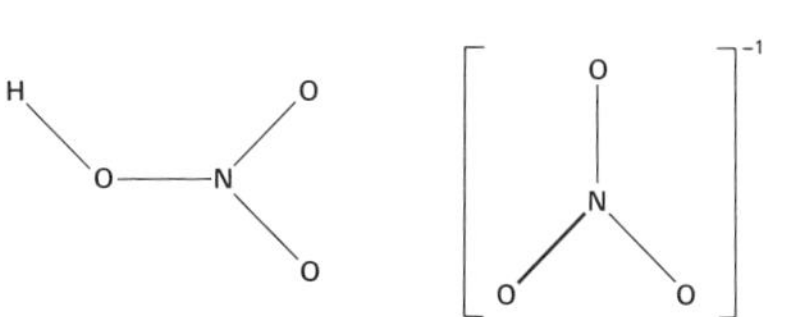

Figure 8-4 The structures of nitric acid and the nitrate ion.

ceuticals. Hexametaphosphates, $(NaPO_3)_6$, are very useful in the treatment of hard water, since they have the property of forming a complex with calcium ions that does not precipitate. Another important sodium salt is tetrasodium pyrophosphate, $Na_4P_2O_7$, which is an integral constituent of soaps and soap powders. It helps prevent the deposition of dirt particles or particles of insoluble calcium and magnesium soaps on the surface of fabrics being washed in hard water. This is accomplished by its property of dispersing solid particles into a fairly permanent suspension. Solutions of this salt have a lower pH than those containing washing powders, such as trisodium phosphate, and can be used for washing silks and woolen goods. Ammonium phosphate is used for fireproofing textiles and wood.

CARBON, SILICON, AND GERMANIUM

The elements of Group IVA include **carbon, silicon, germanium, tin,** and **lead.** Carbon displays physical and chemical properties characteristic of a nonmetal; silicon and germanium behave as metalloids; and tin and lead are metals.

Carbon occurs on earth both as the element and in chemical combination. Elemental carbon occurs as two pure allotropes, **diamond** and **graphite,** and in many impure **amorphous** forms such as coal. The structure of diamond involves carbon atoms each covalently bonded to four adjacent carbon atoms at the corners of a tetrahedron as shown in Figure 8–5. This bonding results in a three dimensional structure with covalent bonds extending through the entire crystal. For this reason diamond crystals are very hard and are broken with difficulty. The graphite structure involves layer after layer of carbon atoms in a plane, each surrounded by three carbon atoms as shown in Figure 8–5. The forces between the planes are comparatively weak, so that they can "slip" relative to one another. For this reason graphite is soft and can act as a lubricant. Graphite is also an electrical conductor while diamond is an insulator.

Graphite has many important uses. In powdered form it is often used as a lubricant, especially under conditions of high temperature or high pressure. When mixed with oil or grease, it increases the temperature at which the lubricant can be safely used. The "lead" in lead pencils is composed of graphite mixed with clay and wax, in varying proportions depending on the hardness of the "lead."

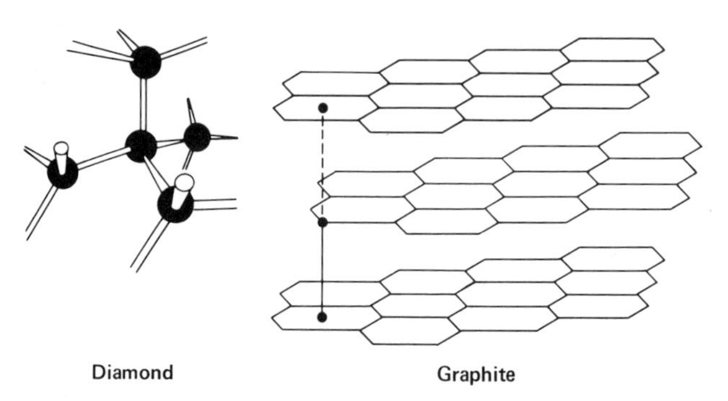

Figure 8-5 The structures of diamond and graphite. The physical properties of the two pure allotropes are strikingly different.

There are also several noncrystalline modifications of elemental carbon called **amorphous carbon** or **charcoal.** When wood, sugar, nutshells, bone, and coal are heated in a container in the absence of air, volatile liquids and gases are given off and a residue of charcoal or impure carbon remains. This process is called destructive distillation and produces many valuable volatile products in addition to charcoal.

Silicon is the second most abundant element in the earth's crust at 26 per cent of the total. It does not occur naturally in the elemental state but rather is found as silicon dioxide or silicates. Sand, flint, and quartz are silicon dioxide, SiO_2. Metallic silicates are the primary constituents of rocks, soil, and clay.

Elementary silicon is usually obtained by reduction of SiO_2 with carbon or calcium carbide, CaC_2, in an electric furnace:

$$SiO_2\,(s) + 2C\,(s) \rightleftharpoons Si\,(s) + 2CO\,(g)$$

The element, which exists only in a form which has the diamond structure, has a gray metallic appearance but does not conduct electricity as well as metals.

Germanium is a relatively rare element. In the elemental state it has the diamond structure and has electrical conductivity properties similar to those of silicon.

Negative Oxidation States. Carbon forms compounds with less electronegative elements in which it assumes negative oxidation numbers. These binary compounds, called **carbides,** can be divided into three classes, **ionic, covalent,** and **interstitial.**

Ionic carbides are formed by the interaction of carbon with active metals at elevated temperatures. Some ionic carbides, including Be_2C and Al_4C_3, contain discrete C^{-4} ions and hydrolyze when added to water to give **methane,** CH_4. Other carbides, including ZnC_2, CdC_2, and Al_2C_6, contain discrete C_2^{-2} ions and yield **acetylene,** C_2H_2, when hydrolyzed.

The compounds SiC and B_4C are typical covalent carbides. These materials, prepared by reducing SiO_2 and B_2O_3 with carbon at very high temperatures, are extremely hard and insoluble in water. Silicon carbide, SiC, finds industrial use as an abrasive.

Metal silicides are formed when elemental silicon is dissolved in metals. These compounds hydrolyze to form silicon hydrogen compounds of the type SiH_4, called **silanes.**

Hydrogen Compounds. The hydrogen compounds of carbon, **hydrocarbons,** can be prepared by hydrolysis of carbides. However, these compounds are abundant in nature and are the primary constituents of petroleum and natural gas. The physical and chemical properties of these compounds will be discussed in the section of this book on organic chemistry.

The **silanes** are most commonly prepared by the hydrolysis of magnesium silicide, Mg_2Si, using sulfuric or phosphoric acid. This hydrolysis gives a mixture of products of the general formula Si_nH_{2n+2}, involving chains of n silicon atoms where n ranges from 1 to 6. Such chains, having bonds between like atoms, are referred to as **catenated structures.** They spontaneously inflame in air (are pyrophoric) to give SiO_2 and water, and are hydrolyzed in basic aqueous solutions to hydrated silicon dioxide:

$$Si_2H_6\ (g) + (4 + 2n)H_2O \rightleftharpoons 2SiO_2 \cdot nH_2O\ (s) + 7H_2\ (g)$$

The thermal stability of silanes decreases as the chain length increases.

The hydrogen compounds of germanium, **germanes,** are prepared by the reduction of **germanium dioxide,** GeO_2. The reduction, performed using **sodium borohydride,** $NaBH_4$, in acidic aqueous solution, gives GeH_4, Ge_2H_6, and Ge_3H_8. Germanes are less pyrophoric than silanes, but are oxidized to GeO_2 by oxygen. They undergo hydrolysis reactions similar to those of the silanes but only in highly basic solutions.

HALOGEN COMPOUNDS. The halogen compounds of carbon will be discussed in the organic chemistry section. Silicon and germanium form tetrahalides of the general formula MX_4 with each of the halogens by direct interaction. Some catenated silicon and germanium halides are also known. All of these undergo hydrolysis reactions when added to water. Except for silicon tetrafluoride, SiF_4, all of the silicon halides hydrolyze rapidly and completely to give hydrogen halides and hydrated silicon dioxide, sometimes called silicic acid:

$$SiX_4\ (X = Cl, Br, I) + (2 + n)H_2O \rightleftharpoons SiO_2 \cdot nH_2O\ (s) + H^+\ (aq) + X^-\ (aq)$$

The silicon halides behave as Lewis acids and combine with a variety of Lewis bases, including organic amines and ethers. In these adducts, silicon assumes a coordination number of five or six.

Germanium tetrachloride, $GeCl_4$, and **tetrabromide,** $GeBr_4$, do not hydrolyze completely in water. The partial hydrolysis products include such species as $[Ge(OH)_4X_2]^{-2}$ and $[Ge(OH)_3X_3]^{-2}$.

OXIDES AND OXYACIDS. Carbon dioxide is formed by the combination of oxygen and carbon. It is a gas that has extensive biological and industrial applications. It occurs in the atmosphere in a concentration of about 0.03 per cent, whereas 20 to 30 times this amount is dissolved in the water of the oceans. Gases emanating from underground pockets or accumulations in volcanic regions are often rich in carbon dioxide.

The preparation of carbon dioxide is readily achieved by burning carbon in the presence of an excess of oxygen. In the laboratory, chunks of marble or other metal carbonates are treated with dilute hydrochloric acid to produce the gas:

$$CaCO_3\ (s) + 2HCl\ (aq) \rightarrow CaCl_2\ (aq) + H_2O + CO_2\ (g)$$

Commercially, large quantities of the gas are formed as a byproduct in the preparation of alcohols by fermentation. After purification, the carbon dioxide is stored in steel cylinders under pressure.

Normally carbon dioxide exists as a colorless, odorless gas with a slightly sharp taste. When the gas is subjected to 60 or 70 atmospheres of pressure, as, for example, in the commercial steel cylinders, most of it liquefies. If this liquid carbon dioxide is allowed to escape into a container, it vaporizes so rapidly that it is quickly cooled to a temperature of $-80°C$ and part of it freezes to a white solid. Commercially, this white carbon dioxide snow is pressed into blocks of **dry ice.** The extremely low temperatures and ease of handling afforded by dry ice have resulted in many applications as a refrigerant.

Carbon dioxide dissolves in water to form **carbonic acid,** H_2CO_3, which is a very weak diprotic acid ($K_{a1} = 4.2 \times 10^{-7}$; $K_{a2} = 9.7 \times 10^{-11}$):

$$CO_2\ (g) + H_2O \rightarrow H_2CO_3\ (aq)$$

If a solution of carbon dioxide and water is prepared under pressure, soda water results, which is the basis for the common carbonated beverages.

When carbon is burned in a limited supply of oxygen, carbon monoxide is formed. This compound may also be formed in a coal furnace when the carbon dioxide from combustion of the coal passes over more hot carbon:

$$CO_2\ (g) + C\ (s) \rightarrow 2CO\ (g)$$

Carbon monoxide is a colorless, odorless gas that differs from carbon dioxide in being combustible in air and not soluble in water. The blue flame of burning carbon monoxide can often be seen over the top layer of unburned coal in a coal furnace. One of the outstanding properties of the gas is its poisonous nature. Continuous inhalation of as low a concentration of 0.05 per cent carbon monoxide will produce headaches, dizziness, and unconsciousness in a few hours. Breathing automobile exhaust gas that contains 7 per cent carbon monoxide will result in death in a few minutes.

Carbon monoxide is used as a reducing agent in the reduction of iron oxide to iron:

$$Fe_2O_3\ (s) + 3CO\ (g) \rightarrow 2Fe\ (s) + 3CO_2\ (g)$$

It can also be used in metallurgical processes to reduce oxides of other metals.

Carbonates occur in abundance in many parts of the world, mainly as calcium carbonate or limestone. The erosion of limestone causes the unusual stalactite and stalagmite formations found in many caves. Other forms of calcium carbonate are marble, sandstone, oyster shells, coral, and chalk. Large quantities of limestone are used in the construction of highways and the manufacture of cement, and marble and limestone are used to produce large quantities of lime and carbon dioxide gas. When limestone is heated in a lime kiln, carbon dioxide is formed, and a residue of lime remains behind, as shown in the following equation:

$$CaCO_3\ (s) \rightarrow CaO\ (s) + CO_2\ (g)$$

The lime produced by this process is used mainly in the production of plaster and as a flux in metallurgy. It is also used in water softening, in tanning hides, and in the manufacture of paper and glass.

Another very important carbonate is sodium carbonate. This salt occurs in nature, often mixed with other salts, such as sodium bicarbonate. Deposits of carbonates are found in several parts of the world. In addition to the natural sources, huge quantities of sodium carbonate are produced by the **Solvay process.**

The physical properties of silicon dioxide differ markedly from those of the carbon oxides. The carbon oxides are gases at room temperatures, whereas SiO_2 is a covalent network solid which melts near 1700°C.

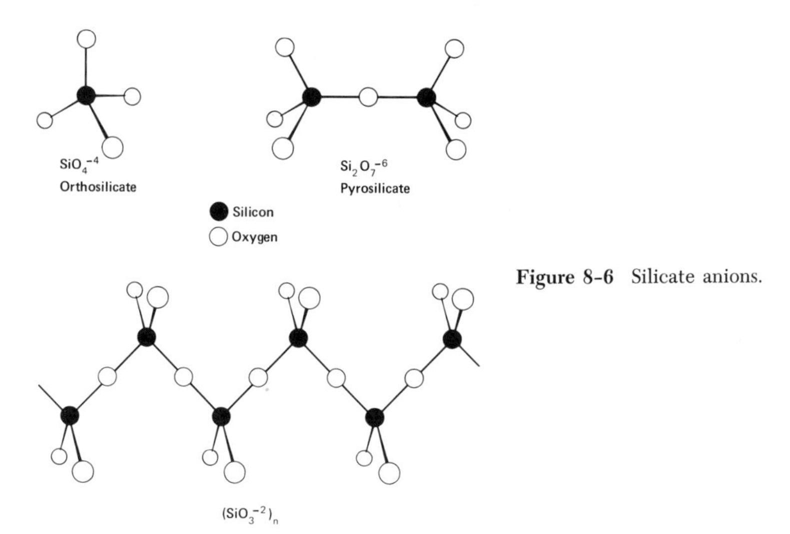

Figure 8-6 Silicate anions.

Many types of silicate anions are derived from hydrated silicon dioxide, or "silicic acid." The simplest of these is **orthosilicate,** SiO_4^{-4}, which is composed of discrete tetrahedral anions. All other silicates can be thought of as being derived from this simple anion by sharing a common oxygen atom. The **pyrosilicate** anion, $Si_2O_7^{-6}$, occurs naturally in many minerals. The combination of many units to generate silicate chains and silicate three-dimensional structures is quite common in many minerals. The structures of orthosilicate, pyrosilicate, and a chain silicate of general formula $(SiO_3^{-2})n$ are found in Figure 8-6.

Glasses, ceramics, and cement are all complex silicates. The properties of these materials are dependent on the type of silicate anion involved and the cations involved.

BORON

Boron is the only member of the Group IIIA elements which behaves as a nonmetal. It does not occur in nature in the elemental state, but is usually obtained as $Na_2B_4O_7 \cdot 10H_2O$, a soluble borate called **borax.** Elemental boron in a semi-pure state can be obtained by magnesium reduction of boric oxide, B_2O_3, at elevated temperatures:

$$B_2O_3\,(s) + 3Mg\,(s) \rightleftharpoons 2B\,(s) + 3MgO\,(s)$$

Pure boron can be obtained by reducing **boron tribromide,** BBr_3, with hydrogen on a tungsten wire at 1300°C. Crystalline boron obtained in this way is metallic black. Its poor electrical conductivity increases with increasing temperature, which is typical of semiconductors. It is very inert and combines only with very strong oxidizing agents, but when finely divided it will burn in oxygen to give **boric oxide,** B_2O_3.

Borides. **Borides,** in which boron assumes negative oxidation numbers, are formed by the interaction of boron and active metals at very high tempera-

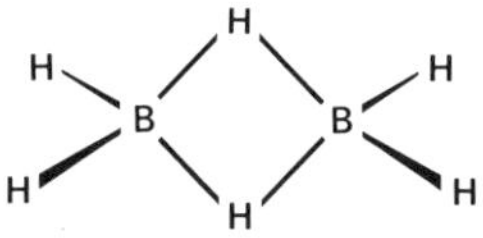

Figure 8-7 The structure of diborane.

tures. In the very hard crystalline products, several structures have been identified, including chains and layers of boron atoms. **Magnesium boride,** MgB_2, which hydrolyzes in water, is used extensively in the synthesis of boron hydrogen compounds called **boranes.**

HYDROGEN COMPOUNDS. The **boranes** are prepared in a number of ways, including the hydrolysis of MgB_2 and the reduction of boron halides with lithium hydride:

$$6LiH\ (s) + 8BF_3\ (g) \rightleftharpoons 6LiBF_4\ (s) + B_2H_6\ (g)$$

The hydrolysis of MgB_2 yields B_2H_6, B_4H_{10}, B_5H_9, B_5H_{11}, B_6H_{10}, $B_{10}H_{14}$, and others. When B_2H_6 is heated, it can be converted into higher boranes. All of these compounds are unusual in that in each case there appear to be more bonds than can be explained by considering the number of electron pairs available. The simplest borane, B_2H_6, **diborane,** has the structure shown in Figure 8-7, in which it is seen that there are two kinds of hydrogens, terminal and bridging. The terminal hydrogens are bonded to boron with classical two-electron bonds, whereas the bridging hydrogens are involved in a bond which contains two electrons but is spread over three atoms. This nonclassical bond is called a **"three-center bond."** In higher boranes the bonding is often more complex and cannot be explained by classical bonding theory.

Diborane is a spontaneously flammable gas at room temperature. It reacts with water to form hydrogen and **boric acid,** H_3BO_3:

$$B_2H_6\ (g) + 6H_2O\ (l) \rightleftharpoons H_3BO_3\ (aq) + 6H_2\ (g)$$

The boranes form **borohydrides** on combining with alkali metal hydrides. The simplest of these, BH_4^-, is prepared from diborane:

$$2NaH\ (s) + B_2H_6\ (g) \rightleftharpoons 2NaBH_4\ (s)$$

The use of borohydrides and combinations of boron, hydrogen, and carbon as improved fuels for jet and rocket engines is receiving considerable attention. For example, **decaborane** ($B_{10}H_{14}$) on burning releases 28,000 B.T.U. per pound, in contrast to the present JP-4 jet fuel, which generates 18,000 B.T.U. per pound. Similar compounds of boron are added to gasoline to improve its performance.

HALOGEN COMPOUNDS. Each of the boron trihalides except BI_3 can be prepared by direct combination of the element. Boron triiodide is prepared by the interaction of iodine and $NaBH_4$. They are relatively volatile compounds, with BF_3 and BCl_3 existing as gases at room temperature. All of the trihalides of boron have planar triangular structures.

All of the boron halides act as Lewis acids and form adducts with ammonia, organic amines, ethers, and many other Lewis bases. In each case the Lewis acidity prevails, because boron is surrounded by only six electrons.

Oxides and Oxyacids. **Boric oxide,** B_2O_3, is the anhydride of **boric acid,** $B(OH)_3$. Pure boric acid, which is a white crystalline solid, dissolves in water and behaves as a weak monoprotic acid ($K_a = 10^{-9}$). It can be obtained as a white crystalline precipitate by acidifying an aqueous solution of any borate.

Borates, like silicates, exist in several forms, but usually do not contain the simplest unit BO_3^{-3}. Complex borates are composed of simple borate units sharing oxygen atoms. In these systems, both planar triangular and tetrahedral borate units are found.

Questions

1. Why does oxygen so readily combine with other elements to form oxides?

2. Describe a laboratory method and an industrial method for preparation of hydrogen.

3. Why is it difficult to show the exact position of hydrogen in the periodic table?

4. Which of the following apply to non-metals?
   (a) high electronegativity
   (b) low electronegativity
   (c) tendency to form anions
   (d) tendency to form cations
   (e) occupies upper left of periodic table
   (f) occupies upper right of periodic table
   (g) many of their oxides react with water to form acids
   (h) many of their oxides react with water to form bases

5. Explain why the halogens differ drastically in reactivity from their immediate periodic table neighbors, the inert gases.

6. The halogens occur in nature in what form? Give an example for each element.

7. What is the laboratory preparation of Cl_2?

8. Write equations for the laboratory preparation of each of the four hydrogen halides?

9. The oxides of fluorine differ in what way from the other halogen oxides?

10. What is disproportionation? Give an illustrative equation.

11. Chalconides refers to what combinations of elements?

12. What types of oxides are formed with sulfur, selenium, and tellurium? What oxidation numbers are assumed by the above?

13. How is H_2SO_4 prepared commercially? Give the reactions involved.

14. What are metalloids? Give examples.

15. Match the nitrogen form on the left with the type of combining elements on the right.

(a) ionic nitrides	(x) nonmetals
(b) covalent nitrides	(y) transition metals
(c) interstitial nitrides	(z) active metals

16. Write an equation for the commercial preparation of ammonia. What is the name of this process?

17. Ammonia, phosphine, arsine, and stibine are products of the Group VA elements assuming the ____________ oxidation state in combination with ____________.

18. Write equations for the formation of phosphorous acid and phosphoric acid.

19. What are the names and formulas of the most important oxyacids of nitrogen?

20. Give equations for the commercial preparation of nitric acid by the Ostwald process.

21. What are the two crystalline allotropic forms of carbon? Describe their distinguishing structures?

22. What are catenated structures?

23. What are some widespread uses of silicates?

24. What is the only member of Group IIIA of periodic table with nonmetallic characteristics?

25. The structure of diborane exemplifies what two kinds of bonding? Describe each.

Suggested Reading

Claasen: The Noble Gases. Lexington, Mass., D. C. Heath & Co., 1966.
Cotton and Wilkinson: Advanced Inorganic Chemistry, 2nd Ed. New York, Interscience Publishers, 1966.
Massey: Boron. Scientific American, Vol. 210, No. 1, p. 88, 1964.
Navratil: Fluorine—A Hostile Element. Chemistry, Vol. 42, No. 2, p. 11, 1970.
Partington: The Discovery of Oxygen. Journal of Chemical Education, Vol. 29, p. 123, 1962.
Sanderson: Inorganic Chemistry. New York, Reinhold Publishing Corp., 1967.
Sanderson: Principles of Halogen Chemistry. Journal of Chemical Education, Vol. 41, p. 361, 1964.
Sanderson: Principles of Hydrogen Chemistry. Journal of Chemical Education, Vol. 41, p. 331, 1964.

CHAPTER 9

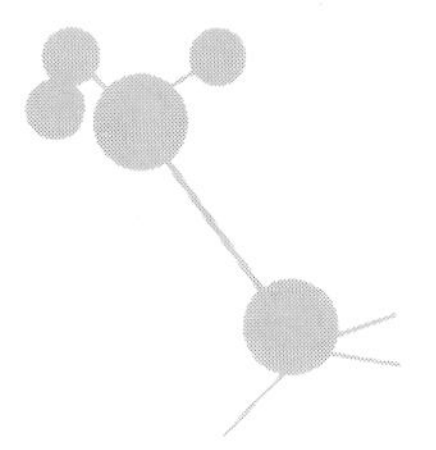

THE METALS

There are 79 elements which have the properties of metals, as shown in Figure 9–1. These are found in every group of the periodic table except those of the oxygen family, the halogens, and the inert gases. The metals are subdivided into the **representative metals,** and **transition metals,** and the **innertransition metals.** The representative metals found in the A groups of the periodic table have either empty or completely filled inner *d*-orbitals. For example, potassium and calcium have empty 3*d*-orbitals, whereas gallium has a completely filled set of 3*d*-orbitals. The transition metals have from one to ten electrons in inner *d*-orbitals and no electrons in the valence shell *p*-orbitals. The elements scandium through nickel contain partially filled 3*d*-orbitals, whereas copper and zinc contain ten electrons in 3*d*-orbitals. None of these elements has electrons in 4*p*-orbitals. The innertransition metals have a partially filled or a completely filled set of *f*-orbitals. For example, the elements cerium, Ce, through thulium, Tm, have a partially filled set of 4*f*-orbitals, whereas ytterbium, Yb, and lutetium, Lu, each has a filled set of 5*f*-orbitals.

It is difficult to give an all-inclusive definition for metals, although there

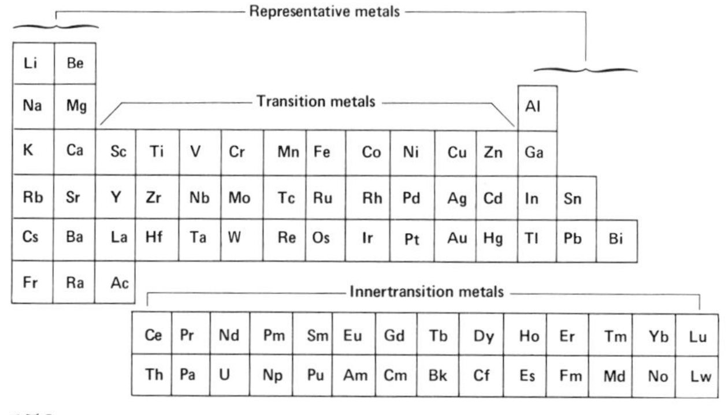

Figure 9-1 The metals in the periodic table.

are many properties that are common to the majority of these elements. For example, most metals exhibit a metallic luster, are good conductors of electricity and heat, and are solids at room temperature. When combined with other elements they always exist in positive oxidation states. Metals usually possess a degree of hardness greater than the nonmetals, although sodium and potassium are very soft metals, and mercury is a liquid. The common metals possess to a varying degree the property of malleability and ductility. A metal is malleable if it can be hammered or rolled into very thin sheets, whereas it is ductile if it can be drawn into a wire without breaking. Gold and silver are the most malleable and ductile of metals. Sheets of gold can be prepared that are less than 0.00001 inch thick, and an ounce of gold can be drawn into a wire over fifty miles long.

Since many of the compounds of metals are ionic, the chemical behavior of the metals is often largely determined by properties associated with the metallic ion. These include ionization potential of the element, atomic radius of the element, ionic radius, and charge on the ion.

THE OCCURRENCE AND RECOVERY OF METALS

Most metals are found in chemical combination in nature. Only the relatively nonreactive metals such as copper, silver, platinum, and gold, are found in nature in their elemental state. Mineral deposits containing chemically combined metals are called **ores.** The most common ores contain metal oxides, sulfides, halides, carbonates, silicates, and sulfates. Although many ores contain only one compound of a metal, they are often mixtures.

The many processes of recovering metals from their ores comprise the field of **metallurgy.** Although unique metallurgical processes are used for almost each metallic element, these processes do have some common features. These include concentration of the ore, reduction to the elemental state, and refinement to the desired purity.

The initial processes in the concentration of an ore usually involve removal of undesirable minerals called **gangue.** This is usually done by crushing and grinding the mineral to a finely divided state and then separating the ore and gangue by taking advantage of some difference in physical property, such as density or the ability to wet.

Sulfide and carbonate ores are often converted to oxides by roasting in air:

$$2MS\ (s) + 3O_2\ (g) \xrightarrow{\text{heat}} 2MO\ (s) + 2SO_2\ (g)$$

$$MCO_3\ (s) \xrightarrow{\text{heat}} MO\ (s) + CO_2\ (g)$$

Reduction of metal ions in ores to elemental metal is performed in a number of ways. Chemical reductions are performed using carbon, carbon monoxide, hydrogen, active metals, and electrolytic processes. Some metallic sulfides, when heated, undergo a self-reduction as follows:

$$MS\ (s) + O_2\ (g) \rightarrow M\ (s) + SO_2\ (g)$$

Impure metals obtained in the reduction of ores are purified most commonly by electrolytic processes. Occasionally, low-melting metals are separated from impurities by melting and recrystallization, or by distillation.

ALLOYS

Although special uses are found for many metals in a pure state, most metals are used as mixtures. These mixtures of metals, usually designed to generate desired physical or chemical properties, are called **alloys.** Alloys are prepared by melting metals together to form a molten homogeneous state and then allowing this state to crystallize. Nonmetals are often included in metals to give special properties.

Alloys are similar to metals in that they are good thermal and electrical conductors. They usually display other physical properties of metals such as malleability, ductility, and metallic luster. The preparation of an alloy is most often performed to adjust hardness, melting point, or chemical reactivity, such as corrosive resistance. Alloys are almost always harder than their major component.

THE REPRESENTATIVE METALS

The representative metals include the alkali metals, the alkaline earth metals, the Group III elements except boron, the Group IV elements tin and lead, and bismuth.

THE ALKALI METALS. The **alkali metals, lithium, sodium, potassium, rubidium,** and **francium,** are the members of Group IA. The atoms of these elements are the largest found in any period. Because of the relatively large distance of the outermost electrons from the nucleus, these elements have the smallest ionization potentials in each period. Consequently, none of the alkali metals is found in nature because of their very high reactivity.

The elemental form of the alkali metals is usually produced by electrolysis of molten salts. The element francium, the heaviest member of the alkali metal family, is produced in natural radioactive decay and is radioactive itself, with a very short half-life. The alkali metals combine directly with most nonmetals to form binary compounds, virtually all of which are ionic and exist as crystalline solids. In these compounds the alkali metals exist as $+1$ ions. On combining with oxygen, lithium forms the oxide Li_2O, sodium forms the peroxide Na_2O_2, and potassium, rubidium, and cesium give superoxides of formula MO_2. Each of these compounds hydrolyzes when added to water to give basic solutions. In the case of Na_2O_2, hydrogen peroxide, H_2O_2, is produced. The superoxides also result in the evolution of oxygen.

$$Li_2O\,(s) + H_2O \rightleftharpoons 2Li^+\,(aq) + 2OH^-\,(aq)$$
$$Na_2O_2\,(s) + 2H_2O \rightleftharpoons 2Na^+\,(aq) + 2OH^-\,(aq) + H_2O_2\,(aq)$$
$$2KO_2\,(s) + 2H_2O \rightleftharpoons 2K^+\,(aq) + 2OH^-\,(aq) + O_2\,(g) + H_2O_2\,(aq)$$

Crystalline alkali metal hydroxides, when dissolved in water, behave as strong electrolytes.

Compounds of alkali metals, including salts of oxyacids, are highly soluble in water. However, many compounds of lithium show much lower solubility than corresponding compounds of the other elements.

The alkali metals react violently with water to give hydrogen and basic solutions:

$$2M\ (s) + 2H_2O \rightleftharpoons H_2\ (g) + 2M^+\ (aq) + OH^-\ (aq)$$

THE ALKALINE EARTH METALS. The **alkaline earth metals, beryllium, magnesium, calcium, strontium,** and **radium,** are the members of Group IIA. These elements are slightly less reactive than the alkali metals, but are too reactive to occur naturally in the elemental state. They do occur in many minerals, primarily as halides, carbonates, and sulfates. The elements are usually recovered from these minerals by converting to halides and electrolyzing the fused halide. The radioactive Group IIA element **radium,** which is the heaviest member of the alkaline earth family, is formed in the natural radioactive decay series of $^{238}_{92}U$.

The chemical and physical properties of the alkaline earth metals can usually be predicted by considering the properties of the neighboring alkali metals and how they change with decreased atomic and ionic radii.

In their reactions, the alkaline earth metals tend to lose two electrons to generate M^{+2} ions. These ions have the same electronic configurations as inert gases and the M^{+1} ions of the neighboring alkali metal. There are no stable compounds in which a Group IIA element exhibits an oxidation number of +1.

Each of the alkaline earth elements has an atomic and ionic radius smaller than the Group IA metal of the same period. This is due to larger nuclear charges in the case of the alkaline earth metals. The effect of variation of atomic and ionic sizes is evident in comparing the properties of similar Group IA and Group IIA compounds. These effects are most pronounced in the cases of beryllium and magnesium compounds.

Binary and ternary compounds of calcium, strontium, and barium are highly ionic and exist as typical crystalline ionic solids which conduct electricity when in the molten state. Beryllium, and to a much smaller extent magnesium, have a pronounced tendency to form compounds with covalent bond characteristics.

Group IIA elements are quite reactive and combine with most nonmetals to form binary compounds. Each of the elements combines with oxygen to give the oxide MO. Barium also gives the peroxide, BaO_2. All of the oxides, except beryllium oxide, BeO, hydrolyze on addition to water to form the hydroxides, $M(OH)_2$, which are relatively insoluble but are strong bases. Beryllium oxide is insoluble in water but dissolves slowly in concentrated acid solution to give $[Be(H_2O)_4]^{+2}$. It also dissolves in concentrated basic solution to give $[Be(OH)_4]^{-2}$. These reactions are represented as follows:

acidic solution $\quad BeO\ (s) + 3H_2O + 2H^+\ (aq) \rightleftharpoons [Be(H_2O)_4]^{+2}\ (aq)$

basic solution $\quad BeO\ (s) + H_2O + 2OH^-\ (aq) \rightleftharpoons [Be(OH)_4]^{-2}\ (aq)$

Thus, beryllium oxide acts as a base when added to an acid, and as an acid when added to a base. Such behavior is referred to as **amphoterism,** and beryllium oxide is called **amphoteric.**

The relative reducing abilities of the alkaline earth metals are indicated by their reactions with water. Calcium, strontium, and barium react with water as do the alkali metals to give hydrogen and basic solutions:

$$M\ (s) + 2H_2O \rightleftharpoons M^{+2}\ (aq) + 2OH^-\ (aq) + H_2\ (g)$$

Magnesium undergoes the same reaction at elevated temperatures (steam), but beryllium will not react with water even at much higher temperatures.

The binary and ternary compounds of the Group IIA metals have much lower water solubility than corresponding alkali metal compounds. The solubilities of alkaline earth sulfates and carbonates are quite low and decrease going down the family. Precipitations of these salts are often used in separation and qualitative analysis schemes.

The Group IIIA Metals. All of the Group IIIA elements except boron behave as metals. These include **aluminum, gallium, indium,** and **thallium.** Aluminum, the third most abundant element in the earth's crust, at 7.5 per cent, is the most abundant metal. It is a common component of many silicates, but the most valuable ores are **bauxite,** $Al_2O_3 \cdot 2H_2O$, and **cryolite,** Na_3AlF_6. The element is isolated by electrolyzing a molten bauxite-cryolite mixture in an electrolytic cell. In this process the carbon anodes are consumed and the electrode reactions are as follows:

$$\text{(cathode)}\quad Al^{+3} + 3e^- \rightarrow Al$$
$$\text{(anode)}\quad C + 2O^{-2} \rightarrow CO_2 + 4e^-$$

Gallium, indium, and thallium are often found in small quantities in ores of aluminum and zinc. The elements are obtained by electrolysis of aqueous solutions of the salts.

Aluminum is a very hard, strong, lustrous metal, which in air always has a surface film of aluminum oxide, Al_2O_3. This film protects the metal from corrosion. Aluminum is used extensively as a structural material, but usually as an alloy. Although the +3 oxidation state is most common in compounds of the Group IIIA metals, the +1 oxidation state is known in each case. The stability of the +1 oxidation state increases going down the family and is quite important in the chemistry of thallium. The Group III metals react completely with most nonmetals, including the halogens, oxygen, and sulfur. In these compounds, all of the metals assume the +3 oxidation state, but thallium also forms compounds such as TlCl, Tl_2O, and Tl_2S in which it assumes the +1 oxidation state.

Aluminum oxide is insoluble in water and, in fact, can be precipitated in a hydrated form from aqueous solution containing Al^{+3} by adding OH^-:

$$2Al^{+3}\ (aq) + 6OH^- \rightleftharpoons Al_2O_3 \cdot 3H_2O\ (s)$$

This material, which if heated loses water to form aluminum oxide, Al_2O_3, can be thought of as aluminum hydroxide, $Al(OH)_3$. It behaves like beryllium hydroxide in that it is amphoteric. In basic solutions aluminum oxide and aluminum hydroxide dissolve because of the combination with OH^- to form

$[Al(OH)_4]^{-1}$, the **aluminate ion:**

$$Al_2O_3\,(s) + 3H_2O + 2OH^-\,(aq) \rightleftharpoons 2[Al(OH)_4]^{-1}\,(aq)$$
$$Al(OH)_3\,(s) + OH^-\,(aq) \rightleftharpoons [Al(OH)_4]^{-1}\,(aq)$$

In acidic solution they dissolve to form the hydrated aluminum ion $[Al(H_2O)_6]^{+3}$:

$$Al_2O_3\,(s) + 6H^+\,(aq) + 9H_2O \rightleftharpoons 2[Al(H_2O)_6]^{+3}\,(aq)$$
$$Al(OH)_3\,(s) + 3H^+\,(aq) + 3H_2O \rightleftharpoons 2[Al(H_2O)_6]^{+3}\,(aq)$$

Gallium oxide and hydroxide are also amphoteric, but the indium and thallium compounds are basic. The oxides In_2O_3, Tl_2O_3, and Tl_2O, when dissolved in water, give basic solutions.

Many binary compounds of the Group IIIA metals hydrolyze when added to water to give acidic solutions. This is explained by considering the M^{+3} ions as cations of weak bases which hydrolyze as follows:

$$M^{+3}\,(aq) + H_2O \rightleftharpoons MOH^{+2}\,(aq) + H^+\,(aq)$$

The Group IVA Metals. **Tin** and **lead** are the only Group IVA elements which behave as metals. **Galena,** PbS, the most common ore of lead, is converted to metallic lead by roasting and reduction with coke (carbon). Tin, obtained as the oxide SnO_2 from its ore, is reduced to the metal with coke.

Tin, which is known to exist as three temperature-dependent allotropes, is at room temperature a malleable, lustrous metal, but a poor electrical conductor. Elemental tin is used extensively in alloys. Some of the important alloys of tin are **solder** (33% Sn, 67% Pb), **babbitt metal** (90% Sn, 7% Sb, 3% Cu), **pewter** (85% Sn, 7% Cu, 6% Bi, 2% Sb), and several **bronzes** which contain from 10 to 20% tin.

Lead, which appears as dull gray, does have a lustrous surface when freshly exposed. It is very soft and has a relatively low melting point. Elemental lead is used extensively in the pure state in storage battery plates, lead pipes, bullets, and cable covering. Many alloys of lead are also used, including **solder** and **type metal** (82% Pb, 15% Sb, 3% Sn).

In the chemistry of the Group IVA metals, the +2 and +4 oxidation states are both prevalent. The +2 oxidation state is more important in lead chemistry.

THE TRANSITION METALS

The **transition metals** are those elements which have partially filled sets of *d*-orbitals or filled sets of *d*-orbitals with empty outer *p*-orbitals. The **inner-transition metals** are really a subgroup of the transition metals. These elements have a partially filled set of *f*-orbitals or a filled set of *f*-orbitals.

The Innertransition Series. The innertransition metal series, **cerium,** Ce, through **lutetium,** Lu, which is referred to as the **lanthanides,** has a partially or totally filled set of 4*f*-orbitals. These orbitals are located considerably inside of the 6*s*-orbital. The physical and chemical behavior of the members of this series is consequently not significantly related to the 4*f*-orbital electron configurations. The major differences in the properties observed within the series are due to a smooth decrease in atomic or ionic radius with increasing atomic number. This effect is called the **"lanthanide contraction."**

The second row of innertransition metals, called the **actinides,** contains only six elements found in nature. The other elements of the series have all been synthesized by nuclear processes.

GENERAL PROPERTIES OF THE TRANSITION METALS. With the exception of the Group IIB metals **zinc, cadmium,** and **mercury,** all of the transition metals have relatively high melting and boiling points. **Mercury** is a liquid which freezes at $-38.9°C$ and cadmium and zinc are solids which melt at $321°C$ and $420°C$, respectively. All of the transition metals are good thermal and electrical conductors. The Group IB elements, **copper, silver,** and **gold** are outstanding electrical conductors.

Some of the metals form oxide films which are not attacked by acids and therefore can be used as protective films on other metals.

The elements of the first transition metal series, as well as the second and third series, typically display several oxidation states. Unlike the representative metals, many of the oxidation states of the transition metals leave an odd number of electrons on the metal ion. As a result of this, many transition metal compounds are **paramagnetic** (i.e., attracted to a magnetic field). The transition metals form many simple binary or ternary compounds. However, the chemistry of transition metal ions is dominated by the tendency of the ion to surround itself with the maximum possible number of Lewis bases, whether molecules, atoms, or ions. This behavior results in the formation of coordination compounds which are discussed in a later section.

THE SCANDIUM FAMILY. The **scandium family** includes **scandium, yttrium,** and **lanthanum,** and a subgroup composed of the **lanthanides.** Scandium, rather rare and relatively unimportant, displays only the +3 oxidation state in its compounds. In a number of respects the chemistry of Sc^{+3} is like that of Al^{+3}. However, scandium oxide, Sc_2O_3, is basic rather than amphoteric like aluminum oxide, Al_2O_3. Yttrium and lanthanum display only the +3 oxidation state, and this is the principal oxidation state displayed by all of the lanthanides.

THE TITANIUM FAMILY. The titanium family includes **titanium, hafnium,** and **zirconium.** Titanium, the seventh most abundant metal in the earth's crust is obtained in metallic form by treating its ores with hot chlorine gas and carbon to give **titanium tetrachloride,** $TiCl_4$, which is then reduced to the metal with molten magnesium in an inert atmosphere. Titanium metal is very hard, although malleable and ductile. Because of its low density, very high strength, and high melting point, this metal has extensive uses in the aviation industry.

Zirconium and hafnium, comparatively rare elements, are usually obtained from their ores by magnesium reduction of the tetrachlorides.

The +4 oxidation state of titanium is the most stable and most important. Titanium dioxide, TiO_2, is extensively used as a white pigment in paints. This oxide is insoluble in water, but displays amphoterism by dissolving in basic and acidic solutions. Titanium tetrachloride is a liquid (b.p. $137°C$) which is hydrolyzed in moist air to TiO_2 and HCl:

$$TiCl_4 (l) + 2H_2O (l) \rightarrow TiO_2 (s) + 4HCl (g)$$

This reaction is used in making dense white smoke screens.

The Vanadium Family. The vanadium family includes **vanadium, niobium,** and **tantalum.** Vanadium ores, which occur widely but in small abundance, are used most extensively to produce an alloy with iron, called **ferrovanadium.** This alloy is produced by the reduction of a mixture of vanadium pentoxide, V_2O_5, and ferric oxide with carbon in an electric furnace.

The most important compound of vanadium, V_2O_5, is obtained as an orange powder when finely divided metallic vanadium is burned in excess oxygen. This oxide, which is used as a catalyst in many processes, is the catalyst employed in converting SO_2 to SO_3 in the contact process for making sulfuric acid. Although V_2O_5 is only slightly soluble in water, it is amphoteric.

The Chromium Family. The chromium family includes **chromium, molybdenum,** and **tungsten.** Metallic chromium is extremely resistant to atmospheric corrosion and consequently is used extensively as a protective coating. It is also used as an alloying component of steel to increase strength and toughness.

The heavy metals of the chromium family are obtained by converting the ores to the oxides MoO_3 and WO_3, which are reduced to the metal with hydrogen. These metals are important because of their very high melting points and extreme hardness. Molybdenum increases the strength and hardness of steels, and tungsten increases the hardness of steels at higher temperatures.

Metallic chromium is covered by a thin layer of the oxide Cr_2O_3. This oxide is relatively insoluble in water and is amphoteric. It is obtained in the hydrated form when solutions containing Cr^{+3} are made basic. With the addition of excess base, the amphoteric oxide dissolves by formation of **chromite ion,** $[Cr(OH)_4]^-$. In basic solution, $[Cr(OH)_4]^-$ is readily oxidized to the **chromate ion,** $[CrO_4]^{-2}$, in which chromium is in the +6 oxidation state. When acidified $[CrO_4]^{-2}$, which is yellow, is converted to orange **dichromate ion,** $[Cr_2O_7]^{-2}$, a powerful oxidizing agent.

Addition of $[CrO_4]^{-2}$ or $[Cr_2O_7]^{-2}$ salts to concentrated sulfuric acid solution produces the bright red oxide CrO_3. This highly acidic oxide, soluble in water, is a very strong oxidizing agent. Its solutions are commonly used as cleaning agents to remove greases from laboratory glassware.

The Manganese Family. The manganese family contains the elements **manganese, technetium,** and **rhenium.** Technetium does not occur in nature, but has been artificially produced. Rhenium, which is very rare, is of little significance compared to manganese which ranks eighth in abundance among the metals in the earth's crust.

Metallic manganese is used most extensively in making steel alloys. Small quantities added to steel remove traces of oxygen and sulfur by the formation of manganese oxides and sulfides, which enter the slag. When added in quantities of 10 per cent or more, very tough, hard steel is formed.

Manganese dioxide is relatively insoluble in water but is amphoteric. In acidic media it is a strong oxidizing agent being reduced to Mn^{+2}. Manganese exhibits the oxidation number +7 in $[MnO_4]^-$, the **permanganate** anion which is a strong oxidizing agent in basic or acidic solution. In basic solution the reduction product is MnO_2, whereas Mn^{+2} is formed in acidic solutions:

$$MnO_4^- (aq) + 2H_2O + 3e^- \rightleftharpoons MnO_2 (s) + 4OH^- (aq) \quad \text{(basic)}$$
$$MnO_4^- (aq) + 8H^+ (aq) + 5e^- \rightleftharpoons Mn^{+2} (aq) + 4H_2O \quad \text{(acidic)}$$

Iron, Cobalt, and Nickel. **Iron, cobalt,** and **nickel** have many very similar properties and, consequently, these three elements are considered to be members of the same group in the periodic table, Group VIII. The heavier members of this group also display similar properties and will be discussed in the next section.

Iron, cobalt, and nickel are hard metals with very high melting points. They are all strongly attracted by a magnetic field **(ferromagnetic).** The metals are moderately reactive, and the important chemistry primarily involves the +2 and +3 oxidation states.

Iron is the second most abundant metal, constituting 4.7 per cent of the earth's crust. The major ores are **hematite,** Fe_2O_3, **limonite,** $FeO(OH)$, **siderite,** $FeCO_3$, and **magnetite,** Fe_3O_4. In the earth's crust cobalt and nickel are rare compared to iron, and commonly occur as sulfide and arsenide impurities in ores of iron, silver, and copper.

To recover metallic iron from its ores, the oxides are reduced by heating in the presence of coke and limestone. The burning coke or carbon combines with oxygen to form carbon dioxide and carbon monoxide:

$$C (s) + O_2 (g) \rightarrow CO (g) + CO_2 (g)$$

The carbon monoxide formed reacts with the iron oxide in the ore to form iron and carbon dioxide:

$$3CO (g) + Fe_2O_3 (s) \rightarrow 3CO_2 (g) + 2Fe (l)$$

Pure iron is oxidized in moist air, whereas cobalt and nickel are attacked only at elevated temperatures. The most important oxidation states of iron and cobalt are +2 and +3, whereas most of the chemistry of nickel involves the +2 state.

The Platinum Metals. The six heavy members of Group VIII, **ruthenium, rhodium, palladium, osmium, iridium,** and **platinum,** are referred to as the **platinum metals.** All of these metals have relatively low reactivity. Rhodium, palladium, and platinum are softer than the other members of this group. In fact, ruthenium, osmium, and iridium are brittle and hard. Platinum is used as a catalyst in many processes, including the hydrogenation of organic compounds and the oxidation of ammonia in the Ostwald processes for making nitric acid. The pleasing luster, the chemical inertness, and the high price of platinum have resulted in its use in the manufacture of jewelry.

The Copper Family. The copper family includes **copper, silver,** and **gold.** All of these elements occur in the free and combined states in nature.

Metallic copper is obtained from sulfide ores by heating in air, whereas carbon is often used to reduce oxide and carbonate ores at higher temperatures.

Impure copper obtained in this way is usually refined in two steps. Air is passed over molten impure copper to oxidize impurities. The partially refined copper is finally purified by electrolysis, using impure copper as the anode and pure copper as the cathode:

$$\text{(anode)} \quad Cu\,(s)\,(\text{impure}) \rightleftharpoons Cu^{+2}\,(aq) + 2e^-$$
$$\text{(cathode)} \quad Cu^{+2}\,(aq) + 2e^- \rightleftharpoons Cu\,(s)\,(\text{pure})$$

All of the metals of the copper family are soft, malleable, and ductile, and are very good thermal and electrical conductors. One of the major uses of copper is in the manufacture of copper wire for the electrical industry. Copper is also used extensively in alloys such as brass and bronze. Silver and gold are considered precious metals and are used in the manufacture of jewelry. Silver is also used in manufacturing tableware and plated cutlery. One third of the silver produced in the United States is used in the photographic industry for AgBr and AgCl emulsions on films.

The +2 oxidation state is the most important for copper, although the +1 oxidation state is also well characterized. Copper (I) compounds usually decompose slowly to copper (II) compounds at room temperatures.

The Zinc Family. The zinc family contains the elements **zinc, cadmium,** and **mercury.** Zinc, never found in the elemental state, is obtained primarily from the ores **zinc blende,** ZnS, **calamine,** $ZnCO_3$, **willemite,** Zn_2SiO_4, and **zincite,** ZnO. Cadmium is usually obtained as an impurity in zinc ores. Mercury, although found in the elemental state alloyed with other metals, is obtained primarily from the ore **cinnabar,** HgS.

Zinc is usually obtained from its ores by roasting followed by carbon reduction at high temperatures. The metallic zinc thus generated contains metallic cadmium in small quantities. Metallic zinc and cadmium are separated by distillation. Mercury is obtained from cinnabar simply by heating to 500°C, whereby elemental mercury distills from the solid.

Zinc and cadmium are lustrous metals which are corroded slowly by the atmosphere. Mercury is a liquid at room temperature and is much less reactive than zinc or cadmium. Acidic solutions dissolve zinc and cadmium to give the +2 oxidation state, which is the most common oxidation state for these two elements.

Mercury displays both the +1 and the +2 oxidation states. In the +1 oxidation state, a diatomic metallic ion, Hg_2^{+2}, has been established. This oxidation state is less stable than the +2 state.

COORDINATION COMPOUNDS

A **coordination compound** consists of a central metal ion bonded to surrounding atoms, molecules, or ions by coordinate covalent bonds. Transition metal ions have a high tendency to form coordination compounds, because they have partially filled *d*-orbitals which can accept electron pairs. The representative metal ions also form coordination compounds.

The atoms or ions surrounding the central metal ion in a coordination

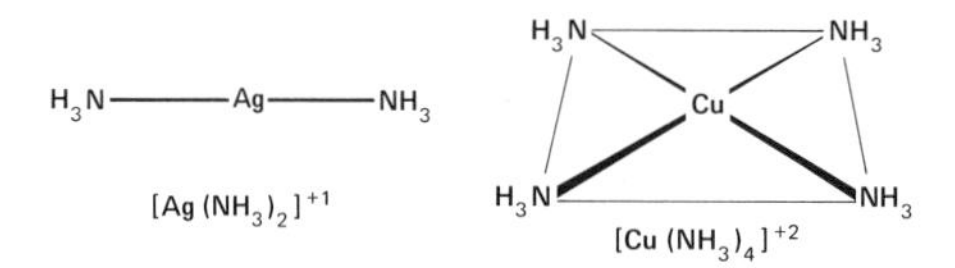

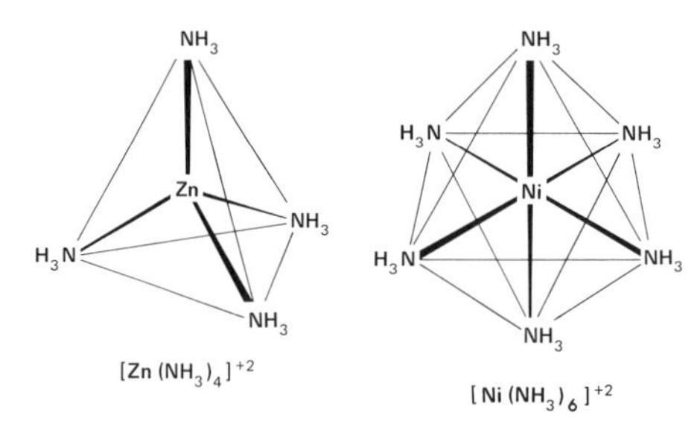

Figure 9–2 The common geometries of two, four and six coordinate complex ions.

compound are called **ligands,** and they constitute the **primary coordination sphere.** The number of ligands in the primary coordination sphere is the **coordination number.** In the complex ion $[PtCl_4]^{-2}$, Pt^{+2} is the central metal ion, and the primary coordination sphere consists of four ligands which are chloride ions. The coordination number is four. In the complex compound $[Pt(NH_3)_3Cl]Cl$, the primary coordination sphere contains three ammonia molecules and a chloride ion. A second chloride ion is part of the compound, but it is outside of the primary coordination sphere. In writing the formulas of complex compounds, the central metal ion and the contents of the primary coordination sphere are enclosed in brackets.

The most common coordination numbers found in coordination compounds of the transition metal ions are two, four, and six. The geometries of typical complex ions are shown in Figure 9–2. Species of coordination number two, such as $[Ag(NH_3)_2]^+$, are linear. Species of coordination number four are either tetrahedral or square planar, and six coordinate compounds are octahedral.

Molecules, ions, and atoms which have lone pairs of electrons available for donation can act as ligands. Ligands which bond through one lone pair of electrons are called **monodentates.** Many ligands have more than a single site which can furnish a lone pair of electrons. These are called **chelates** (Greek, *chele*-claw) or **polydentates.** Those which bond to a metal ion through two sites are called **bidentate ligands.** Common bidentates include **ethylenediamine** (en), **oxalate** (ox), and **acetylacetonate** (acac), as shown in Figure 9–3. Polydentate ligands having up to six and more donor sites are known. When chelates coordinate to metal ions, they form closed rings which include the metal ion.

$H_2N-CH_2-CH_2-NH_2$

(en)

(ox)

(acac)

Figure 9–3 Some bidentate ligands.

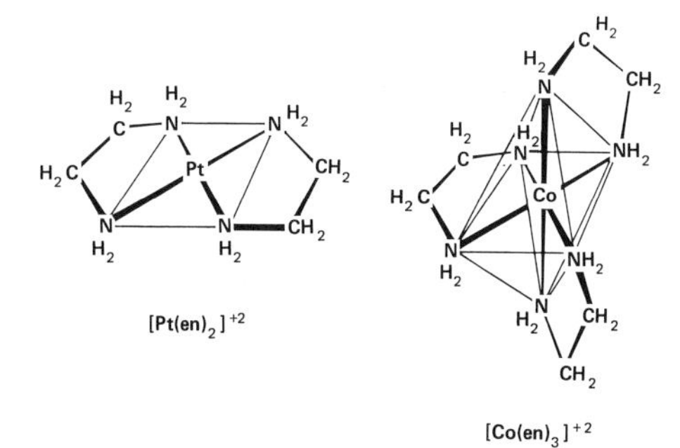

Figure 9-4 Closed rings in chelates.

In the compounds $[Pt(en)_2]^{+2}$ and $[Co(en)_3]^{+2}$, two and three rings are formed as seen in Figure 9–4.

Isomerism. When a coordination compound has two or more of each different ligand on the central metal ion, it is possible to have more than one structural formula. This phenomenon is referred to as **stereoisomerism.** This is illustrated by the two possible structures for $[Pt(NH_3)_2Cl_2]$ shown in Figure 9–5.

These two structures are called **isomers** of the compound. The isomer with like ligands occupying adjacent positions in the coordination sphere is called the ***cis-* isomer.** The other isomer is the ***trans-* isomer.** In the tetrahedral complex $[Zn(Nh_3)_2Cl_2]$, isomerism of the type mentioned cannot prevail because there is only one possible structure. In the octahedral complex $[Co(en)_2Cl_2]$, *cis-* and *trans-* isomers prevail as seen in Figure 9–6. In addition there are two types of *cis-* isomers called **optical isomers** which are mirror images as shown in Figure 9–6. Optical isomerism will be discussed in Chapter 20.

Nomenclature of Complexes. The multitudes of coordination compounds and complex ions are named by the rules suggested by the International Union of Pure and Applied Chemistry (IUPAC). These are as follows:

1. In a salt, cations are named first, whether complex or not.
2. In naming a complex, cationic or neutral ligands are named first, then neutral molecule ligands, followed by the metal ion with its oxidation state in Roman numerals in parentheses.
3. If the complex is an anion, the suffix -ate is attached to the metal followed by the oxidation state of the metal in Roman numerals.
4. Anionic ligands have the suffix -o added to the stem; Cl^-, chlor; Br^-, bromo; SO_4^{-2}, sulfato; OH^-, hydroxo; NO_3^-, nitrato; and CN^-, cyano.
5. Neutral ligands have the name of the ligand itself, except for H_2O (aquo) and NH_3 (ammine).

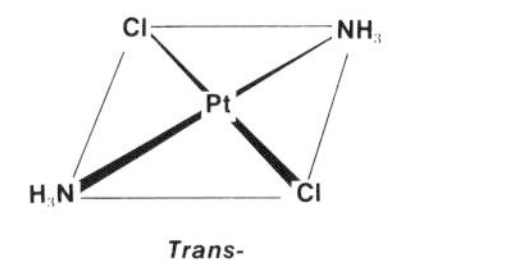

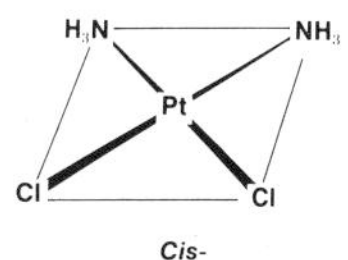

Figure 9-5 *Cis-* and *trans-*$[Pt(NH_3)Cl_2]$.

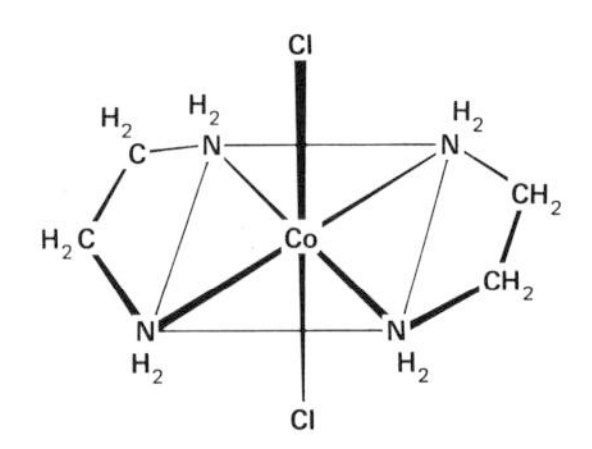

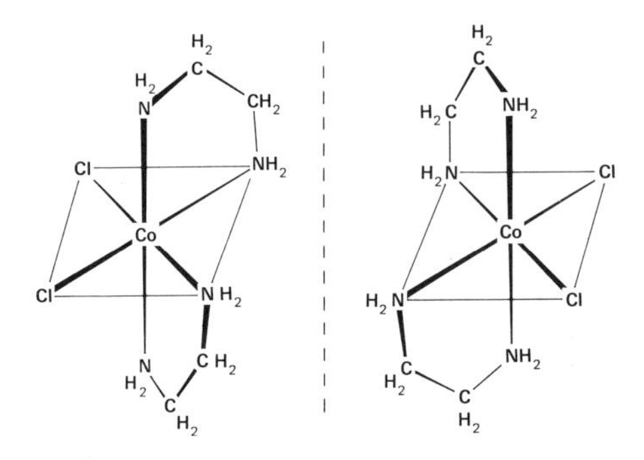

Figure 9-6 The *trans-* isomers of $[Co(en)_2Cl_2]$ (above), and the optical isomers of *cis*-$[Co(en)_2Cl_2]$ (below).

6. The number of each type of monodentate ligand is specified by the Greek prefixes **di-, tri-, tetra-, penta-,** and **hexa-,** whereas the number of chelate or complicated ligands is indicated by the prefixes **bis-, tris-, tetrakis-, pentakis-,** and **hexakis-.**

The following examples illustrate the use of these rules:

$K_2[PtCl_4]$	potassium tetrachloroplatinate (II)
$[Pt(NH_3)_3Cl]Cl$	chlorotriammineplatinum (II) chloride
$[Ag(NH_3)_2]NO_3$	diamminesilver (I) nitrate
$[Ni(H_2O)_6][NiCl_4]$	hexaquonickel (II) tetrachloronickelate (II)
$[Pt(en)_2]Cl_2$	bisethylenediamineplatinum (II) chloride
$[Co(en)_2Cl_2]$	dichlorobisethylenediaminecobalt (II)
$[Pt(PCl_3)_2Cl_2]$	dichlorobistrichlorophosphineplatinum (II)
$[Sn(OH)_2(H_2O)_4]^{+2}$	dihydroxotetraquotin (IV) ion
$[Al(OH)_4(H_2O)_2]^{-1}$	tetrahydroxodiaquoaluminate (III) ion

Aquo Complexes. Metal ions in aqueous solution, in the absence of other strongly coordinating ligands, have water molecules in their primary coordination spheres. The coordination number of the metal ion in water is dependent upon the charge on the ion and the ionic radius. Most aquo complexes have a coordination number of six. Some of those which have been established include:

$$[Al(H_2O)_6]^{+3};\ [Cr(H_2O)_6]^{+3};\ [Fe(H_2O)_6]^{+3};\ [Cu(H_2O)_6]^{+2}$$

In aqueous solution the formation of coordination compounds other than aquo complexes is best visualized as a metathesis reaction in which a ligand replaces water in the primary coordination sphere. This is exemplified by the reaction of Cu^{+2} ion in water on the addition of ammonia:

$$[Cu(H_2O)_6]^{+2}\,(aq) + 4NH_3\,(aq) \rightleftharpoons [Cu(NH_3)_4(H_2O)_2]^{+2}\,(aq) + 4H_2O$$

THE IMPORTANCE OF COORDINATION COMPOUNDS. Coordination compounds and complex ions play an important role in many processes. Various chelating ligands are often used to remove unwanted ions from aqueous solution. This is a common procedure in water softening. In the photographic process, **thiosulfate ion,** $S_2O_3^{-2}$, is used as a liquid to remove silver ions from the film in the developing process. Many complex ions play an important role in the catalysis of organic reactions, including hydrogenations.

Undoubtedly, one of the most striking roles of a coordination compound is the oxygen transport behavior of hemoglobin, an iron coordination compound. There are many other important biochemical processes in which coordination compounds play an important role, including the chemistry of vitamins and the hydrolysis of proteins. These will be treated more extensively in the biochemistry section of this book.

Questions

1. Name the three subdivisions of the metals, indicating the differences according to electron configuration.

2. What groups of the periodic table do *not* contain metals?

3. How do most metals occur in nature? Give exceptions.

4. Considering the electron configuration and the first and second ionization potentials, explain why alkali metals always occur in chemical combination as +1 ions.

5. Which group of the representative metals reacts violently with water? Illustrate with an equation.

6. Group the following characteristics under the proper heading: A)Alkali metals; B)Alkaline earth metals; C)Group IIIA metals; D)Group IV metals
   (a) atoms are the largest found in any period
   (b) members of Group IIA
   (c) +3 oxidation state is most common
   (d) electron configuration shows one electron in *s*-orbital
   (e) contains strontium and beryllium
   (f) contains lithium and cesium
   (g) contains thallium and gallium
   (h) tend to lose 2 electrons to generate M^{+2} ions
   (i) contains tin and lead
   (j) electron configuration of ns^2np^2

7. Write an equation illustrating the hydrolysis of an alkaline earth metal oxide in water.

8. What is amphoterism?

9. Do binary halides of Group IIIA metals hydrolyze to form acidic or basic solutions? Illustrate with an equation.

10. Explain the "lanthanide contraction."

11. What is the most important compound of vanadium? Write the equation for its preparation.

12. In what form does chromium occur in nature? Write the equation for reduction with carbon to the elemental metal.

13. What are the most important oxidation states of iron and cobalt?

14. Match the following columns:

(a) coordination compound	(w) ions or atoms surrounding central metal ion
(b) ligand	(x) all ligands directly bonded to the metal ion
(c) coordination number	(y) central metal with atoms, ions, or molecules bonded by coordinate covalent bonds
(d) primary coordination sphere	(z) number of ions or atoms surrounding central metal ion

15. Write chemical reactions which represent dissolving an amphoteric metal oxide, MO_2, in acidic solution and in basic solution.

16. Give the coordination number and name of each of the following and predict the structure.

(a) $[Co(NH_3)_4Cl_2]^+$ (c) $[Pt(NH_3)_2(H_2O)_2]^{+2}$
(b) $[Pt(NH_3)_6]^{+4}$ (d) $[AgCl_2]^-$

17. What is the difference between *cis-* and *trans-* isomers? Illustrate your answer by drawing the *cis-* and *trans-* isomers of square planar $[Pd(H_2O)_2I_2]$.

Suggested Reading

Cotton and Wilkinson: Advanced Inorganic Chemistry, 2nd Ed. New York, Interscience Publisher, 1966.
Larsen: Transition Elements. New York, W. A. Benjamin, Inc., 1965.
Moeller: The Chemistry of the Lanthanides. New York, Reinhold Publishing Corp., 1963.
Murmann: Inorganic Complex Compounds. New York, Reinhold Publishing Corp., 1964.
Sanderson: Inorganic Chemistry. New York, Reinhold Publishing Corp., 1967.

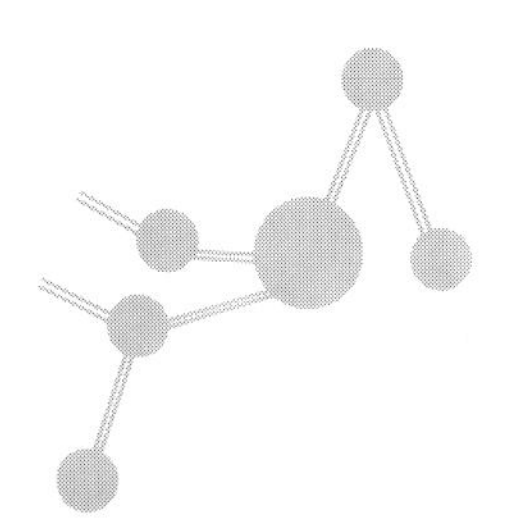

CHAPTER 10

INTRODUCTION AND GENERAL PRINCIPLES OF ORGANIC CHEMISTRY

At the beginning of the nineteenth century considerable evidence had been accumulated concerning the nature, properties, and reactions of inorganic compounds. In contrast to this large body of knowledge about inorganic compounds, relatively little was known about organic compounds. It was known at that time that organic compounds were composed of only a few elements, such as carbon, hydrogen, oxygen, nitrogen, and sulfur; and that, in contrast to inorganic materials, the organic compounds were easily combustible and many of them were sensitive to heat and strong acids and bases.

In the first quarter of the nineteenth century many simple organic compounds were studied by chemists such as Lavoisier, Dalton, Gay-Lussac, Berzelius, and Gerhardt. The compounds investigated by these early chemists, however, were all obtained from either vegetable or animal sources and are what are commonly known as "natural products." Since these initial organic compounds had been isolated as part of the life process, the early theory of organic chemistry postulated that organic compounds could only arise through the operation of a "vital force" inherent in living cells. Consequently, the chemical synthesis of organic compounds in the laboratory seemed impossible to these early chemists.

In 1828, however, a German chemist, Friedrich Wöhler, discovered that by heating an aqueous solution of the inorganic salt ammonium cyanate, urea was produced. This material isolated by Wöhler was identical to urea isolated from urine.

$$NH_4OCN \rightarrow CO(NH_2)_2$$

Ammonium cyanate   Urea

This discovery, of course, dealt a severe blow to the "vital force" theory for the synthesis of organic compounds. Although it still took many years and further experiments for this "vital force" theory to be completely abandoned, Wöhler's experiments marked the beginning of the end of this theory and set the stage for a new era in organic chemistry. Since that time the organic compounds which have been synthesized in the laboratory by organic chemists far outnumber those isolated and identified from natural sources, although the branch of organic chemistry known as "natural products" is still an important area of research in organic chemistry. In fact, most natural products now known have been prepared by independent synthesis in the laboratory, and many of these compounds make up the vitamins and medicinals that people use daily.

Although Wöhler's conversion of ammonium isocyanate to urea dispelled the "vital force" idea and established a link between inorganic and organic chemistry, the designation "organic" has persisted as a convenient means of classifying groups of compounds having some features in common. Analyses of many of the early investigated organic compounds revealed that most of them contained carbon and hydrogen, and that many contained oxygen, nitrogen, halogen, sulfur, phosphorus, and other elements as well. Today, it is recognized that the common feature among compounds classified as "organic" is that they all contain the element carbon. Hence, the present day definition of organic chemistry is **the chemistry of carbon compounds.** Conversely, compounds not containing carbon are designated as inorganic compounds.*

The importance of organic chemistry to modern society is evident if we consider our surroundings and environment. Most of the clothes that one wears are made of rayon, dacron, nylon, orlon, or some other synthetic fiber developed in the organic chemistry laboratory. Not only the fiber used in making clothing, but even the dyes employed to color these fabrics are the result of organic research. Although the past thirty years have seen startling developments in man-made fibers, certainly the future will deliver even greater developments in this area.

Modern plastics have not only revolutionized many of our commonplace utensils, such as cups, plates, glasses, and so forth, but have replaced even wood and metal in many of our household furnishings, building materials, and even toys. In today's world plastic materials are as commonplace as wood materials were a hundred years ago.

Even common, naturally occurring organic materials such as petroleum have felt the improving hand of the organic chemist. The chemist and chemical engineer have designed catalytic cracking methods for transforming raw petroleum into improved gasoline for automotive and aviation fuels, and have developed methods for utilizing the non-gasoline fractions of petroleum to make raw materials for synthetic rubber and plastics. In fact, much of our modern chemical technology has been developed from by-products of the petroleum industry.

In addition to these few mentioned items, other materials such as sulfa

---

*The student will recognize that several carbon-containing compounds such as CO_2, CO, H_2CO_3, and other carbonates have been previously classified as inorganic. These compounds were classified as inorganic before Wöhler's experiments and before carbon was recognized as the common feature of organic compounds. For convenience, their classification has remained the same and they are studied as part of inorganic chemistry.

drugs, penicillin, cortisone, perfumes, detergents, vitamins, pesticides, anesthetics, and many of the recent antibiotics are among the contributions made by organic chemistry to our present society. Throughout the remainder of the text, the nature of many of these substances, including what their chemical structures are, how they are made, and what their chemical behavior is, will be presented. However, in order to be able to understand these more complex compounds, a knowledge of certain basic reactions and properties of organic molecules must be understood. In the following pages the more common chemistry of the fundamental types of organic compounds will be presented, and a thorough understanding of these basic concepts and reactions will serve as a basis for understanding the more complex materials.

COMPARISON OF ORGANIC AND INORGANIC COMPOUNDS

Although inorganic chemistry preceded the study of organic chemistry by many years, the number of known organic compounds now far exceeds the number of known inorganic compounds. Today, there are well over 1,500,000 known organic compounds, and it should be obvious that some sort of logical and consistent theory of reactions is necessary if one is to study such a multitude of chemical structures.

Before embarking on a detailed analysis of organic compounds and their reactions, certain fundamental principles need to be learned. In some cases we can extend certain basic principles that one has learned in inorganic chemistry directly to organic chemistry, such as the concepts of acidity and basicity,[*] but in many cases it is necessary to introduce new principles and new nomenclature to be able to understand organic compounds.

In contrast to inorganic compounds, which can contain any of the elements in various combinations, organic compounds must contain the element carbon. In addition to carbon, other elements such as hydrogen, oxygen, fluorine, chlorine, bromine, iodine, nitrogen, phosphorus, and sulfur are commonly found in organic compounds. The number of atoms present in an organic compound is often large (for example, $C_{20}H_{40}$) in contrast to many inorganic compounds (for example, H_3PO_4 and $K_2Cr_2O_7$) which generally contain few atoms. In addition, the structures of organic molecules are often complex, such as cholesterol, and may contain ring (cyclic) structures as well as carbon-carbon chains.

Simple carbon structures, such as C_2H_6O, may exist as more than one structure. The two compounds dimethyl ether and ethyl alcohol are known as **isomers.** Isomers can be defined as compounds which have the same molecular formula, but which have different atomic arrangements. The physical properties and reactions of isomers can be quite similar or quite different

```
    H     H              H  H
    |     |              |  |
  H-C--O--C-H          H-C--C-O-H
    |     |              |  |
    H     H              H  H
```

Dimethyl ether — Ethyl alcohol

[*] The student should review Chapter 7 to refresh his memory of the concepts of acids and bases.

depending on how the atoms are arranged in the isomeric structures. In the previous example, dimethyl ether and ethyl alcohol exhibit quite different physical properties. Dimethyl ether is a gas at room temperature, whereas ethyl alcohol is a liquid (b.p. 78°C). These two compounds are also quite different chemically, the alcohol being quite reactive with many reagents such as HCl, H_2SO_4, whereas dimethyl ether is unreactive with these reagents.*

If the natures of the bonds and the organic groups in the isomers are quite similar, similar properties and reactivities can be expected. For example, C_4H_{10} can exist as the following isomers:

```
    H  H  H  H               H       H       H
    |  |  |  |               |       |       |
 H—C—C—C—C—H           H—C———C———C—H
    |  |  |  |               |       |       |
    H  H  H  H               H   H—C—H   H
                                     |
                                     H

     n-Butane               2-Methylpropane
    (b.p. 0°C)               (b.p. −10°C)
```

These two isomeric structures have similar boiling points and react with similar chemical reagents. As the number of carbon atoms in the molecule increases, the number of isomeric structures also increases, and it is the ability of carbon to form many isomeric structures that accounts in part for the large number of known organic compounds. Most inorganic compounds, with the exception of complex ions, do not form isomeric structures. Hence, for a particular combination of atoms in an inorganic molecule often only one structure is possible.

The types of bonding in organic and inorganic compounds also differ and account for the large difference in some of the physical properties of organic and inorganic compounds. Whereas many inorganic compounds are composed of ions and held together by strong electrostatic forces, most organic compounds are composed of weak covalently bonded atoms and are relatively nonpolar materials.† This difference in bonding is reflected in the physical properties such as boiling point, melting point, and solubility. Most ionic inorganic compounds have high melting points and high boiling points (generally > 1000°C), whereas most organic compounds melt at temperatures less than 300°C and boil at temperatures less than 500°C. The high temperatures required to volatilize inorganic compounds indirectly measure the polarity of the bonds in the molecule.

```
                            O
                            ‖
    Na⁺Cl⁻              CH3C—NH2

Sodium chloride          Acetamide
 (m.p. 801°C)           (m.p. 81°C)
 (b.p. 1413°C)          (b.p. 222°C)
```

Since most inorganic compounds are made up of ions held together electrostatically, it would be expected that inorganic compounds should be soluble in polar solvents and, as expected, most inorganic compounds are soluble in

* As will be obvious from later discussions of alcohols and ethers, the presence of the —OH group in the alcohol will account for its extensive reactivity compared to the ether which has no —OH group.

† The student is encouraged to review Chapter 3 and the concepts of ionic and covalent bonding.

the polar solvent water. Water breaks the bond between the ions in the ionic inorganic crystal and hydrates the individual ions. It is also found that these hydrated ions conduct an electric current and behave as good electrolytes. On the other hand, most organic compounds are insoluble in a polar solvent like water but are quite soluble in nonpolar solvents like ether, benzene, and hydrocarbons. Since dissolution of an organic compound into an organic solvent does not produce ions, most solutions of organic compounds do not conduct an electric current and are classified as nonelectrolytes.*

THE ROLE OF CARBON IN ORGANIC CHEMISTRY

Since organic chemistry is defined as "the chemistry of carbon compounds," the obvious question is—why define and separate a branch of chemistry for one element, such as carbon, and classify the chemistry of the other hundred or so elements as another branch of chemistry, namely, inorganic chemistry?

The fact that carbon can form four covalent bonds cannot be the complete answer, since other Group IV elements, such as silicon and germanium, also possess this property. The unique character of carbon is its ability not only to bond with other carbon atoms but also at the same time to form strong bonds with other elements as well. One carbon nucleus can form a covalent bond with another nucleus, which in turn can covalently bond with still another carbon nucleus, and so on, so as to form chains of carbon atoms, as shown below for a four carbon atom chain.

$$\cdot\underset{\cdot}{\dot{C}}:\underset{\cdot}{\dot{C}}:\underset{\cdot}{\dot{C}}:\underset{\cdot}{\dot{C}}\cdot$$

Each carbon atom still contains unshared electrons which can form covalent bonds with other elements. If hydrogen nuclei form covalent bonds with these unshared electrons on carbon, the following compound will result.† In addition

$$\begin{array}{c}H\;H\;H\;H\\ H:\ddot{\underset{\cdot\cdot}{C}}:\ddot{\underset{\cdot\cdot}{C}}:\ddot{\underset{\cdot\cdot}{C}}:\ddot{\underset{\cdot\cdot}{C}}:H\\ H\;H\;H\;H\end{array} \quad \text{or} \quad \begin{array}{c}H\quad H\quad H\quad H\\ |\quad\;\; |\quad\;\; |\quad\;\; |\\ H—C—C—C—C—H\\ |\quad\;\; |\quad\;\; |\quad\;\; |\\ H\quad H\quad H\quad H\end{array} \quad \text{or} \quad CH_3CH_2CH_2CH_3$$

(A) *n*-Butane (B) *n*-Butane (C) *n*-Butane

to the ability to bond with itself to form carbon chains, carbon also has the unique ability (because of its small atomic radius and the strength of the carbon-carbon bond) to form multiple bonds with itself. If carbon were to share two electrons between each carbon atom, the following situation would result:

$$\cdot\underset{\cdot}{\dot{C}}\cdot + \cdot\underset{\cdot}{\dot{C}}\cdot \rightarrow \cdot\dot{C}::\dot{C}\cdot \equiv \cdot\dot{C}{=}\dot{C}\cdot$$

*Although an organic compound such as trimethyl amine, $(CH_3)_3\ddot{N}$, does not conduct an electric current, treatment of this amine with HCl produces a salt, $[(CH_3)_3\overset{+}{N}H]Cl^-$, which does conduct an electric current. Consequently, some organic compounds, which can be converted into ions by the appropriate acid or base reaction, can behave as conductors, but this is not the normal behavior of most organic compounds.

†Note that a covalent bond is conveniently represented by a dash (—) as in structure (B). In some cases, even the single bonds connecting carbon are omitted, such as in (C)—in structures of this type, single bonds are understood.

Now there are two covalent bonds between the two carbon atoms and only four valence electrons remain for additional bonding. If hydrogen atoms are used to complete the bonding capacity of carbon in this system, the molecule ethylene, C_2H_4, results. In ethylene, there is a double (or multiple) bond between

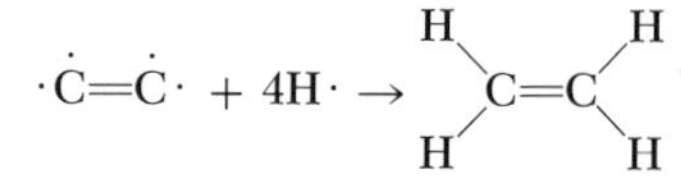

$$\cdot\dot{C}{=}\dot{C}\cdot + 4H\cdot \rightarrow H_2C{=}CH_2$$

Ethylene

the carbon atoms. If this same sort of process is used to share three electrons between carbon, and then to use hydrogen atoms to complete any unused bonding capacity, the molecule acetylene, C_2H_2, results, as follows:

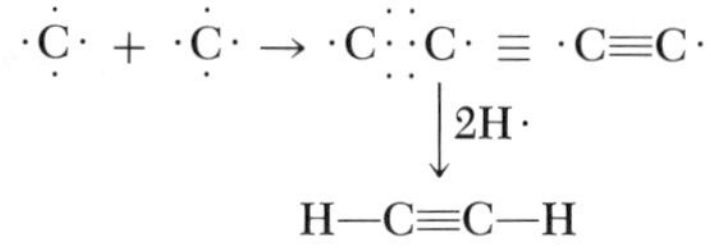

$$\cdot\dot{\underset{\cdot}{C}}\cdot + \cdot\dot{\underset{\cdot}{C}}\cdot \rightarrow \cdot C\vdots C\cdot \equiv \cdot C{\equiv}C\cdot \xrightarrow{2H\cdot} H{-}C{\equiv}C{-}H$$

Acetylene

Acetylene again contains a multiple bond between the carbon atoms, and in this case there is a triple bond between carbon atoms. Applying this same process further, the molecule C_2, C≣C, could be formed by sharing all four valence electrons between two carbon atoms. However, four bonds between two carbon atoms has not been observed, and only single, double, and triple bonded carbon atoms have been found in organic compounds.

The process used in the preceding paragraphs could be repeated again and again using additional carbon and hydrogen atoms to give even longer carbon chains. In addition, carbon can share electrons, not only with itself and with hydrogen, but with many other simple elements to form cyclic organic compounds as well as linear-chain compounds. Some examples of these various types of compounds are as follows:

$CH_3CH_2CH_3$ Propane

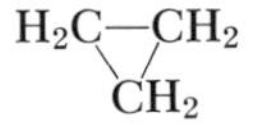

$H_2C{-}CH_2$ / CH_2 (ring) Cyclopropane

CCl_4 Carbon tetrachloride

$HCCl_3$ Chloroform

$H_3C{-}O{-}CH_3$ Dimethyl ether

$H_3C{-}NH_2$ Methyl amine

$H_3C{-}CH_2I$ Ethyl iodide

THE SHAPES OF ORGANIC MOLECULES

Ionic compounds, such as those commonly found among inorganic compounds, are held together by electrostatic forces between positive and negative ions. Electrostatic forces of this type, such as in Na^+Cl^-, are exerted symmetrically in all directions, and the ions can be thought of as a point charge, or a sphere of unit charge on which the charge is distributed equally over the surface of the sphere.

In contrast to the nondirectional nature of electrostatic forces, covalent bonds are directional in nature and give a definite shape to the molecule which depends on the type of covalent bond. In the simple examples methane, ethane, ethylene, and acetylene, which were considered in the previous section, different shapes and bond angles are found in each case. In methane, the carbon atom is considered to be at the center of a regular tetrahedron and the four bonds to hydrogen are directed to the corners of the tetrahedron (cf. Fig. 10–1). The molecule can be pictured as follows:

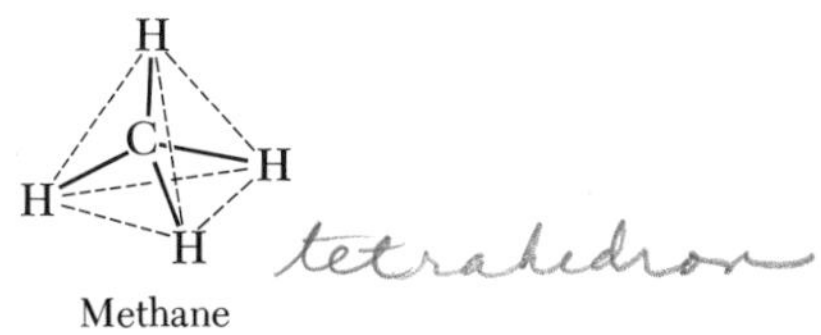

Methane

Other methods of defining chemical shapes and geometry, such as x-ray and electron-diffraction, have confirmed the regular tetrahedron shape of molecules such as CH_4 and CCl_4. The bond angles in molecules of this shape are 109.5°. When all the bonds to carbon are not identical, such as in chloroform, $CHCl_3$, the shape of the molecule is still tetrahedral. It is, however, no longer a regular tetrahedron, but a distorted tetrahedron with bond angles slightly different than 109.5°. (See Figures 10–2 and 10–3 for stick models and space models of ethane and *n*-butane.)

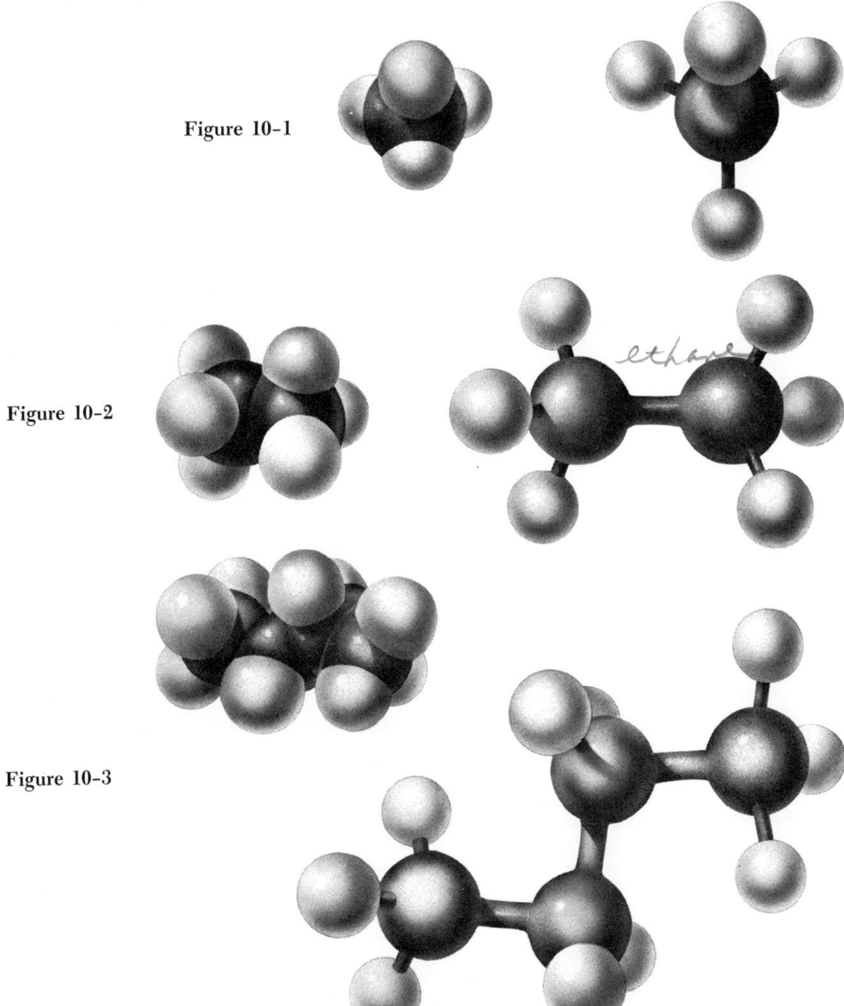

Figure 10–1

Figure 10–2

Figure 10–3

When carbon-carbon bonds are linked together in the formation of more complex molecules, such as ethane, propane, and so forth, the shape of the molecules is a series of tetrahedrons which share a common corner. The normal carbon-carbon single bond distance in molecules such as this is 1.54 Å.* Longer linear-chain molecules can be assembled by adding on additional tetrahedrons which share a common corner.

In a compound, such as ethylene, C_2H_4, the formation of the carbon-carbon double bond in the molecule imposes certain geometric requirements on the shape of the molecule. First, the introduction of the double bond limits rotation around the carbon-carbon bond. In compounds, such as ethane, which share a corner of a tetrahedron, there is free rotation† around the carbon-carbon single bond. However, introduction of the carbon-carbon double bond restricts any free rotation (360°), and for all practical purposes no rotation is allowed in this molecule unless the carbon-carbon double bond is broken. Secondly, a molecule containing a carbon-carbon double bond has been shown to be a coplanar molecule (all the carbon and hydrogen atoms in ethylene lie in the same plane) with bond angles of 120° and a carbon-carbon bond length of 1.34 Å. The bond angles and bond distances again may vary slightly depending upon what atoms are attached to carbon, but the gross overall features of the molecule will not change.

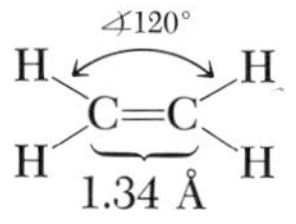

∢120°

H₂C=CH₂ (H and H on each carbon)

1.34 Å

Ethylene

In acetylene, C_2H_2, and other molecules containing a carbon-carbon triple bond, even greater deviations from the simple tetrahedral structures occur. X-ray and electron-diffraction methods have shown that compounds containing a —C≡C— linkage are linear molecules with a carbon-carbon bond length of 1.21 Å, as illustrated below for acetylene:

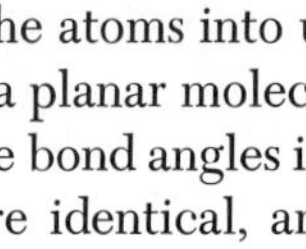

∢180°

H—C≡C—H

1.21 Å

Acetylene

In cyclic organic compounds some deviation from the normal bond angles illustrated above may be expected, as the constraining of the carbon atoms into rings of certain sizes will force the atoms into unusual and strained shapes. For example, cyclopropane must be a planar molecule, since three points (the three carbon atoms) define a plane. The bond angles in cyclopropane must necessarily be equal, since all the atoms are identical, and have been shown to be 60°.

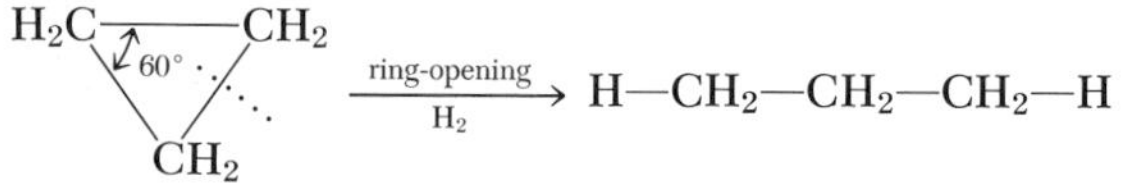

$$\text{Cyclopropane } (H_2C-CH_2-CH_2 \text{ ring}, \angle 60^\circ) \xrightarrow[H_2]{\text{ring-opening}} H-CH_2-CH_2-CH_2-H \text{ (Propane)}$$

Cyclopropane Propane

* 1 Å = 10^{-8} cm.

† In actuality there is a small energy barrier to rotation, since the atoms do occupy space and must pass each other on rotation. However, in most compounds containing carbon-carbon single bonds this barrier is very small.

Since the normal bond angle of a single carbon-carbon bond is 109.5° (cf. Fig. 10–1), to constrain or compress these bond angles from 109.5° to 60° is going to introduce lots of strain into the molecule. Consequently, we might expect cyclopropane and any other highly strained compound to be particularly susceptible to ring-opening reactions, since after ring-opening the bond-angles become approximately 109.5° again.

Experimentally, it is commonly found that the chemical susceptibility to ring-opening reactions does increase with increasing amount of ring strain, and cyclopropane does undergo ring-opening reactions with many chemical reagents. As the size of the ring increases, the bond angles increase, and the amount of strain decreases. Consequently, ring systems higher than cyclopropane are less prone to undergo ring-opening reactions.

Larger ring systems are not as common as five and six membered rings, but they have been found in many natural products, and in recent years synthetic methods have been developed in the laboratory for preparing many larger ring systems.

ORGANIC FUNCTIONAL GROUPS

Most of the compounds used as illustrations in this chapter thus far have contained only carbon and hydrogen atoms. Experimentally, compounds containing exclusively carbon and hydrogen make up only a small number of organic compounds. In addition, the chemical reactivity of compounds containing only carbon and hydrogen is much less in most cases than compounds containing atoms such as oxygen, nitrogen, halogen, and sulfur, as well as carbon and hydrogen.

When the chemistry of organic compounds is considered in detail in the following chapters, it will become apparent that only certain bonds and organic groups participate in the chemical reaction, and that most of the carbon chain structure of the molecule remains unchanged in going from reactants to products. The following reaction of *n*-butyl alcohol is illustrative:

$$\underset{n\text{-Butyl alcohol}}{CH_3CH_2CH_2CH_2OH} + \underset{\text{Hydrogen bromide}}{HBr} \rightarrow \underset{n\text{-Butyl bromide}}{CH_3CH_2CH_2CH_2Br} + \underset{\text{Water}}{HOH}$$

In this reaction the —OH group of the alcohol is lost, and its place on the carbon chain is taken by the bromine atom of the hydrogen bromide. In turn, the —OH lost by the alcohol combines with the hydrogen of the hydrogen bromide to produce water as the other product of this reaction.* The important point to consider here is that only a small portion of the organic molecule undergoes change. The bond between the carbon atom and the —OH group is broken, and a new bond between carbon and bromine is formed. Other than these simple changes, all the remaining bonds (all carbon and hydrogen in this case) in the organic compound remain unchanged. This is a characteristic

*The student should be aware that this explanation is very simplified, and that the details in converting the alcohol to the bromide are more involved than merely exchanging groups or atoms. This will become apparent when we look at a mechanism for this type of reaction.

feature of organic molecules, that only certain atoms or groups of atoms in an organic molecule determine the chemistry of the class of compounds containing that particular atom or group of atoms. *The atom or group of atoms that defines the structure of a particular class of organic compounds and determines its properties is called the* ***functional group.*** In the particular example above, the functional group in the alcohol is the —OH group, and the functional group in the product is the halogen atom, Br.

A large portion of organic chemistry is concerned with the transformation of one functional group into another. A basic understanding of organic chemistry is a mastery of the properties and reactivities of each type of functional group, and how one functional group can be transformed into another functional group of different properties and reactivity. In later chapters, using simple molecules, one shall learn to associate a particular set of properties with a particular functional group. When encountering a more complicated molecule, one may expect the properties of this molecule to roughly approximate the properties of the various functional groups contained in the molecule. For example, by understanding the simple properties of an alcohol like *n*-butyl alcohol, one can then extrapolate this knowledge to a more complicated alcohol, such as cholesterol, which contains an alcohol functional group and undergoes many of the same chemical reactions as *n*-butyl alcohol.*

A summary of the typical functional groups to be taken up in later chapters is presented in the accompanying table with representative examples. Although it is not assumed that the student understands anything about the chemistry

*It should be pointed out that the properties of a complicated molecule containing several different functional groups may be modified relative to a simple monofunctional compound, and a complete extrapolation of chemical behavior from simple to complicated compounds should not be expected. However, in most cases this simple extension from simple to complicated molecules using functional group chemistry yields surprisingly excellent results.

TABLE 10-1 SIMPLE FUNCTIONAL CLASSES OF ORGANIC COMPOUNDS

Illustrative Example	*Name of Functional Class*	*Functional Group*
$CH_3CH_2CH_3$	Alkanes	—
$CH_3CH{=}CH_2$	Alkenes	$C{=}C$
$CH_3C{\equiv}CH$	Alkynes	$C{\equiv}C$
$CH_3CH_2CH_2OH$	Alcohols	$-OH$
$CH_3CH_2\overset{H}{\overset{\vert}{C}}{=}O$	Aldehydes	$-\overset{H}{\overset{\vert}{C}}{=}O$
$CH_3\overset{O}{\overset{\Vert}{C}}CH_3$	Ketones	$-\overset{O}{\overset{\Vert}{C}}-$
$CH_3CH_2\overset{O}{\overset{\Vert}{C}}OH$	Acids	$-\overset{O}{\overset{\Vert}{C}}-OH$
$CH_3CH_2CH_2NH_2$	Amines	$-NH_2$
$CH_3OCH_2CH_3$	Ethers	$-C-O-C-$
$CH_3CH_2CH_2Br$	Halides	$-Br$

of these functional groups, it is extremely helpful at this stage to learn the names of some of the simple functional classes, and especially to learn to associate a particular atom or group of atoms with a particular functional class.

REACTION MECHANISMS AND REACTION INTERMEDIATES

To really understand an organic chemical reaction, it is important to know not only what happens (to know the reactant and products) but also how it happens (to know something about the intermediates formed, how they are formed, and how they recombine to form the products). The answer to this question—how does the reaction occur?—is called a reaction mechanism. It is a stepwise description which is advanced to account for the facts. In most cases it is difficult to prove a mechanism outright, since it is very difficult to detect or trap many of the transient intermediates or compounds that may make up some of the steps which lead to the final product. Nevertheless, by investigating many reactions using various chemical, spectroscopic, and stereochemical techniques, the organic chemist has learned that a few simple chemical intermediates can be used to account for many of the observed facts of organic chemistry. Also, by using these intermediates to explain chemical reactions, it becomes apparent that many unrelated reactions appear to proceed by the same or a similar mechanistic pathway; so that by having some knowledge of the more common mechanisms of organic chemistry the student can relate and correlate many facts that appear on the surface to be totally unrelated. Obviously, if one has a knowledge of basic mechanisms, this will facilitate his understanding of organic chemistry; it is much better than trying to merely memorize all the facts.

Three fundamental types of reaction intermediates—carbonium ions, carbanions, and free radicals—can be used to explain many of the more common organic reactions. These intermediates are trivalent carbon atoms bearing a positive, negative, or zero charge, respectively. Examples of each type are given below:

$CH_3CH_2CH_2\overset{+}{C}H_2$	$CH_3CH_2CH_2\overset{-}{C}H_2$	$CH_3CH_2CH_2\dot{C}H_2$
Carbonium ion	Carbanion	Free radical

The monochlorination of methane will illustrate the detailed steps involved in a reaction mechanism.

$$:\ddot{\underset{..}{Cl}}:\ddot{\underset{..}{Cl}}: \xrightarrow{h\nu} 2:\ddot{\underset{..}{Cl}}\cdot$$

Chlorine Chlorine atoms

$$:\ddot{\underset{..}{Cl}}\cdot + CH_4 \rightarrow H:\ddot{\underset{..}{Cl}}: + \cdot CH_3$$

$$:\ddot{\underset{..}{Cl}}:\ddot{\underset{..}{Cl}}: + \cdot CH_3 \rightarrow H_3C:Cl + \cdot\ddot{\underset{..}{Cl}}:$$

In the first step, a chlorine molecule is irradiated ($h\nu$ is the symbol used for photochemical or radiant energy) by light of the appropriate wavelength, which dissociates* some of the chlorine molecules into chlorine atoms (free radicals). The chlorine atom then abstracts a hydrogen atom from the methane to give hydrogen chloride and a methyl free radical ($\cdot CH_3$). The methyl free radical then abstracts a chlorine atom from a chlorine molecule to give methyl chloride and regenerates a chlorine atom, which can go back and repeat this cycle.† The overall reaction then can be summarized in the following equation:

$$\underset{\text{Methane}}{CH_4} + Cl_2 \xrightarrow{h\nu} \underset{\text{Methyl chloride}}{CH_3Cl} + HCl$$

The preceding example illustrates that much more detail about a chemical reaction can be learned by investigating its reaction mechanism, rather than merely writing an equation which represents the reactants and products. Of course, it is not always easy to decide experimentally whether a reaction is proceeding through a carbonium ion, a carbanion, or a free-radical type of intermediate. In later chapters, the current reaction mechanisms known for some of the functional group reactions will be employed as an aid to understanding the fundamental concepts of an organic reaction. The criterion which is used to arrive at a particular mechanism will generally not be included, as this type of material is beyond the scope of this book. However, in most cases good theoretical and experimental reasoning is available for writing a particular mechanism for a reaction, and the student is referred to more advanced texts for more mechanistic detail.

---

* Only a small percentage of the chlorine molecules are actually dissociated by this irradiation process, as recombination of chlorine atoms can occur to regenerate the chlorine molecule.

† Most free-radical reactions proceed via a cyclic process. Consequently, in many of these types of reaction, only a small amount of an initiator (compound which dissociates easily to free radicals) is required to get the reaction going.

Questions

1. Which of the following compounds are isomers?

(a) $CH_3CH_2CH_2OH$
(b) $CH_3CHClCH_3$
(c) $CH_3CH_2CH_3$
(d) $CH_3\overset{O}{\overset{\|}{C}}CH_2CH_3$
(e) $CH_3CH_2CH_2Cl$

2. Which of the following compounds are identical?

```
     H  H  H  H              H   H   H                 H  H  H  Cl
     |  |  |  |              |   |   |                 |  |  |  |
(a) H—C—C—C—C—H      (d) H—C———C———C—H          (g) H—C—C—C—C—H
     |  |  |  |              |   |   |                 |  |  |  |
     H  H  H  Cl             H  HCH  Cl                H  H  H  H
                                 H

     H  H  Cl H              H  H  H  H
     |  |  |  |              |  |  |  |
(b) H—C—C—C—C—H      (e) Cl—C—C—C—C—H
     |  |  |  |              |  |  |  |
     H  H  H  H              H  H  H  H

     H  H  H  H              H  H  H  H  H
     |  |  |  |              |  |  |  |  |
(c) H—C—C—C—C—H      (f) H—C—C—C—C—C—Cl
     |  |  |  |              |  |  |  |  |
     H  Cl H  H              H  H  H  H  H
```

3. Draw all the possible structures for compounds having the following molecular formula.

(a) C_3H_8 (c) C_3H_4
(b) C_4H_8 (d) C_3H_5Cl

4. Classify each of the following compounds with respect to the shape of the molecule.

(a) C_2Cl_4 (d) CH_2Cl_2
(b) CBr_4 (e) C_2F_6
(c) C_2Cl_2

5. In each of the following compounds, pick out the functional group and classify each compound into a functional class.

(b) $(CH_3)_2CHCH_2C(CH_3)_3$

(c) $CH_3CH_2OCH_2CH_3$

(d) $CH_3CH{=}CHCH_3$

(e) $CH_3CH_2NH_2$

(f) $CH_3CH_2CH_2CH(OH)CH_3$

(g) cyclopropyl–C(=O)–OH

(h) 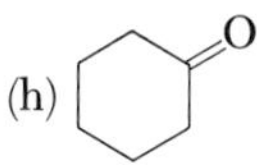

(i) $CH_3CHBrCH_3$

(j) $CH_3C{\equiv}CCH_3$

Suggested Reading

Caserio: Reaction Mechanisms In Organic Chemistry. Journal of Chemical Education, Vol. 42, pp. 570 and 627, 1965.

DeLaMare and Vaughan: Detection and Reactions of Free Alkyl Radicals. Journal of Chemical Education, Vol. 34, p. 10, 1957.

Kurzer and Sanderson: Urea In the History of Organic Chemistry. Journal of Chemical Education, Vol. 33, p. 452, 1956.

Stewart: The Reactive Intermediates of Organic Chemistry. Journal of Chemical Education, Vol. 38, p. 308, 1961.

CHAPTER 11

HYDROCARBONS

The simplest types of organic compounds are those which are composed only of the elements carbon and hydrogen, and compounds containing only these two elements are referred to as **hydrocarbons.** Replacement of a carbon-hydrogen bond by a functional group, such as those given in Chapter 10, gives rise to the various classes of organic compounds, and these functional classes may be thought of as derivatives of hydrocarbons.

Although hydrocarbons contain only two elements, these elements may be combined in several different ways. For example, in the following compounds the carbon atoms may be linked together to form a linear chain or a cyclic

$CH_3CH_2CH_2CH_2CH_2CH_3$
n-Hexane
C_6H_{14}

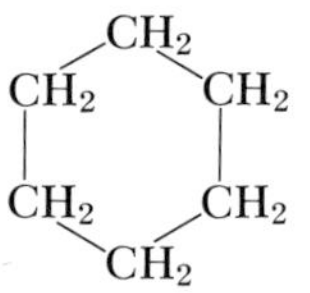

Cyclohexane (ring of six CH_2 groups)
C_6H_{12}

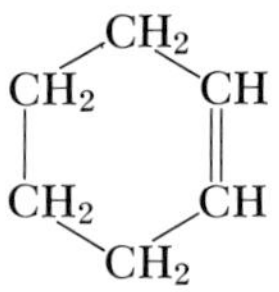

Cyclohexene (ring of four CH_2 groups and two CH groups joined by a double bond, $CH=CH$)
C_6H_{10}

$CH_3CH_2CH_2CH_2CH{=}CH_2$
1-Hexene
C_6H_{12}

$CH_3C{\equiv}CH$
Propyne
C_3H_4

ring. Also, the molecule may contain only carbon-carbon single bonds or may contain carbon-carbon multiple bonds. Even within a similar type of compound, such as *n*-hexane and cyclohexane which do not contain any multiple bonds, the ratio of carbon to hydrogen is not constant, and hence the molecular formula is not an indication of the type of hydrocarbon structure. For example, the molecular formula C_6H_{12} as shown above could refer to either cyclohexane or 1-hexene. The chemistry of these two compounds is quite different even though they have the same ratio of carbon to hydrogen atoms.

TABLE 11-1 HYDROCARBONS

Class	*Distinguishing Feature*	*Simplest Example*
Alkanes	—C—C— (with single bonds)	CH_4, Methane
Alkenes	>C=C<	C_2H_4, Ethylene
Alkynes	—C≡C—	C_2H_2, Acetylene

Consequently, within the class of compounds known as hydrocarbons there are different degrees and types of chemical reactivity, and in order to classify the properties and chemical reactions of hydrocarbons, it is convenient to divide hydrocarbons into several subclasses. The basis for classification is the number of covalent bonds formed between the carbon atoms in the compounds. If only carbon-carbon single bonds are involved in the compound, the class is known as **alkanes,** or **saturated hydrocarbons.** The term "saturated" means that only one pair of electrons is shared covalently between any two bonded atoms in the molecule. Therefore, *n*-hexane and cyclohexane are alkanes by this definition. If the molecule contains multiple carbon-carbon bonds (more than one pair of electrons is shared convalently between any two bonded atoms), the compounds are classified on the basis of the number of multiple bonds between any two bonded atoms in the molecules. For example, cyclohexene and 1-hexene both contain a carbon-carbon double bond and would therefore be classed in the same category. However, propyne contains a carbon-carbon triple bond and would not be classed with cyclohexene or 1-hexene. Compounds containing carbon-carbon double bonds are known as **alkenes,** and compounds containing carbon-carbon triple bonds are known as **alkynes.** Hence, cyclohexene and 1-hexene are alkenes and acetylene is an alkyne. These classifications of hydrocarbons are summarized in Table 11-1.

Although only simple examples of the classes of hydrocarbons are illustrated, it should be obvious that more complex hydrocarbons are possible either by introducing more carbon atoms or by introducing more than one alkene or alkyne linkage in the molecule. For example, 1,4-pentadiene contains two carbon-carbon double bonds per molecule, and diacetylene contains two car-

$CH_2{=}CHCH_2CH{=}CH_2$	$HC{\equiv}C{-}C{\equiv}CH$	$CH_2{=}CH{-}C{\equiv}CH$
1,4-Pentadiene	Diacetylene	Vinyl acetylene

bon-carbon triple bonds per molecule. These compounds, however, behave chemically similar to simple alkenes and alkynes, except that instead of one mole of reagent, they require two moles of reagents in a chemical reaction. On the basis of our previous definition, they are also classified as alkenes and alkynes, respectively. It is not the number of multiple centers in the molecule which determines its classification, but the *number of multiple bonds between any two bonded atoms.* In some special cases the molecule may contain two different types of multiple bonds, as illustrated in vinyl acetylene. This compound can behave as either an alkene or an alkyne, and its classification will depend on the type of reaction. Complex cases of this type are beyond the scope of the present treatment and will not be considered here, but the student should recognize that such compounds do exist and that in many cases the behavior of the compound can be predicted by reference to the chemistry of simple alkenes and alkynes.

ALKANES

The simplest known compound containing only carbon and hydrogen is methane. Higher members of this series can be developed by replacing one or more of the carbon-hydrogen bonds by one or more carbon atoms and by completing the valence requirements of the added carbon atom with hydrogen atoms. Several examples of this stepwise development of a carbon chain are illustrated below:

$$\begin{array}{c} \text{H} \\ | \\ \text{H—C—H} \\ | \\ \text{H} \\ \text{Methane} \end{array} \xrightarrow[\text{of 1 carbon atom}]{\text{addition}} \begin{array}{c} \text{H H} \\ |\ \ | \\ \text{H—C—C—H} \\ |\ \ | \\ \text{H H} \\ \text{Ethane} \end{array} \text{ or } CH_3CH_3 \text{ or } C_2H_6$$

$$\begin{array}{c} \text{H} \\ | \\ \text{H—C—H} \\ | \\ \text{H} \\ \text{Methane} \end{array} \xrightarrow[\text{of 2 carbon atoms}]{\text{addition}} \begin{array}{c} \text{H H H} \\ |\ \ |\ \ | \\ \text{H—C—C—C—H} \\ |\ \ |\ \ | \\ \text{H H H} \\ \text{Propane} \end{array} \text{ or } CH_3CH_2CH_3 \text{ or } C_3H_8$$

$$\begin{array}{c} \text{H} \\ | \\ \text{H—C—H} \\ | \\ \text{H} \\ \text{Methane} \end{array} \xrightarrow[\text{of 3 carbon atoms}]{\text{addition}} \begin{array}{c} \text{H H H H} \\ |\ \ |\ \ |\ \ | \\ \text{H—C—C—C—C—H} \\ |\ \ |\ \ |\ \ | \\ \text{H H H H} \\ \text{Normal butane} \end{array} \text{ or } CH_3CH_2CH_2CH_3 \text{ or } C_4H_{10}$$

Another and easier way to view this process of building a carbon chain structure is that the propane carbon chain is developed from ethane by replacing a carbon-hydrogen bond of ethane by a $—CH_3$ group. Similarly, normal[*] butane is developed from propane by replacement of the terminal carbon-hydrogen bond with a $—CH_3$ group. Higher members of this series are built up in an analogous manner as shown below:

$$\begin{array}{c} \text{H H H H} \\ |\ \ |\ \ |\ \ | \\ \text{H—C—C—C—C—H} \\ |\ \ |\ \ |\ \ | \\ \text{H H H H} \\ \text{Normal butane} \end{array} \xrightarrow[\text{of C—H by —CH}_3]{\text{replacement}} \begin{array}{c} \text{H H H H H} \\ |\ \ |\ \ |\ \ |\ \ | \\ \text{H—C—C—C—C—C—H} \\ |\ \ |\ \ |\ \ |\ \ | \\ \text{H H H H H} \\ \text{Normal pentane} \end{array}$$

In the compounds illustrated above, the molecular formulas conform to the general formula C_nH_{2n+2}, in which n is the number of carbon atoms in the molecule. This general formula fits all linear (straight chain) alkanes and can be used to predict the molecular formula of any linear alkane. For example, an alkane containing seven carbon atoms (heptane) would have a molecular

[*] **Normal** refers to the fact that all of the carbon atoms in the chain are arranged in a linear manner. This is in contrast to other isomers having the same molecular formula, but in which the carbon atoms are arranged in other than a linear chain.

formula C_7H_{16}; one with eight carbons (octane), C_8H_{18}; one with nine carbons (nonane), C_9H_{20}; one with ten carbons (decane), $C_{10}H_{22}$, and so on. Careful inspection of each of these structures shows that propane differs from butane by a $—CH_2$ unit; pentane differs from hexane by a $—CH_2$ unit; hexane differs from heptane by a $—CH_2$ unit, and so on. This $—CH_2$ unit is known as a methylene group. Each higher member in this series differs from the next lower or higher one by one more or one less methylene unit. A series of compounds of this type in which each member differs from the next higher or lower member by a constant increment is known as a **homologous series.** In general, members of a homologous series have closely related physical and chemical properties.

NOMENCLATURE

The nomenclature of the simple straight chain alkanes is straightforward. The first four members have common names (methane, ethane, propane, and butane), but the root names of the higher members are derived from the number of carbon atoms in the chain, and the -ane ending is added to the root name. Therefore, the names hexane, heptane, octane, and so forth are used for alkanes containing six, seven, and eight carbons in the chain.

In the preceding section normal butane was developed by replacing a terminal carbon-hydrogen bond in propane by a $—CH_3$ group. Since both ends of the propane molecule are equivalent, it doesn't matter what terminal C—H

$$\underset{\text{Propane}}{H{-}\overset{H}{\underset{H}{C}}{-}\overset{H}{\underset{H}{C}}{-}\overset{H}{\underset{H}{C}}{-}H} \xrightarrow[\text{group}]{+CH_3} \underset{\text{Normal butane}}{H{-}\overset{H}{\underset{H}{C}}{-}\overset{H}{\underset{H}{C}}{-}\overset{H}{\underset{H}{C}}{-}\overset{H}{\underset{H}{C}}{-}H} \xleftarrow{+CH_3} \underset{\text{Propane}}{H{-}\overset{H}{\underset{H}{C}}{-}\overset{H}{\underset{H}{C}}{-}\overset{H}{\underset{H}{C}}{-}H}$$

bond is replaced by the $—CH_3$ group. However, another possibility still exists for developing the carbon chain by this process. If the carbon-hydrogen bond of the $—CH_2—$ group in propane is replaced by a $—CH_3$ group, an isomer of normal butane is formed, namely, isobutane.

$$H{-}\overset{H}{\underset{H}{C}}{-}\overset{H}{\underset{H}{C}}{-}\overset{H}{\underset{H}{C}}{-}H \xrightarrow[\text{group}]{+CH_3} \underset{\text{Isobutane}}{H{-}\overset{H}{\underset{H}{C}}{-}\overset{H{-}\overset{H}{C}{-}H}{\underset{H}{C}}{-}\overset{H}{\underset{H}{C}}{-}H} \text{ or } CH_3\overset{CH_3}{C}HCH_3 \text{ or } C_4H_{10}$$

In the case of propane, only two isomers of butane can be developed by replacing carbon-hydrogen bonds by a $—CH_3$ group. With higher members of this series, many isomers are possible by replacement of the different C—H bonds. For example, in the octane series, C_8H_{18}, eighteen possible structural isomers are possible. Of course, when the number of isomers of a particular carbon chain structure becomes high, the naming of these isomers becomes difficult, and an unambiguous nomenclature is necessary to avoid confusion.

The naming system selected should be simple, but must provide a name that fits one and only one structure. The International Union of Pure and Applied Chemistry (IUPAC) has recommended a system which is used universally by organic chemists to name organic compounds. The rules of this system are as follows:

1. The characteristic ending -ane is applied to the root stem name to obtain the name of a linear saturated hydrocarbon.
2. For branched chain alkanes, the compound is named as a derivative of the hydrocarbon corresponding to the longest continuous carbon chain in the molecule.
3. Substituents (atoms or groups of atoms) are indicated by a suitable prefix and a number to indicate their position on the carbon chain.
4. Numbering of the longest continuous carbon chain (Rule 2) must be done in such a way that the numbers giving the position of the substituents are kept as low as possible.

For example, the following compound can be named in two different ways:

$$\overset{1}{C}H_3\overset{2}{C}H(CH_3)\overset{3}{C}H_2\overset{4}{C}H_2\overset{5}{C}H_3 \quad \text{or} \quad \overset{5}{C}H_3\overset{4}{C}H(CH_3)\overset{3}{C}H_2\overset{2}{C}H_2\overset{1}{C}H_3$$

(A) 2-Methylpentane (B) 4-Methylpentane

In both structures, (A) and (B), the longest continuous carbon chain contains five carbons; hence, the compound will be named as a derivative of pentane. The $—CH_3$ group (methyl group) is attached at position 2 in structure (A) or at position (4) in structure (B). Rule 4 demands that the lowest number be used in numbering substituents; hence, 2-methylpentane is the correct name for this compound—not 4-methylpentane.

When more than one substituent is present on the carbon chain, each substituent is given a number. For example, structure (C) is correctly named as 2,2-dimethylbutane—not 2-dimethylbutane. Even though the substituents are identical, each must be given a number, and the number of identical substituents

$$CH_3C(CH_3)(CH_3)CH_2CH_3 \quad \text{or} \quad CH_3C(CH_3)_2CH_2CH_3$$

(C) 2,2-dimethylbutane

is also indicated by the appropriate prefix. The following examples further illustrate this nomenclature method*:

$$CH_3CH_2CH(CH_3)CH_2CH_2CH(CH_3)CH_3 \quad \text{2,5-Dimethylheptane}$$

* It should be evident that by counting the number of carbon atoms indicated in the name given to the longest continuous chain and adding to this the number of carbons indicated in the substituent names, the total number of carbons in the compound is obtained. This is a good way to check that no carbons have been omitted in naming the compound.

$$\begin{array}{c} \quad CH_2CH_3 \\ | \\ CH_3CHCHCHCH_2CH_3 \\ |\quad\; | \\ CH_3\;\; CH_3 \end{array}$$ 2,4-Dimethyl-3-ethylhexane

In these previous examples, the substituents (other than hydrogen atoms) were designated as methyl or ethyl. The names of these substituents were derived from the alkane containing the same number of carbon atoms by changing the **-ane** ending to **-yl.** These groups are derived from the parent alkane by removing one of the hydrogen atoms and are known as **alkyl** groups (from alk**ane** → alk**yl**). Therefore, methyl and ethyl groups can be formulated from methane and ethane as follows:

$$H{-}\overset{H}{\underset{H}{C}}{-}H \rightarrow H{-}\overset{H}{\underset{H}{C}}{-} \quad \text{or} \quad H_3C{-}$$

Methane Methyl group

$$H{-}\overset{H}{\underset{H}{C}}{-}\overset{H}{\underset{H}{C}}{-}H \rightarrow H{-}\overset{H}{\underset{H}{C}}{-}\overset{H}{\underset{H}{C}}{-} \quad \text{or} \quad CH_3CH_2{-} \quad \text{or} \quad C_2H_5{-}$$

Ethane Ethyl group

In compounds containing more than two carbon atoms, the possible number of alkyl groups will depend on the number of different types of C—H bonds in the molecule. From propane, for example, two different alkyl groups are possible:

(1)

$$H{-}\overset{H}{\underset{H}{C}}{-}\overset{H}{\underset{H}{C}}{-}\overset{H}{\underset{H}{C}}{-}H \rightarrow H{-}\overset{H}{\underset{H}{C}}{-}\overset{H}{\underset{H}{C}}{-}\overset{H}{\underset{H}{C}}{-} \quad \text{or} \quad CH_3CH_2CH_2{-} \quad \text{or} \quad n{-}C_3H_7{-}$$

Propane Normal propyl (*n*-propyl)

(2)

$$H{-}\overset{H}{\underset{H}{C}}{-}\overset{H}{\underset{H}{C}}{-}\overset{H}{\underset{H}{C}}{-}H \rightarrow H{-}\overset{H}{\underset{H}{C}}{-}\overset{H}{\underset{|}{C}}{-}\overset{H}{\underset{H}{C}}{-}H \quad \text{or} \quad CH_3\underset{|}{C}HCH_3 \quad \text{or} \quad \text{iso}{-}C_3H_7{-}$$

2-Propyl or isopropyl

From higher homologues, even more possibilities can be formulated. In general, naming of alkyl groups by this method is only reasonable for groups-containing a small number of carbons. This nomenclature is usually employed for compounds containing 1 to 4 carbon atoms. Table 11–2 summarizes some of the more important alkyl groups.*

Cycloalkanes are analogous to straight chain alkanes, except that the ends

* Use of alkyl groups in naming compounds provides another method of nomenclature which has much current usage for simple compounds, and the student should become acquainted with both the IUPAC and alkyl group systems, although it should be kept in mind that the IUPAC is the correct systematic method.

TABLE 11-2 ALKYL GROUPS

Alkyl Group	*IUPAC Name*	*Common Name*
$CH_3—$	Methyl	Methyl
$CH_3CH_2—$	Ethyl	Ethyl
$CH_3CH_2CH_2—$	Propyl	*n*-Propyl
$CH_3\overset{}{C}HCH_3$ (with bond down from central C)	Methylethyl	Isopropyl*
$CH_3CH_2CH_2CH_2—$	Butyl	*n*-Butyl
$CH_3CH_2CHCH_3$ (with bond down from the CH)	1-Methylpropyl	*s*-Butyl**
$CH_3CHCH_2—$ with CH_3 on the CH	2-Methylpropyl	Isobutyl*
CH_3CCH_3 with bond up and CH_3 down on the central C	Dimethylethyl	*t*-Butyl**
$CH_3CH_2CH_2CH_2CH_2—$	Pentyl	*n*-Pentyl (Amyl)

* iso- refers to any structure having a terminal CH_3CHCH_3 (bond up from the CH) grouping.

** *s*- and *t*- refer to secondary and tertiary. In this system of nomenclature, the carbon having the unsatisfied valence is the focal point. Whether this carbon is referred to as primary, secondary, or tertiary depends upon whether it is attached to 1, 2, or 3 additional carbon atoms.

of the carbon chain are joined together in a ring. This process of linking the ends of the chain into a ring requires the use of one additional valence from each terminal carbon atom. Consequently, two less C—H bonds are formed, and the general formula for these compounds is C_nH_{2n}, where n equals the number of carbon atoms. The parent name of the cyclic hydrocarbon is derived by adding the prefix **cyclo-** to the linear alkane having the same number of carbon atoms as shown below:

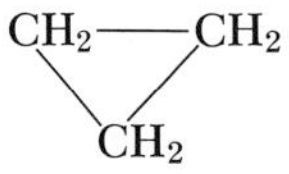

Cyclopropane	Cyclobutane	Cyclopentane	Cyclohexane	Cycloheptane
C_3H_6	C_4H_8	C_5H_{10}	C_6H_{12}	C_7H_{14}

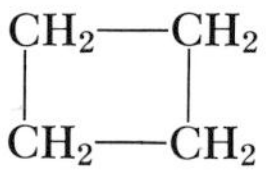

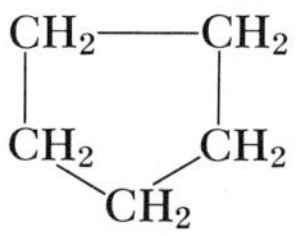

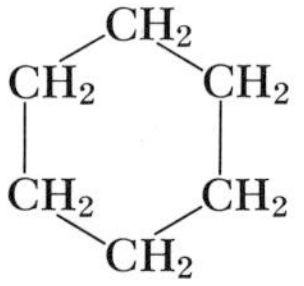

(Each drawn as a ring of CH_2 groups.)

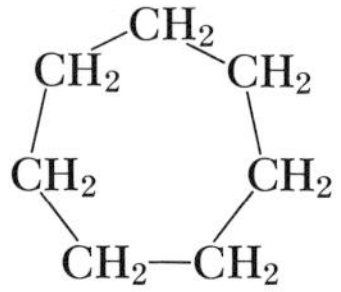

For substituted cycloalkanes, the IUPAC rules again require that each substituent be given the lowest possible number, and that each substituent be named and indicated by the appropriate prefix. Some further examples are illustrated below:

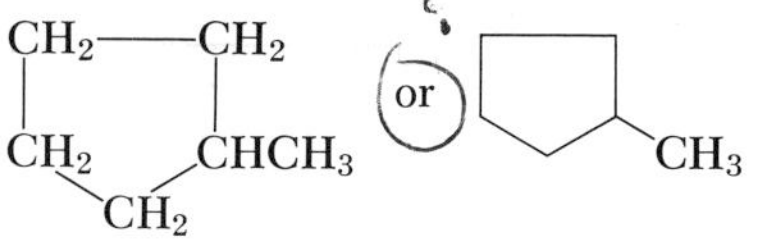

$CH_2—CH_2$, CH_2, $CHCH_3$, CH_2 ring or pentagon with CH_3 — 1-Methylcyclopentane

$$\begin{array}{c} CH_2 \\ CH_2 \quad C(CH_3)_2 \\ | \qquad\quad | \\ CH_2 \quad CH_2 \\ CH \\ | \\ CH_3 \end{array} \quad \text{or} \quad \text{(cyclohexane ring with two } CH_3 \text{ on C-1 and one } CH_3 \text{ on C-3)}$$

1,1,3-Trimethylcyclohexane—not
1,5,5-trimethylcyclohexane

SOURCES OF HYDROCARBONS

Natural gas and petroleum are the most important natural sources of hydrocarbons. Large deposits of these substances have been formed over the years by the gradual decomposition of marine life and other biological materials. These deposits usually accumulate under a dome-shaped layer of rock several thousand feet under the earth's surface (Fig. 11–1). When a hole is drilled through the rock layer, the pressure under the dome forces the gas or oil to the surface. After the pressure is released, pumps are required to bring the remaining oil to the surface.

Natural Gas. Natural gas is an excellent source for low molecular weight alkanes. Natural gas occurs in most parts of the United States, but most of it is produced in the southwest. In recent years a vast network of pipelines has been installed to carry natural gas from Texas to other parts of the United States.

The typical composition of natural gas is shown in Table 11–3. The propane and butane are removed by liquefaction before the gaseous fuel is introduced into the pipelines for distribution. The liquid propane and butane are stored under pressure in steel cylinders from which they are released as a gaseous fuel to be used in rural areas and in locations that are not supplied by natural gas mains. Large quantities of **carbon black,** also called gas black or lamp black,

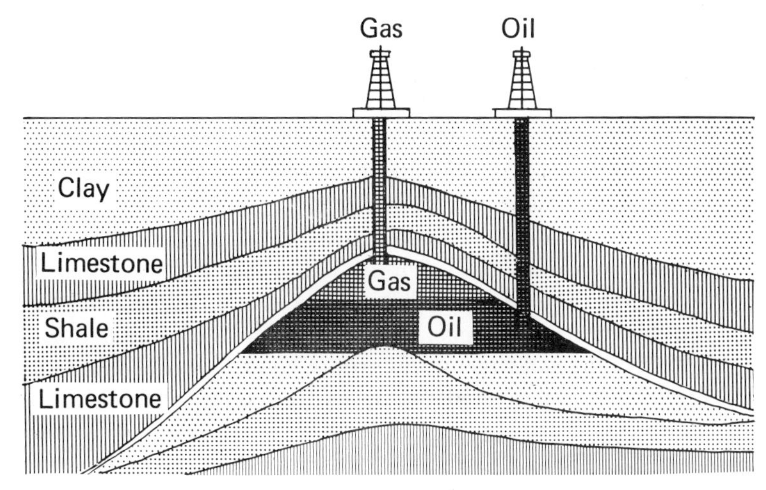

Figure 11–1 An illustration of an oil dome, or anticline, showing deposits of petroleum and natural gas.

TABLE 11-3 COMPOSITION OF NATURAL GAS

Hydrocarbon	*Typical Composition (per cent)*
Methane	82
Ethane	10
Propane	4
Butane	2
Higher hydrocarbons	2

are produced from natural gas. A large proportion of the carbon black produced by this process is used in the manufacture of rubber tires. Natural gas is also used for the production of hydrogen and for the preparation of many organic compounds, including alcohols and acids.

PETROLEUM. The crude oil, or petroleum, obtained from oil wells is another rich source of hydrocarbons. The hydrocarbons in petroleum, in contrast to those of natural gas, are of higher molecular weight. Petroleum has been known for several centuries and has been used for many purposes, most particularly as a fuel. It was not until recent years, however, that petroleum was separated into its hydrocarbon components. It is a very complex mixture of hydrocarbons, and its composition varies with the location of the oil field from which it is obtained. It contains mainly a mixture of alkanes, cycloalkanes, and aromatic hydrocarbons. In addition to the hydrocarbons, petroleum contains about 10 per cent by weight of sulfur, nitrogen, and oxygen compounds.

Petroleum is separated into its hydrocarbon fractions by the process of distillation. Since the forces of attraction between the individual hydrocarbon molecules are small, they may be converted into the gaseous state without decomposition at a fairly low temperature (boiling point). When the hydrocarbon vapors are cooled, they condense to a liquid form. This process is called distillation and can be used to separate hydrocarbons, because the forces of attraction between the molecules of one compound differ from those between other molecules and result in different distillation temperatures. Fractional distillation is commonly used in the petroleum industry. In this process, petroleum is separated by distillation into several fractions possessing different distilling temperatures. By the use of a fractionating column, more efficient continuous fractional distillation of petroleum can be achieved. A fractionating column consists of a tall column containing perforated plates or irregular-shaped glass or ceramic pieces designed to promote intimate contact between the distilling vapors and the refluxing liquid that condenses and runs back down the column. The effect of such a column is to concentrate the lower-boiling constituents in the vapors as they rise, and to enrich the reflux with the higher-boiling constituents. By proper construction and operation, the various petroleum fractions can be removed from different levels in the fractionating column (Fig. 11–2). In the petroleum industry, these distillation procedures are called the refining process. The distillation fractions from a typical petroleum are shown in Table 11–4.

Gases from Petroleum. The first products given off during the distillation of petroleum are the gaseous hydrocarbons containing from 1 to 5 carbon atoms.

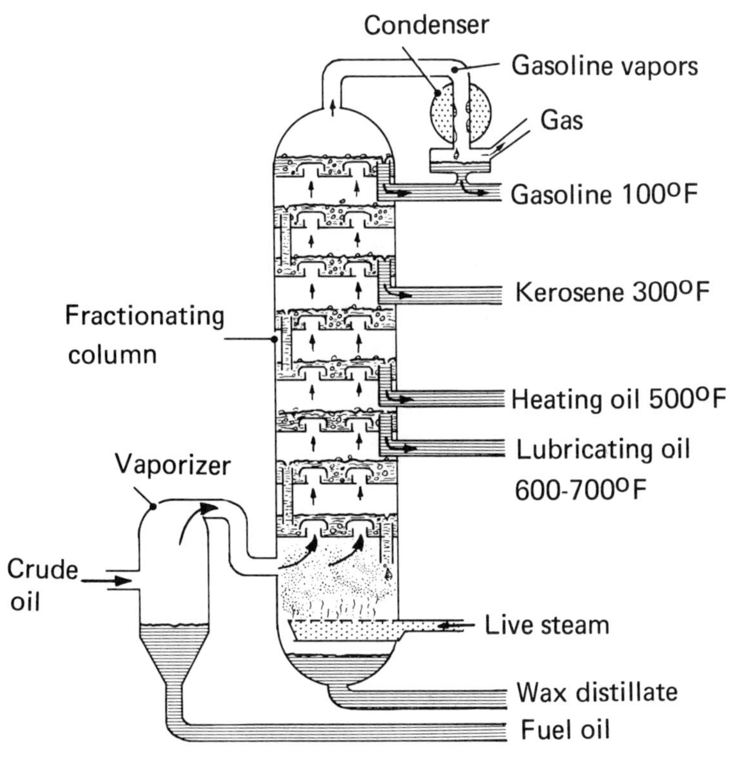

Figure 11-2 A diagram of a fractionating column, showing the various levels from which the petroleum fractions are removed.

These hydrocarbons are both saturated and unsaturated, and are usually separated from each other by chemical methods. The unsaturated gases are used in the production of aviation gasoline, synthetic rubber, and other organic compounds. The saturated hydrocarbons, especially propane and butane, are liquefied and sold as **bottled gas.** Over two billion gallons of these liquefied gases from both petroleum and natural gas are used in the United States every year.

Petroleum Ether. The second fraction that is distilled from petroleum is called petroleum ether. This consists mainly of pentanes, hexanes, and heptanes, and is used extensively as fat solvents; paint, varnish, and enamel thinner; and dry-cleaning agents. This fraction may also be separated into its compo-

TABLE 11-4 DISTILLATION FRACTIONS FROM A TYPICAL PETROLEUM

Name	*Composition (per cent)*	*Molecular Size*	*Boiling Range (°C)*	*Uses*
Gases	2	C_1–C_5	0	Fuel
Petroleum ethers	2	C_5–C_7	30–110	Solvents
Gasoline	32	C_6–C_{12}	30–200	Motor fuel
Kerosene	18	C_{12}–C_{15}	175–275	Diesel and jet fuel
Gas oil (fuel oil)	20	C_{15}——	250–400	Fuel
Lubricating oils and residue		C_{19}——	300——	Lubricants, paraffin wax, petrolatum, and asphalt

nents, which are called **ligroin** and **naphtha,** depending on the temperature of the boiling range.

Gasolines. In the early days of the automobile industry simple distillation of petroleum gave more than enough gasoline to supply the demands. This type of gasoline is called **straight-run gasoline** and is composed essentially of alkanes plus minor amounts of cycloalkanes and aromatic hydrocarbons. The composition of gasoline varied considerably and depended on the source of petroleum from which it was distilled.

As the number of automobiles increased, the supply of straight-run gasoline was insufficient to meet the demands. In addition to the shortage of gasoline, the petroleum industry was faced with an excess of other fractions distilled from petroleum. It was found as early as 1912 that lower molecular weight hydrocarbons could be produced from the higher fractions by heating them to a temperature of from 400 to 500°C. At this temperature some of the bonds of the larger hydrocarbon molecules break to form lower boiling range fractions. If the temperature is increased beyond 500°C, small molecular weight gaseous hydrocarbons are produced. This process, called **thermo-cracking,** not only produces smaller hydrocarbons, but also yields unsaturated and aromatic hydrocarbons. More gasoline with a higher octane number* was therefore produced by the cracking process. Research by petroleum engineers has resulted in the development of several cracking processes that employ petroleum fractions of lower molecular weight and higher molecular weight than those in gasoline. Catalysts have been found that increase the yield of high octane gasoline from these petroleum fractions. In fact, straight-run gasoline can have its octane number increased by passing it through a **catalytic cracking unit** (Fig. 11–3). After the methods for the measurement of the octane number of a gasoline were devised, several hydrocarbons were found that possessed an octane number of over 100. By the proper combination of hydrocarbons in a cracking process, **aviation gasoline** of an octane number over 100 can be produced. During World War II many catalytic cracking units were built to produce aviation gasoline.

Kerosene. In the early days of the petroleum industry the most important fraction from petroleum was kerosene. It was used for lighting purposes, for cooking, and for heating. With the advent of the electric light and the automobile, the demand for kerosene decreased and that for gasoline increased. **Kerosene** is composed of a mixture of saturated, unsaturated, and aromatic hydrocarbons containing from 12 to 15 carbon atoms. It is distilled from petroleum at a temperature of 175 to 275°C. Since the unsaturated hydrocarbons produce an inferior flame when kerosene is used for lighting purposes, they are usually removed by a refining process.

Gas Oil and Fuel Oil. The next higher boiling fraction after kerosene contains a mixture of hydrocarbons whose smallest members have 15 carbon atoms. This fraction contains gas oil, fuel oil, and diesel oil. The name **gas oil** is derived from the fact that this fraction was used originally to enrich water gas for use as a fuel. Large quantities of **fuel oil** are used in furnaces that burn

---

*The performance of a gasoline is expressed as its octane number, compared with an octane number of zero for *n*-heptane and 100 for isooctane. Mixtures of these two hydrocarbons are prepared that exactly match the knocking characteristics of a gasoline under test. The octane number of the gasoline is then equal to the percentage of isooctane in the mixture.

Figure 11-3 One of the catalytic cracking plants used by the petroleum industry. (Reprinted from The Lamp, Standard Oil Company, N. J.)

oil, whereas **diesel oil** is used in diesel engines. This fraction may also be cracked to produce gasoline.

Lubricating Oils. Lubricating oils are produced from the fraction of petroleum that distills at the highest temperature, usually over 300°C. This fraction consists of hydrocarbons with 20 or more carbon atoms and can be separated into oils of different viscosity by fractional distillation. The **viscosity,** or consistency, of lubricating oils is directly related to the structure of their constituent hydrocarbons. For example, an increase in the length of the carbon chain results in increased viscosity and the higher boiling fractions have a higher viscosity than the lower boiling fractions. It is also well known that the viscosity of a lubricating oil increases as the temperature of the oil is decreased. Such processes as redistillation, refining, and dewaxing are used to prepare lubricating oils with different viscosities. One example of the importance of viscosity in oils is the switch from an oil of relatively high viscosity to one of low viscosity when preparing an automobile for winter driving. Recently, by the proper combination of oils and substances such as detergents, a superior multirange viscosity automobile oil has been produced. Such oils remain fluid at low temperature and possess a greater viscosity at high temperatures than regular motor oil. Lubricating greases are composed of a mixture of metallic soaps and lubricating oils.

The Residual Fraction. The residual material that is left after the removal of the distillable fractions of petroleum usually contains either asphalt or paraffin types of hydrocarbons. The nature of the residue has given rise to the terminology of **paraffin base motor oils, asphalt base motor oils,** and **mixed base oils. Paraffin wax** is prepared from the residue of paraffin base oil and consists of straight chained alkanes with 26 to 30 carbon atoms. This residue also yields petrolatum, which is commonly known as petroleum jelly or Vaseline. **Petrolatum** is a semisolid substance that is used as a pharmaceutical base for many salves and ointments. The asphalt type crude oil produces a residue containing pitch or asphalt that is used in roofing material, protective coatings, paving, and asphalt tiles for floors.

PHYSICAL PROPERTIES

The lower molecular weight alkanes, methane through butane, are gases at ordinary temperatures and pressures. The C_5 to C_{17} alkanes (m.p. $< 20°C$) are liquids at room temperature, and compounds containing more than C_{17} are solids at room temperature. The alkanes in the C_{26} to C_{36} range make up the substance known as paraffin wax, and these alkanes are sometimes referred to as paraffin hydrocarbons. Table 11–5 summarizes some of the physical properties of the more common alkanes. A close look at the physical properties of the alkanes reveals that the melting points of the even-numbered alkanes are higher than the next higher odd-numbered homologues. If the melting points of the alkanes are plotted against the number of carbon atoms, a zigzag shaped curve is obtained. The zigzag shape of the melting point curve is probably due to the shape of the molecules. The alkanes with an odd number of carbon atoms have their end carbon atoms on the same side of the molecule, and the even-numbered carbon atom alkanes have their end carbon atoms on opposite sides of the molecule as shown below:

```
   C   C   C            C   C   C
  / \ / \ / \          / \ / \ /
 C   C   C   C        C   C   C
      Odd                 Even
```

Apparently, the alkanes having an even number of atoms can pack in the crystal so as to permit greater intermolecular attraction between molecules, and therefore, have slightly higher melting points.

In contrast to the zigzag shape of the melting point curve, the boiling points of the alkanes show a regular increase of approximately 20 to 30°C with the introduction of each new methylene group. Branching of the carbon chain, however, lowers the boiling point of the alkane, and, with compounds that can exist as isomers, the straight-chain isomer is always the highest boiling isomer. For example, there are three isomers having the molecular formula, C_5H_{12}, as illustrated below:

		B.P. °C
n-Pentane	$CH_3CH_2CH_2CH_2CH_3$	36
Isopentane	$CH_3CHCH_2CH_3$ with CH_3 on C-2	28
Neopentane	CH_3CCH_3 with two CH_3 on C-2	9.5

As the amount of branching increases, the boiling point decreases.

TABLE 11-5 ALKANES

Name	*M.P. °C*	*B.P. °C*	*Mol. Formula*	*Sp. Gravity (as liquids)*
Methane	−183	−162	CH_4	—
Ethane	−172	−89	C_2H_6	—
Propane	−187	−42	C_3H_8	—
n-Butane	−135	−0.5	C_4H_{10}	—
n-Pentane	−130	+36	C_5H_{12}	0.626
n-Hexane	−94	69	C_6H_{14}	0.659
n-Heptane	−90	98	C_7H_{16}	0.683
n-Octane	−57	126	C_8H_{18}	0.703
n-Nonane	−54	151	C_9H_{20}	0.718
n-Decane	−30	174	$C_{10}H_{22}$	0.729
n-Undecane	−26	196	$C_{11}H_{24}$	0.740
n-Dodecane	−10	216	$C_{12}H_{26}$	0.749
n-Tridecane	−6	235	$C_{13}H_{28}$	0.757
n-Tetradecane	+6	251	$C_{14}H_{30}$	0.764
n-Pentadecane	10	268	$C_{15}H_{32}$	0.769
n-Hexadecane	18	280	$C_{16}H_{34}$	0.775
n-Heptadecane	22	303	$C_{17}H_{36}$	0.777
n-Octadecane	28	308	$C_{18}H_{38}$	0.777
n-Nonadecane	32	330	$C_{19}H_{40}$	0.778
n-Eicosane	36	343	$C_{20}H_{42}$	0.778

The saturated hydrocarbons are almost completely insoluble in water, but are soluble in organic solvents. Because of their stability and resistance to many chemical reagents, many of the alkanes are useful as solvents for organic reactions.

CHEMICAL PROPERTIES

Many of the alkanes are actually obtained from petroleum and natural gas, either directly or by fractional distillation of petroleum products. In many cases, however, a particular compound or isomer may be required that is not available or easily obtained commercially from petroleum. Then the compound must be synthesized in the laboratory by the organic chemist. For the preparation of most organic compounds there are generally several possible routes whereby one functional group may be converted to another. The method selected by the organic chemist takes into consideration the yield and purity of the reaction product, the cost of the reagents, the difficulty in carrying out the reaction, and the ease of isolating the product. These factors will vary from compound to compound, and a method of synthesis for one type of compound may not be the best method for another. Consequently, some knowledge of these factors is necessary in order to select the best possible method to prepare a specific compound. The following are several of the more common methods to prepare alkanes.

REDUCTION OF ALKENES. The carbon-carbon double bond present in alkenes can add a mole of hydrogen (H_2) in the presence of a catalyst to give an alkane as the final product. The catalysts most commonly used are palladium

$$\rangle C{=}C\langle + H_2 \xrightarrow{\text{catalyst}} -\underset{H}{\overset{|}{C}}-\underset{H}{\overset{|}{C}}-$$

(Pd), platinum (Pt), and nickel (Ni). For most small-scale laboratory work the reaction is generally carried out in the **liquid phase** at temperatures below 100°C and pressures of 1 to 10 atmospheres of hydrogen. The reaction is a general one and can be employed in the preparation of both straight chain and cyclic alkanes. Some specific examples are illustrated below:

$$\underset{\text{Propene}}{CH_3CH{=}CH_2} + H_2 \xrightarrow{Pt} \underset{\text{Propane}}{CH_3CH_2CH_3}$$

$$\underset{\text{Cyclohexene}}{C_6H_{10}} + H_2 \xrightarrow{Pd} \underset{\text{Cyclohexane}}{C_6H_{12}}$$

Wurtz Reaction*. The French chemist, Adolphe Wurtz, discovered that alkyl halides react with metallic sodium to yield hydrocarbons. The hydrocarbon obtained was a dimer of the alkyl groups contained in the original alkyl halide. Since the product is a dimer, only even-numbered alkanes are obtained as final products. To prepare an odd-numbered alkane by this method requires two

$$\underset{\text{Ethyl iodide}}{2CH_3CH_2I} + 2Na \rightarrow \underset{n\text{-Butane}}{CH_3CH_2CH_2CH_3} + 2NaI$$

different alkyl halides, an even-numbered and an odd-numbered one. However, self-dimerization of each alkyl halide occurs as well as intermolecular dimerization. Therefore, the method is generally not useful for preparing odd-numbered alkanes.

$$\underset{\text{Methyl bromide}}{CH_3Br} + \underset{\text{Ethyl bromide}}{CH_3CH_2Br} \xrightarrow{2Na} \underset{\text{Ethane}}{CH_3CH_3} + \underset{\text{Propane}}{CH_3CH_2CH_3} + \underset{n\text{-Butane}}{CH_3CH_2CH_2CH_3}$$

CHEMICAL REACTIONS

The alkanes are the least reactive organic compounds. They are resistant to strong acids and alkalies, and under ordinary conditions are resistant to oxidizing agents. For these reasons they are commonly employed as inert solvents in chemical reactions of the other functional groups. They do, however, react with halogens, nitric acid, and oxygen under forced conditions.

Combustion. Hydrocarbons will react with oxygen when ignited and in the presence of excess oxygen. The products of the combustion reaction are carbon dioxide and water. This reaction may be expressed as:

* Many synthetic organic reactions have been named after the discoverer of the reaction. Although memorization of the reaction name is unimportant in understanding the chemistry involved, many of these reactions are commonly known by these names, and the student will find it useful to become acquainted with some of them.

$$C_nH_{2n+2} + \frac{3n+1}{2} O_2 \rightarrow nCO_2 + (n+1)\ H_2O + \text{heat}$$

The heat given off in this reaction is called the heat of combustion. It is the utilization of this heat that accounts for much of the commercial use of hydrocarbons as heating fuels. Gasoline, which is composed of hydrocarbons containing 6 to 12 carbon atoms per molecule, burns similarly with oxygen in an automobile engine. The gases CO_2 and H_2O power the pistons of the engine, and the heat evolved is the heat carried away by the cooling system. Some specific examples are shown below:

$$CH_4 + 2O_2 \rightarrow CO_2 + 2H_2O + 213\ \text{Kcal.}$$
$$C_2H_6 + 3\tfrac{1}{2}O_2 \rightarrow 2CO_2 + 3H_2O + 373\ \text{Kcal.}$$

If combustion occurs in the absence of sufficient oxygen, carbon monoxide is formed. Carbon monoxide is the toxic gas found in the exhaust fumes of a car.

$$C_2H_6 + 2\tfrac{1}{2}O_2 \rightarrow 2CO + 3H_2O$$

Halogenation. The halogens, fluorine, chlorine, and bromine, will react under the proper conditions with alkanes. Iodine does not react with alkanes. In contrast to chlorine and bromine which react very slowly at room temperature, fluorine reacts with alkanes almost explosively at room temperature. Fluorine undergoes a controlled reaction with alkanes only under very carefully controlled conditions, and because of its extreme reactivity it is generally not too useful for ordinary laboratory conditions.

Consequently, chlorine and bromine are the only practical halogens useful in the laboratory for reaction with alkanes. At higher temperatures or in the presence of sunlight, a chlorine or bromine atom can substitute for the hydrogen of an alkane. These reactions are known as **halogenation** reactions, or more specifically as **chlorination** or **bromination**, depending on whether a chlorine or a bromine atom substitutes for the hydrogen. The mechanism for this reaction is a free radical one, as was outlined in Chapter 10 for the formation of methyl chloride from methane. Chlorine molecules absorb energy (either heat or light) and dissociate into chlorine atoms, which attack the alkane and remove a hydrogen atom and generate an alkyl free radical. The free alkyl radical then abstracts a chlorine atom from a chlorine molecule to form the halogenated product and a chlorine atom which can repeat the cycle to generate more halogenated alkane. The mechanism can be represented as follows:

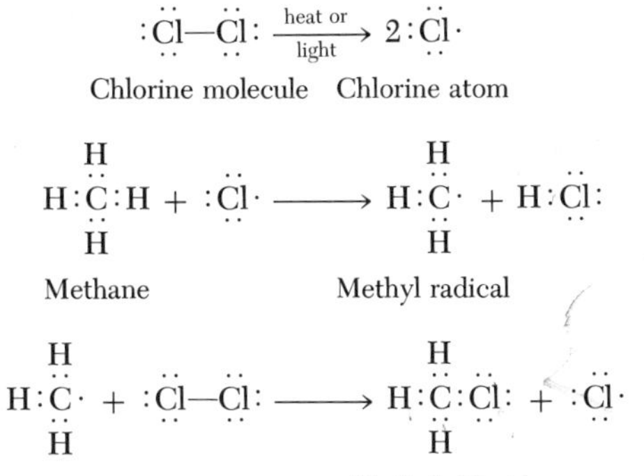

$$:\ddot{\underset{..}{Cl}}—\ddot{\underset{..}{Cl}}: \xrightarrow[\text{light}]{\text{heat or}} 2:\ddot{\underset{..}{Cl}}\cdot$$

Chlorine molecule   Chlorine atom

$$\begin{array}{c} H \\ H:\ddot{\underset{..}{C}}:H \\ H \end{array} + :\ddot{\underset{..}{Cl}}\cdot \longrightarrow \begin{array}{c} H \\ H:\ddot{\underset{..}{C}}\cdot \\ H \end{array} + H:\ddot{\underset{..}{Cl}}:$$

Methane   Methyl radical

$$\begin{array}{c} H \\ H:\ddot{\underset{..}{C}}\cdot \\ H \end{array} + :\ddot{\underset{..}{Cl}}—\ddot{\underset{..}{Cl}}: \longrightarrow \begin{array}{c} H \\ H:\ddot{\underset{..}{C}}:\ddot{\underset{..}{Cl}}: \\ H \end{array} + :\ddot{\underset{..}{Cl}}\cdot$$

Methyl chloride

The overall reaction can be summarized as:

$$Cl_2 + CH_4 \xrightarrow[\text{sunlight}]{\text{heat or}} CH_3Cl + HCl$$

Only half of the halogen ends up in the organic product; the other half ends up as hydrogen chloride. Substitution reactions always give two products, whereas addition reactions (such as the hydrogenation of alkenes to give alkanes) give only one product.

Substitution reactions in many instances give rise to more than two products. In the case of the chlorination of methane to give methyl chloride, the product, methyl chloride, may react further (via the same type of free-radical mechanism) to give methylene chloride (CH_2Cl_2), which can react further to give chloroform ($CHCl_3$), which reacts further to give carbon tetrachloride (CCl_4). The reactions may be summarized as follows:

$$CH_3Cl + Cl_2 \rightarrow CH_2Cl_2 + HCl$$

Methyl chloride (chloromethane) — Methylene chloride (dichloromethane)

$$CH_2Cl_2 + Cl_2 \rightarrow CHCl_3 + HCl$$

Chloroform (trichloromethane)

$$CHCl_3 + Cl_2 \rightarrow CCl_4 + HCl$$

Carbon tetrachloride (tetrachloromethane)

In actual practice, chlorination of methane gives all four products. The amounts of each depend on the amount of chlorine used and the reaction conditions.

Higher homologues increase the complexity of even the simple monohalogenation step. For example, in the chlorination of propane, two possible monochlorinated products are possible, depending upon which of the two different types of hydrogens chlorine substitutes for. In practice, both products are obtained. Further halogenation of these monochlorinated products produces

$$Cl_2 + CH_3CH_2CH_3 \xrightarrow[\text{sunlight}]{\text{heat or}} CH_3CH_2CH_2Cl + CH_3CHClCH_3 + HCl$$

n-Propyl chloride (1-chloropropane) — Isopropyl chloride (2-chloropropane)

an even more complex mixture. Because of this multiplicity of products, halogenation of alkanes is not a particularly useful laboratory reaction. However, in many commercial products where ultra pure products are not always required, the mixtures obtained are useful, and this type of reaction is commercially important for preparing halogenated alkanes used in such products as dry cleaning agents and fire extinguishers.

Pyrolysis (Cracking) of Alkanes. Although alkanes are generally some of the most stable organic compounds, they can be broken down (cracked into smaller fragments) by heating to high temperatures (400–700°C) in the absence of air (to avoid combustion). This is an important reaction in the petroleum industry for converting high molecular weight alkanes into fragments in the gasoline range, thereby increasing the amount of gasoline obtainable from crude petroleum. In the petroleum industry many catalysts have been developed which affect the "cracking" reaction at much lower temperatures than simple pyrolysis. A simple illustrative example of pyrolytic cracking is shown below:

$$CH_3CH_2CH_2CH_3 \xrightarrow{700^\circ C} CH_2{=}CHCH_2CH_3 + CH_3CH{=}CHCH_3 + CH_3CH_3 + CH_2{=}CH_2 + CH_3CH{=}CH_2 + CH_4$$

Both carbon-carbon and carbon-hydrogen bonds are broken at these temperatures, and a complex mixture of products is obtained.

Questions

1. Draw out all the possible isomers for the compound having a molecular formula, C_6H_{14}. Give the correct IUPAC name to each of the compounds.

2. Which of the isomeric compounds in Question 1
   (a) contains a tertiary butyl group?
   (b) is isohexane?
   (c) contains an isopropyl group?
   (d) has the lowest boiling point?
   (e) has the highest boiling point?

3. Draw a correct structure for each of the following compounds:
   (a) 1,2-dimethylcyclohexane
   (b) 4-isopropyl-5-methyldecane
   (c) *n*-heptane
   (d) tertiary butyl bromide
   (e) 3-methyl-4-ethyl-5-isopropyloctane

4. What alkanes would be expected from the reaction of the following alkyl halides with sodium? Give the IUPAC name for each of the products.
   (a) methyl iodide
   (b) ethyl bromide
   (c) isopropyl bromide
   (d) isobutyl bromide
   (e) tertiary butyl bromide

5. An alkane was found to have a molecular weight of 58. On photochemical chlorination, two isomeric monochlorinated products were obtained. What is the structure of the original alkane? Name the substitution products by the IUPAC system. Write a reasonable mechanism to form the substitution products.

Suggested Reading

Conrad and Sabin: Motor Fuel Quality As Related to Refinery Processing and Antiknock Compounds. Journal of Chemical Education, Vol. 34, p. 262, 1957.

Hurd: The General Philosophy of Organic Nomenclature. Journal of Chemical Education, Vol. 38, p. 43, 1961.

Kimberlin: Chemistry In The Manufacture of Modern Gasoline. Journal of Chemical Education, Vol. 34, p. 569, 1957.

Nelson: The Origin of Petroleum. Journal of Chemical Education, Vol. 31, p. 399, 1954.

Rossini: Hydrocarbons in Petroleum. Journal of Chemical Education, Vol. 37, p. 554, 1960.

Shoemaker, d'Ouville, and Marschner: Recent Advances In Petroleum Refining. Journal of Chemical Education, Vol. 32, p. 30, 1955.

CHAPTER 12

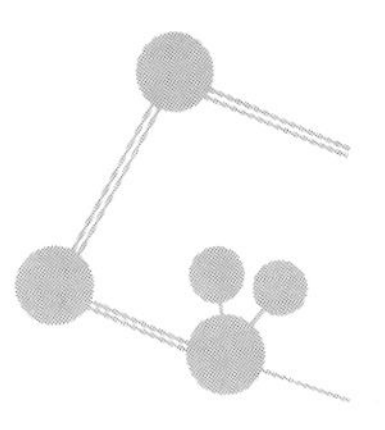

ALKENES

Alkenes are distinguished from alkanes by the presence of the carbon-carbon double bond in the molecule. Other terms used to denote alkenes are **olefins** (from olefiant gas—the old name for ethylene) and **unsaturated hydrocarbons** (to denote the presence of a multiple bond in the molecules). The presence of the carbon-carbon double bond in the molecule makes these kinds of compounds highly reactive compared to alkanes.

In Chapter 10 the differences in bonding between alkanes and alkenes were briefly noted. In alkanes, which contain only carbon-carbon single bonds, essentially free rotation is allowed around the axis of the bond between the two carbon atoms. The orbitals used in the bonding of alkanes were sp^3 and gave a tetrahedral structure. However, in alkenes three sp^2 orbitals are used to form the carbon-hydrogen bonds and the carbon-carbon single bond. The use of sp^2 orbitals in bonding gives a planar arrangement of the carbon and hydrogen atoms involved.* (See also Fig. 3–11.) The remaining carbon-carbon bond is formed by overlap of the $2p$ orbital remaining on each carbon atom as illustrated in Figure 12–1. Hence, the first carbon-carbon bond is formed by the overlap of two sp^2 orbitals (called a sigma [σ] bond) and the second carbon-carbon bond is formed by overlap of $2p$ orbitals (called a pi [π] bond).

Figure 12–1 The carbon-carbon double bond.

*The student is encouraged to use models to confirm this coplanar arrangement of the atoms and the restricted rotation in alkenes.

The π-bond is much weaker than the σ-bond and is the first bond broken in the chemical reactions of most alkenes.

Resultant features of this multiple bonding in alkenes are: (1) The bond distance between the two carbon atoms of the multiple bond is 1.34 Å compared to 1.54 Å in an alkane. This is because two pairs of electrons pull the two nuclei closer together than the one pair in alkanes; (2) Rotation about the carbon-carbon double bond is restricted. Although the atoms or groups of atoms attached to the double bond can vibrate or twist within small angles, no free rotation is allowed, since the π-bond would have to be broken to allow this rotation; (3) Because of restricted rotation around the multiple bond, a new type of isomerism is possible—namely, geometrical isomerism in which the two isomers differ only in the arrangement of the four atoms or groups of atoms that are attached to the multiple bond. This point will be discussed more fully later in this chapter.

NOMENCLATURE AND STEREOCHEMISTRY

Alkenes have the general formula C_nH_{2n}, and cycloalkenes have the general formula C_nH_{2n-2}. Two systems again are most commonly used to name these compounds. The correct nomenclature system is the IUPAC system. In this system the characteristic ending **-ene** is added to the root stem name (formed by dropping the **-ane** ending from the hydrocarbon having the same number of carbon atoms); the carbon atoms in the *longest continuous chain containing the double bond* are numbered so that the carbon atoms of the double bond have the lowest possible numbers. Therefore the location of the double bond is indicated by the lowest possible number, and this number is placed in front of the name. Substituents are listed in front of the alkene name with the appropriate number to indicate their position. The other system of nomenclature is the use of common names for the lower members of this series. The common names are formed by dropping the **-ane** ending of the parent alkane and adding the ending **-ylene** to the remaining root stem. Some examples of these nomenclature systems are illustrated below with the common name in parentheses:

$$H_2C{=}CH_2 \text{ or } H_2C{=}CH_2$$

Ethene
(ethylene)

$$H_3C{-}CH{=}CH_2 \text{ or } CH_3CH{=}CH_2$$

Propene
(propylene)

In olefins containing four carbon atoms or more, not only are isomers possible by varying the arrangement of the carbon atoms, as in 1-butene and isobutylene, but isomerism can also result from a shift in the position of the double bond without changing the carbon skeleton itself, as illustrated below for the pentenes:

$CH_2{=}CHCH_2CH_2CH_3$	1-Pentene
$CH_3CH{=}CHCH_2CH_3$	2-Pentene

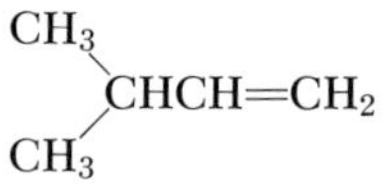

$$\begin{array}{l} CH_3 \\ \quad \diagdown \\ \quad\; CHCH{=}CH_2 \\ \quad \diagup \\ CH_3 \end{array}$$ 3-Methyl-1-butene

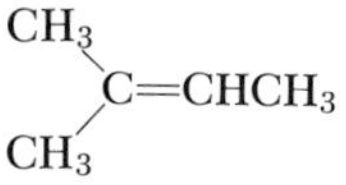

$$\begin{array}{l} CH_3 \\ \quad \diagdown \\ \quad\; C{=}CHCH_3 \\ \quad \diagup \\ CH_3 \end{array}$$ 2-Methyl-2-butene

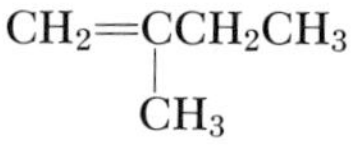

$$\begin{array}{l} CH_2{=}\underset{\displaystyle CH_3}{\underset{|}{C}}CH_2CH_3 \end{array}$$ 2-Methyl-1-butene

The only difference between 1-pentene and 2-pentene is the position of the double bond. Further examination of the pentene isomers has revealed that there are two different pentenes which have the structure $CH_3CH{=}CHCH_2CH_3$. The structure of these two pentenes has been shown to be:

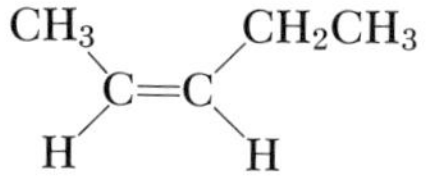

$$\begin{array}{ccc} CH_3 & & CH_2CH_3 \\ & C{=}C & \\ H & & H \end{array} \quad \text{and} \quad \begin{array}{ccc} CH_3 & & H \\ & C{=}C & \\ H & & CH_2CH_3 \end{array}$$

cis-2-Pentene (A) *trans*-2-Pentene (B)

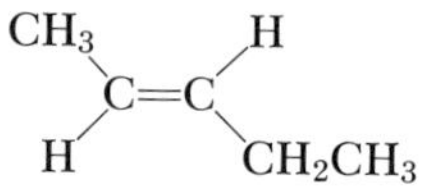

The only difference between these two isomers is that in one structure (A) the two hydrogens attached to the double bond are on the same side of the double bond (this is called a *cis* arrangement) and the two alkyl groups are also on the same (but opposite to the hydrogen) side of the double bond. In structure (B), the two hydrogens attached to the double bond are on opposite sides of the double bond (this is called a *trans*-arrangement), and the two alkyl groups are also *trans* across the double bond. Because of the restricted rotation around the double bond, these two compounds are fixed in respect to the arrangement of the atoms and alkyl groups around the double bond. Both the *cis*- and *trans*-compounds do exist, and both have been isolated and their structures unambiguously proved.

The necessary requirements for geometrical isomerism are: (1) that restricted rotation of some kind be present in the molecule. This may be either a double bond in olefins or the presence of a ring system which also prevents free rotation. Consequently, cyclic alkanes also exhibit geometric isomerism; (2) that neither carbon atom involved in the restricted rotation may hold identical groups. Some examples of geometric isomerism are shown below.

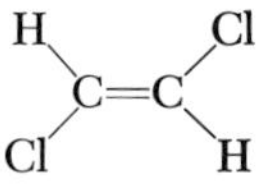

$$\begin{array}{ccc} H & & Cl \\ & C{=}C & \\ Cl & & H \end{array} \qquad \begin{array}{ccc} H & & H \\ & C{=}C & \\ Cl & & Cl \end{array} \qquad \begin{array}{ccc} H & & Cl \\ & C{=}C & \\ H & & Cl \end{array}$$

trans-1,2-Dichloroethene 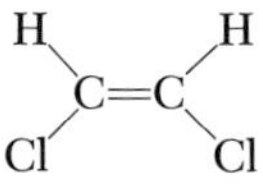 *cis*-1,2-Dichloroethene 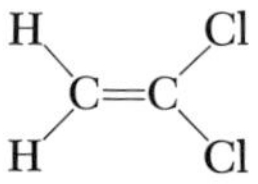 1,1-Dichloroethene (no geometric isomers)

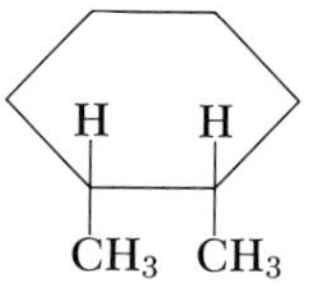

(cyclohexane ring: H H above; CH_3 CH_3 below)

cis-1,2-Dimethylcyclohexane

(cyclohexane ring: H CH_3 above; CH_3 H below)

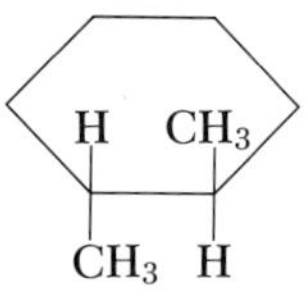

trans-1,2-Dimethylcyclohexane

In **cycloalkenes,** the possibility of *cis-trans* isomerism is limited by the constraints of the ring. Cycloalkenes of less than seven carbons in the ring exist only in the *cis*-arrangement. The corresponding *trans*-isomer in the smaller rings is too highly strained and is not capable of being isolated.

PHYSICAL PROPERTIES OF ALKENES

The alkenes are quite similar to the alkanes in physical properties. Alkenes containing 2 to 4 carbon atoms are gases; those containing 5 to 18 carbon atoms are liquids; and those containing more than 18 carbons are solids. They are relatively insoluble in water but are soluble in concentrated sulfuric acid. The physical properties of some of the more common alkenes are summarized in Table 12–1.

PREPARATION OF ALKENES

The more general laboratory synthesis of alkenes involves a reaction known as an **elimination** reaction in which a molecule of water, hydrogen halide, or halogen is removed from adjacent carbon atoms in a saturated compound. The generalized reaction may be summarized as:

$$-\underset{\mathrm{x}}{\overset{|}{\underset{|}{C}}}-\underset{\mathrm{y}}{\overset{|}{\underset{|}{C}}}- \rightarrow \;\rangle C{=}C\langle \; + \mathrm{xy}$$

x = H or halogen
y = OH or halogen

In this process the group —x is removed in such a manner that it leaves its electron pair which then forms an additional bond (the double bond) between the carbon atoms involved.

TABLE 12–1 PHYSICAL PROPERTIES OF ALKENES AND DIENES

Compound	*Structure*	*M.P. °C*	*B.P. °C*
Ethylene	$CH_2{=}CH_2$	−169	−104
Propylene	$CH_3CH{=}CH_2$	−185	−48
1-Butene	$CH_3CH_2CH{=}CH_2$	−185	−6
2-Butene (*cis*)	$CH_3CH{=}CHCH_3$	−139	+4
2-Butene (*trans*)	$CH_3CH{=}CHCH_3$	−106	+1
1-Pentene	$CH_3CH_2CH_2CH{=}CH_2$	−165	+30
1-Hexene	$CH_3(CH_2)_3CH{=}CH_2$	−140	+64
Cyclobutene	(square ring with one double bond)	—	+2
Cyclopentene	(pentagon ring with one double bond)	−135	+44
Cyclohexene	(hexagon ring with one double bond)	−104	+83
Allene	$CH_2{=}C{=}CH_2$	−136	−35
1,3-Butadiene	$CH_2{=}CH{-}CH{=}CH_2$	−109	−4

Dehydration of Alcohols. When the molecule (xy) lost in an elimination reaction is water (HOH), the reaction is called a **dehydration reaction.** The generalized equation can then be represented as shown below:

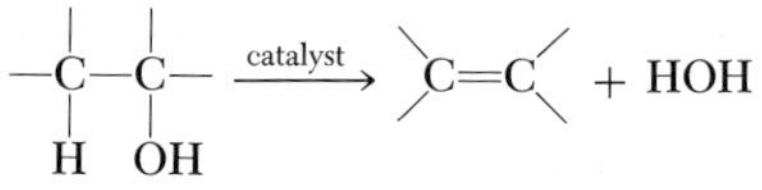

$$-\underset{H}{\overset{|}{\underset{|}{C}}}-\underset{OH}{\overset{|}{\underset{|}{C}}}- \xrightarrow{\text{catalyst}} \rangle C{=}C\langle + HOH$$

The starting material which is dehydrated is an alcohol, and the alcohol can be either a straight chain or a cyclic alcohol. The catalysts employed for dehydration reactions are acidic, and all have a strong affinity for water. Acids such as concentrated sulfuric acid and phosphoric acid, or aluminum oxide and phosphorus pentoxide at high temperatures, are generally employed. For example, propene can be prepared from 1-propanol and cyclohexene can be prepared from cyclohexanol by this method. In some alcohols the elements of

$$\underset{\text{1-Propanol}}{CH_3CH_2CH_2OH} \xrightarrow{H_2SO_4} \underset{\text{Propene}}{CH_3CH{=}CH_2}$$

H_2O may be able to be lost in more than one way. For example, in 2-butanol either 1-butene or 2-butene may be formed by the elimination of water.

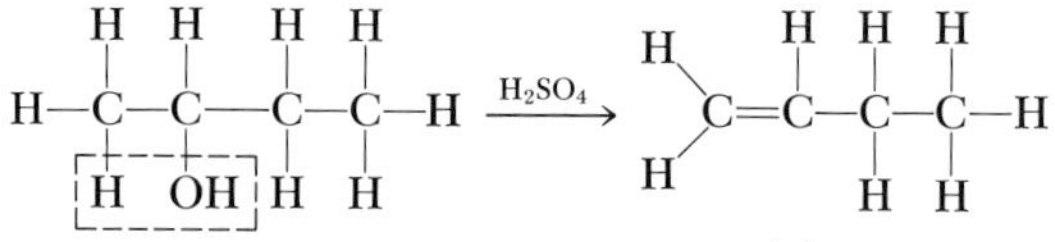

$$H-\underset{H}{\overset{H}{C}}-\underset{OH}{\overset{H}{C}}-\underset{H}{\overset{H}{C}}-\underset{H}{\overset{H}{C}}-H \xrightarrow{H_2SO_4} \underset{\text{1-Butene}}{\overset{H}{\underset{H}{\rangle}}C{=}\overset{H}{C}-\underset{H}{\overset{H}{C}}-\underset{H}{\overset{H}{C}}-H}$$

$$H-\underset{H}{\overset{H}{C}}-\underset{OH}{\overset{H}{C}}-\underset{H}{\overset{H}{C}}-\underset{H}{\overset{H}{C}}-H \xrightarrow{H_2SO_4} \underset{\text{2-Butene}}{H-\underset{H}{\overset{H}{C}}-\overset{H}{C}{=}\overset{H}{C}-\underset{H}{\overset{H}{C}}-H}$$

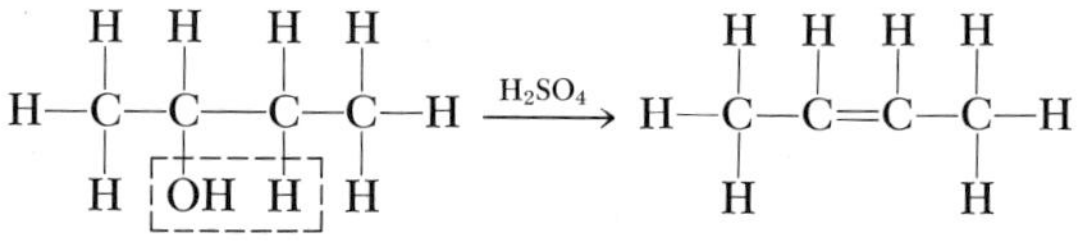

In alcohols of this type, mixtures of both possible olefins are generally obtained. However, one olefin is usually formed as the major product, and in nearly all cases the predominant product is the *most highly substituted olefin.** The ratio of 2-butene to 1-butene in this reaction is 4:1.

The ease of dehydration of alcohols depends upon the type of alcohol used. Alcohols are classified as primary, secondary, and tertiary by counting the number of alkyl groups bonded to the carbon atom bearing the —OH group, as follows:

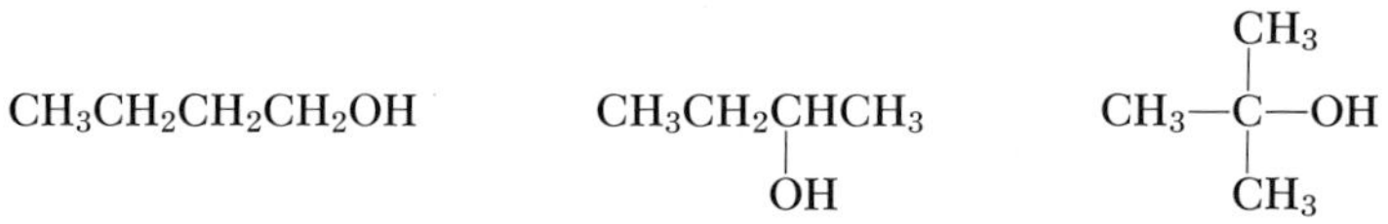

$CH_3CH_2CH_2CH_2OH$	$CH_3CH_2\underset{OH}{\underset{\mid}{C}}HCH_3$	$CH_3-\underset{CH_3}{\overset{CH_3}{C}}-OH$
Primary alcohol 1-Butanol	Secondary alcohol 2-Butanol	Tertiary alcohol 2-Methyl-2-propanol

* In the previous example, 1-butene is a monosubstituted olefin (one hydrogen of ethylene has been substituted by an ethyl group), and 2-butene is a disubstituted olefin (two hydrogens of ethylene have been substituted by methyl groups).

The ease of dehydration has been proved to be: tertiary > secondary > primary. Thus, it is easier to dehydrate 2-methyl-2-propanol to isobutylene than it is to dehydrate 1-butanol to 1-butene, or 2-butanol to 2-butene.

The following mechanism for the acid catalyzed dehydration of alcohols has been proposed:

$$CH_3CH_2\underset{\underset{\cdot\cdot}{:OH}}{\underset{|}{C}}HCH_3 + H_2SO_4 \rightleftharpoons CH_3CH_2\underset{\underset{\cdot\cdot}{^{+}OH_2}}{\underset{|}{C}}HCH_3 + HSO_4^-$$

Oxonium ion

$$CH_3CH_2\underset{\underset{\cdot\cdot}{^{+}OH_2}}{\underset{|}{C}}HCH_3 \rightleftharpoons CH_3CH_2\overset{+}{C}HCH_3 + H_2O$$

Carbonium ion

$$CH_3\underset{\underset{H}{|}}{CH}-\overset{+}{C}HCH_3 + HSO_4^- \rightarrow CH_3CH{=}CHCH_3 + H_2SO_4$$

$$CH_3CH_2\overset{+}{C}H-\underset{\underset{H}{|}}{CH_2} + HSO_4^- \rightarrow CH_3CH_2CH{=}CH_2 + H_2SO_4$$

The first step involves protonation of the oxygen of the hydroxyl group to generate an oxonium ion, followed by loss of water to generate a carbonium ion. The carbonium ion can lose a proton (H^+) from either the adjacent $—CH_2$ group or $—CH_3$ group. The proton does not simply dissociate off but is lost to a base. In this case, the bisulfate ion can act as the base and abstract the proton. The electron pair left behind by the proton then forms the double bond of the olefin. In the process, the acid catalyst, H_2SO_4, is regenerated and can then repeat this cycle.

Dehydrohalogenation of Alkyl Halides. An elimination reaction, similar to the dehydration of alcohols, can occur with **alkyl halides** (compounds containing a halogen bonded to an alkyl group) in the presence of base to eliminate the elements of —HX (where X = halogen) and form an alkene. For example, if isopropyl chloride is treated with potassium hydroxide in alcohol, propene, potassium chloride, and water are formed as products.

$$CH_3CHClCH_3 + KOH \xrightarrow{\text{alcohol}} CH_3CH{=}CH_2 + K^+Cl^- + H_2O$$

Isopropyl chloride (2-chloropropane) — Propene

The ease of removal of halogen is in the following order: tertiary alkyl halides > secondary > primary. Also, the ease of dehydrohalogenation for a particular type of carbon structure increases as the halogen is varied from fluorine through iodine. Consequently, it is easier to dehydrohalogenate isopropyl iodide than it is to dehydrohalogenate isopropyl chloride.

The generally accepted mechanism for dehydrohalogenation is as follows:

$$\underset{\underset{H}{|}}{CH_2}-\overset{\overset{Cl}{|}}{C}H-CH_3 + OH^- \rightarrow CH_2{=}CHCH_3 + H_2O + Cl^-$$

The base, OH^- in the example, abstracts a proton from the adjacent carbon atom to form water. The electron pair left behind by the proton being abstracted then forms the carbon-carbon double bond with the ejection of a chloride ion. This type of reaction is known as a β-elimination, since the proton is being abstracted from the carbon atom which is *beta* to the group being eliminated (Cl^-). In this example, it doesn't matter whether a proton is abstracted from either methyl group, since both groups are identical, and the same product would be formed in either case. However, in compounds which are not symmetrical, two different olefins can be produced. In 2-chlorobutane, both 1-butene and *cis*- and *trans*-2-butene are formed, but the more highly substituted alkene (2-butene) is the predominant product. In this particular compound, 80% of the 2-butene and 20% of the 1-butene are produced.

REACTIONS OF ALKENES

In the previous chapter the reactions of alkanes were shown to be reactions of a substitution type in which the carbon-hydrogen bond is broken and a new bond is formed between carbon and a new atom or group of atoms. Two products are formed in reactions of this type; for example, methyl chloride and hydrogen chloride are formed in the chlorination of methane.

In contrast to this behavior of alkanes, *the most characteristic reaction of alkenes is addition to the double bond.* The π-bond of the olefin is broken and two new single bonds are formed in this process. *Only one product is formed in addition reactions of this type.* A generalized reaction scheme for an addition reaction can be represented by the following equation:

$$>C{=}C< + xy \rightarrow -\underset{x}{\overset{|}{\underset{|}{C}}}-\underset{y}{\overset{|}{\underset{|}{C}}}-$$

Addition reaction

The π-bond (the weaker of the two bonds) is broken and a new bond is formed between carbon and atom or group —x, and between carbon and atom or group —y, to yield the addition product, which now contains only single bonds between all atoms or groups of atoms. The molecule or compound xy can be one of many types of materials added to olefins. We shall only consider several simple types of compounds which have been added to olefins, but it should be kept in mind that this is a very general and widely used reaction by the organic chemist and one of the more important types of reactions in organic chemistry. In later chapters we shall again encounter addition reactions to other multiple bonds, such as $>C{=}O$, or $-C{\equiv}N$. The necessary requirement for an addition reaction is the presence of a multiple bond, and addition reactions are not necessarily limited to carbon-carbon multiple bonds.

Hydrogenation of Alkenes. In Chapter 11, the preparation of alkanes was shown to be possible by the catalytic hydrogenation of alkenes. This is a general reaction of alkenes and is important both in the laboratory and commercially in the preparation of many organic compounds. Dienes, which

$$\rangle C{=}C\langle + H_2 \xrightarrow{\text{catalyst}} -\overset{|}{\underset{H}{C}}-\overset{|}{\underset{H}{C}}-$$

are discussed later in this chapter, also will undergo this reaction, but will add twice as much hydrogen.

ADDITION OF HALOGEN. Halogens behave similar to hydrogen and undergo 1,2-addition across the double bond. In practice, chlorine and bromine are the halogens generally employed. Fluorine can be added to olefins under special conditions but is too reactive for general laboratory use. Iodine adds reversibly to olefins, and most 1,2-vicinal iodides are unstable; consequently, iodine addition is not of much practical value. An example of halogen addition is illustrated as follows:

$$\underset{\text{Propene}}{CH_2{=}CHCH_3} + Cl_2 \rightarrow \underset{\text{1,2-Dichloropropane}}{CH_2ClCHClCH_3}$$

ADDITION OF ACIDS. Acids such as sulfuric acid and the hydrogen halides (HF, HCl, HBr, HI) can be added across a carbon-carbon double bond to give either alkyl hydrogen sulfates or alkyl halides, as illustrated below for ethylene:

$$\underset{\text{Ethylene}}{CH_2{=}CH_2} + \underset{\text{Sulfuric acid}}{H_2SO_4(HOSO_2OH)} \rightarrow \underset{\text{Ethyl hydrogen sulfate}}{CH_3CH_2OSO_2OH}$$

$$CH_2{=}CH_2 + HBr \rightarrow CH_3CH_2Br$$

A mechanism opposite to that for dehydration can be written in which the proton (H^+) of the acid adds to the double bond to give a carbonium ion which then picks up an anion (either OSO_2OH^-, F^-, Cl^-, Br^-, or I^-) to give the addition product. In ethylene, addition of the proton to either carbon gives

$$CH_2{=}CH_2 + HOSO_2OH \rightleftharpoons CH_3\overset{+}{C}H_2 + \overset{-}{O}SO_2OH \rightarrow CH_3CH_2OSO_2OH$$

the ethyl carbonium ion ($CH_3CH_2^+$), and only one product can result. However, in unsymmetrical olefins, the proton could add to give two different carbonium ions,* and two possible products could result. In actuality, the proton will add to give the *more stable carbonium ion,* and the product resulting from the more stable carbonium ion will be the product produced. For example, in the addition of hydrogen iodide to propene, either isopropyl iodide or *n*-propyl iodide could be produced. The product actually formed is isopropyl iodide

$$\underset{\text{Propene}}{CH_3CH{=}CH_2} + HI \rightleftharpoons \underset{\text{Secondary}}{CH_3\overset{+}{C}HCH_3} \quad \text{or} \quad \underset{\text{Primary}}{CH_3CH_2\overset{+}{C}H_2}$$

$$CH_3\overset{+}{C}HCH_3 + I^- \rightarrow CH_3CHICH_3 \quad \text{Isopropyl iodide}$$

$$CH_3CH_2\overset{+}{C}H_2 + I^- \rightarrow CH_3CH_2CH_2I \quad n\text{-Propyl iodide}$$

formed from the more stable secondary carbonium ion. An empirical rule, called **Markownikoff's rule,** generalizes the addition of unsymmetrical reagents to unsymmetrical olefins as follows: *When an unsymmetrical reagent adds to an*

* The stability of carbonium ions is: tertiary > secondary > primary.

unsymmetrical olefin, the positive part of the unsymmetrical reagent becomes attached to the carbon atom of the double bond which bears the greatest number of hydrogen atoms. This rule also predicts isopropyl iodide as the product in the above reaction. Other unsymmetrical reagents add similarly.

The mechanism of ionic addition to a carbon-carbon double bond has been shown to be similar for the addition of halogen, sulfuric acid, and the hydrogen halides. However, the attacking positive species was different in several of these reactions and involved either a proton (H^+) or a chloronium (Cl^+), bromonium (Br^+) or iodonium ion (I^+). A general name to include all species of this type is the term **electrophilic reagent.** An electrophilic reagent is defined as an electron-seeking reagent. Therefore, it is either positively charged or is electron deficient. Some common examples of electrophilic reagents are: acids (hydrogen halides, sulfuric acids), Ag^+, SO_2, and oxidizing agents (MnO_4^-, $Cr_2O_7^=$). The term used to define the opposite of an electrophilic reagent is the term **nucleophilic reagent.** A nucleophilic reagent can be defined as a nucleus-seeking (positive center seeking) reagent. Some common examples of nucleophilic reagents are H_2O:, $\ddot{N}H_3$, $\bar{O}H$, $\bar{C}N$, and reducing agents such as Zn and Na.

Many common organic mechanisms can be viewed as reactions between electrophilic and nucleophilic reagents. For example, in the addition of hydrogen chloride to propene, the hydrogen chloride is an electrophilic reagent and the olefin is a nucleophilic reagent. Since the emphasis in the mechanistic interpretation of these addition reactions of acids is on the addition of the proton (the electrophilic reagent), *the characteristic reaction of olefins may be defined even more explicitly as electrophilic addition reactions.* These terms, electrophilic and nucleophilic, are useful in describing mechanistic features of chemical reactions and will be used as needed throughout the organic section of the book.

DIENES

Dienes are organic compounds containing two double bonds. The characteristic molecular formula which describes dienes is C_nH_{2n-2}. The nomenclature of these compounds is similar to alkenes, except that the term **-diene** is added to the parent hydrocarbon prefix instead of **-ene** with simple alkenes. In addition, the position of the two double bonds must be specified using the lowest possible numbers.

Two double bonds may be incorporated into an organic compound in three possible ways, as illustrated for pentadiene:

$CH_2{=}CH{-}CH{=}CHCH_3$ (1,3-Pentadiene) $CH_2{=}C{=}CHCH_2CH_3$ (1,2-Pentadiene) $CH_2{=}CH{-}CH_2{-}CH{=}CH_2$ (1,4-Pentadiene)

In 1,4-pentadiene, the two double bonds behave chemically as if the other double bond were not present in the molecule. The second double bond has no influence on the first double bond, and the compound behaves like a simple alkene except that it requires twice as much reagent to saturate the molecule. Consequently, it will add two moles of hydrogen, two moles of hydrogen halide, two moles of halogen, and so forth.

Compounds containing a 1,2-diene system are known as **allenes.** Most of their chemistry is similar to compounds containing only one double bond; but again they add two moles of reagent per allene molecule. They are generally more reactive than simple olefins and do undergo some reactions that are the result of the molecule containing the allenic double bond. However, most of these reactions are quite complex and beyond the scope of this treatment.

Compounds containing a 1,3-diene system are known as **conjugated dienes. Conjugated systems** are compounds containing *alternating double and single bonds*. Therefore, 1,3-butadiene; 1,3-pentadiene; and cyclopentadiene are examples of a conjugated olefin, but 1,2-butadiene; 1,4-pentadiene; and 1,4-cyclohexadiene are not conjugated systems.

Conjugated Olefins	*Nonconjugated Olefins*
$CH_2{=}CH{-}CH{=}CH_2$ 1,3-Butadiene	$CH_2{=}C{=}CHCH_3$ 1,2-Butadiene
$CH_2{=}CH{-}CH{=}CHCH_3$ 1,3-Pentadiene	$CH_2{=}CHCH_2CH{=}CH_2$ 1,4-Pentadiene
Cyclopentadiene	1,4-Cyclohexadiene

Most diene systems can be prepared by methods similar to those employed for introducing a double bond into a simple olefin, except that the starting material must be polyfunctional (contain two or more functional groups).

Conjugated dienes differ from other diene systems, not in the types of reagents that they react with, but in the mode of addition of these reagents. Whereas most dienes undergo 1,2-addition of electrophilic reagents similar to ethylene, conjugated dienes undergo 1,4-addition of electrophilic reagents. For example, if the amount of bromine is controlled, 1,3-butadiene will add only one mole of bromine to give a dibromide. Two dibromides are possible for this addition product. If 1,2-addition occurs, 3,4-dibromo-1-butene is the expected

$$\underset{\text{1,3-Butadiene}}{CH_2{=}CH{-}CH{=}CH_2} + Br_2\ (1\ \text{mole}) \xrightarrow[\text{addition}]{\text{1,2-}} \underset{\text{3,4-Dibromo-1-butene}}{CH_2{=}CHCHBrCH_2Br}$$

$$\underset{\text{1,3-Butadiene}}{CH_2{=}CH{-}CH{=}CH_2} + Br_2\ (1\ \text{mole}) \xrightarrow[\text{addition}]{\text{1,4-}} \underset{\text{1,4-Dibromo-2-butene}}{CH_2BrCH{=}CHCH_2Br}$$

product. However, the product is actually found to be the 1,4-addition product, 1,4-dibromo-2-butene. In addition to the bromine adding in a 1,4-manner, the double bond is shifted to the 2,3- position in the carbon chain. This type of conjugative addition is common for conjugated dienes, and differentiates this kind of diene from other dienes.

ALKYNES

Alkynes are organic compounds that contain a carbon-carbon triple bond. The general molecular formula which describes these compounds is C_nH_{2n-2}— the same as dienes, since both types of compounds contain the same degree of unsaturation. This class of compounds is also referred to as "acetylenes," after the first member of this series, acetylene itself, $HC{\equiv}CH$.

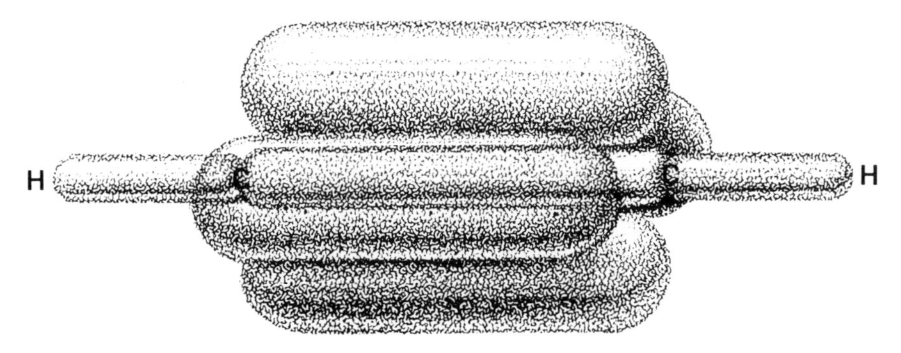

Figure 12–2 Acetylene.

The carbon-carbon triple bond in alkynes is composed of one sigma bond formed by using an *sp*-orbital (see Fig. 3–11) from each carbon, and two π-bonds formed from the remaining *p*-orbitals on the respective carbon atoms. The two π-bonds are perpendicular to one another and enclose the carbon-carbon sigma bond in a cylinder of electron density (Fig. 12–2).

The alkynes are similar to alkanes and alkenes in physical properties. They are insoluble in water and soluble in organic solvents. Their boiling points are similar to the corresponding alkanes and alkenes, and the physical properties of some of the more common alkynes are listed in Table 12–2. Since the alkynes involve bonding using *sp*-orbitals, they are linear molecules. Consequently, *cis-trans* isomers are not possible for compounds such as 2-butyne ($CH_3C{\equiv}CCH_3$) as was found for the 2-butenes.

NOMENCLATURE

In the IUPAC nomenclature system the ending **-yne** is added to the stem name, and the position of the triple bond is given the lowest number. Other substituents are named as before. Some of the lower homologues of acetylene are also named as derivatives of acetylene in which the —C≡C— group is named as acetylene, and the substituents attached to this group are denoted and attached to the acetylene name. Some typical examples are cited on page 213.

TABLE 12–2 PHYSICAL PROPERTIES OF ALKYNES

Compound	*Structure*	*M.P. °C*	*B.P. °C*
Acetylene	$HC{\equiv}CH$	−82	−84 (subl.)
Propyne	$CH_3C{\equiv}CH$	−103	−23
1-Butyne	$CH_3CH_2C{\equiv}CH$	−126	+8
2-Butyne	$CH_3C{\equiv}CCH_3$	−32	+27
1-Pentyne	$CH_3CH_2CH_2C{\equiv}CH$	−106	+40
2-Pentyne	$CH_3C{\equiv}CCH_2CH_3$	−109	+56
1-Hexyne	$CH_3CH_2CH_2CH_2C{\equiv}CH$	−132	+71
2-Hexyne	$CH_3C{\equiv}CCH_2CH_2CH_3$	—	+84

Compound	*IUPAC Name*	*Common Name*
$HC\equiv CH$	Ethyne	Acetylene
$CH_3C\equiv CH$	Propyne	Methylacetylene
$CH_3CH_2C\equiv CH$	1-Butyne	Ethylacetylene
$CH_3C\equiv CCH_3$	2-Butyne	Dimethylacetylene

METHODS OF PREPARATION

Acetylene is prepared industrially by the hydrolysis of calcium carbide, which is prepared from lime and coke. This two-step process is outlined below:

$$\underset{\text{Coke}}{3C} + \underset{\text{Lime}}{CaO} \xrightarrow[\text{temps.}]{\text{high}} \underset{\text{Calcium carbide}}{CaC_2} + \underset{\text{Carbon monoxide}}{CO}$$

$$\underset{\text{Calcium carbide}}{CaC_2 + 2H_2O} \longrightarrow \underset{\text{Acetylene}}{HC\equiv CH} + \underset{\text{Calcium hydroxide}}{Ca(OH)_2}$$

Laboratory preparations of acetylenes are similar to those used for the preparation of alkenes and dienes.

Dehydrohalogenation.

$$\underset{\text{2,2-Dichlorobutane}}{CH_3CH_2CCl_2CH_3} + 2KOH \xrightarrow{\text{alcohol}} \underset{\text{2-Butyne}}{CH_3C\equiv CCH_3} + 2KCl + 2H_2O$$

$$\underset{\text{2,3-Dichlorobutane}}{CH_3CHClCHClCH_3} + 2KOH \xrightarrow{\text{alcohol}} \underset{\text{2-Butyne}}{CH_3C\equiv CCH_3} + 2KCl + 2H_2O$$

Vicinal dihalides required in the previous reaction are conveniently prepared via the addition of halogens to alkenes. Thus, the synthesis of propyne may be envisioned as a two-step process starting from propene:

$$\underset{\text{Propene}}{CH_3CH{=}CH_2} + Br_2 \longrightarrow \underset{\text{1,2-Dibromopropane}}{CH_3CHBrCH_2Br}$$

$$\underset{\text{1,2-Dibromopropane}}{CH_3CHBrCH_2Br} + 2KOH \xrightarrow{\text{alcohol}} \underset{\text{Propyne}}{CH_3C\equiv CH}$$

Via Other Acetylenes. The carbon-hydrogen bonds in alkanes and alkenes are generally very stable to basic reagents. In contrast to this behavior, terminal acetylenes (those containing at least one hydrogen attached to the carbon-carbon triple bond) exhibit an acidic character. The hydrogen attached to the triple bond can be replaced when the terminal acetylene is treated with a strong base, such as metallic sodium. The acetylene is converted into an acetylide salt. Acetylides of this type react with alkyl halides to give a new acetylene as shown below:

$$\underset{\text{Propyne}}{CH_3C\equiv CH} + Na \xrightarrow{\text{liq. } NH_3} CH_3C\equiv C^-Na^+ + \tfrac{1}{2}H_2$$

$$CH_3C\equiv C^-Na^+ + \underset{\text{Ethyl bromide}}{CH_3CH_2Br} \longrightarrow \underset{\text{2-Pentyne}}{CH_3C\equiv CCH_2CH_3} + Na^+Br^-$$

The halide ion is displaced by the acetylide anion to produce the alkyne. For best results, the alkyl halide must be a primary alkyl halide. Secondary and tertiary halides undergo side reactions with the acetylide anion, and low yields of the alkyne are produced. This method is, however, quite versatile for preparing many homologues of acetylene.

REACTIONS OF ALKYNES

Since acetylenes contain two π-bonds, it might be expected that the characteristic reaction of acetylenes would also be an addition reaction. In fact, this prediction is realized experimentally, and alkynes undergo electrophilic addition reactions like olefins, except that two moles of reagent are required to saturate the two π-bonds. In many cases, addition reactions to acetylenes can be controlled to give an olefin derivative, which is formed via addition of one mole of reagent to a triple bond. Some of the more characteristic reactions of alkynes are summarized in the following sections.

Hydrogenation. Exhaustive catalytic (Pt, Pd, Ni) hydrogenation of alkynes gives alkanes as the final product. Partial hydrogenation gives an alkene. In compounds which can exhibit *cis-trans* isomerism, the *cis*-isomer is the predominant isomer formed on partial catalytic hydrogenation of an acetylene.

$$CH_3C{\equiv}CCH_3 + H_2 \xrightarrow{Pt} \underset{H}{\overset{CH_3}{}}\!\!\!\!\!\;C{=}C\!\!\!\!\!\;\underset{H}{\overset{CH_3}{}} \xrightarrow[Pt]{H_2} CH_3CH_2CH_2CH_3$$

2-Butyne — *cis*-2-Butene — *n*-Butane

Halogenation. Similarly to alkenes, chlorine and bromine add easily to a triple bond. Fluorine is generally too vigorous, and iodine generally does not form stable addition products. Again, the reaction may be carried out stepwise, as follows:

$$HC{\equiv}CH + Cl_2 \longrightarrow ClCH{=}CHCl \xrightarrow{Cl_2} CHCl_2CHCl_2$$

Acetylene — 1,2-Dichloroethylene — 1,1,2,2-Tetrachloroethane

Addition of Hydrogen Halides. The addition of hydrogen halides (HF, HCl, HBr, and HI) leads first to vinyl* halides and then to 1,1-dihalides. These addition reactions follow Markownikoff's rule.

$$HC{\equiv}CH + HBr \longrightarrow CH_2{=}CHBr \xrightarrow{HBr} CH_3CHBr_2$$

Acetylene — Vinyl bromide (1-bromoethene) — 1,1-Dibromoethane

---

* The $CH_2{=}CH{-}$ is known as the vinyl group. $CH_2{=}CHCl$ (vinyl chloride) is an important olefin in preparing the commercial polyvinyl chloride.

Questions

1. Draw a structural formula for each of the following compounds:
   (a) *trans*-2-hexene
   (b) *cis*-2,3-dichloro-2-butene
   (c) 1-methylcyclopentene
   (d) *trans*-1,2-dibromocyclohexane
   (e) 4-ethyl-1-octene
   (f) 3-hexyne
   (g) *cis*-diiodoethylene
   (h) 2-methyl-2-butene
   (i) 2-chloro-1,3-cyclohexadiene
   (j) 2-bromo-1,3-butadiene

2. Name each of the following compounds by the IUPAC system:
   (a) $CH_3CH_2C{\equiv}CH$
   (b) $CH_2{=}CH{-}CBr{=}CH{-}CH_3$
   (c) [cyclohexene ring bearing CH_3]
   (d) $(CH_3)_2C{=}CHCH_3$
   (e) $CH_3CH{=}CCl_2$
   (f) $CH_2{=}CF_2$
   (g) $CH_3(H)C{=}C(Cl)CH_3$ — CH_3 and Cl on top, H and CH_3 on bottom
   (h) $CH_2{=}CHCH_2CH_2CH{=}CH_2$
   (i) $(CH_3)_2C{=}CHCH_2CH(CH_3)_2$
   (j) $CH_3CHClC{\equiv}CCH_3$

3. Draw out all the possible structural isomers for the compound having a molecular formula, C_5H_{10}. Name each isomer according to the IUPAC system.

4. Which of the isomeric compounds in Question 3
   (a) contain(s) no geometric isomers?
   (b) has(ve) no double bonds?
   (c) is (are) symmetrical?
   (d) has the highest boiling point?
   (e) contain(s) geometric isomers, but no double bonds?

5. Write equations for the reactions of 1-pentene with the following reagents:
   (a) Br_2
   (b) H_2SO_4
   (c) HI
   (d) H_2/Ni

6. Write equations for the reactions of 1-butyne with the following reagents:
   (a) excess Cl_2
   (b) 1 mole Br_2
   (c) excess HI
   (d) $Ag(NH_3)_2^+$
   (e) excess H_2/Ni

7. Using reactions discussed in this chapter and the preceding chapter, show how each of the following compounds could be prepared from isopropyl iodide:
   (a) propene
   (b) 2,3-dimethylbutane
   (c) propyne
   (d) isopropyl chloride
   (e) 2,2-dibromopropane

Suggested Reading

Bonner and Castro, Essentials of Modern Organic Chemistry. Rheinhold Publishing Corp., 1965, pp. 143–194.

Ihde: The Unraveling of Geometric Isomerism and Tautomerism. Journal of Chemical Education, Vol. 36, p. 330, 1959.

Tucker: Catalytic Hydrogenation Using Raney Nickel. Journal of Chemical Education, Vol. 27, p. 489, 1950.

CHAPTER 13

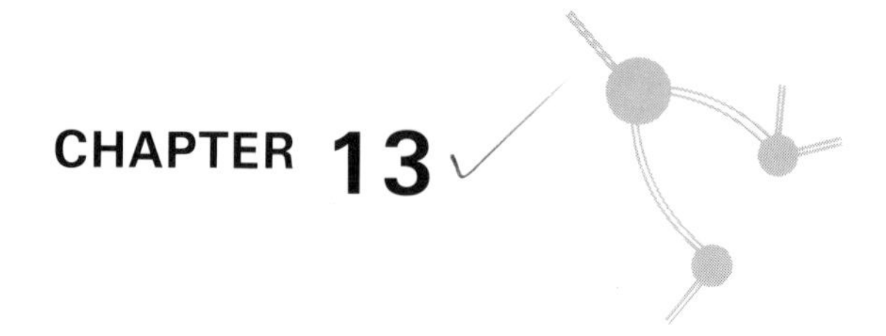

AROMATIC HYDROCARBONS

Early in the nineteenth century a class of organic compounds was isolated from aromatic substances such as oils of cloves, vanilla, wintergreen, cinnamon, bitter almonds, and benzoin. These compounds were pleasant smelling substances, and the term **aromatic** was given to this class of compounds to denote their aroma. In fact, many of these "aromatics" are still used in the perfumery and flavor extract industries because of their distinctive and pleasant odors.

The investigation of the chemistry of this class of compounds soon made it evident that the aromatic compounds were not related in an obvious manner to alkanes, alkenes, or alkynes and constituted a new class of hydrocarbons. Further chemical investigation also made it evident that all of the members of this class of hydrocarbons were structurally related to a cyclic hydrocarbon, **benzene,** which has the molecular formula C_6H_6. Benzene is not a pleasant smelling substance like many of its derivatives, and the original meaning of the term aromatic can only be loosely applied to benzene. However, the term has been carried over to include the chemistry of benzene and benzene derivatives; hence, this class of hydrocarbons is still called **aromatic hydrocarbons.**

Benzene, itself, was first isolated from an oily distillate of illuminating gas by Michael Faraday in 1825. Later, it was also shown to be a constituent of coal tar. Coal tar is a heavy black liquid obtained from the destructive distillation of coal at high temperatures. Coal tar has been the chief source of aromatic compounds and today still serves as the commercial source of many aromatics, although some aromatic hydrocarbons are now obtained from the petroleum industry by aromatization of petroleum hydrocarbons.

The initial preparation of benzene from illuminating gas by Faraday was carried out by pyrolyzing benzoic acid with lime. From the high ratio of carbon to hydrogen, it was expected that benzene would be highly unsaturated. Den-

$$\underset{\text{Sodium benzoate}}{C_6H_5CO_2Na} + NaOH \xrightarrow[\Delta]{CaO} \underset{\text{Benzene}}{C_6H_6} + Na_2CO_3$$

sity, combustion analysis, and molecular weight studies established the molecular formula as C_6H_6. If a straight-chain structure is written for a compound of this molecular formula, several double bonds or triple bonds must be included in the carbon chain to attain the molecular formula C_6H_6. Surprisingly, however, it was found that benzene did not decolorize a solution of bromine in carbon tetrachloride (therefore, did not undergo addition of bromine), and it was not oxidized by potassium permanganate, a reagent which easily oxidizes alkenes and alkynes. Therefore, structures containing several double and/or triple bonds were ruled out for the structure of benzene.

Further investigation of benzene established the following facts: (1) catalytic hydrogenation of benzene indicates that benzene absorbs three moles of hydrogen and gives cyclohexane, C_6H_{12}, as the final product. Therefore, benzene must contain a cyclic six-membered ring of carbon atoms. Similarly, benzene adds three moles of chlorine in sunlight to give hexachlorocyclohexane, again consistent with a cyclic six-membered ring. These addition reactions also suggest that benzene contains an unsaturation which is equivalent to three double bonds. (2) benzene can be chlorinated or brominated in the presence of a catalyst to give only one compound (C_6H_5X). Experimental evidence indicates that all six hydrogens in benzene are equivalent. If they were not equivalent, more than

$$C_6H_6 + X_2 \xrightarrow{Fe} C_6H_5X + HX$$
$$(X = Cl, Br)$$

one isomer would be possible from this kind of halogenation experiment. In addition, the halogenation reaction gives two products, indicating that this reaction is a **substitution** reaction and not an **addition** reaction.

To account for the above experimental facts, Friedrich Kekulé in 1865 proposed the following structure for benzene. He suggested that benzene is composed of a hexagon of carbon atoms with alternating single and double

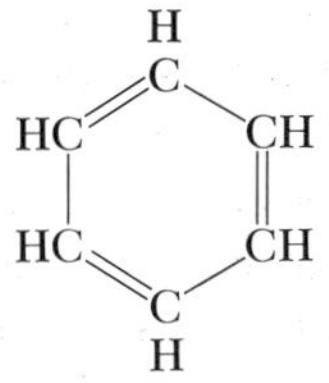

H
C
HC CH
HC CH
C
H

Kekulé benzene

bonds between the carbon atoms. This structure could account quite nicely for the addition of three moles of hydrogen and chlorine to give cyclohexanes and for the formation of a monohalogenated derivative on halogenation with catalysis.

However, the **Kekulé benzene structure** does not account for the stability of benzene to oxidizing agents like potassium permanganate. Also, this structure predicts two isomers for a 1,2-disubstituted benzene (I and II). In isomer I, the two carbon atoms holding the chlorine atoms are bonded with a double bond, whereas in II they are bonded with a single bond. However, only one 1,2-dichlorobenzene has ever been isolated.

```
         H                          H
         C                          C
    HC       CCl               HC       CCl
                      and
    HC       CCl               HC       CCl
         C                          C
         H                          H
        (I)                        (II)
```

To circumvent these difficulties, Kekulé further proposed that benzene is composed of a dynamic equilibrium between two equivalent structures (as

```
         H                          H
         C                          C
    HC       CH                HC       CH
                      ⇌
    HC       CH                HC       CH
         C                          C
         H                          H
```

shown) in which there is a rapid oscillation of the double and single bonds. Kekulé suggested that if the equilibrium were rapid enough, the two isomeric 1,2-disubstituted benzenes would be converted into each other at too fast a rate, so that only one isomer could be isolated. However, this rapid equilibrium still does not account for the stability of benzene to oxidizing agents and for its ease in undergoing substitution reactions rather than addition reactions.

The modern concept of the structure of benzene has evolved from this early chemical investigation and more recently from the use of x-ray analysis. X-ray diffraction has shown that benzene is a **planar hexagon** and that all the carbon-carbon bond distances are identical and equal to 1.39 Å. Consequently, the idea of alternating single and double bonds is unacceptable, since structures of this type cannot account for identical carbon-carbon bond lengths.

The modern structure of benzene is depicted as the formation of a six-membered ring of carbon atoms using the sp^2 orbitals of carbon. The resulting p orbitals (one on each carbon) then overlap to form a π-bond (similar to a π-bond in an alkene, except that the π-bond can be formed with one of two near neighbors). This π-bond is formed by the overlap of six p orbitals which results in a region of π-electron density above and below the hexagon structure, such that the hexagonal carbon skeleton is encased in a donut shape π-electron cloud. Thus, benzene can be depicted as shown in the following diagram:

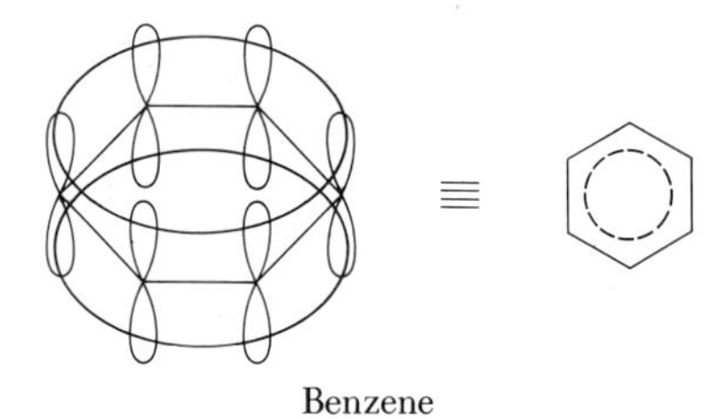

Benzene

The dotted line indicates that the π-electron density is distributed evenly over the six-carbon ring.

Another representation commonly used is a double headed arrow between the two Kekulé structures with the understanding that the positions of the double bonds are not fixed. It has been demonstrated that compounds like

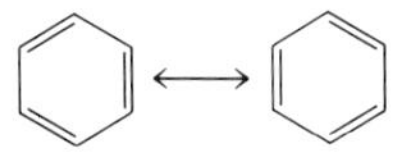

benzene, which have a continuous π-electron cloud encompassing all the carbon atoms of the ring system, exhibit unusual stability compared to analogous systems in which the electrons would be fixed in double bonds. In benzene this stability amounts to 36 Kcal./mole compared to a cyclohexatriene having fixed double bonds. Consequently, any chemical reaction which occurs to disrupt

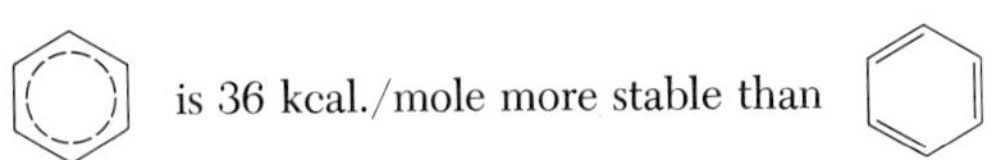

this continuous π-electron cloud will be unfavored, since it will take 36 Kcal./mole of energy more to carry out this reaction on benzene than on a cyclohexatriene type structure. Therefore, oxidation or addition reactions which would disrupt the continuous π-cloud are unfavorable energetically, and do not occur under ordinary conditions.

In addition to benzene and substituted benzenes, other aromatic hydrocarbons are known which have one or more **fused** or **condensed** rings. Again, the π-electron system is continuous over the entire carbon skeleton, and these compounds are more stable than would be expected for similar compounds containing fixed double bonds. Some of the more common condensed ring systems are shown in Table 13–1. Several systems containing condensed rings have been shown to be carcinogenic (will cause cancer) and should be handled with caution.

NOMENCLATURE

Many aromatic compounds are named by common names, or as derivatives of the parent hydrocarbon by naming the substituent attached to the ring followed by the name of the aromatic hydrocarbon.

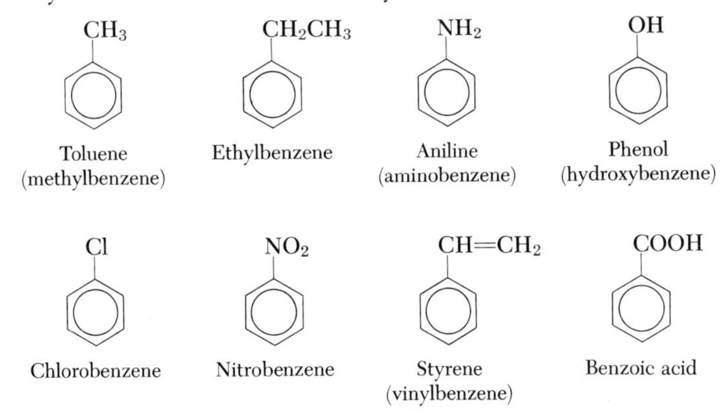

The C_6H_5 group is known as the **phenyl** group.

TABLE 13-1 PHYSICAL PROPERTIES OF AROMATIC HYDROCARBONS

Compound	*Structure*	*M.P. °C*	*B.P. °C*
Benzene		+6	+80
Toluene	CH_3	−95	+111
Ethylbenzene	CH_2CH_3	−95	+136
Isopropylbenzene	$CH(CH_3)_2$	−96	+152
Naphthalene		+80	+218
Anthracene		+217	+355
Phenanthrene		+100	+340
o-Xylene	CH_3 CH_3	−25	+144
m-Xylene	CH_3 CH_3	−48	+139
p-Xylene	CH_3 CH_3	+13	+138

When two substituents are attached to the benzene ring, two systems of nomenclature are used. The position and number of each substituent can be indicated by the appropriate number and prefix. Alternatively, the relative positions of the two substituents can be indicated by the prefixes *ortho-* (*o*-), *meta-* (*m*-), and *para-* (*p*-) to indicate either a 1,2-; 1,3-; or 1,4- position of the substituents relative to each other. For aromatic hydrocarbons containing more than two substituents, the numbering system is used. The manner in which the numbers are applied is not always consistent. Sometimes it is done alphabetically, and sometimes it is done by assigning the most important substituent the lowest number, and numbering the other substituents accordingly. Some examples of polysubstituted benzenes are shown below:

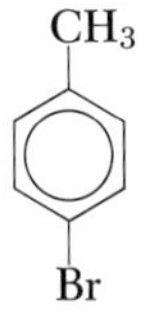

CH_3 / Br — *p*-Bromotoluene (4-bromo-1-methylbenzene)

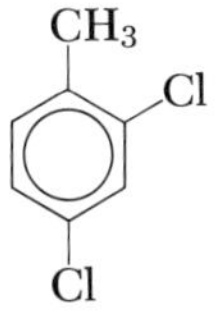

CH_3 / Cl / Cl — 2,4-Dichlorotoluene

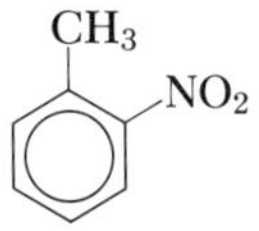

CH_3 / NO_2 — *o*-Nitrotoluene (2-nitrotoluene)

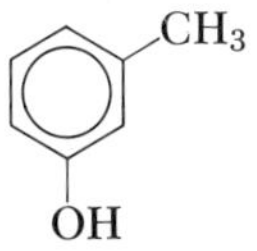

CH_3 / OH — *m*-Hydroxytoluene (3-hydroxytoluene)

PHYSICAL PROPERTIES OF AROMATIC HYDROCARBONS

Benzene and its homologues are similar to other types of hydrocarbons with respect to their physical properties. They are insoluble in water but soluble in organic solvents. The boiling points of the aromatic hydrocarbons are slightly higher than those of the alkanes of similar carbon content. For example, *n*-hexane, C_6H_{14}, boils at 69°C, whereas benzene, C_6H_6, boils at 80°C. The planar structure and highly delocalized electron density in the aromatic hydrocarbon increase the forces acting between molecules and result in a higher boiling point. Also, the symmetrical structure of benzene permits better packing in the crystal, resulting in a higher melting point than the straight-chain alkane of similar carbon content. Benzene melts at +5.5°C, whcreas *n*-hexane melts at −95°C. The physical properties of some of the common aromatic hydrocarbons are summarized in Table 13–1.

Aromatic hydrocarbons are quite flammable and should be handled with caution. Benzene is toxic when taken internally and must be used with proper precautions in any commercial process. Prolonged inhalation of its vapors results in a decreased production of red and white corpuscles in the blood which may prove fatal. Consequently, compounds of this class should only be handled under well-ventilated conditions.

METHODS OF PREPARATION

Wurtz-Fittig Reaction. The Wurtz coupling reaction of alkyl halides with sodium was illustrated earlier in Chapter 11 as a preparative method for even-numbered alkanes. Attempts to employ this reaction for the preparation of odd-numbered alkanes gives a complex mixture of all possible alkanes. However, if one of the halides is an aromatic (aryl) halide, suitable coupling

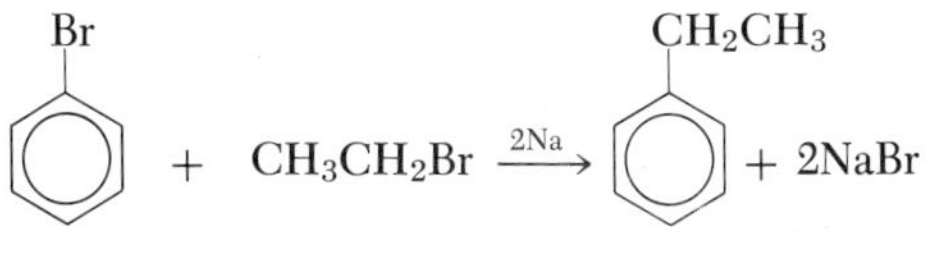

$$C_6H_5Br + CH_3CH_2Br \xrightarrow{2Na} C_6H_5CH_2CH_3 + 2NaBr$$

Bromobenzene — Ethyl bromide — Ethylbenzene

does take place to give substituted aromatic hydrocarbons. Although some symmetrical coupling of the halides does take place to give *n*-butane ($CH_3CH_2CH_2CH_3$) and biphenyl ($C_6H_5—C_6H_5$), these products are easily separated from the ethylbenzene produced. Consequently, the Wurtz coupling of two different halides can be useful if one of the halides is an aromatic halide. This kind of reaction was particularly valuable in the early history of organic chemistry.

Friedel-Crafts Reaction. The French chemist, Charles Friedel, and the American chemist, James Crafts, discovered an alkylation reaction which could be used to prepare benzene derivatives. They found that when benzene is treated with an alkyl halide in the presence of a Lewis acid like aluminum chloride, an alkylated benzene is produced. For example, benzene and methyl chloride in the presence of aluminum chloride gives toluene. Ethylbenzene could be similarly prepared from benzene, ethyl bromide, and aluminum bromide.* The major limitation on this kind of reaction is that in many cases

$$C_6H_6 + CH_3Cl \xrightarrow[\Delta]{AlCl_3} C_6H_5CH_3 + HCl$$

Benzene Methyl chloride Toluene

the aromatic hydrocarbon obtained contains an alkyl substituent that has a rearranged carbon skeleton. For example, the reaction of benzene, *n*-propyl chloride, and aluminum chloride would be expected to yield *n*-propylbenzene. However, cumene is the major product of this reaction.

$$C_6H_6 + CH_3CH_2CH_2Cl \xrightarrow[\Delta]{AlCl_3} C_6H_5CH(CH_3)_2 + HCl$$

Benzene *n*-Propyl chloride Cumene

Rearrangement of the carbon skeleton of the alkyl halide is usually observed with longer chain alkyl halides, and hence, this reaction is of limited use for the preparation of aromatic hydrocarbons containing a straight-chain alkyl substituent.

Although the Friedel-Crafts reaction historically involved alkyl halides as reactants, it has since been found that alkenes and alcohols also can be reacted with benzene in the presence of a suitable acid catalyst to give similar derivatives of aromatic hydrocarbons. Several examples are cited below:

$$C_6H_6 + CH_2{=}CHCH_3 \xrightarrow[\Delta]{AlCl_3} C_6H_5CH(CH_3)_2$$

Benzene Propene Cumene

---

*Although the aluminum halides are generally used, other acid catalysts such as H_2SO_4, BF_3, HF, and $SnCl_4$ have been used, depending on the type of starting material and its reactivity.

$$C_6H_6 + CH_3CH_2CH_2OH \xrightarrow[\Delta]{BF_3} C_6H_5CH(CH_3)_2$$

Benzene n-Propyl alcohol Cumene

Again, isomerization of the alkyl group occurs in the appropriate cases, and the same limitation exists for these reactants.

REACTIONS OF AROMATIC HYDROCARBONS

The most characteristic reactions of aromatic hydrocarbons are substitution reactions rather than addition reactions. Consequently, two products are formed in the reaction. One is an organic product in which an atom or group of atoms has substituted for the hydrogen atom of the aromatic ring. The second product is usually an inorganic acid or water.

Halogenation. With chlorine or bromine, halogenation occurs readily in the presence of a Lewis acid catalyst. Usually, iron or an iron halide containing the same halide atom as the halogenating agent is used as the catalyst. Fluorine is generally too reactive to be used in substitution reactions of this

$$C_6H_6 + Br_2 \xrightarrow{FeBr_3} C_6H_5Br + HBr$$

Benzene Bromobenzene

type, and iodine is generally too unreactive to be successful by this method. The introduction of the first halogen (chlorine or bromine) into the aromatic ring makes the ring less susceptible to further attack and deactivates it, and the introduction of a second or third halogen atom becomes progressively more difficult.

Nitration. The introduction of a nitro group ($—NO_2$) into an aromatic ring can be readily carried out using a mixture of concentrated nitric and

$$C_6H_6 + HONO_2(HNO_3) \xrightarrow[50\text{–}60^\circ C]{H_2SO_4} C_6H_5NO_2 + H_2O$$

Benzene Nitric acid Nitrobenzene

sulfuric acids. The introduction of a nitro group into the ring also deactivates the aromatic ring to further substitution, and more vigorous conditions must be used to introduce a second or third nitro group. Nitrobenzene has a harmful physiological effect on the red blood corpuscles and on the liver. Therefore, caution should be used in handling this material, and inhalation of its vapor should be avoided.

Sulfonation. The introduction of a sulfonic acid ($—SO_3H$) group can be accomplished by treating benzene with concentrated sulfuric acid at elevated temperatures, or with fuming sulfuric acid (sulfuric acid containing SO_3) at moderate temperatures.

$$C_6H_6 + HOSO_2OH + SO_3 \rightarrow C_6H_5SO_3H$$

Benzene Fuming sulfuric acid Benzenesulfonic acid

Sulfonic acids are strong acids and are usually water soluble. Their acidity is generally comparable to that of the mineral acids.

Although sulfonic acids themselves are important, derivatives of sulfonic acids have received extensive treatment, since many of them were found to exhibit physiological action. Sulfanilamide, the forerunner of the sulfa drugs, is a sulfonic acid derivative.

$$H_2N{-}C_6H_4{-}SO_2NH_2$$

Sulfanilamide

Sulfanilamide was found to be effective in the treatment of streptococcus infections, pneumonia, puerperal fever, gonorrhea, and gas gangrene. Unfortunately, sulfanilamide is only slightly soluble in water and tends to crystallize from aqueous solutions. When administered orally, the drug is absorbed and eventually carried to the kidney for excretion. When the dose is large, or under prolonged therapy, the kidneys are damaged by the accumulated sulfanilamide. Other toxic reactions, including methemoglobinemia, caused a search for derivatives that were less toxic. Sulfathiazole, sulfapyridine, sulfaguanidine, and sulfadiazine are among these derivatives.

$$H_2N{-}C_6H_4{-}SO_2NH{-}C_3H_2NS \qquad H_2N{-}C_6H_4{-}SO_2NH{-}C_5H_4N \qquad H_2N{-}C_6H_4{-}SO_2NH{-}C({=}NH)NH_2 \qquad H_2N{-}C_6H_4{-}SO_2NH{-}C_4H_3N_2$$

Sulfathiazole Sulfapyridine Sulfaguanidine Sulfadiazine

Extensive studies of the therapeutic properties of each of the sulfa drugs were carried out following the initial therapeutic discovery of sulfanilamide in 1936. This resulted in better treatment and control of various infectious diseases. Sulfadiazine, for example, is less toxic than the other sulfa drugs, yet is one of the most effective in the treatment of pneumonia and staphylococcus infections. These studies also disclosed that some of the sulfa drugs are very poorly absorbed from the intestinal tract and can be used as intestinal antiseptics. However, in destroying organisms in the intestine, these drugs interfere with the synthesis of vitamin K and members of the Vitamin B complex, such as p-aminobenzoic acid, biotin, and folic acid, and may produce vitamin deficiencies.

MECHANISM OF AROMATIC SUBSTITUTION REACTIONS

The generally accepted mechanism for aromatic substitution reactions is one involving electrophilic attack on the aromatic hydrocarbon. The π-electron system of the aromatic hydrocarbon is attacked by an electrophilic reagent to give a charged intermediate (A), which then loses a proton to a base to reform the aromatic π-system. Thus, the additional stability of the aromatic system gained from the continuous π-electron system is not lost. A generalized scheme which represents the main features of this mechanism is outlined below:

$$C_6H_6 + X^+ \rightarrow (A) \xrightarrow{\text{base}} C_6H_5X + \text{base } H^+$$

Benzene Electrophilic reagent (A)

The electrophilic reagent X^+ can be any of the electrophiles formed by the following reactions:

Halogenation:

$$Y_2 + FeY_3 \rightleftarrows Y^+[FeY_4]^-$$

$Y = Cl, Br$

Nitration:

$$HONO_2 + H_2SO_4 \rightleftharpoons H_2O + NO_2^+[HSO_4]^-$$

Nitric acid Sulfuric acid

Sulfonation:

$$H_2SO_4 \rightleftharpoons H^+ + HSO_4^-$$

Friedel-Crafts:

Alkylation:

$$RX + AlX_3 \rightleftharpoons R^+[AlX_4]^-$$

$X = \text{Halogen}$

Thus, attack by a halonium ion (Cl^+, Br^+), a nitronium ion (NO_2^+), a proton (H^+), or a carbonium ion (R^+) gives the charged intermediate (A). The proton is removed from (A) by the bases $[FeY_4]^-$, $[HSO_4]^-$, or $[AlX_4]^-$. The electron pair left behind on removal of the proton is used to reform the aromatic sextet of electrons.

ORIENTATION IN AROMATIC SUBSTITUTION REACTIONS

The introduction of an atom or a group of atoms into an unsubstituted benzene ring presents no problem as to the position taken by the new atom or groups of atoms. Since all the carbon-hydrogen bonds in benzene are equivalent, substitution of a halogen atom, a nitro group, a sulfonic acid group, an alkyl group, or an acyl group can give only one product. However, if a second group is substituted on the ring, more than one isomer is possible, and the position of the new atom or group of atoms becomes important. For example, nitration of toluene could give any or all of the following products:

$$\text{Toluene} + HNO_3 \xrightarrow{H_2SO_4} o\text{-Nitrotoluene}, \quad m\text{-Nitrotoluene}, \quad p\text{-Nitrotoluene}$$

From an extensive study of many electrophilic reactions of substituted benzene derivatives, a set of simple rules has been deduced which can be used to predict the expected product in reactions of this type.

Orientation Rules. The predominant products are predicted by the following orientation rules:

1. The position of the second substituent is determined by the group already present on the ring.
2. The atom or group of atoms already present on the ring may be divided into two classes.

   Class A: Atoms or groups of atoms which orient the new group predominantly into the *ortho-* and *para-* positions. The members of this class are called *ortho-para* directors and include atoms or groups such as $—NH_2$, $—OH$, OCH_3, alkyl groups (CH_3, C_2H_5, and so forth), Cl, Br, and I.

   Class B: Atoms or groups of atoms which orient the new group predominantly into the *meta-* position. The members of this class are called *meta* directors and include groups such as NO_2, COOH, CHO, CN, CO_2CH_3, and SO_3H.

Though these rules predict the major products, in most cases small amounts of the other isomers are also formed, but generally only to a minor extent.

Using these orientation rules, the nitration reaction of toluene would be expected to give mainly *ortho* and *para* nitrotoluenes, since the methyl group is an *ortho-para* director. Some other applications of these rules are outlined below:

$$\text{Nitrobenzene} + HNO_3 \xrightarrow[\Delta]{H_2SO_4} m\text{-Dinitrobenzene}$$

$$\text{Toluene} + H_2SO_4 \xrightarrow{SO_3} o\text{-Methylbenzene sulfonic acid} + p\text{-Methylbenzene sulfonic acid}$$

SIDE-CHAIN REACTIONS OF AROMATIC HYDROCARBONS

A substituted aromatic hydrocarbon can undergo two possible modes of chemical attack. It can undergo electrophilic aromatic substitution on the aromatic ring itself, or reaction can take place in the group (side-chain) attached to the aromatic ring. The mode of chemical reaction will be dependent on the type of side chain, the kind of chemical reagent, and the conditions used for carrying out the reaction. For example, toluene undergoes chlorination in the ring under ionic conditions; but under free-radical conditions (sunlight), chlorination occurs in the methyl group similarly to the free-radical substitution reaction of carbon-hydrogen bonds in alkanes (see Chapter 11).

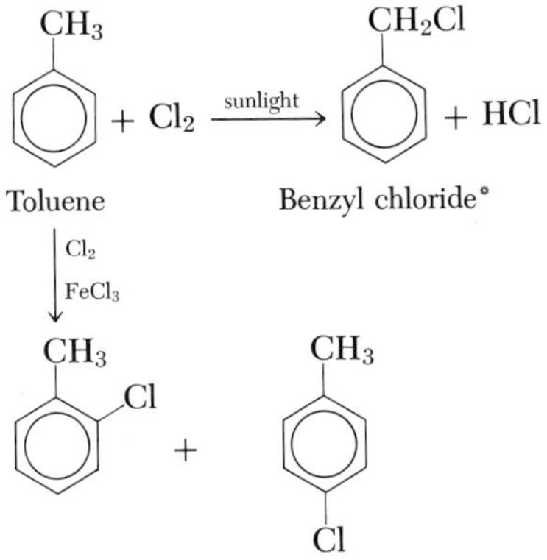

$$C_6H_5CH_3 + Cl_2 \xrightarrow{\text{sunlight}} C_6H_5CH_2Cl + HCl$$

Toluene → Benzyl chloride*

$$C_6H_5CH_3 \xrightarrow[FeCl_3]{Cl_2} o\text{-}ClC_6H_4CH_3 + p\text{-}ClC_6H_4CH_3$$

o-Chlorotoluene *p*-Chlorotoluene

In many cases, the reagent and conditions used will affect only the side-chain, and the aromatic ring will be unaffected by the chemical reaction. For example, aromatic hydrocarbons are resistant to oxidizing agents. Consequently, an alkyl group side-chain can be oxidized to a carboxylic acid without oxidizing the

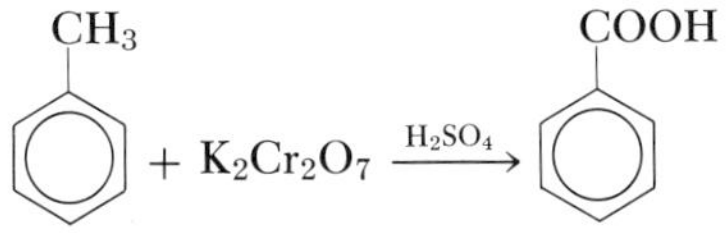

$$C_6H_5CH_3 + K_2Cr_2O_7 \xrightarrow{H_2SO_4} C_6H_5COOH$$

Toluene → Benzoic acid

aromatic ring. The alkyl group, regardless of its length, is degraded to the —COOH group. If more than one alkyl group is attached to the ring, a polyfunctional acid results on oxidation. Thus, *para*-xylene gives terephthalic acid, (cf. Chapter 18) an important compound in the synthesis of Dacron.

Similarly, the aromatic ring is stable to chemical reduction. Thus, nitro groups can be reduced by tin and hydrochloric acid to give the amino group. Nitrobenzene gives aniline as a reduction product.

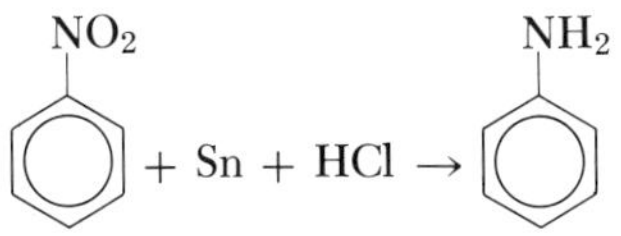

$$C_6H_5NO_2 + Sn + HCl \rightarrow C_6H_5NH_2$$

Nitrobenzene → Aniline

*The $C_6H_5CH_2$ is commonly known as the benzyl group.

Derivatives of aromatic amines have important medicinal properties and are used as drugs. Nitration of an aromatic hydrocarbon followed by chemical reduction of the nitro group to the amino group provides a facile route to the intermediates used in many of these medicinal compounds.

HETEROCYCLIC COMPOUNDS

As has been stated previously, the two major types of organic compounds are the aliphatic and the cyclic. If the cyclic compounds are composed of rings of carbon atoms only, as in benzene and its derivatives, they are called **carbocyclic.** When atoms other than carbon are also included in the ring the compounds are termed **heterocyclic.** The most commonly occurring elements other than carbon in these ring structures are oxygen, nitrogen, and sulfur.

The basic ring structure of a heterocyclic compound is called the **heterocyclic nucleus.** In the majority of heterocyclic compounds the nucleus consists of five- or six-membered rings that contain either one or two elements other than carbon. Examples of the important heterocyclic nuclei and their derivatives will be considered in the following sections.

Five-Membered Rings. Several important heterocyclic compounds are derived from a heterocyclic ring made up of 1 oxygen and 4 carbon atoms. This ring is known as **furan,** and one of its most important derivatives is the α-aldehyde, **furfural.** For purposes of nomenclature, the rings are numbered counterclockwise starting with the element other than carbon. In the furan nucleus illustrated here, the carbon atoms adjacent to the oxygen are α, the next ones are called β carbons.

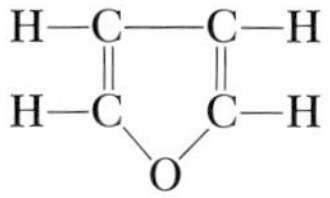

```
H—C——C—H        H—C——C—H            β'C4——3C β
  ||   ||          ||   ||               ||    ||
H—C    C—H       H—C    C—CHO        α'C5    2C α
   \  /             \  /                 \  /
    O                O                    O

  Furan            Furfural
```

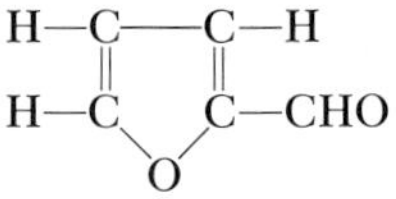

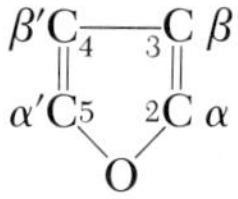

Another important five-membered heterocyclic ring, containing nitrogen in place of oxygen, is **pyrrole.** It was originally obtained from animal matter but can be readily synthesized from the action of ammonia on a dicarboxylic hydroxy acid called glycaric acid.

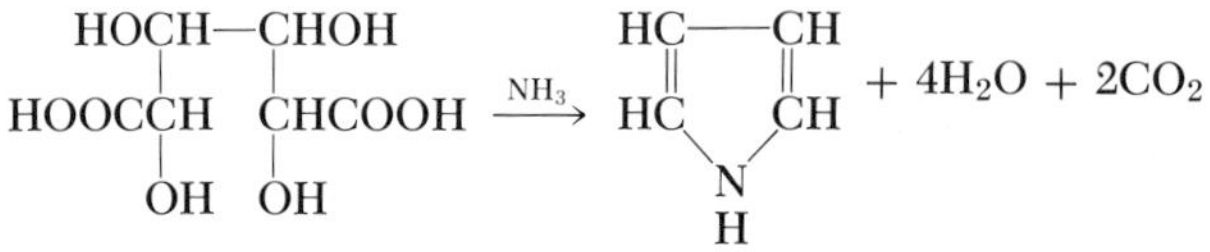

```
   HOCH—CHOH               HC——CH
    |    |                 ||   ||
HOOCCH  CHCOOH  --NH3-->   HC    CH   + 4H2O + 2CO2
    |    |                   \  /
    OH   OH                   N
                              H

   Glycaric acid            Pyrrole
```

Pyrrole is a liquid that gradually forms dark-colored resins on exposure to the air. It is constituent of many important naturally occurring substances such as hemoglobin, chlorophyll, amino acids, and alkaloid drugs.

If pyrrole is condensed with a benzene ring, another type of heterocyclic nucleus called **indole** is produced. Indole and 3-methyl indole, which is called **skatole,** are formed during the putrefaction of proteins in the large intestine.

They are responsible for the characteristic odor of feces. One of the most important derivatives of the indole nucleus is the amino acid **tryptophan.** Tryptophan is present in most proteins and is an essential constituent of the diet of growing animals.

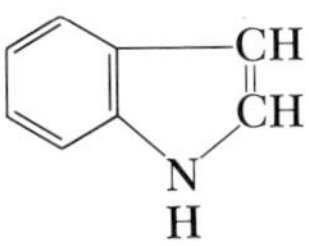

Indole 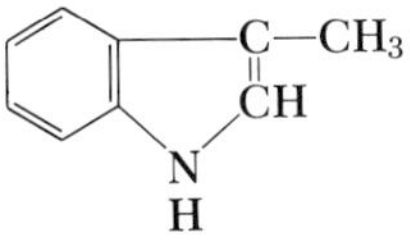   Skatole 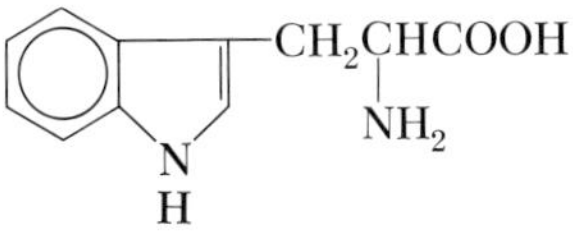   Tryptophan

The other example of a five-membered heterocyclic ring, in which sulfur takes the place of oxygen or nitrogen, is **thiophene.** This compound occurs as an impurity in the benzene obtained from coal tar. It is produced commercially by a reaction of butane and sulfur at high temperatures.

$$CH_3CH_2CH_2CH_3 + 4S \longrightarrow C_4H_4S + 3H_2S$$

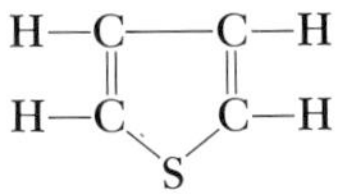

Butane   Thiophene

While many derivatives may be made from thiophene, as yet none of these has outstanding industrial applications.

SIX-MEMBERED RINGS. The most common six-membered rings containing one element other than carbon are pyran and pyridine.

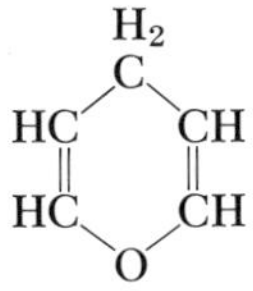

Pyran 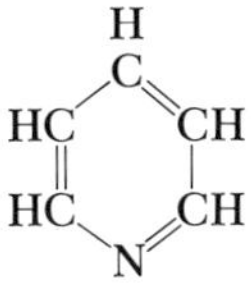   Pyridine

The pyran ring is present in anthocyanin, which is responsible for the color of flowers, and in rotenone, a plant material that is used as an insecticide. The benzopyran ring is found in many plants and is an integral part of the α-tocopherol molecule. Wheat germ oil is especially rich in α-tocopherol, or vitamin E, which is necessary for normal growth and reproduction in animals.

Pyridine is a common heterocyclic compound obtained from coal tar. In recent years the shortage of this compound resulted in the development of synthetic processes for its manufacture. Pyridine is a liquid with a characteristically disagreeable odor. It behaves as a weak base and is a good solvent for both organic and inorganic compounds. It is used to manufacture pharmaceuticals such as sulfa drugs, antihistamines, and steroids. Pyridine serves as a denaturant for ethyl alcohol, as a rubber accelerator, and in the preparation of a waterproofing agent for textiles. The methyl pyridines are known as picolines and may be oxidized to the corresponding picolinic acids. The acid obtained from the oxidation of β-picoline, or 3-methyl pyridine, is known as **nicotinic acid.**

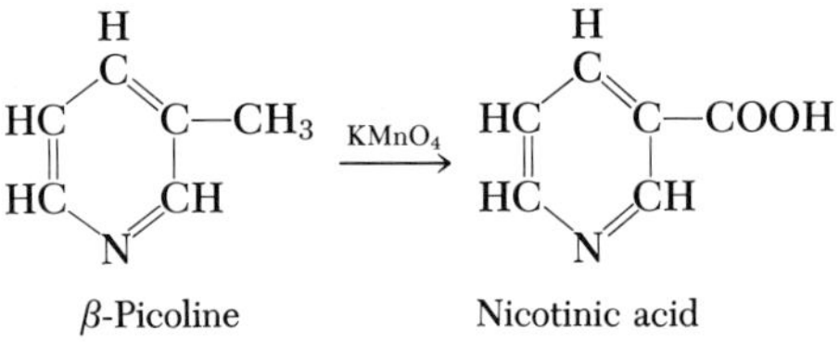

```
      H                            H
      C                            C
  HC     C—CH3    KMnO4       HC      C—COOH
  ||     |       ------>      ||      |
  HC     CH                   HC      CH
      N                            N
```

β-Picoline   Nicotinic acid

Nicotinic acid and its amide are members of the vitamin B complex. Consumption of a diet lacking in these compounds results in a deficiency disease called pellagra. Nicotinamide is used by the body for the manufacture of coenzymes I and II, which are essential for the proper functioning of certain dehydrogenating enzymes.

Questions

1. Draw a structural formula for each of the following compounds:
   (a) toluene
   (b) naphthalene
   (c) *p*-nitrotoluene
   (d) *m*-bromotoluene
   (e) benzyl bromide
   (f) *m*-diethylbenzene
   (g) *o*-dinitrobenzene
   (h) 2,4-dinitrotoluene
   (i) 1,3,5-trimethylbenzene
   (j) 1,4-dichloro-2,5-dibromobenzene

2. Draw out all the possible structural isomers for aromatic compounds with the following molecular formulas and name each of the isomers:
   (a) C_8H_{10}
   (b) C_8H_9Cl
   (c) $C_7H_6Br_2$
   (d) $C_6H_3Cl_2Br$
   (e) C_7H_6ClBr

3. Write equations for each of the following reactions, and name each of the organic products obtained:
   (a) benzene + isopropyl chloride ($AlCl_3$ catalyst)
   (b) toluene + Br_2 (Fe catalyst)
   (c) toluene + H_2SO_4 + SO_3
   (d) nitrobenzene + concentrated nitric acid + sulfuric acid (heat)
   (e) chlorobenzene + bromine (Fe catalyst)
   (f) benzenesulfonic acid + Br_2 (Fe catalyst)
   (g) toluene + nitric acid + sulfuric acid
   (h) ethylbenzene + H_2SO_4 + SO_3
   (i) toluene + bromine (sunlight)

4. Show how the following conversions may be carried out in the laboratory. More than one step may be necessary. Give any necessary catalysts.
   (a) benzene to *m*-chloronitrobenzene
   (b) benzene to benzoic acid
   (c) benzene to *m*-nitroethylbenzene

Suggested Reading

Fedor: Major Aromatics In Transition, 1961–1965. Chemical and Engineering News, Part I, p. 116, March 20, 1961; Part II, p. 130, March 27, 1961.

Ferguson: The Orientation and Mechanism of Electrophilic Aromatic Substitution. Journal of Chemical Education, Vol. 32, p. 42, 1955.

Gero: Kekulé's Theory of Aromaticity. Journal of Chemical Education, Vol. 31, p. 201, 1954.

Hartough: The Chemical Nature of Thiophene and Its Derivatives. Journal of Chemical Education, Vol 27, p. 500, 1950.

Waack: The Stability of the Aromatic Sextet. Journal of Chemical Education, Vol. 39, p. 469, 1962.

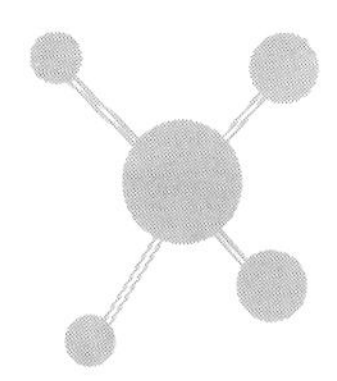

CHAPTER 14

ALCOHOLS

The basic elements of the compounds contained in the preceding chapter were carbon and hydrogen. The addition of a third element, oxygen, greatly extends the number of types of possible organic compounds. Oxygen has six electrons in its valence shell and must therefore share two electrons to attain the stable inert gas configuration. The manner in which oxygen shares electrons with carbon and hydrogen determines the type of organic compound. For example, when oxygen shares one electron with carbon and one electron with hydrogen, the type of organic compound called an **alcohol** results. If each valence electron is shared with a separate carbon atom, a class of compounds called **ethers** is produced. If each electron is shared with the same carbon atom to form a carbon-oxygen double bond, an aldehyde or ketone results, depending upon the other groups or atoms bonded to carbon. The various types of covalent bonding possible for oxygen are illustrated below:

$$\begin{array}{ccccccccc} & & \text{H} & & \text{H} & & & & \\ & & | & & | & & & & \\ \text{H} & — & \text{C} & — & \text{C} & — & \text{O} & — & \text{H} \\ & & | & & | & & & & \\ & & \text{H} & & \text{H} & & & & \end{array}$$

Ethyl alcohol

$$\begin{array}{ccccccccc} & & \text{H} & & & & \text{H} & & \\ & & | & & & & | & & \\ \text{H} & — & \text{C} & — & \text{O} & — & \text{C} & — & \text{H} \\ & & | & & & & | & & \\ & & \text{H} & & & & \text{H} & & \end{array}$$

Dimethyl ether

$$\begin{array}{ccccccc} & & \text{H} & & \text{H} & & \\ & & | & & | & & \\ \text{H} & — & \text{C} & — & \text{C} & = & \text{O} \\ & & | & & & & \\ & & \text{H} & & & & \end{array}$$

Acetaldehyde

$$\begin{array}{ccccccccc} & & \text{H} & & \text{O} & & \text{H} & & \\ & & | & & \| & & | & & \\ \text{H} & — & \text{C} & — & \text{C} & — & \text{C} & — & \text{H} \\ & & | & & & & | & & \\ & & \text{H} & & & & \text{H} & & \end{array}$$

Acetone

Ethers, aldehydes, and ketones will be discussed in detail in subsequent chapters.

Of all the classes of organic compounds, alcohols are probably the best known. For centuries it has been recognized that alcoholic beverages contain ethyl, or grain, alcohol. Also, most temporary automobile antifreezes contain methyl, or wood, alcohol. Most permanent type antifreezes contain ethylene glycol (CH_2OHCH_2OH), an alcohol containing two hydroxyl groups.

Alcohols may be considered derivatives of hydrocarbons in which a carbon-hydrogen bond of the hydrocarbon has been replaced (substituted) by a hydroxyl group. Another way of considering alcohols is to view them as the organic analogs of water, in which one of the hydrogen atoms of water has been replaced by an alkyl or aryl group. If the latter way of viewing alcohols is accepted, the chemical and physical behavior of alcohols might be anticipated to be similar in some respects to water.

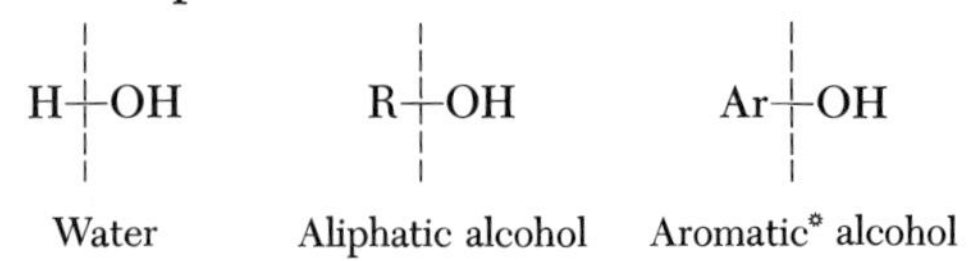

Water Aliphatic alcohol Aromatic* alcohol

Before studying the chemistry of alcohols, it is convenient to subdivide the aliphatic alcohols into three classes, primary, secondary, and tertiary. Each class is dependent upon the hydroxyl group being attached to a carbon that is covalently bonded to one, two, or three other carbon atoms. Many reactions of alcohols can be more easily discussed employing this type of subclassification, and it is only for this reason that the alcohols are classified in the aforesaid manner.

NOMENCLATURE

Alcohols are named in the IUPAC system by replacing the **-e** ending of the corresponding alkane with the characteristic **-ol** ending of the alcohols. Other substituents are named and their positions on the carbon chain indicated by the appropriate number and prefix. Since the hydroxyl group may appear at more than one position on the carbon chain, its position must also be indicated by the number of the carbon atom to which it is attached. Common names are used for the lower members of this series and are formulated by naming the alkyl group attached to the hydroxyl group, followed by the term alcohol. Some typical examples illustrating these various nomenclature and classification systems are illustrated as follows:

Compound	*Name*	*Classification*
CH_3OH	Methanol Methyl alcohol	Primary
CH_3CH_2OH	Ethanol Ethyl alcohol	Primary
$CH_3CH_2CH_2OH$	1-Propanol *n*-Propyl alcohol	Primary
$CH_3CHOHCH_3$	2-Propanol Isopropyl alcohol	Secondary
$CH_3CH_2CH_2CH_2OH$	1-Butanol *n*-Butyl alcohol	Primary
$(CH_3)_3COH$	2-Methyl-2-propanol *t*-Butyl alcohol	Tertiary

*The symbol Ar— is used to denote an aromatic ring. In this case, if Ar— = C_6H_5, the aromatic alcohol would be phenol, C_6H_5OH.

The aromatic alcohols are named as derivatives of the parent compound phenol (or carbolic acid). Some typical examples are represented below:

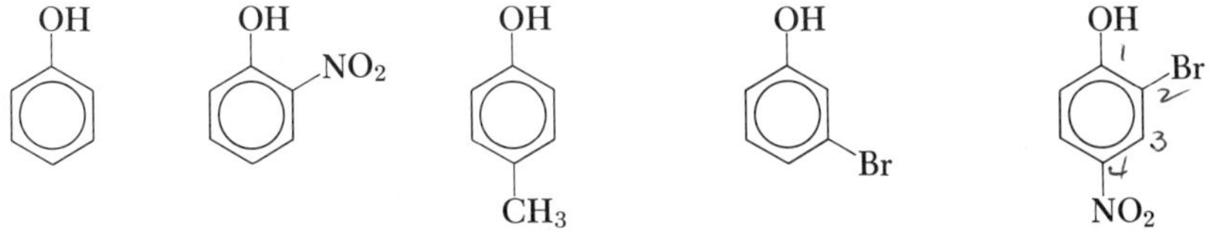

Phenol | *o*-Nitrophenol | *p*-Methylphenol | *m*-Bromophenol | 2-Bromo-4-nitrophenol

The methyl derivatives are also known by the common name, **cresols.**

PHYSICAL PROPERTIES OF ALCOHOLS

The introduction of a hydroxyl group into a hydrocarbon has a pronounced effect on the physical properties of the compound. In contrast to hydrocarbons, which are insoluble in water, the short-chain alcohols (methanol through the butanols) are soluble in water. As the number of carbons in the alcohol increases, the solubility in water decreases, and the physical properties approach those of the saturated hydrocarbons.

The boiling points of alcohols are also abnormally high compared to the saturated hydrocarbons of comparable molecular weight. For example, ethanol, CH_3CH_2OH (mol. wt. 46), has a boiling point of $+78°C$, whereas propane, $CH_3CH_2CH_3$ (mol. wt. 44), boils at $-42°C$. The large increase in boiling point is due to the presence of the hydroxyl group. That the increase is not due merely to the presence of the oxygen atom can be proved by comparing ethanol to its isomer, dimethyl ether, CH_3OCH_3 (mol. wt. 46), which boils at $-24°C$. Therefore, the change in boiling point must be due to the presence of the $-OH$ group. The effect of this group has been attributed to the presence of hydrogen bonds formed between alcohol molecules in the liquid state, similar to the type of hydrogen bonding encountered in water. This type of association can also

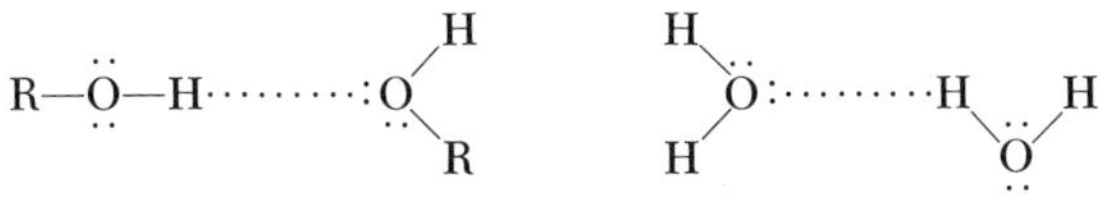

Hydrogen-bonded alcohol | Hydrogen-bonded water

be used to explain the solubility of the short-chain alcohols in water. When these alcohols are dissolved in water, hydrogen bonding occurs between the hydroxyl group of the alcohol and the hydroxyl group of the water.

The aromatic alcohol, **phenol,** is slightly soluble in water, but very soluble in alcohol, ether, and other organic solvents. It has strong antiseptic properties, but is a poison when taken internally. It is very caustic and causes blisters and deep burns when it comes into contact with the skin. If it is accidentally spilled on the skin, it should be washed off immediately. A dilute solution (3%) of phenol is an effective disinfectant for use in hospitals. Similarly, dilute solutions of the cresols, known by the name Lysol, are used as disinfectants in hospitals. The caustic action of phenol and its derivatives, in contrast to the aliphatic alcohols which have no caustic properties, is due to the difference in acidity of these

TABLE 14-1 PHYSICAL PROPERTIES OF ALCOHOLS AND PHENOLS

Compound	*Structure*	*M.P. °C*	*B.P. °C*
Methyl alcohol	CH_3OH	−97	+65
Ethyl alcohol	CH_3CH_2OH	−114	+78
n-Propyl alcohol	$CH_3CH_2CH_2OH$	−126	+97
Isopropyl alcohol	$(CH_3)_2CHOH$	−89	+82
n-Butyl alcohol	$CH_3CH_2CH_2CH_2OH$	−90	+118
Isobutyl alcohol	$(CH_3)_2CHCH_2OH$	−108	+108
t-Butyl alcohol	$(CH_3)_3COH$	+25	+83
Cyclohexanol	(cyclohexane ring)—OH	+24	+162
Phenol	(benzene ring)—OH	+41	+182

two types of alcohols. The aliphatic alcohols, ROH (Ka $\cong 10^{-18}$), are weaker in acidity than water (Ka $= 10^{-14}$). Phenol (Ka $\cong 10^{-10}$), on the other hand, is a much stronger acid than water. This difference in acidity between aliphatic alcohols and phenols is also reflected in their behavior with bases. Alcohols do not form salts with aqueous sodium hydroxide, whereas phenol does form salts with aqueous alkalis. Although phenols are stronger acids than water or the aliphatic alcohols, they are still weak acids when compared to the carboxylic acids (RCOOH, Ka $\sim 10^{-5}$) and the inorganic mineral acids.

METHODS OF PREPARATION OF ALCOHOLS

HYDROLYSIS OF ALKYL HALIDES. Alkaline hydrolysis of alkyl halides, in which the halogen atom of the alkyl halide is displaced by a hydroxyl group, is a useful method for preparing alcohols. A generalized representation of this reaction is shown as follows:

$$R{-}X + OH^- \xrightarrow{H_2O} R{-}OH + X^-$$

X = Cl, Br, I
R = Alkyl group

Aqueous sodium or potassium hydroxide, or aqueous silver oxide* (Ag_2O), is generally used as the hydrolyzing agent. The reactivity of the halides in this reaction is RI > RBr > RCl. Aromatic halides, in which the halogen is directly attached to the ring, are inert to this type of displacement reaction. However, halides such as benzyl halides ($C_6H_5CH_2Cl$) do undergo this reaction. A competing side reaction in this method of preparation is the elimination of hydrogen halide from the alkyl halide to give an olefin. The best yields of the alcohols are obtained from primary alkyl halides using silver oxide at moderate temperatures. Some typical examples are outlined on the following page:

*The aqueous solution of silver oxide may be represented as AgOH (silver hydroxide). The silver oxide method promotes less elimination than the aqueous alkali method.

$$CH_3CH_2Br + Na^+OH^- \xrightarrow{H_2O} CH_3CH_2OH + Na^+Br^-$$
Ethyl bromide — Ethyl alcohol

$$CH_3CH_2CH_2I + Ag_2O \xrightarrow{H_2O} CH_3CH_2CH_2OH + AgI\downarrow$$
n-Propyl iodide — *n*-Propyl alcohol

Hydration of Olefins. Alkenes undergo hydration in the presence of strong acids such as sulfuric acid to give alcohols as the final product. The hydration reaction follows Markownikoff's rule; therefore, secondary and tertiary alcohols are formed except in the case of ethylene. Some typical examples are shown below:

$$CH_3CH{=}CH_2 + H_2O \xrightarrow{H_2SO_4} CH_3CHOHCH_3$$
Propylene — Isopropyl alcohol

$$(CH_3)_2C{=}CH_2 + H_2O \xrightarrow{H_2SO_4} (CH_3)_3COH$$
Isobutylene — *t*-Butyl alcohol

Since the first step involves the addition of the electrophile H^+, these reactions would be expected to follow Markownikoff's rule.

METHODS OF PREPARATION OF PHENOLS

The methods of preparation of aliphatic alcohols outlined previously are generally unsatisfactory for preparing phenols in the laboratory. The hydrolysis reaction, however, has been used commercially to prepare phenol. The reaction requires high temperatures and pressure. The initially formed salt is subsequently converted to phenol.

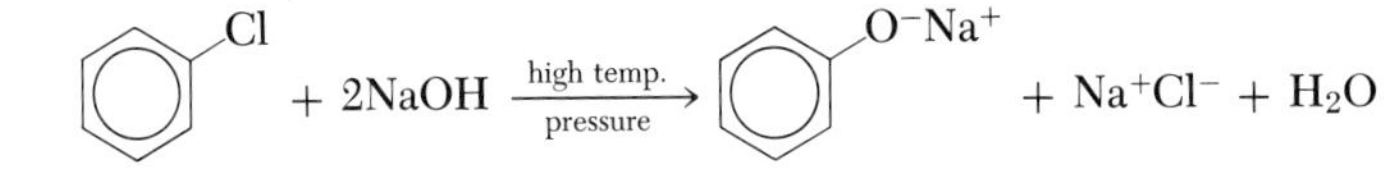

$$C_6H_5Cl + 2NaOH \xrightarrow[\text{pressure}]{\text{high temp.}} C_6H_5O^-Na^+ + Na^+Cl^- + H_2O$$
Chlorobenzene — Sodium phenoxide

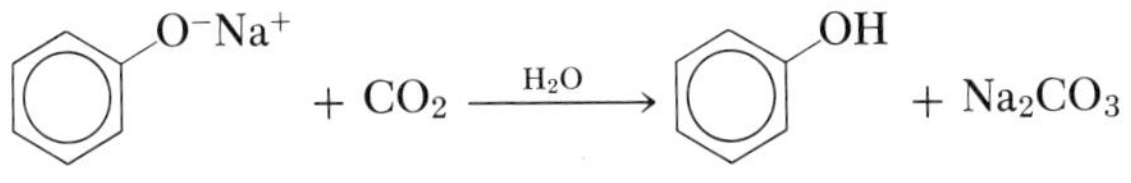

$$C_6H_5O^-Na^+ + CO_2 \xrightarrow{H_2O} C_6H_5OH + Na_2CO_3$$
Sodium phenoxide — Phenol

A similar hydrolysis process, called the Raschig process, was developed in Germany. Chlorobenzene is converted to phenol by treatment with steam at high temperatures in this process.

REACTIONS OF ALCOHOLS AND PHENOLS

Alcohols and phenols can undergo two types of reactions which involve the hydroxyl group. A reaction can occur to cleave the oxygen-hydrogen bond, or a reaction can occur to cleave the carbon-oxygen bond resulting in loss of the hydroxyl group. Reactions which involve oxygen-hydrogen bond cleavage are considered first.

Salt Formation. Because of their increased acidity, phenols will form salts with aqueous alkalies.

$$C_6H_5OH + Na^+OH^- \rightarrow C_6H_5O^-Na^+ + H_2O$$

Phenol — Sodium phenoxide

Aliphatic alcohols will not form salts with aqueous alkali. The free phenol can be regenerated by treatment of the salt with acids such as hydrogen chloride. Although aliphatic alcohols do not form salts with aqueous alkalies, both alcohols and phenols will form salts with active metals such as sodium, potassium, magnesium, and so forth. These reactions are irreversible, and hydrogen gas is liberated as the other product. The reactions are similar to the reaction of water with active metals to give an alkali metal hydroxide and hydrogen.

$$CH_3CH_2OH + Na \rightarrow CH_3CH_2O^-Na^+ + \tfrac{1}{2}H_2\uparrow$$

Ethyl alcohol — Sodium ethoxide

$$C_6H_5OH + K \rightarrow C_6H_5O^-K^+ + \tfrac{1}{2}H_2\uparrow$$

Phenol — Potassium phenoxide

$$H_2O + Na \rightarrow HO^-Na^+ + \tfrac{1}{2}H_2\uparrow$$

It is interesting to note that the salts of alcohols (called alkoxides) are strong bases when used in a nonaqueous solvent. Just as HO^- is a strong inorganic base, alkoxide ions (RO^-, where R is an alkyl group) are strong organic bases. In aqueous solution, the alkoxides are hydrolyzed back to the alcohol; therefore, they must be used in a nonaqueous medium.

$$CH_3CH_2O^-Na^+ + H_2O \rightarrow CH_3CH_2OH + Na^+OH^-$$

Sodium ethoxide — Ethyl alcohol

Ester Formation. In the presence of an acid catalyst, both alcohols and phenols will react with a carboxylic acid to form an ester. Experiments carried out using an alcohol containing oxygen-18 instead of the normal oxygen-16 have shown that the oxygen of the alcohol ends up as one of the oxygen atoms in the ester. Therefore, only the oxygen-hydrogen bond of the alcohol is broken, and the hydroxyl group used in forming the by-product water comes from the carboxylic acid and not the alcohol. This reaction is outlined schematically below:

$$R{-}O^{18}{-}H + HO{-}\overset{O}{\overset{\|}{C}}{-}R' \xrightarrow{H^+} R{-}O^{18}{-}\overset{O}{\overset{\|}{C}}{-}R' + HOH$$

Alcohol — Carboxylic acid — Ester

Some typical examples of ester formation are illustrated as follows:

$$CH_3CH_2OH + CH_3CO_2H \xrightarrow{H^+} CH_3\overset{O}{\overset{\|}{C}}{-}OCH_2CH_3$$

Ethyl alcohol — Acetic acid — Ethyl acetate

$$CH_3OH + C_6H_5\text{—}\overset{O}{\overset{\|}{C}}\text{—}OH \xrightarrow{H^+} C_6H_5\text{—}\overset{O}{\overset{\|}{C}}\text{—}OCH_3$$

Methyl alcohol   Benzoic acid   Methyl benzoate

REPLACEMENT OF THE HYDROXYL GROUP

Dehydration. Alcohols undergo loss of water (dehydration) in the presence of acid catalysts under the proper conditions. These reactions have been considered earlier under the preparation of olefins.

$$CH_3CH_2CH_2OH + Al_2O_3 \xrightarrow{\Delta} CH_3CH{=}CH_2$$

n-Propyl alcohol   Propylene

Conversion to Alkyl Halides. The hydroxyl group of alcohols can be replaced with a halogen atom by several types of reagents. With hydrogen chloride, the reaction is generally carried out using zinc chloride ($ZnCl_2$) as a catalyst. The order of reactivity is tertiary > secondary > primary alcohols.

$$ROH + HCl \xrightarrow{ZnCl_2} RCl + H_2O$$

The order of reactivity of the hydrogen halides in this type of reaction is HI > HBr > HCl. The mechanism of this reaction is dependent on the type of alcohol undergoing displacement of the hydroxyl group. With primary and secondary halides the mechanism involves protonation of the hydroxyl group and subsequent displacement of water by a halide ion, as outlined below:

$$R\ddot{O}H + HX \rightarrow R\overset{+}{\ddot{O}}H_2 + X^-$$

X = halogen
R = primary or secondary

$$X^- \curvearrowright R\text{—}\overset{+}{\ddot{O}}H_2 \rightarrow X\text{—}R + H_2O$$

It is much easier to displace water than hydroxide ion (HO^-), hence the function of the proton catalyst is to facilitate the displacement reaction. With most tertiary halides, the protonated alcohol undergoes ionization to form a carbonium ion, which then picks up a halide ion to form the alkyl halide. In this case ionization gives the much more stable tertiary carbonium ion.

$$R\text{—}\underset{R''}{\overset{R'}{C}}\text{—}\ddot{O}H + HX \rightarrow R\text{—}\underset{R''}{\overset{R'}{C}}\text{—}\overset{+}{\ddot{O}}H_2 \xrightarrow{-H_2O} R\text{—}\underset{R''}{\overset{R'}{C^+}} \xrightarrow{X^-} R\text{—}\underset{R''}{\overset{R'}{C}}\text{—}X$$

$$+ X^-$$

X = halogen
R,R′,R″ = Alkyl groups

With secondary or primary alcohols, ionization would generate the less stable secondary or primary carbonium ions. The energy required for this type of ionization process is higher for generating the secondary and primary carbonium ions than it is for the displacement of water from the protonated alcohol; consequently, the reaction proceeds via the lowest energy pathway, which is displacement of water by the halide anion. Phenols generally do not undergo displacement of the hydroxyl group attached to the aromatic ring with hydrogen halides.

OTHER REACTIONS OF ALCOHOLS AND PHENOLS

In addition to the reactions involving cleavage of the oxygen-hydrogen bond of the hydroxyl group and displacement of the hydroxyl group by carbon-oxygen bond cleavage, alcohols and phenols can undergo other kinds of reactions which involve the carbon-hydrogen bonds of the alcohol or the phenol. In phenols most of these reactions involve electrophilic aromatic substitution reactions similar to those observed earlier for benzene. The hydroxyl group attached to the aromatic ring activates the aromatic ring toward electrophilic attack, and phenol undergoes substitution by electrophilic reagents much more easily than either benzene or toluene. The hydroxyl group is an *ortho-para* director. Some typical reactions of phenol involving substitution on the aromatic ring are illustrated below:

HALOGENATION.

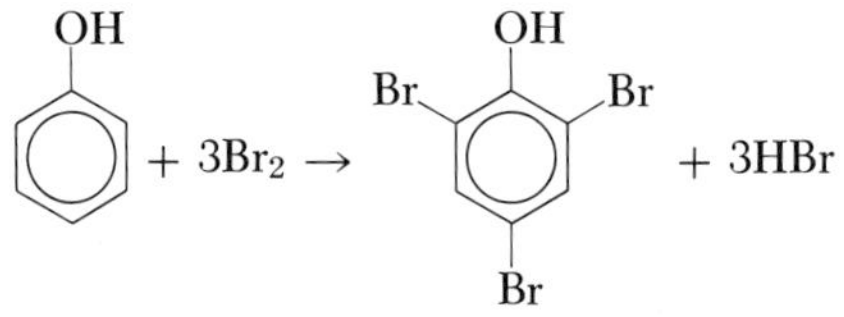

$$C_6H_5OH + 3Br_2 \rightarrow C_6H_2Br_3OH + 3HBr$$

Phenol — 2,4,6-Tribromophenol

NITRATION.

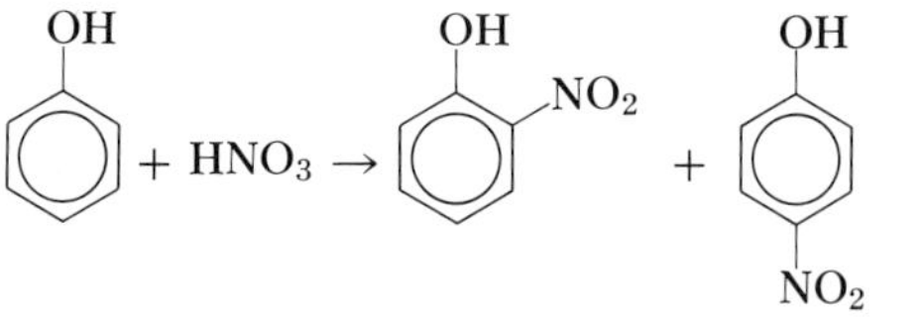

$$C_6H_5OH + HNO_3 \rightarrow o\text{-}C_6H_4(NO_2)OH + p\text{-}C_6H_4(NO_2)OH$$

Phenol — *o*-Nitrophenol — *p*-Nitrophenol

OXIDATION REACTIONS. A reaction involving dehydrogenation (overall loss of H_2) is also possible for alcohols. This type of reaction is an **oxidation reaction,** and the usual type of oxidizing agents used are alkaline potassium permanganate or acidic potassium dichromate. Primary alcohols give carboxylic acids and secondary alcohols give ketones on oxidation by these reagents. Under the normal conditions of this reaction, tertiary alcohols are not oxidized. Some typical examples are given below:

$$\underset{\text{Ethyl alcohol}}{CH_3CH_2OH} + K_2Cr_2O_7 \xrightarrow{H_2SO_4} \underset{\text{Acetaldehyde}}{CH_3CHO} \rightarrow \underset{\text{Acetic acid}}{CH_3{-}\overset{\overset{O}{\|}}{C}{-}OH}$$

$$CH_3CHOHCH_3 + KMnO_4 \xrightarrow{NaOH} CH_3{-}\overset{\overset{\displaystyle O}{\|}}{C}{-}CH_3$$

Isopropyl alcohol — Acetone

Oxidation of primary and secondary alcohols serves as a general method of preparation of carboxylic acids and ketones, both in the laboratory and in industry.

IMPORTANT ALCOHOLS AND PHENOLS

Many important laboratory, industrial, and biologically related compounds contain more than one functional group per molecule. Some typical examples containing a hydroxyl group are:

Salicylaldehyde (OH, CHO on benzene ring) — Ethylene glycol $HOCH_2CH_2OH$ — Salicylic acid (OH, COOH on benzene ring) — Glycerol $HOCH_2CHOHCH_2OH$

Some of the more important alcohols and their uses are discussed in the following pages of this chapter.

IMPORTANT ALCOHOLS. ***Methyl Alcohol.*** Methyl alcohol is commonly called wood alcohol because it was once exclusively produced by the destructive distillation of wood. When hard wood is heated in a retort at temperatures up to 400°C, the vapors that are given off contain methyl alcohol, acetic acid, acetone, and other organic compounds. A synthetic process for the production of methyl alcohol from carbon monoxide and hydrogen was developed in 1923 and has largely supplanted the wood distillation method. In this process, carbon monoxide and hydrogen react under a high pressure and a temperature of 350°C in the presence of zinc and chromium oxide catalysts.

$$CO + 2H_2 \xrightarrow[Cr_2O_3]{ZnO} CH_3OH$$

Carbon monoxide — Methyl alcohol

Recently a method has been developed for the production of methyl alcohol by the controlled oxidation of natural gas. Since ethyl and propyl alcohols and other important organic compounds are produced by this same reaction, this method is becoming increasingly important.

Methyl alcohol is a colorless, volatile liquid with a characteristic odor. It is used as a denaturant for ethyl alcohol, as a solvent for shellac in the varnish industry, as an antifreeze for automobile radiators, and as the raw material for the synthesis of other organic compounds. About 40 per cent of all the methyl alcohol that is made in this country is used in the preparation of formaldehyde, which is a starting material for many plastics. When taken internally, methyl alcohol is poisonous, small doses producing blindness by degeneration of the optic nerve, whereas large doses are fatal. The taste, odor, and poisonous properties of wood alcohol make it a desirable denaturing agent to be added to ethyl alcohol to prevent its use in beverages. During the prohibition era in the United States, many persons were blinded and others died after drinking ethyl alcohol denatured in this fashion. An individual who was blinded temporarily after drinking a small amount of methyl alcohol was said to be "blind drunk."

Ethyl Alcohol. Ethyl alcohol is commonly known as alcohol, or as grain alcohol since it may be made by fermentation of various grains. It is prepared commercially by two major methods. One involves the fermentation of the sugars and starch of common grains, potatoes, or black-strap molasses. The yeast used in fermentation contains enzymes that catalyze the transformation of more complex sugars into simple sugars, and then into alcohol and carbon dioxide, as shown in the following reactions:

$$\underset{\text{Complex sugar}}{C_{12}H_{22}O_{11}} + H_2O \xrightarrow{\text{enzymes}} \underset{\text{Simple sugar}}{2\ C_6H_{12}O_6} \xrightarrow{\text{enzymes}} \underset{\text{Ethyl alcohol}}{4\ C_2H_5OH} + \underset{\text{Carbon dioxide}}{4\ CO_2}$$

Enzymes and fermentation will be studied more completely in the section on biochemistry. The other method is a synthetic method that makes use of the reaction of the unsaturated hydrocarbon ethylene with sulfuric acid, followed by a hydrolysis reaction. This reaction is illustrated using semistructural formulas for the organic compounds involved:

$$\underset{\text{Ethylene}}{H_2C{=}CH_2} + H_2SO_4 \rightarrow \underset{\text{Ethyl sulfate}}{CH_3CH_2OSO_3H}$$

$$\underset{\text{Ethyl sulfate}}{CH_3CH_2OSO_3H} + H_2O \rightarrow \underset{\text{Ethyl alcohol}}{CH_3CH_2OH} + H_2SO_4$$

The ethylene used in this process is produced by the cracking of petroleum hydrocarbons.

Ethyl alcohol is a colorless, volatile liquid with a characteristic pleasant odor. Industrial ethyl alcohol contains approximately 95 per cent alcohol and 5 per cent water. It is difficult to remove all the water from alcohol, since in simple distillation processes a constant boiling mixture of 95 per cent alcohol and 5 per cent water is formed. Methods for removing this water have been developed, since the solvent properties of pure, absolute alcohol are considerably different from those of industrial alcohol. For example, when benzene is mixed with 95 per cent alcohol and the mixture is distilled, the final fraction consists of absolute ethyl alcohol. A large proportion of the industrial ethyl alcohol produced each year is used as an antifreeze in automobile radiators. It is an excellent solvent for many substances and is used in the preparation of medicines, flavoring extracts, and perfumes. Large quantities of ethyl alcohol are used in the manufacture of acetaldehyde, ether, and chloroform, and for the synthesis of butadiene for the manufacture of synthetic rubber. Alcohol is widely used in the hospital as an antiseptic, a vehicle for medications, and as a rubbing compound to cleanse the skin and lower a patient's temperature.

The process of fermentation has been used for centuries to produce the alcohol found in intoxicating beverages such as beers, liquors, and wines. Many countries levy a high tax on alcohol when used for beverage purposes. This practice has resulted in the production and consumption of large quantities of illegal, or "bootleg," liquors. A comparison of the price of denatured industrial alcohol with that of taxed grain alcohol will readily show why most of the alcohol produced in the United States is denatured before use. Denatured industrial alcohol costs approximately fifty-three cents a gallon, while tax-paid ethyl alcohol costs in excess of twenty dollars a gallon. Alcohol is **denatured,** or rendered unfit for drinking, by the addition of small quantities of methyl alcohol, pyridine, or benzene.

The concentration of alcohol in beverages is usually expressed as per cent or "proof." The relationship between proof and per cent alcohol concentration may be shown in the following examples. The common 100 proof whiskey is 50 per cent alcohol, whereas the standard laboratory 95 per cent alcohol is 190 proof. Beer and wine contain from 3 to 20 per cent, whereas whiskey, rum, vodka, and gin contain from 35 to 45 per cent alcohol.

When ethyl alcohol is taken internally, it is rapidly absorbed and oxidized. It may therefore be used as a readily available source of energy and is often employed to overcome shock or collapse. If large quantities are taken, it causes a depression of the higher nerve centers, mental confusion, lack of muscular coordination, lowering of normal inhibitions, and eventually stupor.

Ethylene Glycol. All the alcohols so far considered have been monohydroxy alcohols. Since it is possible to replace a hydrogen atom on more than one carbon atom in a hydrocarbon with a hydroxyl group, it is possible to have polyhydroxy (polyhydric) alcohols. The simplest polyhydric alcohol would be one formed by replacing a hydrogen on each of the two carbons of ethane by a hydroxyl group. This compound is called **ethylene glycol.** It is prepared by oxidizing ethylene to ethylene oxide, and subsequently hydrolyzing the oxide to ethylene glycol, as shown in the following equations:

$$\underset{\text{Ethylene}}{H_2C{=}CH_2} + O_2 \rightarrow \underset{\text{Ethylene oxide}}{H{-}\overset{H}{\overset{|}{C}}\underset{O}{\diagdown\!\!\diagup}\overset{H}{\overset{|}{C}}{-}H}$$

$$\underset{\text{Ethylene oxide}}{H{-}\overset{H}{\overset{|}{C}}\underset{O}{\diagdown\!\!\diagup}\overset{H}{\overset{|}{C}}{-}H} + H_2O \rightarrow \underset{\text{Ethylene glycol}}{H{-}\overset{H}{\overset{|}{\underset{\underset{H}{O}}{\underset{|}{C}}}}{-}\overset{H}{\overset{|}{\underset{\underset{H}{O}}{\underset{|}{C}}}}{-}H}$$

Ethylene glycol is water soluble and has a very high boiling point compared to those of methyl and ethyl alcohols. These properties make it an excellent permanent, or nonvolatile, type of antifreeze for automobile radiators. Antifreeze preparations such as **Prestone** and **Zerex** consist of ethylene glycol plus a small amount of a dye. A more recent development is the so called extended life antifreeze, which is left in a car's radiator the year round and may be effective for more than two years. The nonvolatile properties of these products, such as **Dowgard** and **Telar,** are also based on their content of ethylene glycol. Large quantities of ethylene glycol are used in the preparation of solvents, paint removers and plasticizers (softeners) used in the paint, varnish, and lacquer industry.

Propylene glycol is one of the constituents of sun-tan lotion. It is also being used in the development of nonaqueous foams in conjunction with fluorocarbon aerosol repellents. These foams are much more stable than those made from aqueous solutions.

If the intermediate product in the above reaction, ethylene oxide, is allowed to react with primary alcohols, a series of useful solvents called **cellosolves** are produced. The cellosolves are excellent organic solvents and yet are completely miscible with water. In industry they are used as solvents for varnishes and lacquers.

Glycerol. The most important trihydric alcohol is glycerol, which is sometimes called glycerin. It is an essential constituent of fat (an ester of glycerol and fatty acids) and may be prepared by the hydrolysis of fat as represented in the following equation:

$$\text{Fat} + \text{Hydrolysis} \rightarrow \begin{array}{c} \text{H} \\ | \\ \text{H—C—OH} \\ | \\ \text{H—C—OH} \\ | \\ \text{H—C—OH} \\ | \\ \text{H} \end{array} + \text{Fatty acids}$$

Glycerol

Glycerol is obtained commercially as a by-product of the manufacture of soap, and from a synthetic process that uses propylene from the catalytic cracking of petroleum as a starting material. Glycerol is a syrupy, sweet-tasting substance that is soluble in all proportions of water and alcohol. It is nontoxic and is often used for the preparation of liquid medications. Since it has the ability to take up moisture from the air, it tends to keep the skin soft and moist when applied in the form of cosmetics and lotions. This property is also used to help maintain the moisture content of products made of tobacco. Large quantities of glycerol are also used in the manufacture of resins and photographic film.

When treated with a mixture of sulfuric and nitric acids at a low temperature, glycerol is converted into glycerol trinitrate, which is commonly known as nitroglycerin. When heated or given a sudden shock, nitroglycerin explodes with considerable force. It is commonly used as an explosive—alone and in the form of dynamite—and in the manufacture of smokeless powder.

Questions

1. Draw out all the possible structural isomers of an alcohol which has a molecular formula $C_5H_{12}O$. Name each of these isomers according to the IUPAC system.

2. Which of the structural isomers in Question 1
   (a) are primary alcohols?
   (b) are secondary alcohols?
   (c) are tertiary alcohols?
   (d) is isoamyl alcohol?
   (e) on dehydration will give 2-pentene?

3. Give each of the following compounds an appropriate name:

(a) 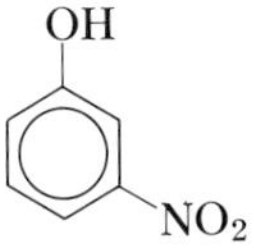 OH, NO_2 (benzene ring)

(d) 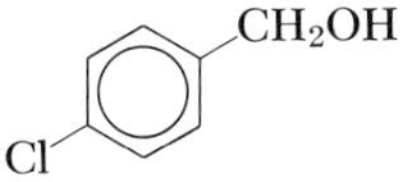 CH_2OH, Cl (benzene ring)

(b) 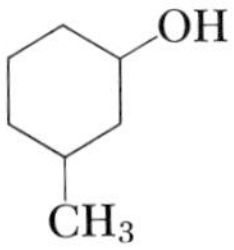 OH, CH_3 (cyclohexane ring)

(e) $(CH_3)_2CHCH_2OH$

(c) $(CH_3)_2CHCHBrCH_2CH_2OH$

4. Write equations for the preparations of the following compounds. Give any necessary catalysts. More than one step may be necessary.

(a) isopropyl alcohol from *n*-propyl bromide
(b) 2-butanol from 2-butyne
(c) 1-phenylethanol from benzene
(d) tertiary butyl alcohol from isobutyl bromide

5. Write equations for the reaction of isopropyl alcohol with the following reagents:

(a) hydrogen iodide
(b) potassium
(c) acetic acid
(d) Al_2O_3/heat
(e) $K_2Cr_2O_7/H_2SO_4$

6. Write equations for the reaction of phenol with the following reagents:

(a) Cl_2
(b) potassium
(c) nitric acid
(d) propionic acid (CH_3CH_2COOH)
(e) sodium hydroxide

7. Arrange the following compounds in order of decreasing acidity:

CH_3CH_2OH, (benzene ring)OH, H_2O, CH_3COOH, CH_3CH_3

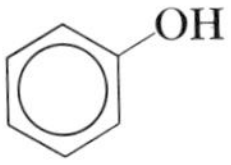

8. Arrange the following compounds in order of decreasing solubility in water:

CH_3CH_2OH, $CH_3(CH_2)_6CH_2OH$, $CH_3(CH_2)_6CH_3$, $HOCH_2CH_2OH$

9. A compound with molecular formula $C_4H_{10}O$ reacts with sodium to liberate hydrogen. It reacts with $ZnCl_2$/HCl to give an oily suspension. When passed over Al_2O_3 at high temperatures, it is converted to a new compound, C_4H_8, which, when treated with H_2SO_4/H_2O, gives the original compound, $C_4H_{10}O$. What are the structures of these compounds?

Suggested Reading

Ferguson: Hydrogen Bonding and Physical Properties of Substances. Journal of Chemical Education, Vol. 33, p. 267, 1956.
Weaver: Glycerol. Journal of Chemical Education, Vol. 29, p. 524, 1952.
Morrison and Boyd: Organic Chemistry, 2nd Ed., Allyn and Bacon, Inc., 1966, pp. 498–546, 789–812.

CHAPTER 15

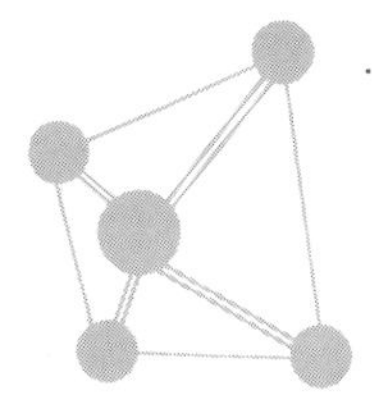

ETHERS

Ethers are closely related to the alcohols and may be considered a derivative of an alcohol in which the hydrogen of the hydroxyl group has been replaced by an alkyl or aryl group. Consequently, ethers may be considered to be organic derivatives of water (HOH) in which both hydrogens have been replaced by alkyl or aryl groups. In addition, cyclic ethers are possible in which the oxygen atom is part of the cyclic structure. The three-membered ring structures containing an oxygen atom as part of the ring are also called **epoxides.** As noted earlier in the chapter on alcohols, ethers are isomeric with the alcohols containing the same number of carbon atoms.

Ethers are named either by common names or by naming the two alkyl or aryl groups linked to the oxygen atom, followed by the word "ether." If one of the alkyl or aryl groups has no simple name, the compound is

TABLE 15-1 PHYSICAL PROPERTIES OF ETHERS

Compound	*Structure*	*M.P. °C*	*B.P. °C*
Dimethyl ether	CH_3OCH_3	−140	−25
Diethyl ether	$CH_3CH_2OCH_2CH_3$	−116 (−123)	+35
Methyl ethyl ether	$CH_3OCH_2CH_3$	—	+8
Di-*n*-propyl ether	$CH_3CH_2CH_2OCH_2CH_2CH_3$	−122	+91
Diisopropyl ether	$(CH_3)_2CH—O—CH(CH_3)_2$	−60	+68
Tetrahydrofuran	(five-membered ring with O)	−108	+66
Anisole	$C_6H_5OCH_3$	−37	+154
Diphenyl ether	$C_6H_5—O—C_6H_5$	+27	+259

named as an **alkoxy** derivative.* Some typical examples of ethers are given in Table 15–1.

Since ethers are essentially hydrocarbons with a single oxygen atom, their physical properties would be expected to parallel those of hydrocarbons. They are colorless, insoluble in water, soluble in acids (whereas alkanes are not), soluble in organic solvents, and in general have densities and boiling points similar to hydrocarbons of corresponding molecular weight. It is interesting to note that dimethyl ether, which is isomeric with ethanol, is a gas at room temperature, whereas ethanol is a liquid at room temperature. Ethers, which have no hydroxyl group, cannot form hydrogen bonds as can alcohols; consequently, the boiling points are not abnormally high like the alcohols. The physical properties of some of the more common ethers are summarized in Table 15–1.

METHODS OF PREPARATION OF ETHERS

Williamson Synthesis of Ethers. A general preparation of both symmetrical and unsymmetrical ethers is a reaction discovered by the British chemist Alexander Williamson. The reaction involves a displacement of halide ion from an alkyl halide by an alkoxide ion (obtained from an alcohol or phenol). The generalized scheme of this reaction, and several specific examples, are illustrated below:

$$\underset{\text{Metal alkoxide}}{RO^-M^+} + \underset{\text{Alkyl halide}}{R'X} \rightarrow \underset{\text{Ether}}{ROR'} + \underset{\text{Salt}}{M^+X^-}$$

M^+ = Na, K (usually)
R = alkyl or aryl

$$\underset{\text{Sodium methoxide}}{CH_3O^-Na^+} + \underset{\text{Methyl bromide}}{CH_3Br} \rightarrow \underset{\text{Dimethyl ether}}{CH_3OCH_3} + Na^+Br^-$$

$$\underset{\text{Sodium methoxide}}{CH_3O^-Na^+} + \underset{\text{Ethyl iodide}}{CH_3CH_2I} \rightarrow \underset{\text{Methyl ethyl ether}}{CH_3OCH_2CH_3} + Na^+I^-$$

$$\underset{\text{Sodium phenoxide}}{C_6H_5\text{—}O^-Na^+} + \underset{\text{Methyl iodide}}{CH_3I} \rightarrow \underset{\text{Anisole}}{C_6H_5\text{—}OCH_3} + Na^+Br^-$$

$$\underset{\text{Sodium methoxide}}{CH_3O^-Na^+} + \underset{\text{Bromobenzene}}{C_6H_5\text{—}Br} \rightarrow \text{No reaction}$$

Compared to the ease of displacement of halide ion from an alkyl halide by alkoxide, it is difficult to displace halide ion from an aromatic ring by alkoxide ion. Therefore, in preparing mixed aromatic-aliphatic ethers, such as anisole, the reaction must be carried out using sodium phenoxide displacement on methyl iodide. Reversal of the types of reagents does not give anisole.

* The —OR group is known as an alkoxy group. The name of this type of group is formed from the hydrocarbon name of the —R group by dropping the **-ane** and adding **-oxy.**

The cyclic three-member ethers (epoxides) can also be prepared by an intramolecular type of Williamson synthesis. Addition of hypohalous acids to alkenes gives halohydrins,* which undergo intramolecular cyclization in the presence of base to give epoxides, as outlined below:

$$CH_2{=}CH_2 + HOCl \rightarrow \underset{OH}{CH_2}{-}\underset{Cl}{CH_2}$$

Ethylene Hypochlorous acid Ethylene chlorohydrin

$$\underset{OH}{CH_2}{-}\overset{Cl}{CH_2} + KOH \rightarrow \left[\underset{O^-}{CH_2}{-}\overset{Cl}{CH_2}\right] \rightarrow H_2C\overset{O}{—}CH_2 + K^+Cl^- + H_2O$$

Ethylene oxide

REACTIONS OF ETHERS AND EPOXIDES

Except for the saturated hydrocarbons, ethers are the most unreactive of any of the simple functional groups. They are stable to dilute acids and bases and are also resistant to many oxidizing and reducing agents. They are similar to alkanes in their lack of chemical reactivity. This lack of chemical reactivity, however, makes ethers quite suitable as solvents for many chemical reactions. Diethyl ether and tetrahydrofuran are two of the most common solvents used today in the organic chemical laboratory. They are, however, quite flammable materials and caution must be exercised when using them. In addition, ethers form nonvolatile peroxides on standing. These peroxides are explosive in the dry state; consequently, ether solutions should never be evaporated or distilled to dryness unless these peroxides have been removed. Under more vigorous reaction conditions, ethers do undergo cleavage reactions with concentrated mineral acids, and the aromatic ethers do undergo ring substitution reactions. Some of the more common reactions of ethers are described in the following pages.

Reaction with Acids. Ethers dissolve in cold concentrated sulfuric acid to form oxonium salts. This is a reversible reaction, and the ether can be regenerated on neutralization of the acid. With the hydrogen halides, ethers can be cleaved at high temperatures.

$$R{-}\ddot{\underset{\cdot\cdot}{O}}{-}R' + H_2SO_4 \rightleftharpoons \left[R{-}\overset{H}{\underset{\cdot\cdot}{O}}{-}R'\right]^+ HSO_4^-$$

Hydrogen iodide and hydrogen bromide are usually used for this purpose. With one equivalent of the hydrogen halide, an alcohol and an alkyl halide are produced. With excess hydrogen halide, two moles of alkyl halide are produced if both groups attached to the oxygen atom of the ether are aliphatic. If one of the groups is aromatic, a mole of the corresponding phenol is formed, since phenols are not converted to aryl halides by hydrogen halides. Some typical cleavage reactions are illustrated in the following examples:

* Hypohalous acid adds to olefins as HO^+Cl^-, according to Markownikoff's Rule.

$$CH_3CH_2OCH_2CH_3 + HI\ (1\ mole) \xrightarrow{\Delta} CH_3CH_2I + CH_3CH_2OH$$

Diethyl ether — Ethyl iodide — Ethyl alcohol

$$CH_3CH_2OCH_2CH_3 + HI\ (excess) \xrightarrow{\Delta} 2CH_3CH_2I$$

$$C_6H_5{-}OCH_3 + HI \xrightarrow{\Delta} CH_3I + C_6H_5{-}OH$$

Anisole — Methyl iodide — Phenol

Aromatic Substitution Reactions of Aromatic Ethers. If one of the groups attached to the ether oxygen atom is an aromatic group, the usual halogenation, nitration, sulfonation, and alkylation reactions can be carried out on the aromatic ring without affecting any cleavage of the ether linkage. The alkoxy group of such a mixed ether is an *ortho-para* directing group and also activates the ring toward electrophilic substitution. Therefore, mixed aromatic-aliphatic ethers undergo ring substitution reactions under very mild conditions compared to benzene and alkyl substituted benzenes. A typical bromination of phenyl ethyl ether (phenetole) is illustrated below, but it should be kept in mind that similar reactions occur for nitration, sulfonation, and Friedel-Crafts alkylation.

$$C_6H_5OCH_2CH_3 + Br_2 \xrightarrow{Fe} o\text{-}BrC_6H_4OCH_2CH_3 + p\text{-}BrC_6H_4OCH_2CH_3$$

Phenetole — *o*-Bromophenetole — *p*-Bromophenetole

Reactions of Epoxides. In contrast to the chemical inertness of simple ethers, epoxides contain a strained three-membered ring and are generally much more chemically reactive than ordinary ethers. As might be expected, the vast majority of reactions of epoxides are ring-opening reactions to relieve the strain in the three-membered ring. Both acid-catalyzed and base-catalyzed ring-opening reactions are known, and two such reactions are illustrated for ethylene oxide:

$$H_2C\overset{O}{-}CH_2 + H_2O \xrightarrow{H^+} \underset{OH}{CH_2}{-}\underset{OH}{CH_2}$$

Ethylene oxide — Ethylene glycol

$$H_2C\overset{O}{-}CH_2 + \ddot{N}H_3 \rightarrow \underset{OH}{CH_2}{-}\underset{NH_2}{CH_2}$$

Ethylene oxide — Ethanolamine

Epoxides, particularly ethylene oxide, undergo a wide variety of ring-opening reactions similar to the ones described above, and ethylene oxide is one of the most important building blocks of the chemical industry.

SOME IMPORTANT ETHERS AND THEIR USES

Diethyl ether, which is often called ethyl ether or simply ether, is extensively used as a general anesthetic. It is easy to administer and causes excellent relaxation of the muscles. Blood pressure, pulse rate, and rate of respiration as a rule are only slightly affected. The main disadvantages are its irritating effect on the respiratory passages and its aftereffect of nausea. More recently, methyl propyl ether has been used as a general anesthetic. It has been claimed that this substance, called Neothyl, is less irritating and more potent than ethyl ether.

Methyl ether is used as a solvent and as a propellant for aerosol sprays. Diethyl ether is also an excellent solvent for fats and is often used in the laboratory for the extraction of fat from foods and animal tissue. In general, ethers are good solvents for fats, oils, gums, resins, and most functional derivatives of hydrocarbons. Ethylene oxide is considered an internal ether. It differs from most ethers in being completely soluble in water. It is used as a fumigating agent for seeds and grains and as the starting material in the preparation of the antifreeze ethylene glycol, the cellosolve solvents ($ROCH_2CH_2OH$, where R is an alkyl group) used in varnishes and lacquers, dioxane, and many other useful solvents and fibers.

Questions

1. Draw a structural formula for each of the following compounds:
   (a) methyl isopropyl ether
   (b) propylene oxide
   (c) *p*-nitroanisole
   (d) ethyl ether
   (e) 2-bromo-4-ethoxyhexane

2. Name each of the following compounds:
   (a) 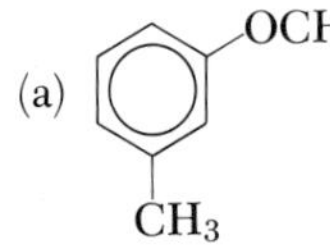

   (b) $(CH_3)_2CH-O-CH(CH_3)_2$
   (c) CH_3CH_2O-
   (d) $HOCH_2CHOHCH_2OH$
   (e) $CH_3OCH_2CH_2OCH_3$

3. Write equations for the reactions of anisole with the following reagents:
   (a) cold conc. H_2SO_4
   (b) excess HI/heat
   (c) $KMnO_4$
   (d) sodium
   (e) HNO_3/H_2SO_4

4. Show by equations how the following conversions may be carried out. More than one step may be necessary.
   (a) methyl *n*-propyl ether from *n*-propyl iodide
   (b) phenyl ethyl ether from phenol
   (c) cyclohexene oxide from cyclohexene
   (d) methyl isopropyl ether from propene
   (e) ethylene glycol from ethyl bromide

Suggested Reading

Bonner and Castro: Essentials of Modern Organic Chemistry. Reinhold Publishing Corp., 1965, pp. 287–294.
Fieser and Fieser: Advanced Organic Chemistry. Reinhold Publishing Corp., 1961, pp. 303–312.

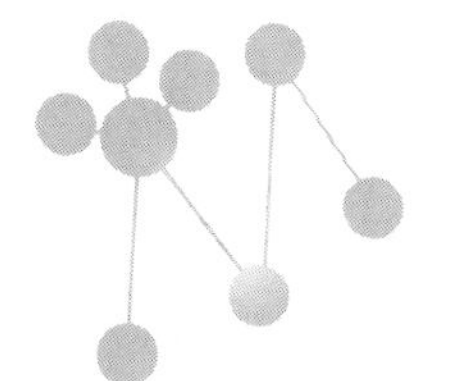

CHAPTER 16

HALOGEN DERIVATIVES OF HYDROCARBONS

In previous chapters compounds have been discussed that contain a carbon-halogen bond. These compounds resulted from the reaction of alcohols with halogenating agents and from the addition of hydrogen halides to olefins. In addition, they were found to be the basic starting materials in some of the preparations of olefins, alcohols, and ethers. The utility of this class of compounds should now be evident, and the purpose of this chapter is to correlate, extend, and amplify somewhat, the properties, reactions, and uses of the halogen derivatives of hydrocarbons.

Halogen derivatives of hydrocarbons can be simply defined as *compounds in which a carbon-hydrogen bond of a hydrocarbon has been replaced by a carbon-halogen bond,* where the halogen can be fluorine, chlorine, bromine, or iodine. The carbon atom bonded to the halogen atom may be a carbon that is part of an alkyl, alkenyl, acetylenic, aromatic, or heterocyclic system. Some examples of the various types of halogen derivatives are shown below:

CH_3CH_2Br	$CH_2{=}CHCl$	$HC{\equiv}CCl$	C_6H_5F	C_5H_4NI (4-position)
Ethyl bromide	Vinyl chloride	Chloroacetylene	Fluorobenzene	4-Iodopyridine

The most widely used types of halogen derivatives are the ones containing the halogen bonded either to an alkyl group (called **alkyl halides**) or to an aromatic ring (called **aryl halides**), and most of the emphasis in this chapter will be on these two types of halides. Although many reactions of alkyl and aryl halides may appear quite similar, in the most important reaction (displacement of the halogen atom by another atom or group of atoms) of halides these two types of halides behave quite differently. Aryl halides are generally quite

TABLE 16-1 PHYSICAL PROPERTIES OF HALIDES

Compound	*Structure*	*M.P. °C*	*B.P. °C*
Methyl fluoride	CH_3F	−142	−78
Methyl chloride	CH_3Cl	−98	−24
Methyl bromide	CH_3Br	−94	+4
Methyl iodide	CH_3I	−66	+42
Methylene chloride	CH_2Cl_2	−95	+40
Chloroform	$CHCl_3$	−64	+62
Carbon tetrachloride	CCl_4	−23	+77
Vinyl chloride	$CH_2{=}CHCl$	−154	−13
Tetrachloroethylene	$CCl_2{=}CCl_2$	−22	+121

resistant to displacement reactions under ordinary conditions, whereas most alkyl halides undergo displacement reactions very easily. This contrasting behavior will be discussed in more detail later in this chapter.

PROPERTIES OF HALIDES

Within any series of alkyl or aryl halides, the boiling points increase with increasing molecular weight; consequently, the boiling points increase in the order: fluorides < chlorides < bromides < iodides, as illustrated in Table 16–1.

The organic halides are insoluble in water and separate from water. The monofluoro and monochloro derivatives are less dense than water, and the monobromo and monoiodo compounds are more dense than water.

METHODS OF PREPARATION OF ORGANIC HALIDES

Several of the preparations of halides have been discussed in earlier chapters and will be noted here to correlate this material, but these previous methods will not be discussed in detail again, and the student is urged to go back and review these preparations in the previous chapters.

DIRECT HALOGENATION. Direct halogenation of alkanes was shown earlier to lead via a free-radical process to a mixture of halogenated hydrocarbons, and, except for a few special cases, is not generally applicable to preparing alkyl halides. Indirect methods are used to prepare the commercially important halides **carbon tetrachloride** and **chloroform.** Carbon tetrachloride is prepared

$$CH_4 + Cl_2 \xrightarrow{h\nu} CH_3Cl + CH_2Cl_2 + CHCl_3 + CCl_4$$

industrially by the chlorination of carbon disulfide using antimony pentachloride as a catalyst. Carbon tetrachloride is a colorless liquid, insoluble in water, soluble in organic solvents, and more dense than water. It is extensively used in the laboratory as an extraction solvent and is used commercially as a solvent for oils and greases. Because of its high solvent power, it has also been used

$$CS_2 + 3Cl_2 \xrightarrow{SbCl_5} CCl_4 + S_2Cl_2$$

Carbon disulfide; Carbon tetrachloride; Sulfur monochloride

as a dry-cleaning agent and as a household cleaning agent. In contrast to the hydrocarbons, alcohols, and ethers, carbon tetrachloride is nonflammable and has found wide use as a fire-extinguishing agent. However, the oxidation of carbon tetrachloride at the temperatures of a fire gives phosgene ($COCl_2$), a toxic gas, and fire extinguishers containing carbon tetrachloride should only be used where adequate ventilation is available.

Chloroform ($CHCl_3$) is obtained commercially by the reduction of carbon tetrachloride using iron and steam as the reducing agent. Chloroform is a

$$CCl_4 + Fe \xrightarrow{H_2O} CHCl_3 + HCl\uparrow$$

Carbon tetrachloride; Chloroform

sweet-smelling volatile liquid (b.p. 62°C) that once was widely used as an anesthetic. Unexplained toxic effects sometimes result from its use, however, and it has been replaced by other anesthetics. Chloroform undergoes photochemical (sunlight) oxidation to give phosgene unless stabilized, and commercially available chloroform contains $\frac{3}{4}$% ethanol as a stabilizer to prevent air oxidation.

$$CHCl_3 + O_2 \xrightarrow{\text{sunlight}} COCl_2$$

Chloroform; Phosgene

Although direct halogenation of alkanes is unsuitable for preparing alkyl halides, it is the most widely used method for preparing aryl halides (see reactions of aromatic hydrocarbons) containing chlorine or bromine. Substituted aryl halides can also be prepared satisfactorily by this method.

$$C_6H_5R + Br_2 \xrightarrow{Fe} o\text{-}BrC_6H_4R + p\text{-}BrC_6H_4R$$

R = alkyl group or alkoxy group

Addition of Hydrogen Halides to Olefins. The addition of hydrogen halides (HF, HCl, HBr, and HI) to olefins has been shown previously (see reactions of alkenes) to give alkyl halides. The addition reaction follows Markownikoff's rule.

$$CH_3CH{=}CH_2 + HCl \rightarrow CH_3CHClCH_3$$

Propene; Isopropyl chloride

REPLACEMENT OF OTHER FUNCTIONAL GROUPS

Hydroxyl Group of Alcohols. The hydroxyl group of alcohols can be replaced by a halide atom using either the hydrogen halides, phosphorus halides (PCl_3 and PCl_5), or thionyl chloride. Fluorides are generally not prepared by this method, but chlorides, bromides, and iodides are easily attainable. Phenols do not usually undergo this type of reaction easily.

$$ROH + HX \rightarrow RX + H_2O$$
$$X = Cl, Br, I$$

REACTIONS OF HALIDES

The most common type of reaction of alkyl halides involves displacement of the halogen atom. This type of reaction may be rationalized as the attack of a negative ion (nucleophile) on an alkyl halide to bring about displacement of a halide ion. Because of the great variety of nucleophiles that will cause displacement of a halide ion from an alkyl halide, numerous types of functional derivatives can be prepared by this kind of reaction. Consequently, alkyl halides are one of the basic starting materials for preparing many of the general classes of hydrocarbon derivatives. Some of the more important of such displacement reactions are summarized in the following equations. The general reaction may be summarized as follows:

$$N^- + R{-}X \rightarrow N{-}R + X^-$$
$$N = \text{anion}$$
$$X = \text{halogen}$$

Nucleophilic Displacement of Halide Ion.

Alcohol formation:	$R{-}X + OH^- \rightarrow R{-}OH + X^-$
Ether formation:	$R'{-}X + OR^- \rightarrow R'{-}O{-}R + X^-$
Amine formation:	$R{-}X + NH_2^- \rightarrow R{-}NH_2 + X^-$
Ester formation:	$R'{-}X + RCO_2^- \rightarrow R'{-}O{-}\overset{\overset{\large O}{\|}}{C}{-}R + X^-$
Acetylene formation:	$R'{-}X + RC{\equiv}C^- \rightarrow R'{-}C{\equiv}CR + X^-$
Nitrile formation:	$R{-}X + CN^- \rightarrow R{-}CN + X^-$
Mercaptan formation:	$R{-}X + SH^- \rightarrow R{-}SH + X^-$

In this type of displacement reaction, the reactivity is dependent on the type of nucleophile and the kind of alkyl group. For any given anion, the tendency toward displacement decreases in the order: primary $>$ secondary $>$ tertiary. Aromatic halides, unless activated by the presence of one or more nitro groups located *ortho* or *para* to the halogen atom, do not generally undergo this type of displacement reaction under normal conditions. Benzyl halides, of course, behave like alkyl halides and undergo this type of displacement reaction easily. Vinyl halides are likewise inert to this kind of displacement reaction under normal conditions.

Formation of Organometallic Reagents. Compounds in which a metal is bonded to carbon are called organometallic compounds. Depending upon the metal involved, these compounds may be mainly ionic or mainly covalent. In recent years, the preparation of organometallic compounds and the study of their chemistry has been one of the most fascinating and productive areas of chemistry. Compounds containing most of the known metals have been prepared and studied. Several of these organometallic compounds are well known to most people. For example, tetraethyl lead, $(C_2H_5)_4Pb$, is an important antiknock ingredient added to most gasolines. Mercurochrome and Merthiolate are two mercury derivatives used as antiseptics.

Of the numerous and diversified kinds of organometallic compounds, the magnesium compounds have been the most useful to the organic chemist in the laboratory. These reagents are usually prepared by an exchange reaction in an inert solvent such as ether or tetrahydrofuran. Generally, they are not isolated, but are generated *in situ* and used as intermediates for preparing other compounds.

The preparation and use of organomagnesium compounds were developed by the French chemist, Victor Grignard, and this type of compound is known as a **Grignard reagent.** Both alkyl and aryl halides form a Grignard reagent when reacted with metallic magnesium in anhydrous ether.

$$\underset{\text{Ethyl iodide}}{CH_3CH_2I} + Mg \xrightarrow{\text{ether}} \underset{\text{Ethyl magnesium iodide}}{[CH_3CH_2]^{-}\overset{+}{Mg}I}$$

These types of reagents are sensitive to hydrolysis and are hydrolyzed by water or acids to hydrocarbons.

Grignard reagents are highly reactive compounds and undergo reaction with many functional groups (particularly carbonyl-containing functional groups) to give useful products. Applications of these organometallic reagents will be introduced in subsequent chapters.

Electrophilic Substitution Reactions of Aryl Halides. Although aromatic halides undergo nucleophilic substitution reactions only with extreme difficulty, they undergo the normal type of electrophilic substitution reactions without too much difficulty. The halogen atoms are *ortho-para* directing, and because of the inductive effect of the halogen compared to hydrogen, aryl halides are deactivating as compared to benzene. Therefore, bromobenzene gives *ortho-* and *para*-dibromobenzene on bromination, but the rate of bromination of bromobenzene is slower than the rate of bromination of benzene. Nitration, sulfonation, and alkylation proceed similarly.

$$C_6H_5Br + Br_2 \xrightarrow{FeBr_3} o\text{-}C_6H_4Br_2 + p\text{-}C_6H_4Br_2$$

Bromobenzene o-Dibromobenzene p-Dibromobenzene

FLUORINE COMPOUNDS

Because of the great reactivity of fluorine, it is not normally useful as a general purpose fluorinating agent except under special conditions. Consequently, organic compounds containing fluorine are usually prepared by special methods. The most important fluorine compounds are those in which all the hydrogen atoms have either been replaced by fluorine (this class is known as fluorocarbons) or by a combination of fluorine and other halogens. Substances containing many fluorine atoms are generally inert to oxidizing and reducing agents and to most acids. Consequently, they find extensive use in refrigeration, aerosols, lubricants, electrical insulators, and in the plastics industry.

The most widely used fluorine compounds are the methane and ethane derivatives sold under the trade names Freon, Genetron, and Ucon. Some typical examples are shown below:

CF_2Cl_2	$CFCl_3$	$CFCl_2CF_2Cl$	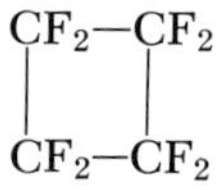 $CF_2—CF_2$ / $CF_2—CF_2$ (ring)
Freon-12 Dichlorodifluoromethane	Freon-11 Trichlorofluoromethane	Freon-113 1,1,2-Trifluoro-trichloroethane	Freon-C318 Octafluorocyclobutane

Freon-12 is volatile (b.p. $-28°C$) and nontoxic, and is used as a refrigerant in the majority of household and commercial refrigeration units and air conditioners. Freon-C318 (octafluorocyclobutane) is odorless, tasteless, nontoxic, and extremely stable to hydrolysis, and has application in the aerosol industry. Trifluorobromoethane, CF_3Br, is a nontoxic fire extinguishing agent and has replaced many of the carbon tetrachloride extinguishers which produce the toxic gas phosgene when used to extinguish fires.

Probably the best known of the fluorine containing compounds is the polymer known as **Teflon,** which is made by polymerization of tetrafluoroethylene, $CF_2{=}CF_2$. Teflon contains the repeating unit $(—CF_2—CF_2—)_n$. This fluorocarbon polymer is inert to most chemical reagents, has excellent electrical insulating properties, maintains its lubricating properties over a wide range (-50 to $+300°C$), and has "non-sticking" properties for most materials. This latter characteristic has been exploited in making Teflon coated frying pans, cookie sheets, pie tins, and so forth, which can be used for cooking and baking without grease, as foodstuffs do not adhere to Teflon.

Another important property of fluorocarbons is their nonwettability. Materials and fabrics coated with a fluorocarbon become water and oil repellant. Fabrics treated with fluorochemicals become stain resistant, and this particular property has been put to practical use in the manufacture of "Scotch Guard" coated fabrics.

MISCELLANEOUS HALOGEN COMPOUNDS

Halogen compounds play an important role in the chemical industry as synthetic starting materials. In addition, many halogen containing compounds find application as fumigants (CH_3Br) for controlling various insects and rodents, and as fungicides and pesticides. Some of the more common halogen-containing pesticides are shown below:

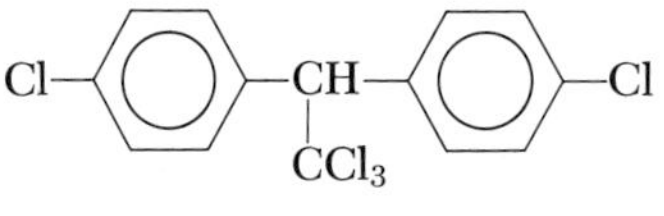

Cl—C6H4—CH(CCl3)—C6H4—Cl

DDT

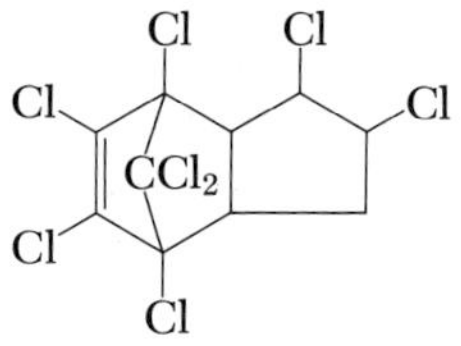

Cl Cl Cl Cl CCl2 Cl Cl

Chlordan

Cl Cl Cl Cl Cl Cl

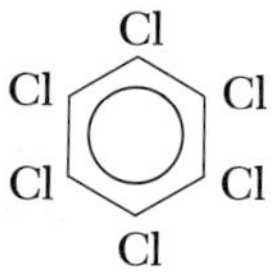

Lindane

Questions

1. Draw a structural formula for each of the following compounds:
   (a) tertiary butyl bromide
   (b) *p*-iodotoluene
   (c) cyclohexyl bromide
   (d) 1,1,2-trichlorotrifluoroethane
   (e) carbon tetrabromide

2. Name each of the following compounds:
   (a) $CHCl_3$
   (b) $CHCl{=}CCl_2$
   (c) Benzene ring–CH_2Cl
   (d) Benzene ring–Cl

   (e) CH_3CHICH_3

3. Write equations for the reactions of *n*-propyl bromide with the following reagents:
   (a) sodium ethoxide
   (b) potassium cyanide
   (c) Mg/ether
   (d) silver oxide/H_2O
   (e) sodium acetylide

4. Write equations for the reactions of chlorobenzene with the following reagents:
   (a) HNO_3/H_2SO_4
   (b) Mg/ether
   (c) sodium methoxide
   (d) sodium cyanide

5. Write equations for the preparation of the following compounds. Give any necessary catalysts. More than one step may be necessary.
   (a) ethane from ethyl iodide
   (b) *n*-propyl fluoride from *n*-propyl alcohol
   (c) cyclopentyl bromide from cyclopentanol
   (d) benzyl cyanide from benzyl alcohol
   (e) 2,2-dibromopropane from propene

Suggested Reading

C. & E. News; Special Report: Fluorocarbons. Chemical and Engineering News, July 18, 1960, p. 92.
Joiner: Pesticide Residue Control Under the Food, Drug, and Cosmetic Act. Journal of Chemical Education, Vol. 38, p. 370, 1961.
McBee and Roberts: Organic Fluorine Chemicals. Journal of Chemical Education, Vol. 32, p. 13, 1955.
Rheinboldt: Fifty Years of the Grignard Reaction. Journal of Chemical Education, Vol. 27, p. 476, 1950.

CHAPTER 17

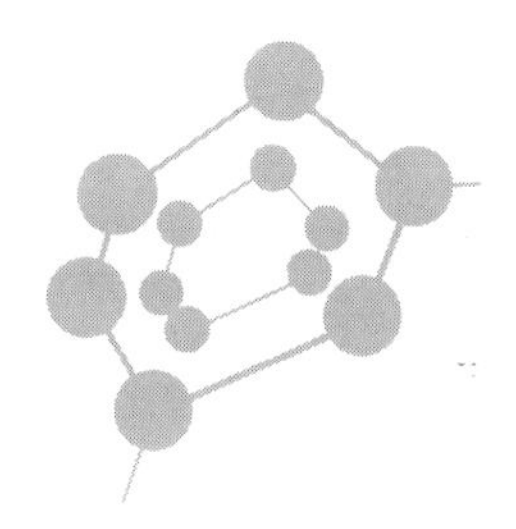

ALDEHYDES AND KETONES

Compounds containing the carbonyl group ($-\overset{O}{\overset{\|}{C}}-$) are known as **aldehydes** or **ketones.** If one of the atoms linked to the carbonyl group is a hydrogen atom, the compound is an aldehyde ($-\overset{O}{\overset{\|}{C}}-H$). The other atom or group attached to the carbonyl group of an aldehyde may be hydrogen, alkyl, or aryl. In the case of ketones, both of the groups attached to the carbonyl group are either alkyl or aryl. Cyclic ketones also exist in which the carbonyl group is part of the ring. Cyclic aldehydes are not possible. Table 17–1 illustrates some of the more common aldehydes and ketones.

NOMENCLATURE

In the IUPAC system, the characteristic ending for aldehydes is **-al,** and the characteristic ending for ketones is **-one.** These endings are added to the root stem name of the hydrocarbon having the same number of carbon atoms. As usual, the compound is named as a derivative of the longest continuous chain of the carbon atoms, including the carbonyl functional group. In the case of aldehydes, the —**CHO** group must always appear at the end of the chain and is always indicated by the number **1** (the lowest number), although this number does not appear in the name. In the case of ketones, however, the carbonyl group may appear at various positions in a carbon chain, and its position must be designated by the lowest possible number. All other substituents are indicated by the appropriate number and prefix to indicate their positions on the carbon chain.

Common names are also used to name the aldehydes and ketones. Alde-

TABLE 17-1 PHYSICAL PROPERTIES OF ALDEHYDES AND KETONES

Compound	*Structure*	*M.P. °C*	*B.P. °C*
Formaldehyde	$HCHO$	−92	−21
Acetaldehyde	CH_3CHO	−123	+21
Propionaldehyde	CH_3CH_2CHO	−81	+49
Chloral	CCl_3CHO	−58	+98
Acetone	CH_3COCH_3	−95	+56
Methyl ethyl ketone	$CH_3COCH_2CH_3$	−86	+80
Cyclohexanone	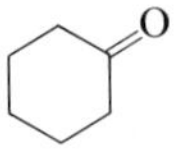 O	−45	+157
Acetophenone	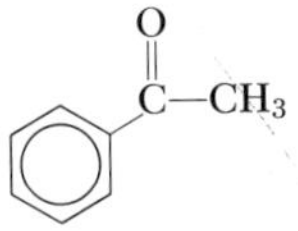 O ‖ C—CH_3	+20	+202
Benzophenone	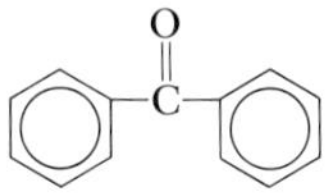 O ‖ C	+48	+306
Acrolein	$CH_2{=}CHCHO$	−88	+53
Crotonaldehyde	$CH_3CH{=}CHCHO$	−77	+104
Benzaldehyde	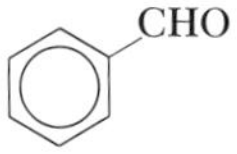 CHO	−26	+179
Salicylaldehyde	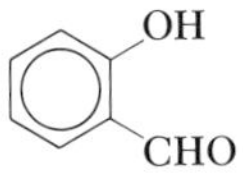 OH, CHO	−7	+197

hydes are generally named as derivatives of the corresponding acid. The **-ic** ending of the acid is dropped and replaced by the term **aldehyde.** Ketones, with the exception of acetone, are named according to the alkyl or aryl groups attached to the carbonyl function, followed by the word **ketone.** Some typical examples are illustrated below (the common name is in parentheses):

$$CH_3\overset{\overset{O}{\|}}{C}CH_2CH_2CH_3$$

2-Pentanone
(methyl *n*-propyl ketone)

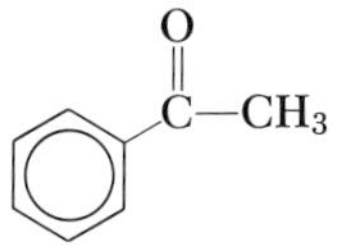

$$\overset{\overset{O}{\|}}{C}{-}CH_3$$

Acetophenone
(methyl phenyl ketone)

$$CH_3CH_2CH_2\overset{\overset{H}{|}}{C}{=}O$$

Butanal
(butyraldehyde)

PHYSICAL PROPERTIES

Formaldehyde is a gas at room temperature (20°C). All other simple aliphatic and aromatic aldehydes are colorless liquids. Most of the lower molecular weight aldehydes have a sharp odor, but the odor becomes more fragrant as the molecular weight increases. The aromatic aldehydes have been used as flavoring agents and perfumes. Benzaldehyde, for example, is a constituent of the seeds of bitter almonds and was once called "oil of bitter almond." It is a colorless liquid with a pleasant almond-like odor.

All of the more common lower molecular weight ketones are liquids. Acetone, the simplest ketone, is a moderately low boiling liquid (b.p. 56°C). It finds extensive use as a solvent. If the hydrogen atoms of acetone are replaced by fluorine atoms, the boiling point is lower. Hence, hexafluoroacetone (CF_3COCF_3) is a gas. With few exceptions all other ketones are either liquids or solids (Table 17–1).

The lower molecular weight aldehydes and ketones are soluble in water. As the length of the carbon chain increases, the water solubility decreases. Limited solubility occurs around 5 to 6 carbon atoms. As expected, aldehydes and ketones dissolve in the normal organic solvents.

The carbonyl group is a polar group (oxygen being more electronegative than carbon). Because of the polarity of the carbonyl group, the boiling points of aldehydes and ketones are higher than either the hydrocarbon or the ether of corresponding molecular weight. However, aldehydes and ketones cannot form hydrogen bonds, and therefore their boiling points are lower than the alcohols and acids of comparable molecular weight.

METHODS OF PREPARATION

Oxidation. Under the proper conditions, primary alcohols can be oxidized to give aldehydes, and secondary alcohols can be oxidized to give ketones. These reactions can be viewed as a "dehydrogenation" process, since the elements of H_2 are lost in going from reactant to product. In the case of aldehydes, the process is somewhat complicated by the fact that aldehydes are more easily oxidized than alcohols. Consequently, as the aldehyde is formed, it undergoes further oxidation to give a carboxylic acid. Therefore, the aldehyde must be removed from the oxidation zone to prevent this additional oxidation to the acid.

$$\underset{\substack{\text{Ethyl alcohol}\\ \text{(B.P. 78°C)}}}{CH_3CH_2OH} + K_2Cr_2O_7 \xrightarrow{H_2SO_4} \underset{\substack{\text{Acetaldehyde}\\ \text{(B.P. 21°C)}}}{CH_3\overset{\overset{\large H}{|}}{C}{=}O}$$

Formaldehyde is prepared industrially by a variation of this type of oxidation reaction.

$$\underset{\text{Methyl alcohol}}{CH_3OH + \tfrac{1}{2}O_2} \xrightarrow[\text{high temps.}]{Cu} \underset{\text{Formaldehyde}}{H{-}\overset{\overset{\large O}{\|}}{C}{-}H} + H_2O$$

Oxygen is used in this process to convert the hydrogen into water. The formaldehyde-water mixture obtained is passed into water, and the solution formed (40% formaldehyde) is sold under the trade name "Formalin."

The problem of further oxidation in the preparation of ketones from secondary alcohols is nonexistent under the normal conditions of the oxidation reactions. Ketones are quite resistant to further oxidation, except under extreme conditions, and the oxidation of the secondary alcohol stops very cleanly at the ketone stage. Some examples to illustrate the scope and versatility of this kind of reaction are shown below:

$$\underset{\text{Isopropyl alcohol}}{CH_3CHOHCH_3} + K_2Cr_2O_7 \xrightarrow{H_2SO_4} \underset{\text{Acetone}}{CH_3\overset{O}{\overset{\|}{C}}CH_3}$$

$$\underset{\text{1-Phenylethanol}}{C_6H_5CHOHCH_3} + KMnO_4 \xrightarrow{OH^-} \underset{\text{Acetophenone}}{C_6H_5\overset{O}{\overset{\|}{C}}{-}CH_3}$$

In the case of compounds containing an alkyl group, the alcohol is obtained by hydration (see reactions of alkenes) of the appropriate olefin. Indirectly then, the ketone is prepared via a two-step synthesis from the olefin, as shown below for the preparation of 2-butanone from 1-butene:

$$\underset{\text{1-Butene}}{CH_3CH_2CH{=}CH_2} + H_2SO_4 \xrightarrow{H_2O} \underset{\text{2-Butanol}}{CH_3CH_2\overset{OH}{\overset{|}{C}}HCH_3} \xrightarrow[H_2SO_4]{K_2Cr_2O_7} \underset{\text{2-Butanone}}{CH_3CH_2\overset{O}{\overset{\|}{C}}CH_3}$$

Friedel-Crafts Acylation. Ketones containing an aromatic ring can be conveniently prepared by the Friedel-Crafts acylation of an aromatic hydrocarbon (see reactions of aromatic hydrocarbons) as illustrated below:

$$\underset{\text{Benzene}}{C_6H_6} + \underset{\text{Acetyl chloride}}{CH_3\overset{O}{\overset{\|}{C}}{-}Cl} \xrightarrow{AlCl_3} \underset{\text{Acetophenone}}{C_6H_5\overset{O}{\overset{\|}{C}}{-}CH_3}$$

REACTIONS OF CARBONYL COMPOUNDS

Oxidation. As noted in the preparation of aldehydes, oxidation of aldehydes occurs very readily to give a carboxylic acid. Even mild oxidizing agents bring about the oxidation of aldehydes. Two such mild oxidizing reagents are **Tollen's reagent,** an ammoniacal solution of silver nitrate, and **Fehling's** or **Benedict's solution,** alkaline solutions of copper sulfate.* Oxidation of an aldehyde with Tollen's reagent produces metallic silver (usually as a silver mirror), whereas Fehling's or Benedict's solution gives a red precipitate of cuprous oxide. These simple visual tests make it easy to distinguish aldehydes from ketones, since ketones are not oxidized by these reagents. Hence, no silver mirror or red precipitate is obtained on treatment of a ketone with either Tollen's reagent or Fehling's and Benedict's solutions. The overall reaction of these reagents with aldehydes is summarized below:

$$\underset{\text{Aldehyde}}{RCHO} + \underset{\substack{\text{Tollen's reagent}\\ \text{(colorless)}}}{2Ag(NH_3)_2^+} + 2OH^- \longrightarrow \underset{\text{Acid salt}}{RCO_2^-NH_4^+} + \underset{\substack{\text{Silver}\\ \text{metal}}}{2Ag\downarrow} + 3NH_3 + H_2O$$

---

* Fehling's solution also contains sodium potassium tartrate, and Benedict's solution contains sodium citrate. These salts form tartrate and citrate complexes with the cupric ion and help keep the cupric ion in solution.

$$\underset{\text{Aldehyde}}{RCHO} + \underset{\text{Fehling's or Benedict's solution (blue)}}{2Cu^{++}\text{ (complex)}} \xrightarrow{NaOH} \underset{\text{Acid salt}}{RCO_2^-Na^+} + \underset{\text{Red precipitate}}{Cu_2O\downarrow}$$

This reaction is the basis of the Fehling and Benedict tests for determining the presence of sugar in urine.

ADDITION REACTIONS

The most characteristic type of reaction of aldehydes and ketones is an addition reaction across the carbon-oxygen double bond. A similar type of reaction is characteristic of compounds containing carbon-carbon multiple bonds (see reactions of alkenes and alkynes). Addition reactions to carbon-carbon double bonds were found to be electrophilic addition reactions, in which the electrophile was added to the carbon atom in the first step of the reaction sequence. With carbonyl groups, however, the polarization of the bond dipole is such that the carbon end of the dipole is the positive end. Therefore, it might be expected that the carbon atom of the carbonyl group would be attacked

$$\underset{\delta+}{>C}=\underset{\delta-}{O}$$

by a nucleophilic type of reagent, and that the addition reactions of aldehydes and ketones would be nucleophilic addition reactions rather than the electrophilic addition reactions characteristic of alkenes.

The investigation of the addition reactions of carbonyl compounds has confirmed this type of prediction, and aldehydes and ketones undergo a great variety of nucleophilic addition reactions. The generalized mechanism for these nucleophilic additions is outlined below:

$$\underset{\text{Nucleophile}}{N:} + >C=\ddot{O}: \rightarrow \underset{\text{(I)}}{N-\overset{|}{\underset{|}{C}}-\ddot{O}:^-} \qquad [1]$$

$$N-\overset{|}{\underset{|}{C}}-\ddot{O}:^- + HN \rightarrow \underset{\text{(II)}}{N-\overset{|}{\underset{|}{C}}-\ddot{O}H} + N: \qquad [2]$$

Overall reaction:

$$HN + >C=O \rightarrow N-\overset{|}{\underset{|}{C}}-OH$$

The first step involves attack by the nucleophile, N:, on the positive end of the carbonyl dipole to give the anion, (I). This anion (I) then abstracts a proton from the nucleophilic agent, HN, to give the final addition product, (II), and the nucleophile, N:, which can then repeat the cycle. The overall reaction (the combination of equations [1] and [2]) is the addition of the nucleophilic reagent, HN, across the carbonyl group. In some cases, the initially formed addition product, (II), can undergo a dehydration reaction (loss of H_2O), so that the actual product isolated may not always appear exactly like (II). Some typical addition reactions of aldehydes and ketones are outlined below.

Addition of Hydrogen Cyanide. The addition of hydrogen cyanide gives an α-hydroxy nitrile,* called a cyanohydrin. These types of compounds are valuable in the preparation of hydroxy acids, amino acids, and sugars.

$$>C{=}O + H^+CN^- \rightarrow >C(OH)(CN)$$

Cyanohydrin

Addition of Organometallic Reagents. Organometallic reagents, particularly Grignard reagents, also add across the carbonyl group. Hydrolysis of the initially formed complex gives alcohols as the final product. Aldehydes, except for formaldehyde, give secondary alcohols, and ketones give tertiary alcohols. This type of reaction provides one of the best laboratory methods for preparing secondary and tertiary alcohols. The generalized scheme of this kind of reaction is outlined below to illustrate the scope and versatility of this reaction:

$$R^{-}\overset{+}{Mg}X + >C{=}O \rightarrow \left[R{-}C{-}OMgX\right] \xrightarrow[H_2O]{H^+} R{-}C{-}OH + Mg(OH)X$$

(not isolated)

$$CH_3\overset{H}{C}{=}O + CH_3CH_2MgBr \rightarrow \left[CH_3\overset{H}{\underset{CH_2CH_3}{C}}{-}OMgBr\right] \xrightarrow[H_2O]{H^+} CH_3\overset{OH}{C}HCH_2CH_3$$

Acetaldehyde — 2-Butanol

Aromatic Substitution Reactions. The carbonyl group is a *meta*-directing group and is also a deactivating group. Aromatic aldehydes and ketones, therefore, undergo electrophilic aromatic substitution reactions less easily than benzene, and more vigorous reaction conditions must be used to bring about ring substitution. The halogenation of a typical aromatic aldehyde and ketone is shown below. Similar reactions occur on nitration, sulfonation, and alkylation.

$$C_6H_5CHO + Br_2 \xrightarrow[\Delta]{Fe} m\text{-}BrC_6H_4CHO$$

Benzaldehyde — *m*-Bromobenzaldehyde

$$C_6H_5COCH_3 + Br_2 \xrightarrow[\Delta]{Fe} m\text{-}BrC_6H_4COCH_3$$

Acetophenone — *m*-Bromoacetophenone

* Organic compounds containing the —C≡N grouping are called either cyanides (relating them to the inorganic analogues) or nitriles.

USES OF IMPORTANT ALDEHYDES AND KETONES

As noted earlier, formaldehyde ($H-\overset{O}{\overset{\|}{C}}-H$) is a colorless gas that readily dissolves in water; a 40% solution is known as **formalin.** Formalin acts as a disinfectant and is used in embalming fluid and as a preservative of various tissues. Formaldehyde gas and the polymer paraformaldehyde are used extensively as insecticides, fumigating agents, and antiseptics. Large quantities of formaldehyde are used in the manufacture of synthetic resins and in the synthesis of other organic compounds.

Acetaldehyde is used in the production of acetic acid, ethyl acetate, synthetic rubber, and other organic compounds. Paraldehyde, a cyclic trimer of acetaldehyde, is more stable than acetaldehyde and serves as a source of the latter compound when heated. It is an effective hypnotic, or sleep producer, but has been replaced by other drugs since it has an irritating odor and an unpleasant taste.

Chloral hydrate is used medically as a sedative or hypnotic. Benzaldehyde is an important intermediate in the preparation of drugs, dyes, and other organic compounds. As noted earlier, it is also used in flavoring agents and perfumes. Cinnamaldehyde, $C_6H_5-CH=CHCHO$, is a constituent of the oil of cinnamon obtained from cinnamon bark. Its main use is as a flavoring agent.

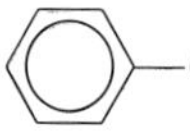

Another important aldehyde is **vanillin,** which may be obtained from the extracts of vanilla beans, and which also occurs in sugar beets, resins, and balsams. Vanillin may be prepared from the phenol derivative eugenol or guaiacol. Eugenol is first rearranged with alcoholic potassium hydroxide to form isoeugenol, which is then oxidized to vanillin. Vanillin is employed as a flavoring agent in many food products, such as candies, cookies, and ice cream.

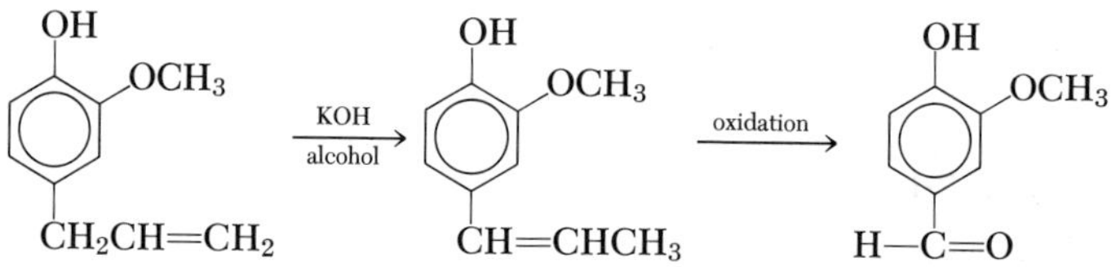

$$\underset{\text{Eugenol}}{C_6H_3(OH)(OCH_3)CH_2CH=CH_2} \xrightarrow[\text{alcohol}]{\text{KOH}} \underset{\text{Isoeugenol}}{C_6H_3(OH)(OCH_3)CH=CHCH_3} \xrightarrow{\text{oxidation}} \underset{\text{Vanillin}}{C_6H_3(OH)(OCH_3)H-C=O}$$

Acetone, the most important ketone, is prepared by the oxidation of isopropyl alcohol, or by the butyl alcohol fermentation process from cornstarch. It is used as a solvent for cellulose derivatives, varnishes, lacquers, resins, and plastics. It is used in the commercial preparation of dyes, acetate rayon, smokeless powders, explosives, and in the manufacture of photographic films.

Methyl ethyl ketone ($CH_3COCH_2CH_3$) is used by the petroleum industry in the dewaxing of lubricating oils. It is an excellent solvent for fingernail polish and is used as a polish remover. It also has found increased use as a solvent.

Acetophenone is used as an intermediate in the preparation of other compounds and in the preparation of perfumes. If one of the hydrogen atoms on the methyl group is replaced by a chlorine atom, chloroacetophenone

($C_6H_5COCH_2Cl$) is formed. This compound is a lachrymator and is commonly used as tear gas.

Benzophenone is a colorless solid with a characteristically pleasant odor. It is used in the manufacture of perfume and soaps and in the preparation of other organic compounds.

An important aromatic ketone that is used extensively in the identification of amino acids and proteins is ninhydrin. This compound also represents one of the few stable hydrates known.

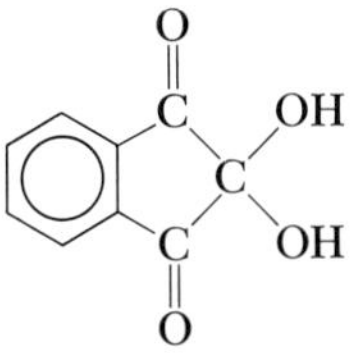

Ninhydrin

Questions

1. Draw a structural formula for each of the following compounds:
   - (a) heptanal
   - (b) 2-hexanone
   - (c) *m*-nitrobenzaldehyde
   - (d) cyclopentanone
   - (e) 5-bromo-3-methylpentanal
   - (f) *p*-chloroacetophenone
   - (g) methyl cyclohexyl ketone
   - (h) *p,p′*-dibromobenzophenone
   - (i) 3-methylcyclohexanone
   - (j) *trans*-3-pentenal

2. Write equations for the preparation of the following compounds. Give any necessary catalysts. More than one step may be necessary.
   - (a) 1-methylcyclohexanol from cyclohexanol
   - (b) 3-hexanone from 3-hexene
   - (c) acetone from acetaldehyde
   - (d) 1-phenylethanol from benzene
   - (e) benzophenone from benzaldehyde

Suggested Reading

Bonner and Castro: Essentials of Modern Organic Chemistry. New York, Reinhold Publishing Corp., 1965, pp. 298–342.

Fieser and Fieser: Advanced Organic Chemistry. New York, Reinhold Publishing Corp., 1961, pp. 395–486, 813–844.

Morrison and Boyd: Organic Chemistry, 2nd Ed. Boston, Allyn & Bacon, Inc., 1966, pp. 615–648, 852–871.

CHAPTER 18

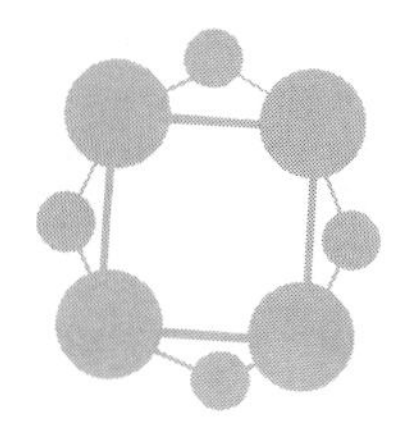

CARBOXYLIC ACIDS AND ACID DERIVATIVES

Organic compounds composed of the three elements carbon, hydrogen, and oxygen have been discussed in the previous chapters. These compounds included alcohols, ethers, aldehydes, and ketones. Further oxidation of primary alcohols or aldehydes eventually leads to the formation of another group of important organic compounds, **carboxylic acids,** which contain the carboxyl functional group $-\overset{O}{\overset{\|}{C}}-OH$. These acids are widely distributed in nature, especially in foodstuffs. Citric acid in citrus fruits, oxalic acid in fruits and vegetables, acetic acid in vinegar, amino acids in proteins, and fatty acids in fats and lipids are some typical examples of naturally occurring organic acids.

Important derivatives of organic acids that will be considered in this chapter include salts, esters, acid halides, acid anhydrides, and amides.

NOMENCLATURE

Carboxylic acids are defined as organic compounds which contain one or more carboxyl groups (—COOH) in the molecule. Some typical acids are illustrated in Table 18–1. Hydrogen, alkyl, and aryl groups may be attached to the carboxyl group, and more than one carboxyl group may be present in the molecule.

Since organic acids* occur so widely in nature either as the free acid or an acid derivative, they were among the earliest organic compounds investi-

---

* The carboxyl group may be written as $-\overset{O}{\overset{\|}{C}}-OH$, —COOH, or $-CO_2H$.

TABLE 18-1 IONIZATION CONSTANTS OF CARBOXYLIC ACIDS

Acid	*Structure*	K_a
Acetic acid	CH_3COOH	1.8×10^{-5}
Propionic acid	CH_3CH_2COOH	1.0×10^{-5}
n-Butyric acid	$CH_3CH_2CH_2COOH$	1.5×10^{-5}
Chloroacetic acid	CH_2ClCO_2H	1.5×10^{-3}
Dichloroacetic acid	$CHCl_2CO_2H$	5.0×10^{-2}
Trichloroacetic acid	CCl_3CO_2H	1×10^{-1}
α-Chloropropionic acid	$CH_3CHClCO_2H$	1.6×10^{-3}
β-Chloropropionic acid	$CH_2ClCH_2CO_2H$	8×10^{-5}
Benzoic acid	C_6H_5—CO_2H	6×10^{-5}
p-Chlorobenzoic acid	Cl—C_6H_4—CO_2H	1×10^{-4}

gated. Common names were given to many of these acids, and these names are still used frequently today. For the aliphatic acids, the IUPAC nomenclature is also used. This system adds the ending **-oic acid** to the root stem name of the hydrocarbon with the corresponding number of carbon atoms. The carboxyl group is given the number **1,** and all the other usual IUPAC rules are then applied. The aromatic acids are named either by common names or as derivatives of the parent aromatic acid, benzoic acid. Some typical illustrations of the use of these nomenclature systems are shown below. The common name (for the aliphatic acids) is given in parentheses.

$H-C(=O)-OH$ — Methanoic acid (formic acid)

$CH_3C(=O)-OH$ — Ethanoic acid (acetic acid)

$CH_3CH_2C(=O)-OH$ — Propanoic acid (propionic acid)

$CH_3CH_2CH_2C(=O)-OH$ — Butanoic acid (butyric acid)

In some cases the position of substituents on the aliphatic chain is indicated by the Greek letters α, β, γ, δ, and so forth.

Di-acids, tri-acids, and so forth, are named in the IUPAC system by adding the ending **-dioic acid, -trioic acid,** and so forth, to the name of the hydrocarbon containing the same number of carbon atoms. In addition, common names are used for many of these compounds. Some examples are shown in the following compounds (the common name is in parentheses):

$HOCH_2CH_2CO_2H$ (β over C3, α over C2; carbons numbered 3 2 1) — 3-Hydroxypropanoic acid (β-hydroxypropionic acid)

$HO-C(=O)-C(=O)-OH$ — Ethanedioic acid (oxalic acid)

$HO-C(=O)CH_2C(=O)-OH$ — 1,3-Propanedioic acid (malonic acid)

$HO-C(=O)(CH_2)_2C(=O)-OH$ — 1,4-Butanedioic acid (succinic acid)

$HO-C(=O)(CH_2)_4C(=O)-OH$ — 1,6-Hexanedioic acid (adipic acid)

$HOOC-C_6H_4-COOH$ (para) — Terephthalic acid

PHYSICAL PROPERTIES OF CARBOXYLIC ACIDS

Carboxylic acids are, as the name indicates, acidic compounds. When dissolved in water, they can ionize slightly to donate a proton to a more basic substance, such as water. Most simple carboxylic acids are only slightly ionized

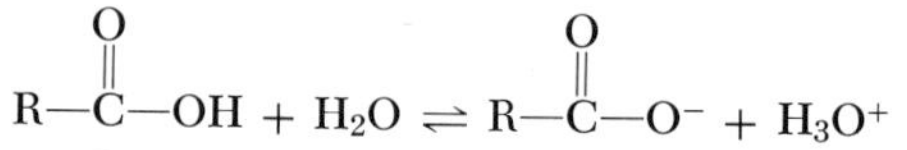

$$\underset{\text{Carboxylic acid}}{R{-}\overset{\overset{O}{\|}}{C}{-}OH} + H_2O \rightleftharpoons \underset{\text{Carboxylate anion}}{R{-}\overset{\overset{O}{\|}}{C}{-}O^-} + H_3O^+$$

in water, and these acids are fairly weak. The ionization constant, K_a, is a measure of the acid strength of these compounds. For the unsubstituted aliphatic acids, the acidity varies with changes in the alkyl group. Substitution of the α-hydrogens by highly electronegative groups, such as the halogens, has a significant effect on the acidity. The effect is an additive one, and the acidity increases as the number of electron-attracting groups in the α-position increases. This effect is only large when the electron-attracting group is in the α-position. When this type of group is further removed from the carboxyl group (β, γ, δ, and so forth), the effect on the acidity of the acid is small.

Aromatic acids are very slightly more acidic than the aliphatic acids. Strongly electron-attracting substituents on the ring also increase the acidity of the aromatic acid, and the effect is generally about the same as in the aliphatic acids. Some typical acids and substituted acids are tabulated in Table 18–1 with their (approximate) K_a to illustrate the preceding points.

Like alcohols, the carboxyl group contains a hydroxyl group which is capable of forming hydrogen bonds with other acid molecules, or with other similar kinds of molecules, such as water. Consequently, the lower molecular weight carboxylic acids are water soluble, the borderline solubility being about 4 to 5 carbon atoms. Branching of the carbon chain increases water solubility, and some branched acids of 5 to 6 carbon atoms are soluble in water.

The carboxylic acids have abnormally high boiling points compared to other functional compounds of similar molecular weight. For example, propionic acid (mol. wt. 74) boils at 141°C, *n*-pentane (mol. wt. 72) boils at 35°C, and *n*-butanol (mol. wt. 74) boils at 118°C. The carboxylic acids have been found to consist largely of dimers; consequently, two hydrogen bonds per molecule are formed, thereby accounting for the high boiling points of this class of compounds.

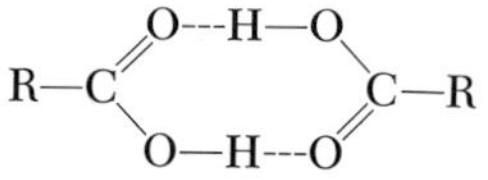

```
       O---H—O
      //       \
R—C             C—R
      \        //
       O—H---O
```

Carboxylic acid dimer

The lower molecular weight aliphatic acids are liquids with sharp, unpleasant odors. As the molecular weight increases, the volatility decreases, and the higher acids ($>C_{10}$) are solids with little odor. The aliphatic acids in the C_{12} to C_{18} range are used in making soaps and candles (Table 18–2).

TABLE 18-2 PHYSICAL PROPERTIES OF ACIDS AND ACID DERIVATIVES

Compound	*Structure*	*M.P. °C*	*B.P. °C*
Formic acid	$H{-}\overset{O}{\overset{\parallel}{C}}{-}OH$	+8	+101
Acetic acid	CH_3CO_2H	+17	+118
Propionic acid	$CH_3CH_2CO_2H$	−21	+141
n-Butyric acid	$CH_3CH_2CH_2CO_2H$	−4	+164
Isobutyric acid	$(CH_3)_2CHCO_2H$	−46	+153
n-Valeric acid	$CH_3(CH_2)_3CO_2H$	−34	+186
Methyl acetate	$CH_3CO_2CH_3$	−98	+57
Ethyl acetate	$CH_3CO_2CH_2CH_3$	−84	+77
Acetyl chloride	CH_3COCl	−112	+52
Acetamide	CH_3CONH_2	+82	+222
Acetic anhydride	$(CH_3CO)_2O$	−73	+140
Benzoic acid	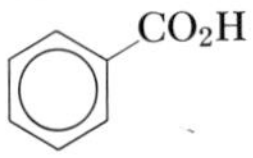 $C_6H_5{-}CO_2H$	+122	+249
Benzoyl chloride	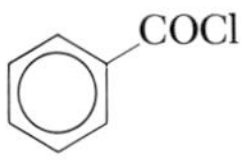 $C_6H_5{-}COCl$	−1	+197
Benzamide	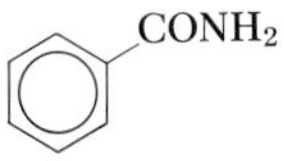 $C_6H_5{-}CONH_2$	+130	+290
Phthalic anhydride	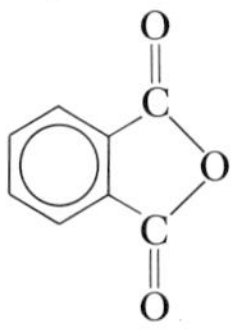 $C_6H_4(CO)_2O$	+130	+285
Methyl benzoate	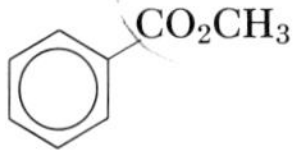 $C_6H_5{-}CO_2CH_3$	−12	+199

PREPARATION OF CARBOXYLIC ACIDS

OXIDATION. Oxidation of primary alcohols and aldehydes (see reactions of aldehydes) gives carboxylic acids. This type of oxidation can be used to prepare both aliphatic and aromatic acids. Oxidizing agents, such as potassium dichromate and sulfuric acid or potassium permanganate, are generally used for these reactions as shown below:

$$\underset{n\text{-Butanol}}{CH_3CH_2CH_2CH_2OH} + K_2Cr_2O_7 \xrightarrow{H_2SO_4} \underset{\text{Butyric acid}}{CH_3CH_2CH_2\overset{O}{\overset{\parallel}{C}}OH}$$

Di-acids can be prepared similarly by oxidation of the corresponding diols or di-aldehydes. This type of oxidation method gives a carboxylic acid with the same number of carbon atoms as the original alcohol or aldehyde.

Benzoic acid and other aromatic acids can also be prepared by the oxidation of alkyl side-chains with either nitric acid or alkaline potassium permanganate. The alkyl group attached to the ring, regardless of its length, is degraded (broken down) until it is converted to the carboxyl group. Since the alkylbenzene can be easily prepared by alkylation reactions (see reactions of aromatic hydrocarbons), this is a useful preparative method for some aromatic acids. Similar types of oxidations of aliphatic hydrocarbons require more vigorous conditions, are not easily controlled, and are not of extensive preparative value. Some typical examples are shown in the following equations:

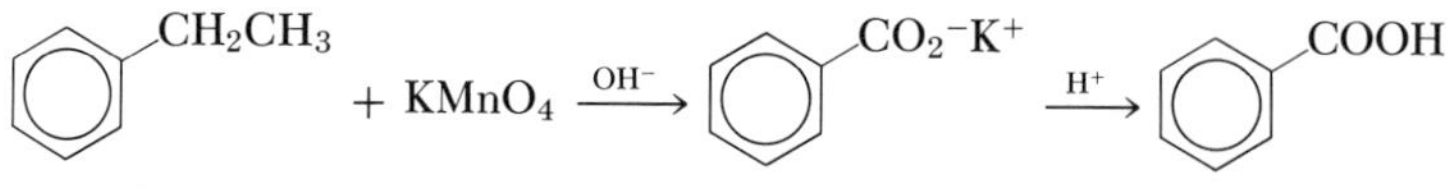

$$C_6H_5CH_2CH_3 + KMnO_4 \xrightarrow{OH^-} C_6H_5CO_2^-K^+ \xrightarrow{H^+} C_6H_5COOH$$

Ethylbenzene — Potassium benzoate — Benzoic acid

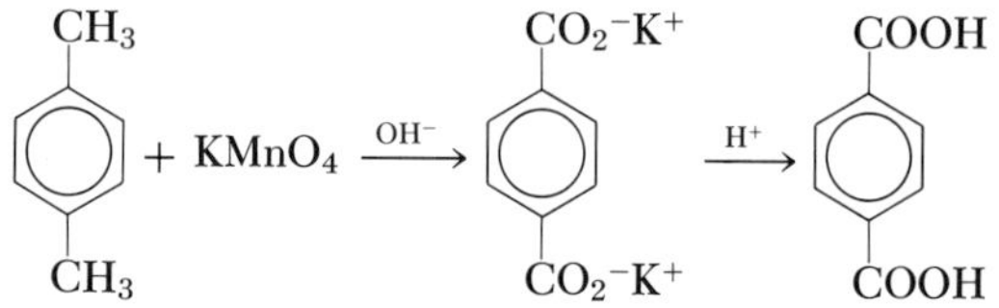

$$CH_3C_6H_4CH_3 + KMnO_4 \xrightarrow{OH^-} K^+{}^-O_2C{-}C_6H_4{-}CO_2^-K^+ \xrightarrow{H^+} HOOC{-}C_6H_4{-}COOH$$

p-Xylene — Potassium terephthalate — Terephthalic acid

Carbonation of Organometallic Reagents. When organometallic reagents, such as Grignard reagents, are treated with carbon dioxide (CO_2), a complex is formed that gives a carboxylic acid on hydrolysis. This reaction may be rationalized as the addition of the organometallic reagent across the carbonyl group (similar to the addition of organometallics to aldehydes and ketones), as shown below:

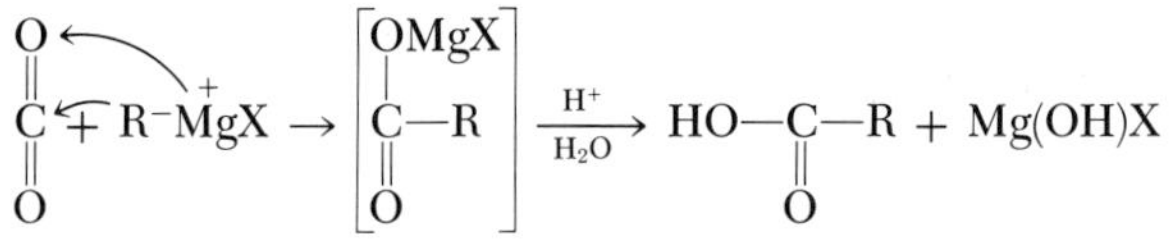

$$O{=}C{=}O + R^-\overset{+}{Mg}X \rightarrow \left[\begin{array}{c} OMgX \\ | \\ C{-}R \\ \| \\ O \end{array} \right] \xrightarrow[H_2O]{H^+} HO{-}\underset{\underset{O}{\|}}{C}{-}R + Mg(OH)X$$

Complex — Carboxylic acid

Both aliphatic and aromatic carboxylic acids can be prepared from the appropriate alkyl or aryl halides by this method. The acids prepared by this method contain one more carbon atom than the original alkyl or aryl halide.

REACTIONS OF CARBOXYLIC ACIDS

Salt Formation. Like their inorganic analogues, carboxylic acids react with bases to form salts. The name of the salt is formed by naming the cation, followed by the name of the acid in which the **-ic** ending (of either the common or IUPAC name) has been changed to **-ate.** Many of these salts are water-soluble. Since these salts are the salts of weak acids, they are readily hydrolyzed, and the acid can be regenerated by treatment of the acid salt with a stronger acid such as HCl or H_2SO_4 as illustrated below:

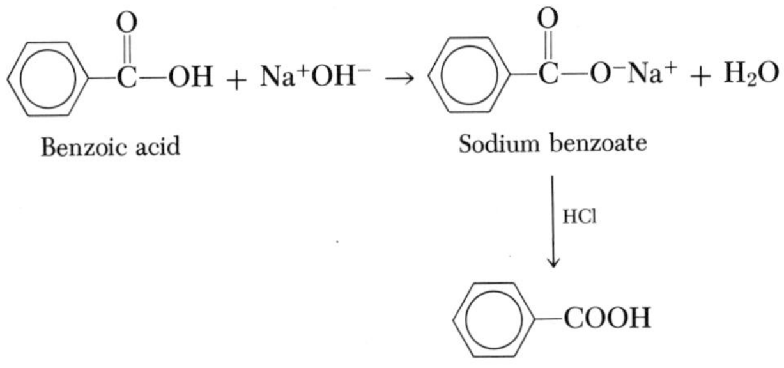

$$C_6H_5\text{—}\overset{\overset{\large O}{\|}}{C}\text{—}OH + Na^+OH^- \rightarrow C_6H_5\text{—}\overset{\overset{\large O}{\|}}{C}\text{—}O^-Na^+ + H_2O$$

Benzoic acid — Sodium benzoate

$$\downarrow HCl$$

$$C_6H_5\text{—}COOH$$

Conversion Into Functional Derivatives. ***Acid Halides (Acyl Halides).*** * The —OH group of a carboxylic acid can be replaced by a halogen atom to give a class of compounds called **acid halides,** RCOX (where X = halogen). This reaction is analogous to the replacement of the —OH group of alcohols and phenols by halogen to give alkyl or aryl halides. Reagents such as phosphorus trichloride, phosphorus pentachloride, and thionyl chloride (see reactions of alcohols) are used to replace the hydroxyl group. The hydrogen halides cannot be used to prepare the acid halides.

$$CH_3\overset{\overset{\large O}{\|}}{C}\text{—}OH + SOCl_2 \rightarrow CH_3\overset{\overset{\large O}{\|}}{C}\text{—}Cl + SO_2\uparrow + HCl\uparrow$$

Acetic acid — Thionyl chloride — Acetyl chloride

$$C_6H_5\text{—}\overset{\overset{\large O}{\|}}{C}\text{—}OH + PCl_5 \rightarrow C_6H_5\text{—}\overset{\overset{\large O}{\|}}{C}\text{—}Cl + POCl_3 + HCl\uparrow$$

Benzoic acid — Benzoyl chloride

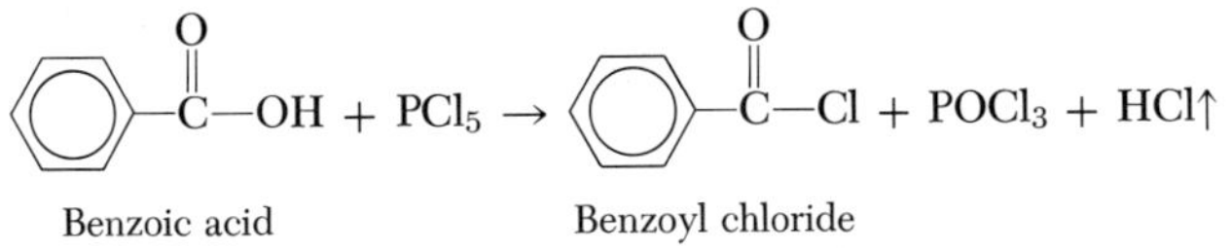

Esters. Carboxylic acids react with alcohols in the presence of an acid to give a class of compounds called **esters,** RCO_2R' (see reactions of alcohols). The name of the ester is formed by changing the **-ic** ending of the acid (either the common or IUPAC name) to **-ate,** preceded by the name of the alkyl or aryl group derived from the alcohol as shown below:

$$CH_3CO_2H + CH_3CH_2OH \overset{H^+}{\rightleftharpoons} CH_3\overset{\overset{\large O}{\|}}{C}\text{—}OCH_2CH_3 + H_2O$$

Acetic acid — Ethyl alcohol — Ethyl acetate

Acid Anhydrides. Carboxylic acids undergo dehydration (loss of a mole of water) when treated with dehydrating agents to give a class of compounds

* The $R\text{—}\overset{\overset{O}{\|}}{C}\text{—}$ group is known as the acyl group. In many reactions of acids, this portion of the acid molecule remains unaffected in the chemical reaction. Many acid derivatives, particularly the halide derivatives, are named according to this group. The name is formed by changing the **-ic** ending of the common acid name to **-yl.**

called **acid anhydrides,** $RC(=O)-O-C(=O)-R'$. Anhydrides are named by adding the name anhydride after the acid from which the anhydride is derived. Cyclic anhydrides, containing five- or six-membered rings, are formed from the appropriate di-acids on dehydration. Some representative examples are shown below:

$$2CH_3C(=O)-OH \xrightarrow{H_2SO_4} CH_3C(=O)-O-C(=O)-CH_3$$

Acetic acid — Acetic anhydride

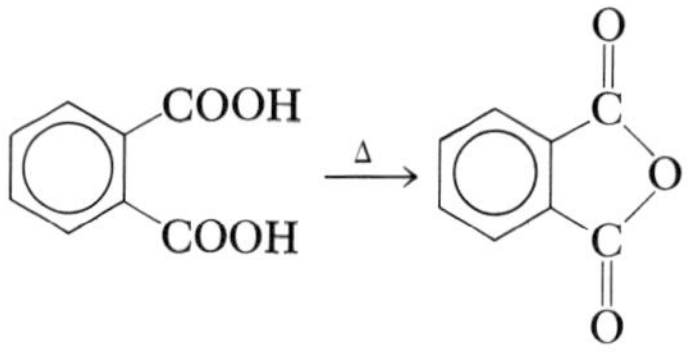

Phthalic acid (benzene ring with two COOH groups) $\xrightarrow{\Delta}$ Phthalic anhydride (benzene ring fused with C(=O)—O—C(=O))

Amides. Carboxylic acids react with ammonia to give salts, which on heating give a class of compounds called **amides,** $RCONH_2$. Substituted amides,

$$CH_3CO_2H + NH_3 \rightarrow CH_3CO_2^-NH_4^+ \xrightarrow{\Delta} CH_3C(=O)-NH_2 + H_2O$$

Acetic acid — Ammonium acetate — Acetamide

$RCONHR'$ or $RCONR_2'$, can be prepared by using amines (RNH_2 or R_2NH) in place of ammonia. The unsubstituted amides ($RCONH_2$) are named by dropping the **-ic** and **-oic** endings of the common and IUPAC names of the acid and adding the term **-amide.** When groups other than hydrogen are present on the nitrogen atom, their position and type is indicated by adding the prefix **N-** to indicate their position, followed by the name of the group attached to the nitrogen atom. Some typical examples are shown below:

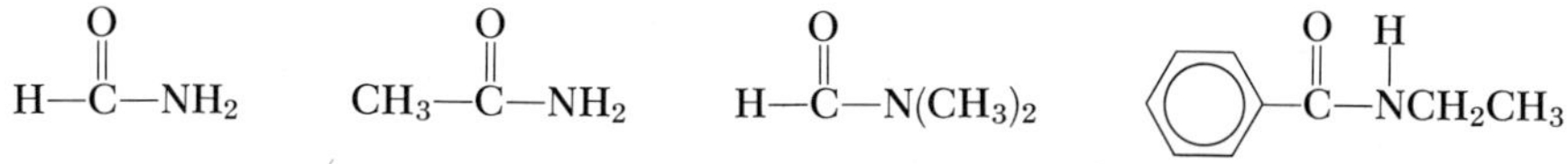

$H-C(=O)-NH_2$ — Formamide

$CH_3-C(=O)-NH_2$ — Acetamide

$H-C(=O)-N(CH_3)_2$ — N,N-Dimethylformamide

$C_6H_5-C(=O)-N(H)CH_2CH_3$ — N-Ethylbenzamide

SUBSTITUTION REACTIONS

Aromatic carboxylic acids undergo the usual electrophilic aromatic substitution reactions. The carboxyl group is a *meta*-director and deactivates relative to benzene. The bromination of benzoic acid represents a typical example of this kind of reaction. Nitration, sulfonation, and alkylation proceed similarly.

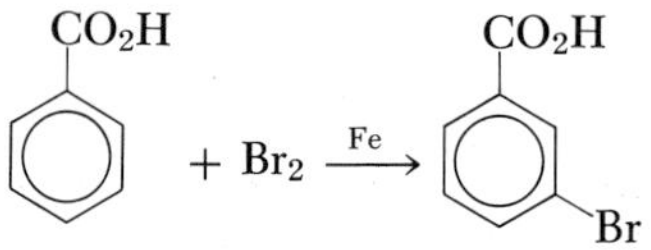

$$C_6H_5CO_2H + Br_2 \xrightarrow{Fe} m\text{-}BrC_6H_4CO_2H$$

Benzoic acid — *m*-Bromobenzoic acid

IMPORTANT ACIDS, ACID DERIVATIVES, AND THEIR USES

Formic acid occurs free in nature in small amounts, and its presence is made known in an unpleasant fashion. Anyone who has been bitten by an ant, stung by a bee, or brushed against stinging nettles has felt the irritating effect of formic acid injected under the skin. Formic acid is prepared commercially by heating powdered sodium hydroxide with carbon monoxide gas under pressure. The sodium formate produced is converted to formic acid by treatment with sulfuric acid.

$$\underset{\text{Carbon monoxide}}{CO} + NaOH \xrightarrow[\text{pressure}]{\Delta} \underset{\text{Sodium formate}}{H-\overset{\overset{\large O}{\|}}{C}-O^-Na^+} \xrightarrow{H_2SO_4} \underset{\text{Formic acid}}{H-\overset{\overset{\large O}{\|}}{C}-OH} + NaHSO_4$$

Formic acid is a colorless liquid with a sharp, irritating odor. It is a slightly stronger acid than most carboxylic acids and produces blisters when it comes in contact with the skin. It is used in the manufacture of esters, salts, plastics, and oxalic acid.

Acetic acid has been known for centuries as an essential component of vinegar. It is formed from the oxidation of ethyl alcohol that is produced from the fermentation of fruit juices in the preparation of the vinegar. Cider vinegar from fruit juices contains about 4% acetic acid in addition to flavoring and coloring agents from the fruit. White vinegar is prepared by diluting acetic acid to the proper concentration with water.

Acetic acid is produced commercially by the oxidation of acetaldehyde, which is obtained by a catalytic oxidation of ethylene. Commercially produced acetic acid is about 99.5% pure and is called "glacial acetic acid," since at temperatures below 17°C it freezes to an ice-like solid. Large quantities of acetic acid are used commercially in the manufacture of white lead, cellulose acetate for the production of rayon, cellophane, plastics, and organic solvents.

Lactic acid, $CH_3CHOHCOOH$, is an important hydroxy acid. It is formed when lactose (milk sugar) is fermented by lactobacillus bacteria. The taste of sour milk and buttermilk is due to the presence of lactic acid. Sour milk is often used in baking for its leavening effect on the dough. The lactic acid reacts with the sodium bicarbonate of baking soda to produce carbon dioxide throughout the dough.

In the process of muscular contraction, lactic acid is formed by the muscle tissues and released into the blood stream. In many of the so-called "cycles" involved in the oxidation of carbohydrates, lipids, and proteins to produce energy in the body, lactic acid is an essential component.

Lactic acid is made commercially by the fermentation of sugar or starch using a strain of lactic acid bacteria. As an acid agent it finds use in the leather industry, in the manufacture of yeast and cheese, and in dye processing. Lactic acid esters are useful as solvents in the preparation of lacquers and plastics. Salts of lactic acid, the lactates, are used in medicine—for example, in calcium therapy—and also in several food products.

Oxalic acid occurs in the leaves of vegetables such as rhubarb and is one of the strongest naturally occurring acids. Its preparation is closely related to that of formic acid, since the sodium salt of oxalic acid is produced by heating sodium formate with alkali. Oxalic acid is used to remove iron stain from fabrics and from porcelain ware, and to bleach straw and leather goods.

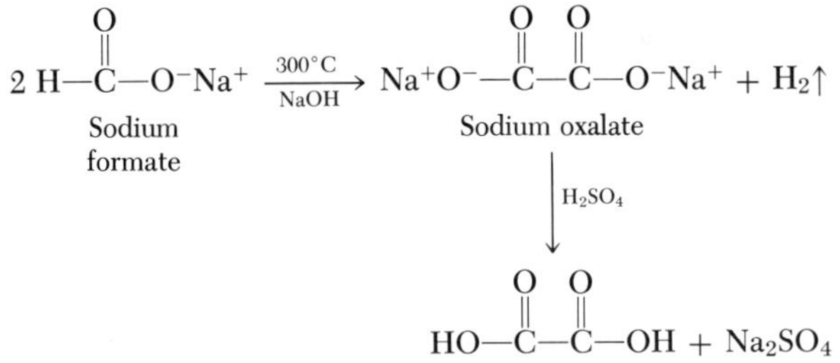

$$2\,H{-}\overset{\overset{\displaystyle O}{\|}}{C}{-}O^-Na^+ \xrightarrow[NaOH]{300^\circ C} Na^+O^-{-}\overset{\overset{\displaystyle O}{\|}}{C}{-}\overset{\overset{\displaystyle O}{\|}}{C}{-}O^-Na^+ + H_2\uparrow$$

Sodium formate — Sodium oxalate

$$\downarrow H_2SO_4$$

$$HO{-}\overset{\overset{\displaystyle O}{\|}}{C}{-}\overset{\overset{\displaystyle O}{\|}}{C}{-}OH + Na_2SO_4$$

Oxalic acid

Citric acid, $HOOCCH_2C(OH)(COOH)CH_2COOH$, is a normal constituent of citrus fruits. Large quantities are produced by a fermentation process from starch or molasses. It is commonly employed to impart a sour taste to food products and beverages. In fact, most of the citric acid produced annually is used by the food and soft drink industries.

Benzoic acid is a colorless solid, slightly soluble in hot water. It is used in the synthesis of organic compounds, and its sodium salt (sodium benzoate) is used as an antiseptic and food preservative. Several substituted benzoic acids are also important commercial materials. ***Para*-aminobenzoic acid** is classified as a vitamin. It is apparently necessary for the proper physiological functioning of chickens, mice, and bacteria, but as yet has not been proved essential in the diet of humans. Certain derivatives of this acid are used as local anesthetics. Procaine, or Novocain, is probably the most important local anesthetic derived from *p*-aminobenzoic acid. Several derivatives of benzene sulfonic acid are used as dyes or indicators, sweetening agents, and antiseptics. Chloramine-T is a sulfonic acid derivative of toluene that is used as an antiseptic. Phenolsulfonphthalein (phenol red) is a derivative of sulfobenzoic acid and phenol; it is used as a pH indicator and as a dye in kidney function tests. Saccharin, a derivative of *o*-sulfobenzoic acid, has a relative sweetness several hundred times that of sucrose. It is used as a sweetening agent when the sugar intake must be restricted. Recently another derivative of *o*-sulfobenzoic acid named Sucaryl has been developed. This sweetening agent has properties similar to saccharin and is available in both liquid and solid form.

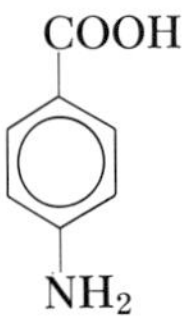

COOH / NH_2 — *p*-Aminobenzoic acid

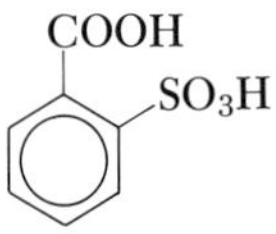

COOH, SO_3H — *o*-Sulfobenzoic acid

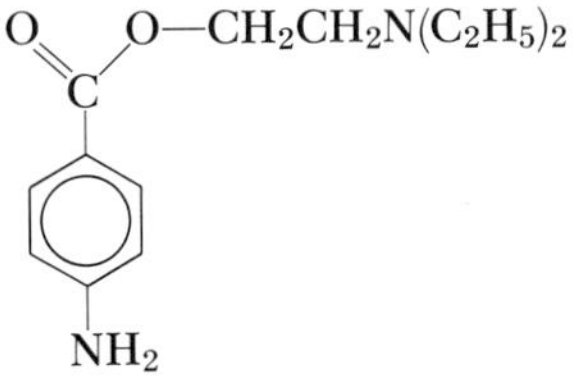

$O{=}C{-}O{-}CH_2CH_2N(C_2H_5)_2$ / NH_2 — Procaine

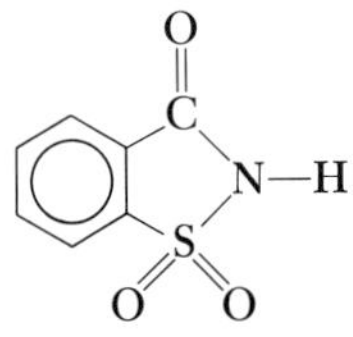

O, C, N—H, S, O O — Saccharin

Acid Derivatives. Organic acid derivatives, such as salts, are often used in place of the acids for many commercial uses. **Sodium acetate** is often used for its buffering effect in reducing the acidity of inorganic acids. It is also used in the preparation of soaps and pharmaceutical agents and in the synthesis of many organic compounds. **Lead acetate,** sometimes called sugar of lead because of its sweet taste, is used externally to treat poison ivy and certain skin diseases. Commercially, it is used in large quantities in printing inks and paper, and in the manufacture of white and pigmented paints. **Paris green** is a complex salt that contains copper acetate and is used as an insecticide. Aluminum acetate is used in certain processes in the dyeing of textile fibers. Calcium propionate is added to bread to prevent molding. The calcium salt of lactic acid is sometimes used to supplement the calcium of the diet. Many infants receive additional iron in their diets from the salt, ferric ammonium citrate. Magnesium citrate has long been used as a saline cathartic. A solution of sodium citrate is employed to prevent the clotting of blood used in transfusions, and potassium oxalate will prevent the clotting of blood specimens drawn for analysis in the clinical laboratory. The sodium salt of salicylic acid is used extensively as an antipyretic or fever-lowering agent, and as an analgesic in the treatment of rheumatism and arthritis.

Commercially, only a few of the esters are produced in large quantity. The two most important are ethyl acetate and butyl acetate, which are used as solvents, especially for nitrocellulose in the preparation of lacquers. Some of the higher molecular weight esters are plastics, whereas others are used in the production of plastics.

Many esters occur in free form in nature and are responsible for the odor of most flowers and fruits. The characteristic taste and odors of different esters find application in the manufacture of artificial flavoring extracts and perfumes. Synthetic esters that are commonly used as food flavors are amyl acetate for banana, octyl acetate for orange, ethyl butyrate for pineapple, amyl butyrate for apricot, isobutyl formate for raspberry, and ethyl formate for rum. Esters used for the manufacture of perfumes are usually esters of aromatic or cyclic acids.

Other esters are commonly used as therapeutic agents in medicine. Ethyl acetate is employed as a stimulant and antispasmodic in colic and bronchial irritations. It is also applied externally in the treatment of skin diseases caused by parasites. Ethyl nitrite, when mixed with alcohol, is called elixir of niter and is used as a diuretic and antispasmodic. Amyl nitrite is used to lower blood pressure temporarily and causes relaxation of muscular spasms in asthma and in the heart condition known as angina pectoris. Glyceryl trinitrate, or nitroglycerin, is a vasodilator that has physiological action similar to amyl nitrate.

Methyl benzoate and methyl salicylate are two of the aromatic esters used in the manufacture of perfumes and flavoring agents. Methyl benzoate has the odor of new-mown hay, whereas methyl salicylate smells and tastes like wintergreen. One of the most important esters of salicylic acid is that formed with acetic acid. **Acetylsalicylic acid,** or aspirin, is the common antipyretic and analgesic drug. An increasing number of medications for the relief of simple headaches and the pain of rheumatism and arthritis contain aspirin combined with other pharmaceutical agents. For example, aspirin combined with antacid and buffering agents is absorbed more rapidly from the intestinal tract and should therefore afford more rapid relief of aches and pains. Several esters of *p*-aminobenzoic acid act as local anesthetics. Butesin is the butyl ester, whereas procaine is the diethyl amino ester of this acid.

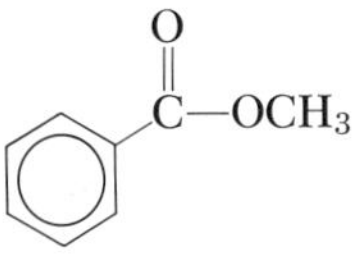

$C_6H_5-C(=O)-OCH_3$ Methyl benzoate

o-$HO-C_6H_4-C(=O)-OCH_3$ Methyl salicylate

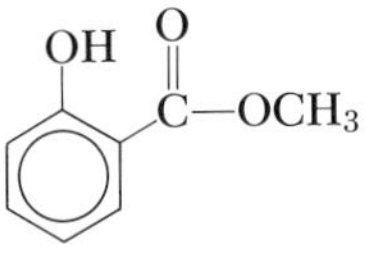

o-$CH_3-C(=O)-O-C_6H_4-COOH$ Acetylsalicylic acid

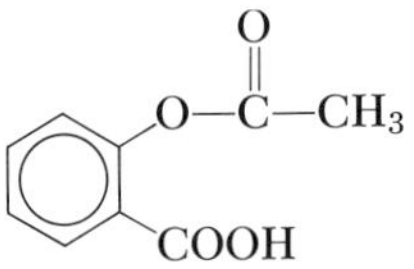

Polyesters make up some of the most important commercial polymers. One of the most interesting and important types of vinyl plastics is the acrylic resins. They are polymers of acrylic acid esters, such as methyl acrylate and methyl methacrylate.

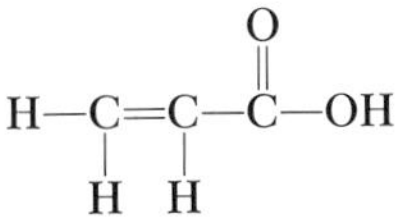

$H-C(H)=C(H)-C(=O)-OH$ Acrylic acid

$H-C(H)=C(H)-C(=O)-OCH_3$ Methyl acrylate

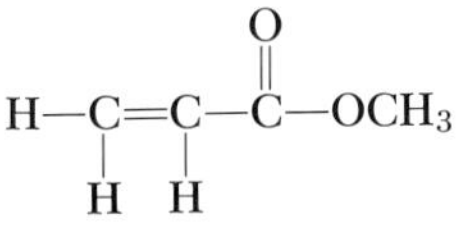

$H-C(H)=C(CH_3)-C(=O)-OCH_3$ Methyl methacrylate

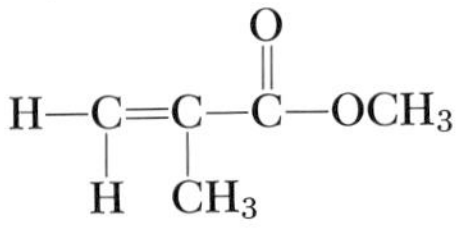

Polymerization of methyl methacrylate results in a clear, transparent plastic that can be used in place of glass. Lucite, Plexiglas, and Perspex are some of the trade names given this polymer. It can be formed into strong, flexible sheets that are highly transparent and lighter than glass, or it can be molded into transparent articles. This plastic rivals nylon in its many applications, since it fills a longstanding need for a clear, transparent substitute for glass.

Another important polyester, Dacron, is prepared from terephthalic acid and ethylene glycol. The initial condensation produces an ester linkage, but the initial condensation product still has two functional groups left which can react further. This type of condensation continues until thousands of the ester units are contained in the chain.

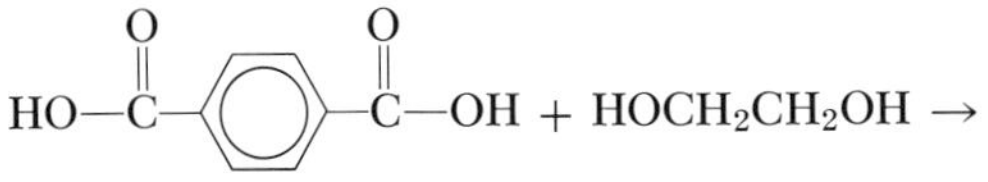

$$HO-C(=O)-C_6H_4-C(=O)-OH + HOCH_2CH_2OH \rightarrow$$

Terephthalic acid; Ethylene glycol

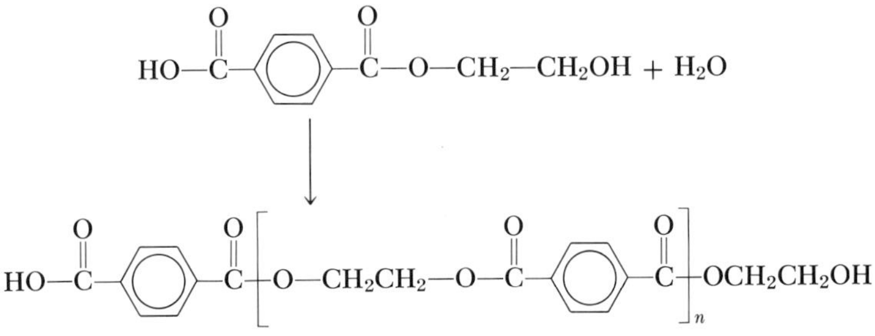

$$HO-C(=O)-C_6H_4-C(=O)-O-CH_2-CH_2OH + H_2O$$

$$\downarrow$$

$$HO-C(=O)-C_6H_4-C(=O)\left[-O-CH_2CH_2-O-C(=O)-C_6H_4-C(=O)-\right]_n OCH_2CH_2OH$$

Dacron

Phthalic anhydride is another very important organic compound that is usually produced commercially by the oxidation of *ortho*-xylene or naphthalene. Large quantities of this material are used as plasticizers of synthetic resins and in the manufacture of the glyptal type of weather-resistant protective coating. Phthalic anhydride is also used for the preparation of anthraquinone dyes and in the manufacture of anthranilic acid, which is used in the manufacture of other types of dyes.

The most important amide is **urea,** $H_2N-\overset{\overset{O}{\|}}{C}-NH_2$, which is a naturally occurring diamide of carbonic acid. The synthesis of this compound by Wöhler in 1828 was the keystone in the development of organic chemistry. Urea is prepared by the action of heat and pressure on carbon dioxide and ammonia.

$$CO_2 + 2NH_3 \xrightarrow[\text{pressure}]{\Delta} \underset{\text{Urea}}{H_2N-\overset{\overset{O}{\|}}{C}-NH_2} + H_2O$$

On heating, urea decomposes to ammonia and isocyanic acid. Alcohols react with isocyanic acid to form carbamates.

$$\underset{\text{Urea}}{H_2N-\overset{\overset{O}{\|}}{C}-NH_2} \xrightarrow{\Delta} \underset{\text{Ammonia}}{NH_3} + \underset{\text{Isocyanic acid}}{HN{=}C{=}O}$$

$$\underset{\text{Isocyanic acid}}{HN{=}C{=}O} + \underset{\text{Ethyl alcohol}}{HOCH_2CH_3} \rightarrow \underset{\text{Ethyl carbamate}}{H_2N-\overset{\overset{O}{\|}}{C}-OCH_2CH_3}$$

Meprobamate (Miltown) is 2-methyl-2-*n*-propyl-1,3-propanediol dicarbamate, a widely used tranquilizing agent.

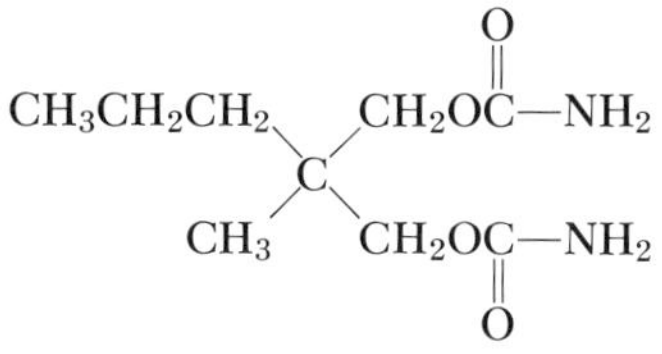

$$(CH_3CH_2CH_2)(CH_3)C(CH_2O\overset{\overset{O}{\|}}{C}-NH_2)(CH_2O\underset{\underset{O}{\|}}{C}-NH_2)$$

Miltown

One of the most important commercial uses of urea is the base catalyzed condensation of urea with substituted diethyl malonates to produce the sedatives known as the **barbiturates.**

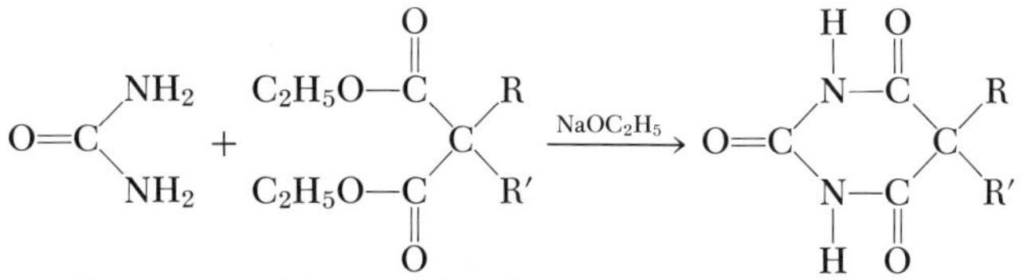

$$\underset{\text{Urea}}{O{=}C(NH_2)_2} + \underset{\text{Substituted diethyl malonates}}{(C_2H_5O-\overset{\overset{O}{\|}}{C})(C_2H_5O-\underset{\underset{O}{\|}}{C})CRR'} \xrightarrow{NaOC_2H_5} \underset{\text{A barbiturate}}{O{=}C\langle\begin{matrix} N(H)-C(=O) \\ N(H)-C(=O) \end{matrix}\rangle CRR'}$$

An important polyamide, nylon, is prepared by the condensation of adipic acid and hexamethylenediamine. The initial reaction involves a neutralization reaction between the acid and the amine to give a nylon salt, which on heating at high temperatures produces nylon.

$$\underset{\text{Adipic acid}}{HO-\overset{O}{\overset{\|}{C}}-(CH_2)_4-\overset{O}{\overset{\|}{C}}-OH} + \underset{\text{Hexamethylenediamine}}{H_2N-(CH_2)_6-NH_2} \rightarrow$$

$$\underset{\text{Nylon salt}}{HO-\overset{O}{\overset{\|}{C}}-(CH_2)_4-\overset{O}{\overset{\|}{C}}-O^{-}H_3\overset{+}{N}-(CH_2)_6-NH_2}$$

$$\downarrow 250°C$$

$$nH_2O + \underset{\text{Nylon}}{HO-\overset{O}{\overset{\|}{C}}-(CH_2)_4-\overset{O}{\overset{\|}{C}}\left[-\overset{H}{N}-(CH_2)_6-\overset{H}{N}-\overset{O}{\overset{\|}{C}}-(CH_2)_4-\overset{O}{\overset{\|}{C}}-\right]_n\overset{H}{N}(CH_2)_6NH_2}$$

Questions

1. Draw a structural formula for each of the following compounds:
   (a) potassium propionate
   (b) isobutyryl chloride
   (c) ethyl benzoate
   (d) succinic anhydride
   (e) diethyl phthalate
   (f) butyramide
   (g) sodium oxalate
   (h) *m*-bromobenzoic acid
   (i) methyl formate
   (j) isohexanoic acid

2. Write equations for the reactions of *n*-butyric acid with the following reagents:
   (a) CH_3CH_2OH/H^+
   (b) $SOCl_2$
   (c) NH_3/heat
   (d) KOH
   (e) H_2SO_4/heat
   (f) PBr_3

3. Write equations for the reactions of benzoic acid with the following reagents:
   (a) $Ca(OH)_2$
   (b) PCl_5
   (c) Br_2/Fe
   (d) HNO_3/H_2SO_4
   (e) benzyl alcohol/H^+

4. Write equations for the preparation of the following compounds. Give any necessary catalysts. More than one step may be necessary.
   (a) *n*-butyric acid from *n*-butyl alcohol
   (b) ethyl propionate from propionic acid
   (c) benzoic acid from benzene
   (d) N-methylpropionamide from ethyl bromide
   (e) benzoyl chloride from benzene
   (f) *n*-propionic acid from ethyl iodide

Suggested Reading

Carter: The History of Barbituric Acid. Journal of Chemical Education, Vol. 28, p. 524, 1951.
Fisher: New Horizons In Elastic Polymers. Journal of Chemical Education, Vol. 37, p. 369, 1960.
Mayo: Contribution of Vinyl Polymerization to Organic Chemistry. Journal of Chemical Education, Vol. 36, p. 157, 1959.
McGrew: Structure of Synthetic High Polymers. Journal of Chemical Education, Vol. 35, p. 178, 1958.
Moncrieff: Linear Polymerization and Synthetic Fibers. Journal of Chemical Education, Vol. 31, p. 233, 1954.
Price: The Effect of Structure On Chemical and Physical Properties of Polymers. Journal of Chemical Education, Vol. 42, p. 13, 1965.

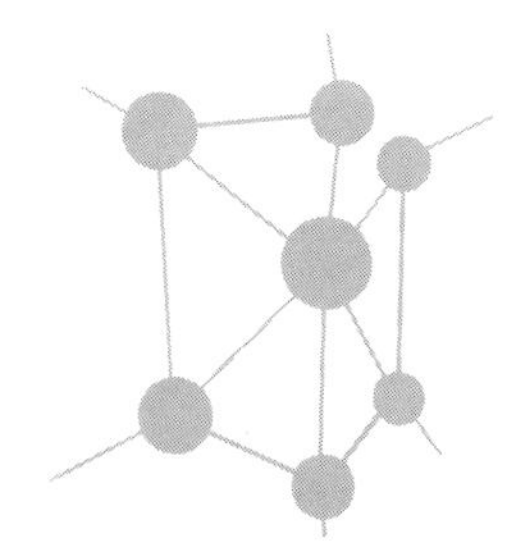

CHAPTER 19

AMINES AND AMINE DERIVATIVES

Amines are organic derivatives of ammonia in which one or more of the hydrogen atoms have been replaced by an alkyl or aryl group. The characteristic functional group present in amines is called the **amino group** and is written as **—NH_2**. Like alcohols, amines are classified as primary, secondary and tertiary amines. In the case of amines the classification is determined by the *number* of alkyl or aryl groups attached to the nitrogen atom. Some representative examples are illustrated below:

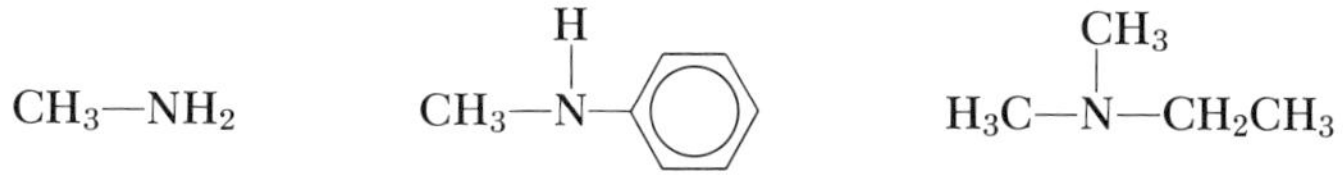

$CH_3—NH_2$ — Primary amine

$CH_3—N(H)—C_6H_5$ — Secondary amine

$H_3C—N(CH_3)—CH_2CH_3$ — Tertiary amine

Another description that classifies amines as primary, secondary, or tertiary describes the nitrogen atom on the basis of its having two, one, or no N—H bonds.

Amines can be named as amino derivatives of hydrocarbons according to the IUPAC system—as amino alkanes for example. This system, however, is little used. The common name nomenclature system is most widely used. It consists of naming the alkyl or aryl groups attached to the nitrogen atom, using the appropriate prefixes if two or more identical substituents are attached to the nitrogen, followed by the word "amine." The following examples illustrate this system of nomenclature (cf. also to Table 19–1).

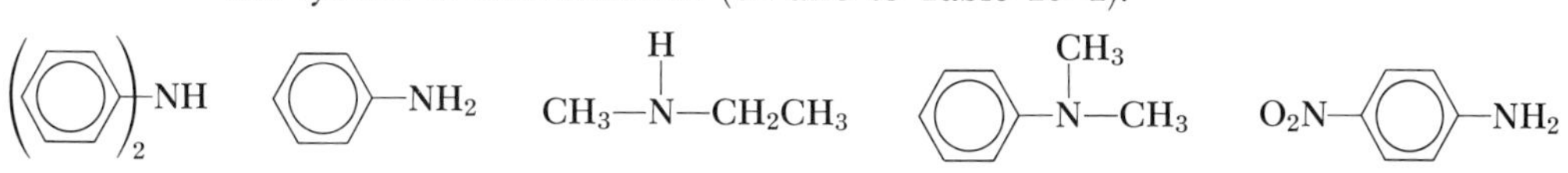

$(C_6H_5)_2NH$ — Diphenylamine

$C_6H_5—NH_2$ — Aniline

$CH_3—N(H)—CH_2CH_3$ — Methyl ethyl amine

$C_6H_5—N(CH_3)—CH_3$ — N,N-Dimethylaniline

$O_2N—C_6H_4—NH_2$ — *p*-Nitroaniline

Simple aromatic amines are named as derivatives of the parent aromatic amine, **aniline.** The nitrogen atom may also be part of a ring system, and both cyclic and aromatic types of amines are known, as illustrated below:

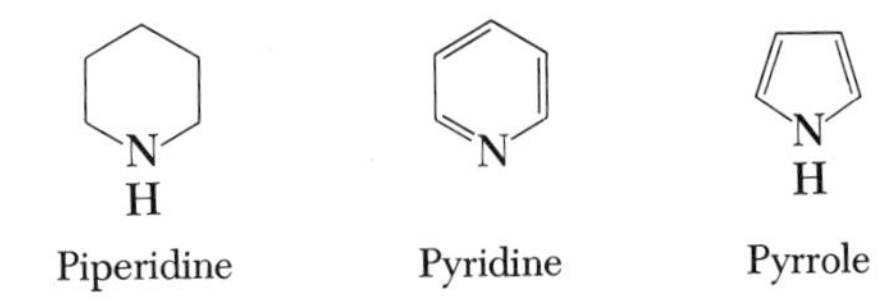

Piperidine Pyridine Pyrrole

PHYSICAL PROPERTIES OF AMINES

The lower molecular weight amines are gases that are soluble in water. As might be anticipated for derivatives of ammonia, the aqueous solutions of the water-soluble amines are alkaline. The volatile amines have unpleasant odors that combine the odor of ammonia with that of decayed fish. The higher molecular weight ($>C_6$) amines are insoluble in water and soluble in organic solvents, and the unpleasant odor decreases with decreasing volatility. The properties of some typical amines are listed in Table 19–1.

Amines are basic like ammonia. Their base strength depends on the type of group (alkyl or aryl) attached to the nitrogen atom. Alkyl groups attached to the nitrogen atom increase the basicity of the amine relative to ammonia, and the alkyl amines are slightly stronger bases than ammonia. On the other hand, aromatic groups attached to the nitrogen atom decrease the basicity of

TABLE 19–1 PHYSICAL PROPERTIES OF AMINES

Compound	*Structure*	*M.P. °C*	*B.P. °C*
Methylamine	CH_3NH_2	−93	−6
Ethylamine	$CH_3CH_2NH_2$	−81	+17
n-Propylamine	$CH_3CH_2CH_2NH_2$	−83	+48
Dimethylamine	$(CH_3)_2NH$	−92	+7
Diethylamine	$(CH_3CH_2)_2NH$	−50	+56
Trimethylamine	$(CH_3)_3N$	−117	+3
Triethylamine	$(CH_3CH_2)_3N$	−115	+90
Aniline	NH_2 (benzene ring)	−6	+184
o-Toluidine	CH_3, NH_2 (benzene ring)	−15	+200
m-Toluidine	CH_3, NH_2 (benzene ring)	−30	+203
p-Toluidine	CH_3, NH_2 (benzene ring)	+44	+201

the amine relative to ammonia. Aniline is a much weaker base than ammonia; diphenylamine is only weakly basic; and triphenylamine is essentially a neutral compound.

The boiling points of primary and secondary amines are higher than might be expected on the basis of molecular weight. Primary and secondary amines can form hydrogen bonds, which accounts for their high boiling points.

$$\begin{array}{ccccc} & \vdots & & & \\ & H & & R & \\ & | & & | & \\ R-&N&-H\cdots\cdots\cdots&N&-H\cdots,\ \text{etc.} \\ & & & | & \\ & & & H & \\ & & & \vdots & \end{array}$$

Tertiary amines, which cannot form hydrogen bonds, have boiling points similar to the hydrocarbons of corresponding molecular weights (compare the b.p. of methyl amine, dimethyl amine, trimethylamine, and *n*-butane).

SALT FORMATION

Ammonia reacts with acids to form salts. The unshared pair of electrons on the nitrogen atom is used to form a new N—H bond, and the ammonia molecule is converted into the ammonium ion.

$$\underset{\text{Ammonia}}{H-\underset{H}{\overset{\cdot\cdot}{N}}-H} + \underset{\text{Hydrogen chloride}}{H^+Cl^-} \rightarrow \underset{\text{Ammonium chloride}}{\left[H-\underset{H}{\overset{H}{N}}-H\right]^+ Cl^-}$$

The organic derivatives of ammonia, the amines, behave similarly. Aqueous mineral acids and carboxylic acids convert amines into salts. The amine salts are typical ionic saltlike compounds.

$$\begin{array}{ccccc} R-\ddot{N}H_2 & & R\overset{+}{N}H_3 & & R\ddot{N}H_2 \\ R_2\ddot{N}H & \xrightarrow{HX} & R_2\overset{+}{N}H_2\ X^- & \xrightarrow{OH^-} & R_2\ddot{N}H + H_2O + X^- \\ R_3\ddot{N} & & R_3\overset{+}{N}H & & R_3\ddot{N} \end{array}$$

They are nonvolatile solids, generally soluble in water and insoluble in nonpolar organic solvents. The free amine can be regenerated by treatment of the salt with a strong base, such as a hydroxide ion.

PREPARATION OF AMINES

Reduction Methods. The reduction of compounds containing the nitro ($-NO_2$) group affords a convenient path to primary amines, particularly the aromatic amines. The aliphatic nitro compounds are more difficult to prepare than the aromatic nitro compounds, and this method is not generally used for preparing aliphatic amines. The aromatic nitro compounds, however, can be easily prepared by nitration of the appropriate aromatic hydrocarbon. Reduction can be carried out catalytically (H_2 + Pt), but the more normal reducing agents used are a metal and a mineral acid, as shown below:

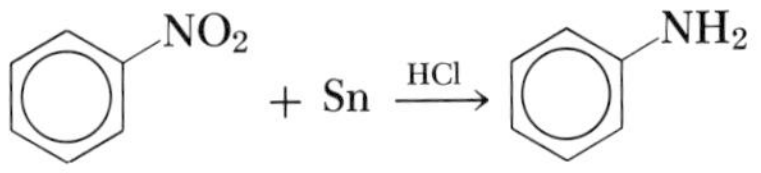

$$C_6H_5NO_2 + Sn \xrightarrow{HCl} C_6H_5NH_2$$

Nitrobenzene Aniline

Alkylation of Alkyl Halides. The use of ammonia or amines in a nucleophilic displacement reaction with alkyl halides (see reactions of alkyl halides) provides a classical route to the aliphatic amines, as shown below:

$$R{-}X + \ddot{N}H_3 \rightarrow [R{-}\overset{+}{N}H_3]X^- \xrightarrow{NaOH} R\ddot{N}H_2 + Na^+X^- + H_2O$$
$$X = Cl, Br, I$$

Unfortunately, the alkylated product (RNH_2) also can react with the alkyl halide to produce some secondary amine, which in turn can give tertiary amine, which in turn can give a quarternary ammonium salt.*

$$R{-}X + \ddot{N}H_3 \rightarrow R\ddot{N}H_2 \xrightarrow{R-X} R_2\ddot{N}H \xrightarrow{R-X} R_3\ddot{N} \xrightarrow{R-X} R_4{}^+NX^-$$

Alkyl halide; Primary amine; Secondary amine; Tertiary amine; Quaternary ammonium salt

In practice, a mixture of all four products results.

REACTIONS OF AMINES

Salt Formation. As noted earlier in this chapter, amines, both aliphatic and aromatic, form salts with mineral acids. The free amines can be regenerated by treatment of the salt with a strong base.

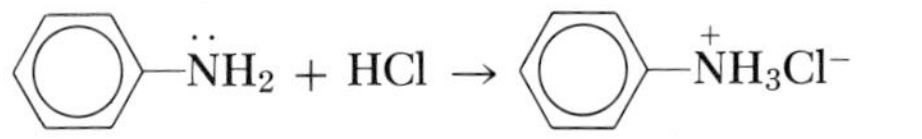

$$C_6H_5{-}\ddot{N}H_2 + HCl \rightarrow C_6H_5{-}\overset{+}{N}H_3Cl^-$$

Aniline; Anilinium chloride (aniline hydrochloride)

Alkylation. As noted in the preparation of amines, alkylation of alkyl halides occurs readily. Since the alkylation is difficult to control, this reaction is most useful in preparing quaternary ammonium salts.

$$(CH_3)_3\ddot{N} + CH_3I \rightarrow (CH_3)_4N^+I^-$$

Acylation. Carboxylic acid amides ($R{-}\overset{O}{\overset{\|}{C}}{-}NH_2$) are prepared via the reaction of ammonia with an acid halide or an acid anhydride. Similar types of acylation reactions occur when primary and secondary amines react with acid halides and acid anhydrides. Since tertiary amines have no available N—H bond, they cannot be acylated.

---

* A quaternary ammonium salt is the organic analog of ammonium halides, $NH_4{}^+X^-$, in which all the hydrogens have been replaced by organic groups.

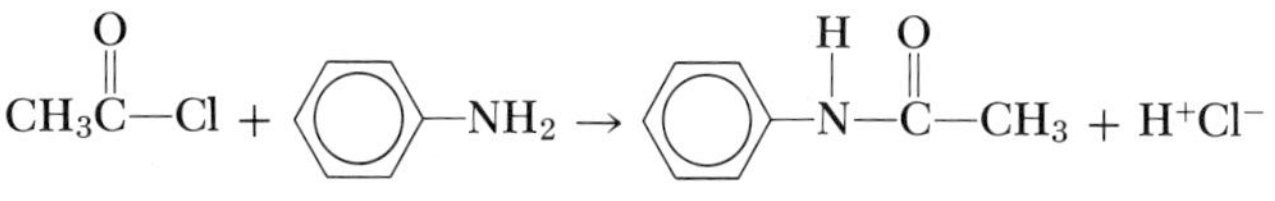

$$CH_3\overset{O}{\overset{\|}{C}}-Cl + C_6H_5-NH_2 \rightarrow C_6H_5-\overset{H}{\overset{|}{N}}-\overset{O}{\overset{\|}{C}}-CH_3 + H^+Cl^-$$

Acetyl chloride Aniline Acetanilide

In addition to being acylated by carboxylic acid derivatives, primary and secondary amines also react with sulfonic acid halides to form sulfonamides, as illustrated below for benzene sulfonyl chloride (the acid chloride of benzene sulfonic acid):

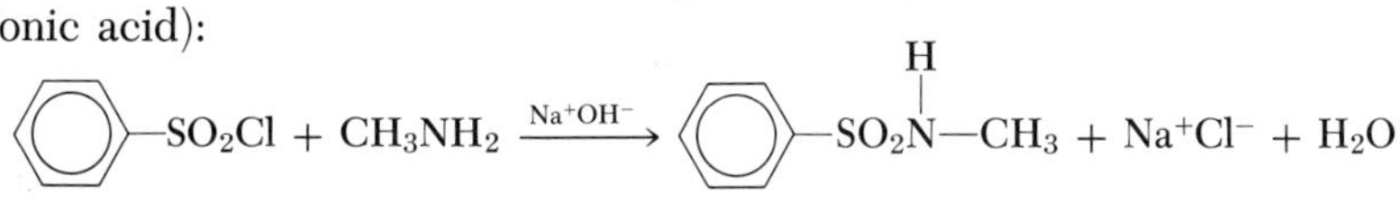

$$C_6H_5-SO_2Cl + CH_3NH_2 \xrightarrow{Na^+OH^-} C_6H_5-SO_2\overset{H}{\overset{|}{N}}-CH_3 + Na^+Cl^- + H_2O$$

Benzene sulfonyl chloride Methyl amine N-Methyl benzenesulfonamide

Sulfanilamide is the parent compound of the important class of chemotherapeutic agents known as the sulfa drugs. The majority of the drugs are prepared by an acylation reaction of a sulfanilamide derivative, as shown below:

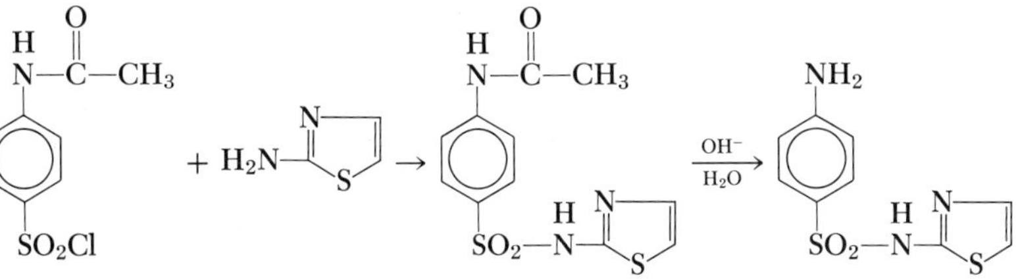

$$p\text{-}(CH_3\overset{O}{\overset{\|}{C}}-\overset{H}{N})-C_6H_4-SO_2Cl + H_2N-C_3H_2NS \rightarrow p\text{-}(CH_3\overset{O}{\overset{\|}{C}}-\overset{H}{N})-C_6H_4-SO_2-\overset{H}{N}-C_3H_2NS \xrightarrow[H_2O]{OH^-} p\text{-}NH_2-C_6H_4-SO_2-\overset{H}{N}-C_3H_2NS$$

2-Aminothiazole Sulfathiazole

The structure of the sulfa drug can be varied by varying the structure of the amine being condensed with the sulfonyl chloride (see reactions of aromatic hydrocarbons).

Aromatic Substitution Reactions. The amino group ($-NH_2$) and amino derivatives ($-\overset{H}{\overset{|}{N}}-\overset{O}{\overset{\|}{C}}-R$, $-\overset{H}{\overset{|}{N}}R$, and $-NR_2$) are *ortho-para* directing groups. They are also activating groups, and compounds containing these functional groups undergo electrophilic aromatic substitution with ease. In practice, the acylated derivative is usually employed, as this group, ($-\overset{H}{\overset{|}{N}}-\overset{O}{\overset{\|}{C}}-R$), is less of an activator than the amine group itself, and the substitution reaction is more easily controlled. The free amine can be regenerated by hydrolysis of the acylated compound. This type of reaction is illustrated below for the bromination of aniline:

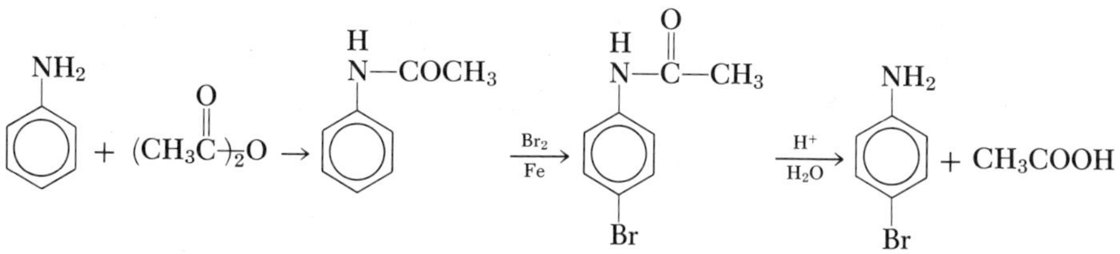

$$C_6H_5NH_2 + (CH_3\overset{O}{\overset{\|}{C}}\!\!\rightarrow_2 O \rightarrow C_6H_5\overset{H}{N}-COCH_3 \xrightarrow[Fe]{Br_2} p\text{-}Br-C_6H_4-\overset{H}{N}-\overset{O}{\overset{\|}{C}}-CH_3 \xrightarrow[H_2O]{H^+} p\text{-}Br-C_6H_4-NH_2 + CH_3COOH$$

Aniline Acetic anhydride Acetanilide Main product *p*-Bromoaniline

Nitration, sulfonation, and Friedel-Crafts reactions proceed similarly.

Reaction with Nitrous Acid. Primary amines react with nitrous acid (HONO) to give a **diazonium** salt, and this type of reaction is known as **diazotization.** Since nitrous acid is not a stable acid, it is generated *in situ* from sodium nitrite and hydrochloric acid.

$$R{-}\ddot{N}H_2 + NaNO_2 + HCl \xrightarrow{0°C} [R{-}\overset{+}{N}{\equiv}N{:}]Cl^-$$

A diazonium salt

Diazonium salts are generally not isolated, as they decompose easily and are explosive in the dry state. Aliphatic diazonium salts (R = alkyl) are usually not stable even at 0°C and are not as useful as the aromatic diazonium salts in synthesis. The aromatic diazonium compounds have a reasonable stability if kept at low temperatures (0° to 5°C), and undergo a variety of displacement and coupling reactions.

The main type of displacement reaction of aromatic diazonium salts is one which involves displacement of molecular nitrogen by a nucleophile. Nucleophiles, such as CN^-, I^-, Br^-, Cl^-, H_2O, and ROH, are effective in this type of reaction. Displacement by some nucleophiles is illustrated below for benzene diazonium chloride:

$$C_6H_5{-}\ddot{N}H_2 + NaNO_2 + HCl \xrightarrow{0°C} [C_6H_5{-}\overset{+}{N}{\equiv}N]\,Cl^-$$

Aniline — Benzenediazonium chloride

$$C_6H_5{-}N_2^+Cl^- + KI \rightarrow C_6H_5{-}I + K^+Cl^- + N_2\uparrow$$

Iodobenzene

$$C_6H_5{-}N_2^+Cl^- + CuCN \rightarrow C_6H_5{-}CN + CuCl + N_2\uparrow$$

Benzonitrile

This type of displacement reaction is particularly useful for introducing cyano and iodo groups into an aromatic ring, as these functional groups cannot usually be introduced by nucleophilic displacement or electrophilic substitution reactions.

Another important reaction of diazonium salts is their **coupling** reaction in alkaline or neutral solution with reactive aromatic compounds such as phenols or aromatic amines. The coupling takes place between the diazonium salt and the *para* position of the aromatic compound, and the nitrogen is retained. The (—N=N—) linkage is known as the **azo** group, and azo compounds of this type are generally colored. Related compounds containing the azo group make up the important group of dyestuffs known as azo dyes.

$$[C_6H_5{-}\overset{+}{N}{\equiv}N{:}]\,Cl^- + C_6H_5{-}OH \rightarrow C_6H_5{-}N{=}N{-}C_6H_4{-}OH$$

Benzene diazonium chloride — Phenol — *p*-Hydroxyazobenzene

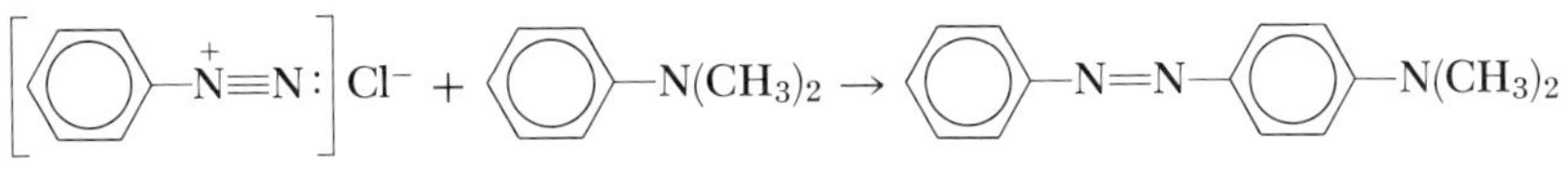

$$\left[C_6H_5-\overset{+}{N}\equiv N:\right]Cl^- + C_6H_5-N(CH_3)_2 \rightarrow C_6H_5-N{=}N-C_6H_4-N(CH_3)_2$$

N,N-Dimethyl aniline   *p*-Dimethylaminoazobenzene

IMPORTANT AMINES, AMINE DERIVATIVES AND THEIR USES

Dimethylamine is an excellent accelerator in the process for removal of hair from skins and hides in the manufacture of leather. It also reacts with carbon disulfide and sodium hydroxide to form an accelerator in the processing of rubber. When allowed to react with nitrous acid, followed by reduction of the product by hydrogen, dimethyl hydrazine, $(CH_3)_2N-NH_2$, is formed. Hydrazines of this type have been used as rocket propellants.

Butyl amine and amyl amines are used as corrosion inhibitors, antioxidants, and in the manufacture of oil-soluble soaps. Several amines are also used as organic bases in the manufacture of pharmaceutical and cosmetic products. Di- and trimethyl amines are essential in the preparation of quaternary ammonium types of anion exchange resins. In general, amines are utilized in the preparation of dyes, drugs, herbicides, fungicides, soaps, disinfectants, insecticides, and photographic developers, and as choline chloride for use in animal feeds.

Among the aromatic amines, aniline is used to synthesize other important organic compounds used as dyes and dye intermediates, antioxidants, and drugs. Dimethyl aniline is used as a dye intermediate and in the manufacture of the explosive known as Tetrol. Other derivatives of aromatic amines, such as *p*-aminosulfonic acid and *p*-toluidine, are used as dye intermediates.

Several derivatives of the aromatic amines have medicinal properties and are used as drugs. **Acetanilide** was used for many years as an antipyretic and analgesic drug. The toxicity of this compound in the body resulted in a search for similar compounds that were less toxic. **Phenacetin,** or **acetophenetidin,** which is closely related to acetanilide, was found to be less toxic and to possess the beneficial effects of acetanilide. Sulfanilamide and other sulfa drugs have been noted earlier for their therapeutic effects.

Another type of important organic compounds that contains amino groups is **amino acids.** Amino acids may be considered as organic acids containing an amino group. Two simple amino acids are illustrated below.

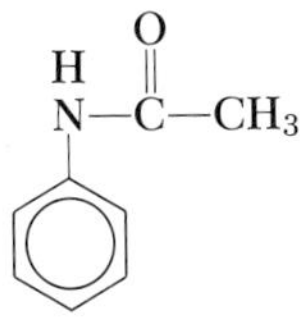

$$C_6H_5-NH-\overset{O}{\overset{\|}{C}}-CH_3 \qquad CH_3CH_2O-C_6H_4-NH-\overset{O}{\overset{\|}{C}}-CH_3 \qquad H-\underset{NH_2}{\underset{|}{\overset{H}{\overset{|}{C}}}}-\overset{O}{\overset{\|}{C}}-OH \qquad CH_3-\underset{NH_2}{\underset{|}{CH}}-\overset{O}{\overset{\|}{C}}-OH$$

Acetanilide 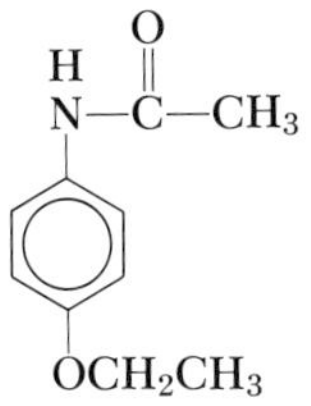   Phenacetin 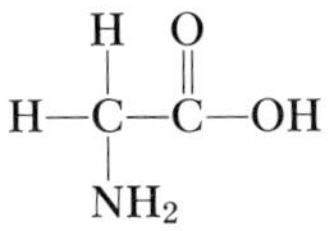   Glycine 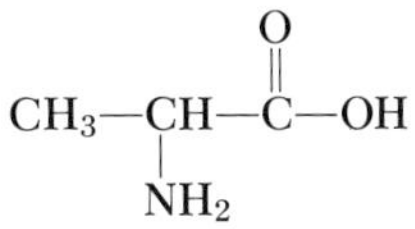   Alanine

Amino acids are the fundamental units in the protein molecule and will be discussed later in detail under the chemistry of proteins.

Questions

1. Draw a structural formula for each of the following compounds:
   (a) diethylamine
   (b) tri-*n*-propyl amine
   (c) aniline
   (d) tetraethylammonium hydroxide
   (e) *m*-nitroaniline
   (f) isohexyl amine
   (g) benzyl amine
   (h) *o*-toluidine
   (i) cyclobutyl amine
   (j) *p*-bromo-N,N-dimethylaniline

2. Give each of the following compounds an appropriate name:

   (a) benzene ring with NH_2 and, meta to it, $\overset{O}{\overset{\|}{C}}—CH_3$

   (b) $CH_3CH_2\overset{H}{N}CH_2CH_2CH_3$

   (c) benzene ring with $\overset{H}{N}—\overset{O}{\overset{\|}{C}}—CH_3$ and, meta to it, Cl

   (d) $(C_6H_5)_3\ddot{N}$

   (e) benzene ring with $N_2^+Cl^-$ and, meta to it, Br

3. Write equations for the reactions of *n*-butyl amine with each of the following reagents:
   (a) acetyl chloride
   (b) HI
   (c) benzene sulfonyl chloride/OH^-
   (d) ethylene oxide

4. Write equations for the reactions of *p*-bromoaniline with the following reagents:
   (a) H_2SO_4
   (b) acetic anhydride
   (c) $NaNO_2/HCl$
   (d) $NaNO_2/HCl$ followed by N,N-dimethylaniline
   (e) benzene sulfonyl chloride/OH^-

5. Write equations for the preparation of the following compounds. Give any necessary catalysts. More than one step may be necessary.
   (a) aniline hydroiodide from nitrobenzene
   (b) acetanilide from nitrobenzene
   (c) *n*-butyl amine from *n*-butyric acid
   (d) *p*-bromobenzonitrile from aniline
   (e) 1,6-hexanediamine from 1,4-butanediol
   (f) *p*-methylaniline from benzene
   (g) *n*-butyl amine from *n*-pentanoic acid
   (h) *p*-nitroaniline from benzene
   (i) *p*-fluorobromobenzene from aniline
   (j) 1,3,5-tribromobenzene from aniline

6. Arrange the following compounds in order of decreasing base strength:

   NH_3, CH_3NH_2, $C_6H_5NH_2$, CH_3COOH, CH_3CH_2OH, H_2O, *p*-$NO_2C_6H_4NH_2$

Suggested Reading

Bonner and Castro: Essentials of Modern Organic Chemistry. New York, Reinhold Publishing Corp., 1965, pp. 397–452.
Fieser and Fieser: Advanced Organic Chemistry. New York, Reinhold Publishing Corp., 1961, pp. 488–529, 706–742.
Morrison and Boyd: Organic Chemistry, 2nd Ed. Allyn & Bacon, Inc., 1966, pp. 718–786.

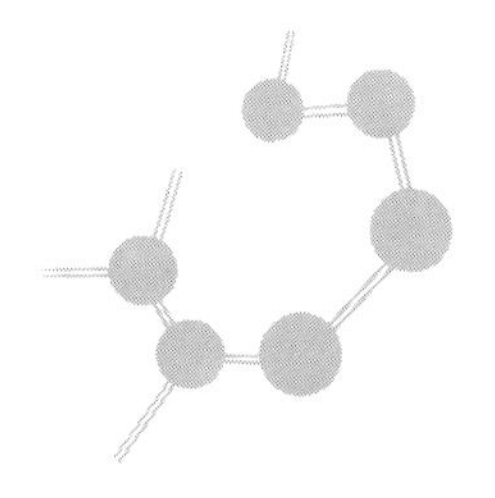

CHAPTER 20

CARBOHYDRATES

In the preceding sections the atomic structure, chemical properties, and reactions of inorganic compounds and organic compounds have been covered. The background gained from this study of the fundamentals of inorganic and organic chemistry, will be a distinct advantage in the consideration of the important section on biochemistry. In the early chapters of the section, the organic chemistry of carbohydrates, lipids, and proteins will be considered. These compounds not only make up the three major types of foods but are also the essential organic constituents of the body.

The carbohydrates will be discussed in the present chapter, since they possess the simplest chemical structure. They are composed of carbon, hydrogen, and oxygen, and the hydrogen and oxygen are usually in the proportion of two to one, the same as in water. As a class of compounds, they include simple sugars, starches, and celluloses. Simple sugars such as glucose, fructose, and sucrose are constituents of many fruits and vegetables. Starches are the storage form of carbohydrates in plants, and cellulose is the main supporting structure of trees and plants. About 75 per cent of the solid matter of plants consists of carbohydrates. In microorganisms and in the cells of higher animals carbohydrates play a fundamental role, serving as an important source of energy for the cell and as a storage form of chemical energy. Specific carbohydrates function as structural units in cell walls and membranes, and in cellular components responsible for function and growth.

OPTICAL ACTIVITY

Stereoisomerism is a common phenomenon in organic chemistry. Stereoisomers are compounds that have the same structure, but different configurations. Examples of **structural** and **geometric isomers** were considered in

Chapters 9 and 12. Optical isomerism is frequently encountered in organic compounds of biochemical interest and is essential in a study of the composition, properties, and reactions of carbohydrates. Many organic molecules, including the carbohydrates, exhibit the phenomenon of **optical activity.** Any optically active compound possesses the property of rotating a plane of polarized light. When ordinary light is passed through a Nicol prism, which consists of two pieces of calcite (natural crystallized $CaCO_3$) cemented together, the resulting beam is traveling in one direction and in one plane, and is called **plane polarized** or just **polarized** light.

Carbohydrates in solution show the property of optical rotation, i.e., a beam of polarized light is rotated when it passes through the solution. The extent to which the beam is rotated, or the angle of rotation, is determined with an instrument called a **polarimeter.**

A substance whose solution rotates the plane of polarized light to the right is said to be **dextrorotatory;** one whose solution rotates the light to the left is called **levorotatory.** The rotation is designated d- for dextrorotatory or l- for levorotatory; for example, d-lactic acid and l-lactic acid. To compare compounds in solution in a polarimeter, the *specific rotation* of a substance has been defined as the rotation in angular degrees produced by a column of solution 1 decimeter long whose concentration is 1 gram per cubic centimeter.

Van't Hoff and La Bel independently advanced the same theory to explain the fundamental reason for the optical activity of a compound. They postulated that the presence of an **asymmetric carbon atom** in a compound was responsible for the optical activity. An asymmetric carbon atom was defined as one which had four different groups attached to it. A simple compound such as lactic acid contains one asymmetric carbon atom, which is marked with an asterisk in the following illustration:

d-Lactic acid	l-Lactic acid
COOH | H—C*—OH | CH_3	COOH | HO—C*—H | CH_3

The d and l forms of lactic acid are mirror images of each other, rotating the plane of polarized light an equal extent in opposite directions. Mirror image isomers form an enantiomeric pair and the d and l forms are called *enantiomers.* A molecular model, constructed as shown in Figure 20–1, in which the asymmetric carbon atom is the central sphere and is joined to four other spheres representing a hydrogen atom, a hydroxyl group, a methyl group, and a carboxyl group, may help to explain optical activity. This model emphasizes the asym-

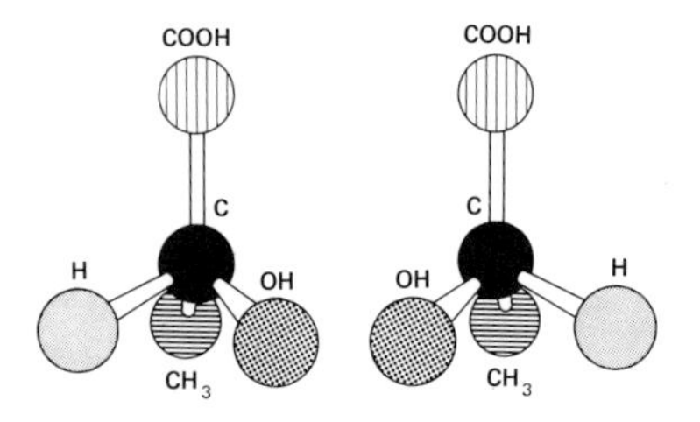

Figure 20–1 The spatial relationship of the groups attached to the asymmetric carbon atom of the d and l forms of lactic acid.

metric carbon atom and the fact that the d and l forms of lactic acid cannot be superimposed upon one another. In general, when a compound that exhibits optical activity is synthesized in the laboratory, a mixture of equal parts of the dextro and levo forms results. Such a mixture is called a **racemic mixture.** Reactions carried out in the body or in the presence of microorganisms often produce optically active isomers, since the reactions are catalyzed by enzymes which themselves are optically active. The enzyme reactions are often specific in producing only one isomer, either for the d or l component of a racemic mixture.

CLASSIFICATION

Carbohydrates are defined as derivatives of polyhydroxyaldehydes or polyhydroxyketones. A sugar that contains an aldehyde group is called an **aldose,** and one that contains a ketone group is termed a **ketose.**

The simplest carbohydrates are known as **monosaccharides,** or simple sugars. Monosaccharides are straight-chain polyhydroxy alcohols and are classified according to the number of carbon atoms in the chain. A sugar with two carbon atoms is called a diose; with three, a triose; with four, a tetrose; with five, a pentose; and with six, a hexose. The ending -ose is characteristic of sugars. When two monosaccharides are linked together by splitting out a molecule of water, the resulting compound is called a **disaccharide.** The combination of three monosaccharides results in a **trisaccharide,** although the general term for carbohydrates composed of two to five monosaccharides is **oligosaccharide.** Polymers composed of several monosaccharides are called **polysaccharides.**

Carbohydrates which will be considered in this chapter may be classified as follows:

I. Monosaccharides
- Trioses—$C_3H_6O_3$
  - Aldose—Glyceraldehyde
  - Ketose—Dihydroxyacetone
- Pentoses—$C_5H_{10}O_5$
  - Aldoses—Arabinose, Xylose, Ribose
- Hexoses—$C_6H_{12}O_6$
  - Aldoses—Glucose, Galactose
  - Ketoses—Fructose
- Ascorbic acid

II. Disaccharides—$C_{12}H_{22}O_{12}$
- Sucrose (glucose + fructose)
- Maltose (glucose + glucose)
- Lactose (glucose + galactose)

III. Polysaccharides
- Hexosans
  - Glucosans—Starch, Glycogen, Dextrin, Cellulose

IV. Mucopolysaccharides
- Hyaluronic acid
- Chondroitin sulfate
- Heparin

Trioses. The trioses are important compounds in muscle metabolism, and are the basic sugars to which all monosaccharides are referred. The definition of a simple sugar may readily be illustrated by the use of the trioses. The polyhydroxy alcohol from which they are derived is glycerol. Oxidation on the end carbon atom produces the aldose sugar known as glyceraldehyde; oxidation on the center carbon yields the keto triose, dihydroxyacetone. It can be seen

H
|
H—C—OH
|
H—C—OH
|
H—C—OH
|
H

Glycerol
(polyhydric alcohol)

→

H
|
C=O
|
H—C—OH
|
H—C—OH
|
H

Glyceraldehyde (aldose)

→

H
|
H—C—OH
|
C=O
|
H—C—OH
|
H

Dihydroxyacetone (ketose)

from the formula of glyceraldehyde that one asymmetric carbon atom is present. Therefore this sugar can exist in two forms, one of which rotates plane polarized light to the right, the other to the left. Originally the two forms were designated d and l for the dextro and levo rotation of polarized light. Modern terminology employs the D and L, written in small capital letters, for structural relationships, and a (+) and (−) for direction of rotation.

H | C=O | H—C*—OH | CH_2OH	H | C=O | HO—C*—H | CH_2OH	H | C=O ⁞ H◄C►OH ⁞ CH_2OH
D (+) Glyceraldehyde	L (−) Glyceraldehyde	Perspective formula

Fischer projection formula

The isomeric forms of sugars are often represented as the **Fischer projection formula.** The asymmetric carbon atom * of glyceraldehyde would represent the central sphere as in the model in Figure 20–1, with the H and OH groups projecting in front of the plane of the paper and the aldehyde and primary alcohol group projecting behind. As ordinarily written, the horizontal bonds are understood to be in front of the plane and the vertical bonds behind the plane of the paper. The **perspective formula** emphasizes the position of the groups using dotted lines to connect those behind the plane and heavy wedges to represent groups in front of the plane. The Fischer projection formulas are always written with the aldehyde or ketone groups at the top of the structure; therefore, all monosaccharides with the hydroxyl group on the right of the carbon atom next to the bottom primary alcohol group are related to D-glyceraldehyde and are called D-sugars.

In like manner, if the hydroxyl group on the carbon atom next to the end primary alcohol group is on the left, it is related to L-glyceraldehyde and is an L-sugar. The direction of rotation of polarized light cannot be ascertained from the formula, but must be determined experimentally, and is designated (+) for dextrorotatory and (−) for levorotatory sugars.

Only two optical isomers of aldotriose exist, since it contains only one asymmetric carbon atom. A sugar with two asymmetric carbon atoms exhibits a total of four isomers, and as each new asymmetric carbon atom is added to the structure the number of isomers is doubled. The total number of optical isomers of a sugar is 2^n, where n is the number of different asymmetric carbon atoms. For example, pentoses contain three asymmetric carbon atoms and can form eight isomers; hexoses contain four asymmetric carbon atoms and can exist as 16 different isomers.

Pentoses. The pentoses are sugars whose molecules contain five carbon atoms and three asymmetric carbon atoms. They occur in nature combined in polysaccharides from which the monosaccharides may be obtained by hydrolysis with acids. Arabinose is obtained from gum arabic and the gum of the cherry tree, and xylose is obtained by hydrolysis of wood, corn cobs, or straw. Ribose and deoxyribose are constituents of the ribose nucleic acids, RNA, and deoxyribose nucleic acids, DNA, that are essential components of the cytoplasm and nuclei of cells.

Hexoses. The hexoses are by far the most important monosaccharides from a nutritional and physiological standpoint. The bulk of the carbohydrates used as foods consist of hexoses free or combined in disaccharides and polysaccharides. Glucose, fructose, and galactose are the hexoses commonly occurring in foods, whereas mannose is a constituent of a vegetable polysaccharide. Glucose, also called **dextrose,** is the normal sugar of the blood and tissue fluids and is utilized by the cells as a source of energy. Fructose often occurs free in fruits and is the sweetest sugar of all the monosaccharides. Galactose is a constituent of **milk sugar** and is found in brain and nervous tissue. All of these monosaccharides are D-sugars. The hydroxyl group on the carbon next to the primary alcohol is on the right. Fructose is a ketose sugar and the others are aldose sugars.

Although glucose and galactose are represented as simple aldehyde structures, this form does not explain all the reactions they undergo. Both of these aldoses, for example, do not give a positive test for aldehyde. Also, when a glucose solution is allowed to stand, a change in its specific rotation may be observed.

D-Glucose (aldose)	D-Fructose (ketose)	D-Galactose (aldose)
H		H
C=O	CH_2OH	C=O
H—C—OH	C=O	H—C—OH
HO—C—H	HO—C—H	HO—C—H
H—C—OH	H—C—OH	HO—C—H
H—C—OH	H—C—OH	H—C—OH
CH_2OH	CH_2OH	CH_2OH

Freshly prepared aqueous solutions of crystalline glucose often yield a specific rotation as high as +113 degrees, whereas glucose crystallized from pyridine exhibits a specific rotation as low as +19 degrees. On standing, both of these solutions change their rotation until an equilibrium value of +52.5 degrees is reached. This change in rotation is called **mutarotation.** Since the specific rotation of an organic compound is related to its structure, as is its melting point, boiling point, and other properties, it may be suspected that glucose exists in two different isomeric forms. This has been shown to be true and is explained by the existence of an **oxygen bridge structure** involving carbon atoms 1 and 5. It can be seen from these formulas that the free aldehyde group no longer exists and a new asymmetric carbon is produced. To indicate the position of the hydroxyl group on the first carbon and to distinguish between the two new isomers, the α-isomer has the OH on the right and the β-isomer on the left as shown. When the α- or β-isomer is dissolved in water, an equilibrium mixture of 37 per cent α and 63 per cent β, with a specific rotation of 52.5 degrees, is formed. This oxygen bridge structure and the phenomenon of mutarotation are common to all aldohexoses, and since the structure contains an additional asymmetric carbon atom, the number of possible isomers is doubled.

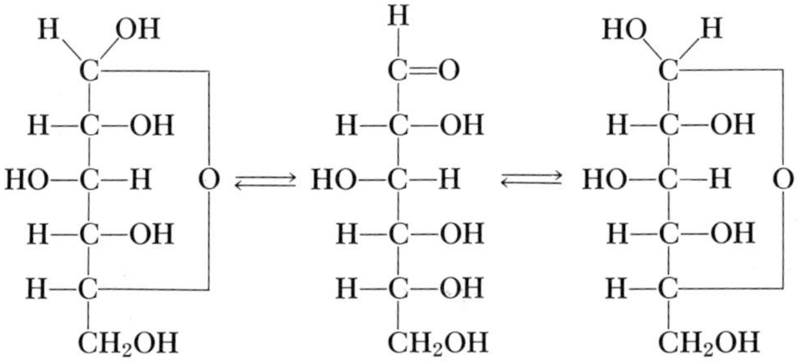

```
  H   OH                   H                HO   H
   \ /                     |                  \ /
    C-------+              C=O                 C-------+
    |       |              |                   |       |
 H--C--OH   |           H--C--OH            H--C--OH   |
    |       |              |                   |       |
HO--C--H    O  <=>    HO--C--H     <=>     HO--C--H    O
    |       |              |                   |       |
 H--C--OH   |           H--C--OH            H--C--OH   |
    |       |              |                   |       |
 H--C-------+           H--C--OH            H--C-------+
    |                      |                   |
   CH2OH                  CH2OH               CH2OH
```

α-D-Glucose +113° | D-Glucose, straight chain form | β-D-Glucose +19°

The oxygen bridge structure in fructose involves carbon atoms 2 and 5, whereas in pentose sugars oxygen joins atoms 1 and 4. These oxygen ring structures suggested a relationship between the pyran and furan rings and the monosaccharides.

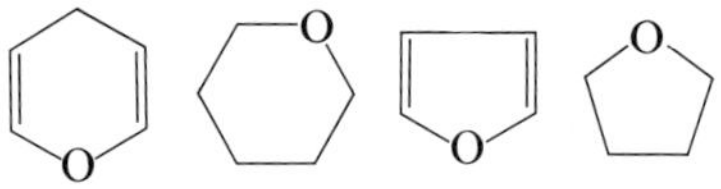

Pyran Furan

Haworth first suggested that the sugars be represented as derivatives of these heterocyclic rings. The relation between the oxygen bridge structure of glucose and Haworth's **glucopyranose** is shown on the top of page 293. The heavy lines represent the base of the ring in which the five carbons and one oxygen are in the same plane perpendicular to the plane of the paper. The thick bonds of the ring extend toward the reader, whereas the thin bonds of the ring are behind the plane of the paper. Groups which are ordinarily written on the right of the oxide ring structure appear below the plane of the pyranose ring, and those to the left of the carbon chain appear above the plane.

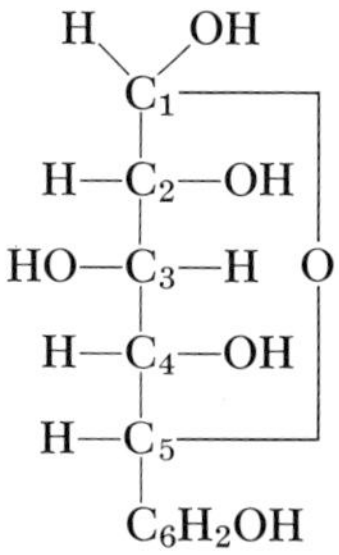

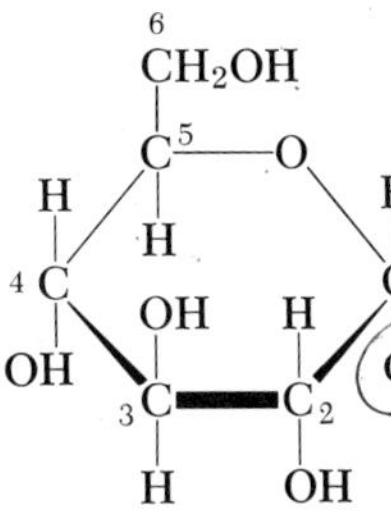

α-D-Glucopyranose

The D-glucopyranose structure has the OH on carbon-1 below the plane of the ring and would therefore be an α-isomer.

Monosaccharides such as the pentoses and fructose, whose oxide rings enclose four carbon atoms, are written as derivatives of furan as shown:

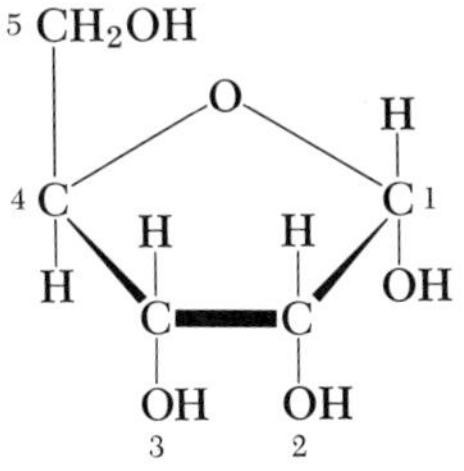

α-D-Ribose 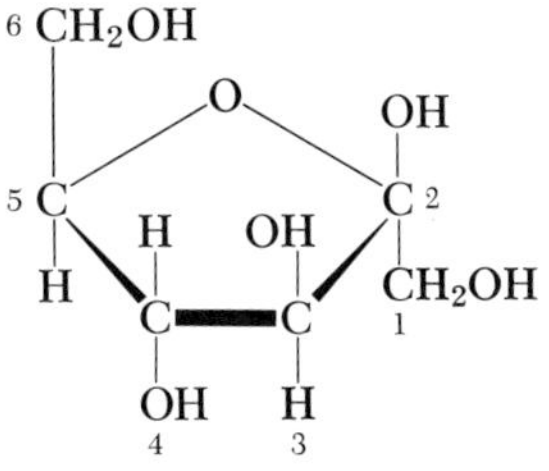 β-D-Fructofuranose

The ring structures of the other pentose and hexose monosaccharides that have been discussed are shown as follows:

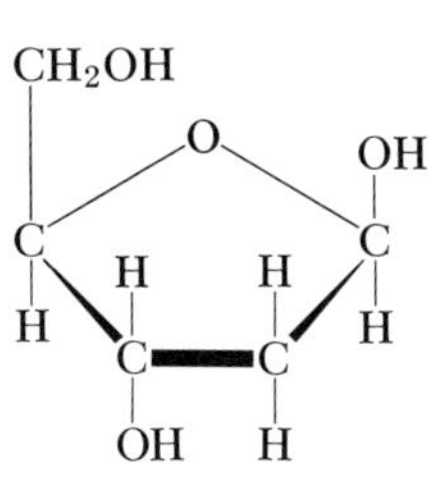

β-D-Deoxyribose 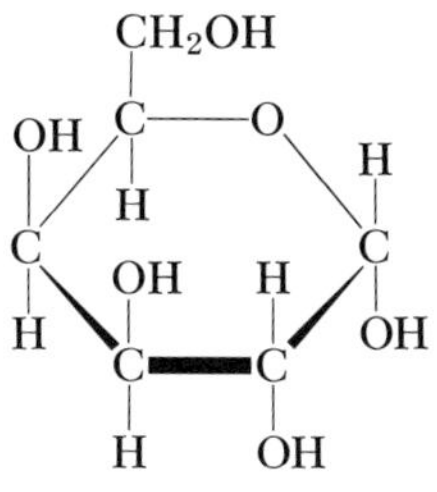 α-D-Galactose 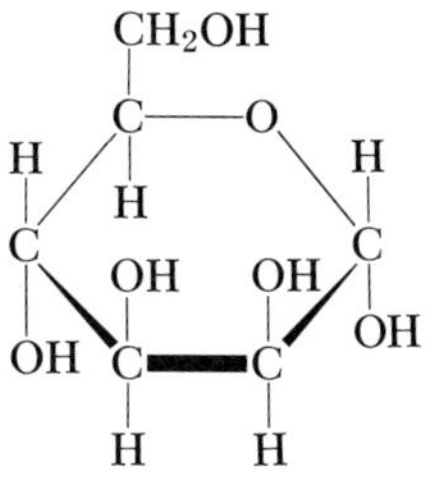 α-D-Mannose

It can readily be observed in the preceding structures that the OH group on carbon-1 (carbon-2 in ketofuranoses) indicates the α or β form of the sugar. The α-isomers have the OH extending below the plane of the ring, whereas in the β-isomers the OH extends above the plane.

REACTIONS OF CARBOHYDRATES

Dehydration. When aldohexoses or aldopentoses are heated with strong acids, they are dehydrated to form furfural derivatives.

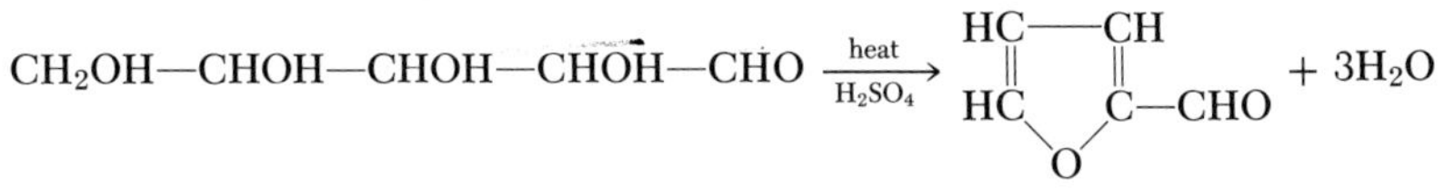

$$CH_2OH{-}CHOH{-}CHOH{-}CHOH{-}CHO \xrightarrow[H_2SO_4]{heat} \text{Furfural} + 3H_2O$$

Pentose Furfural

An aldohexose in this reaction would form 5-hydroxymethyl furfural. The furfural derivatives formed in this reaction combine with α-naphthol to give a purple color. This color is the basis of the Molisch test, a general test for carbohydrate. Furfural reacts with orcinol to yield a green color which constitutes Bial's test for pentoses.

Glycoside Formation. When monosaccharides are treated with an alcohol in a strong acid solution, they form **glycosides.** The alcohol reacts with the hydroxyl group on the first carbon as follows:

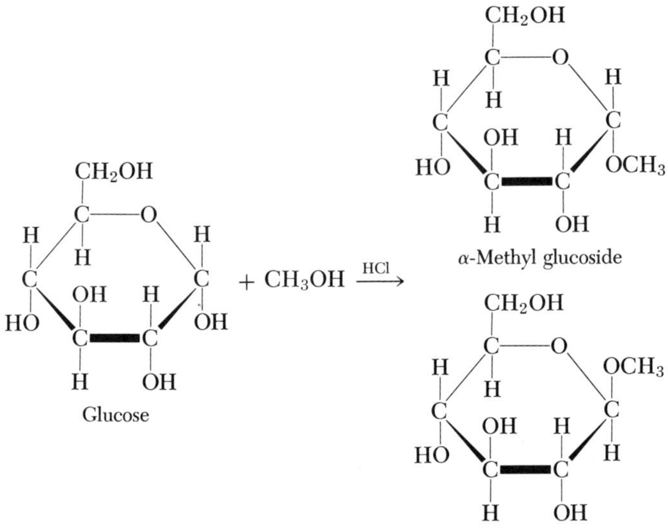

Glucose $+ CH_3OH \xrightarrow{HCl}$ α-Methyl glucoside / β-Methyl glucose

Glucose

α-Methyl glucoside

β-Methyl glucose

The position of the methyl group below or above the plane of the ring indicates α- or β-methyl glucoside in that order. This is a very important reaction since many of the disaccharides and polysaccharides are glycosides in which one of the alcoholic hydroxyl groups in the second monosaccharide reacts with the hydroxyl on carbon 1 of the first monosaccharide.

Oxidation. In Chapter 17 one of the important reactions of carbonyl compounds was the oxidation to carboxylic acids. Sugars that contain *free or potential aldehyde or ketone groups* in the hemiacetal type structure are oxidized in alkaline solution by Cu^{+2}, Ag^{+}, and Bi^{+3}. The reaction of aldehydes in Benedict's, Fehling's, or the silver mirror test has already been described (p. 259). Sugars that undergo oxidation in these reactions are called **reducing sugars.**

All of the reducing sugars that are capable of reducing Cu^{+2} to Cu^{+} are oxidized in the reaction described in the preceding section. If the aldehyde group of glucose is oxidized to a carboxyl group by a weak oxidizing agent, such as NaOBr, gluconic acid is formed. Oxidation of the primary alcohol group, either by chemical agents or enzymes, produces glucuronic acid (see p. 295). Glucuronic acid combines with drugs and toxic compounds in the body, and the glucuronides are excreted in the urine.

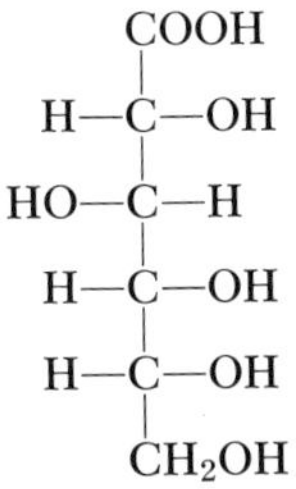

```
   COOH                 H
    |                   |
 H—C—OH                C=O
    |                   |
HO—C—H               H—C—OH
    |                   |
 H—C—OH              HO—C—H
    |                   |
 H—C—OH               H—C—OH
    |                   |
   CH2OH              H—C—OH
                        |
                       COOH
```

Gluconic acid 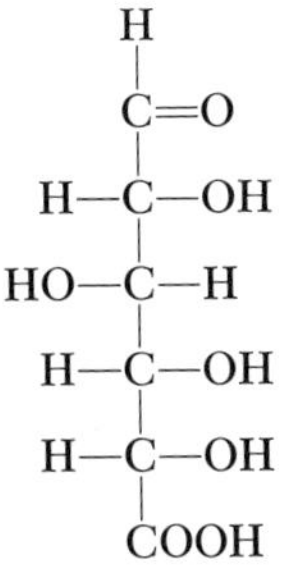   Glucuronic acid

Fermentation. The enzyme mixture called **zymase** present in common bread yeast will act on some of the hexose sugars to produce alcohol and carbon dioxide. The fermentation of glucose may be represented as follows:

$$\underset{\text{Glucose}}{C_6H_{12}O_6} \xrightarrow{\text{zymase}} \underset{\text{Ethyl alcohol}}{2C_2H_5OH} + 2CO_2$$

The common hexoses (with the exception of galactose) ferment readily, but pentoses are not fermented by yeast. Disaccharides must first be converted into their monosaccharide constituents by other enzymes present in yeast before they are susceptible to fermentation by zymase.

There are many other types of fermentation of carbohydrates besides the common alcoholic fermentation. When milk sours, the lactose of milk is converted into lactic acid by a fermentation process. Citric acid, acetic acid, butyric acid, and oxalic acid may all be produced by special fermentation processes.

Ester Formation. Esters formed between a hydroxyl group of phosphoric acid and the hydrogen atom of a hydroxyl group of a monosaccharide are common, and several of these phosphorylated sugars are encountered in carbohydrate metabolism (Chapter 26).

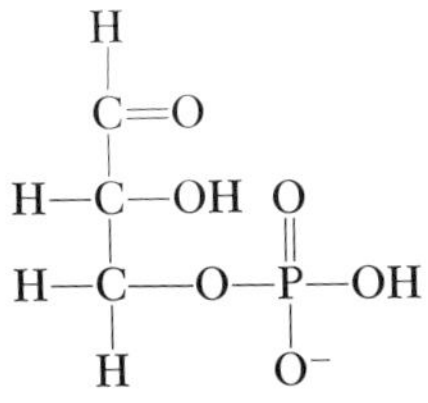

```
      H
      |
     C=O
      |
   H—C—OH  O
      |    ||
   H—C—O—P—OH
      |    |
      H    O⁻
```

D-Glyceraldehyde-3-phosphate

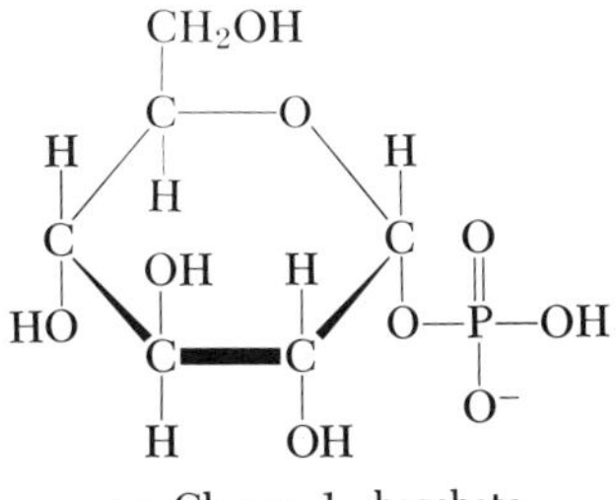

α-D-Glucose-1-phosphate

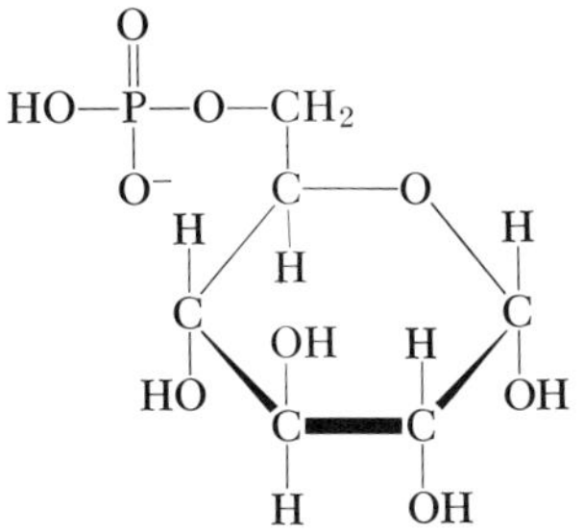

α-D-Glucose-6-phosphate

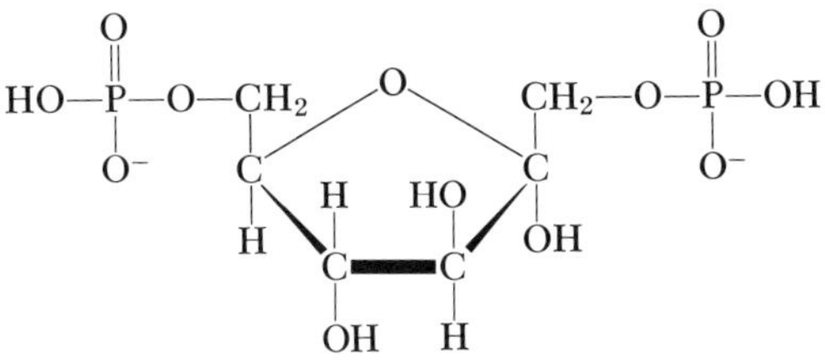

α-D-Fructose-1, 6-diphosphate

Ascorbic Acid (Vitamin C). Ascorbic acid is an oxidation product of a hexose sugar.

```
        O                       O
       //                      //
      C------                 C------
      |      |                |      |
  HO--C      |            O===C      |
      ||     O      <==>      |      O
  HO--C      |            O===C      |
      |      |                |      |
   H--C------              H--C------
      |                       |
  HO--C--H                HO--C--H
      |                       |
     CH2OH                   CH2OH
```

L-Ascorbic acid   Dehydroascorbic acid

The hydrogen atoms on the second and third carbon atoms (from the top) are readily removed by oxidation, forming a ketone called **dehydroascorbic acid.** This reaction is reversible, and because of this oxidation-reduction relationship, ascorbic acid probably functions in oxidations and reductions in the cellular tissues.

A deficiency of ascorbic acid in the diet results in the disease known as **scurvy.** Early symptoms of scurvy are loss of weight, anemia, and fatigue. As the disease progresses, the gums become swollen and bleed readily, and the teeth loosen. The bones become brittle and hemorrhages develop under the skin and in the mucous membrane. Extreme scurvy is not commonly seen today, although many cases of subacute, or latent, scurvy are recognized. Symptoms such as sore receding gums, sores in the mouth, tendency to fatigue, lack of resistance to infections, defective teeth, and pains in the joints are indicative of **subacute scurvy.**

The adrenal cortex contains appreciable amounts of ascorbic acid, which may function in the synthesis of steroid hormones in the adrenal gland. Ascorbic acid is also thought to be involved in hydroxylation reactions and in electron transport in the microsomal region of the cell. The biochemistry of ascorbic acid deficiency in the body is not as yet well understood.

DISACCHARIDES

A disaccharide is composed of two monosaccharides whose combination involves the splitting out of a molecule of water. The hemiacetal linkage is always made from the aldehyde group of one of the sugars to a hydroxyl or ketone group of the second. In order to reduce Benedict's solution, disaccharides must have a potential aldehyde or ketone group that is not involved in the linkage between the two sugars.

Sucrose. Sucrose is commonly called **cane sugar** and is the ordinary sugar that is used for sweetening purposes in the home. It is found in many plants such as sugar beets, sorghum cane, the sap of the sugar maple, and sugar cane. Commercially it is prepared from sugar cane and sugar beets.

α-D-Glucopyranose-β-D-fructofuranoside

Sucrose

Sucrose is composed of a molecule of glucose joined to a molecule of fructose in such a way that the linkage involves the reducing groups of both sugars (carbon-1 of glucose and carbon-2 of fructose). It is the only common mono- or disaccharide that will not reduce Benedict's solution. When sucrose is hydrolyzed, either by the enzyme sucrase or by an acid, a molecule of glucose and a molecule of fructose are formed. The fermentation of sucrose by yeast is possible, since the yeast contains the two enzymes sucrase and zymase. The sucrase first hydrolyzes the sugar, and then the zymase ferments the monosaccharides to form alcohol and carbon dioxide.

Lactose. The disaccharide present in milk is lactose, or **milk sugar.** It is synthesized in the mammary glands of animals from the glucose in the blood. Commercially, it is obtained from milk whey and is used in infant foods and special diets. Lactose, when hydrolyzed by the enzyme lactase or by an acid, forms a molecule of glucose and a molecule of galactose. Lactose will reduce Benedict's solution, but is not fermented by yeast. From its reducing properties, it is obvious that the linkage between its constituent monosaccharides does not involve both potential aldehyde groups (carbon-1 of galactose is connected to carbon-4 of glucose).

β-D-Galactopyranosyl-α-D-glucopyranose

Lactose

Maltose. Maltose is present in germinating grains. Since it is obtained as a product of the hydrolysis of starch by enzymes present in malt, it is often called **malt sugar.** It is also formed in the animal body by the action of enzymes on starch in the process of digestion. Commercially, it is made by the partial hydrolysis of starch by acid in the manufacture of corn syrup. Maltose reduces Benedict's solution and is fermented by yeast. On hydrolysis it forms two molecules of glucose.

α-D-Glucopyranosyl-α-D-glucopyranose
Maltose

POLYSACCHARIDES

The polysaccharides are complex carbohydrates that are made up of many monosaccharide molecules and therefore possess a high molecular weight. They differ from the simple sugars in many ways. They fail to reduce Benedict's solution, do not have a sweet taste, and are usually insoluble in water. Since they are such large molecules, they form colloidal dispersions instead of simple solutions.

There are polysaccharides formed from pentoses or from hexoses, and there are also mixed polysaccharides. Of these, the most important are composed of the hexose glucose and are called **hexosans,** or more specifically, **glucosans.** As in a disaccharide, whenever two molecules of a hexose combine, a molecule of water is split out. For this reason, a hexose polysaccharide may be represented by the formula $(C_6H_{10}O_5)_x$. The x represents the number of hexose molecules in the individual polysaccharide. Because of the complexity of the molecules, the number of glucose units in any one polysaccharide is still an estimate.

STARCH. From a nutritional standpoint, starch is the most important polysaccharide. It is made up of glucose units and is the storage form of carbohydrates in plants. It consists of two types of polysaccharides: **amylose,**

Amylose

composed of a straight chain of glucose molecules connected by α-1,4 linkages, and **amylopectin,** which is a branched chain or polymer of glucose with both α-1,4 and α-1,6 linkages. The repeating structure of glucose molecules in amylose is usually represented as glucopyranose units as shown in the accompanying diagram. Amylose has a molecular weight of about 50,000, compared to about 300,000 for amylopectin. The branching of the glucose chain in amylopectin occurs about every 24 to 30 glucose molecules.

Starch will not reduce Benedict's solution and is not fermented by yeast. When starch is hydrolyzed by enzymes or by an acid, it is split into a series of intermediate compounds possessing smaller numbers of glucose units. The product of complete hydrolysis is the free glucose molecule. A characteristic reaction of starch is the formation of a blue compound with iodine. This test is often used to follow the hydrolysis of starch, since the color changes from blue through red to colorless with decreasing molecular weight:

starch →	amylodextrin →	erythrodextrin →	achroodextrin →	maltose → glucose
blue	blue	red	colorless	colorless with iodine

Dextrins. Dextrins are found in germinating grains, but are usually obtained by the partial hydrolysis of starch. Those formed from amylose have straight chains, whereas those derived from amylopectin exhibit branched chains of glucose molecules. The larger branched chain molecules give a red color with iodine and are the erythrodextrins. They are soluble in water and have a slightly sweet taste. Large quantities of dextrins are used in the manufacture of adhesives because they form sticky solutions when wet. An example of their use is the mucilage on the back of postage stamps.

Glycogen. Glycogen is the storage form of carbohydrate in the animal body and is often called animal starch. It is found in liver and muscle tissue. It is soluble in water, does not reduce Benedict's solution, and gives a red-purple color with iodine. The glycogen molecule is similar to the amylopectin molecule in that it has branched chains of glucose with α-1,4 and α-1,6 linkages that occur at more frequent intervals. The branched chain structure common to both glycogen and amylopectin is shown below:

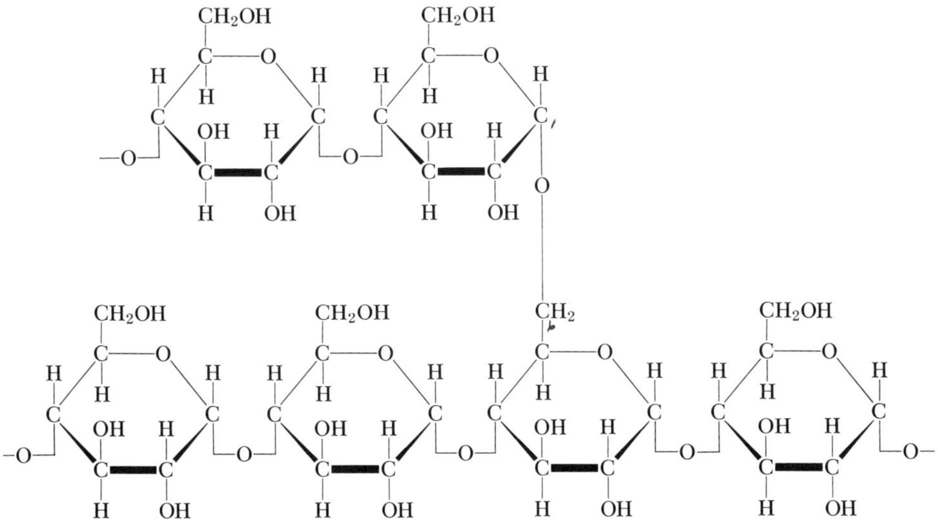

Branched chain of glucose molecules in amylopectin and glycogen

The molecular weight of glycogen is greater than that of amylopectin, varying from 1,000,000 to 5,000,000. When glycogen is hydrolyzed in the animal body, it forms glucose to help maintain the normal sugar content of the blood.

CELLULOSE. Cellulose is a polysaccharide that occurs in the framework, or supporting structure, of plants. It is composed of a straight chain polymer of glucose molecules similar in structure to that pictured for amylose. The major difference concerns the linkage of the glucose molecules. In amylose the linkage is 1,4 α, as in maltose, whereas in cellulose the linkage is 1,4 β, which occurs

Cellulose

in **cellobiose.** The maltose type of structure is hydrolyzed by enzymes and serves as a source of dietary carbohydrate. In contrast, the cellobiose structure is insoluble in water, will not reduce Benedict's solution and is not attacked by enzymes present in the human digestive tract. The molecular weight of cellulose has been estimated to be between 150,000 and 1,000,000. The partial structure of cellulose is represented in the accompanying diagram.

The chemical treatment of cellulose has resulted in several important commercial products. These include mercerized cotton, cellulose nitrates such as gun-cotton and pyroxylin, cellulose acetates used in films, and plastics and **cellophane. Rayon** is produced by treating cellulose with sodium hydroxide and carbon disulfide. The solution resulting from this treatment is forced through small holes into dilute sulfuric acid to make the rayon fibers.

HEPARIN. Heparin is a polysaccharide that possesses anticoagulant properties. It prevents the clotting of blood by inhibiting the conversion of prothrombin to thrombin. Thrombin acts as a catalyst in converting plasma fibrinogen into the fibrin clot. The structure of heparin is still uncertain but it contains a repeating unit of α-1,3 linked glucuronic acid and glucosamine, with sulfate groups on some of the hydroxyl and amino groups.

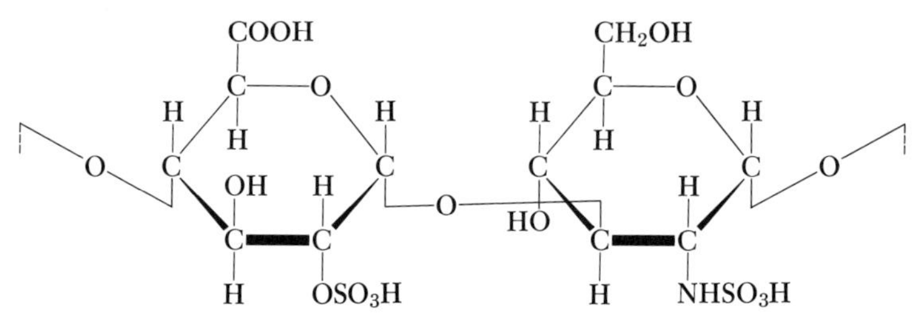

D-Glucuronic acid-2-sulfate D-Glucosamine sulfate

α-1,3 linked repeating unit of heparin

Questions

1. Explain what is meant by the phenomenon of optical activity.

2. Write the formulas for D-fructose and L-fructose. Star the asymmetric carbon atoms in each molecule.

3. Write the formula for lactic acid. Would this compound be classified as a carbohydrate? Explain.

4. What is (1) an aldose, (2) a hexose, (3) a pentose, (4) a ketose, and (5) a disaccharide?

5. Explain fully what is meant by (1) D (+) glucose, and (2) D (−) fructose.

6. How does the phenomenon of mutarotation complicate the representation of the formulas for carbohydrates? Explain with an example.

7. How are the α- and β- isomers of the pyranose and furanose ring forms of the sugars indicated in the structures?

8. Write an equation to illustrate the formation of a glucoside between β-D-galactopyranose and the OH on carbon-4 of α-D-glucopyranose.

9. When a reducing sugar reacts with Benedict's solution, what other products are formed besides Cu_2O? Why is the formation of Cu_2O important in the test?

10. Write the formula for sucrose and use it to explain why sucrose will not reduce Benedict's solution.

11. Write a partial polysaccharide structure that illustrates both α-1,4 and α-1,6 linkages between the monosaccharides.

12. Compare starch, dextrins, glycogen, and cellulose as to size of molecule, chemical composition, color with iodine, and importance.

13. How would you account for the great difference in properties between starch and cellulose?

14. What happens when starch is hydrolyzed?

Suggested Reading

Frohwein: A Simplified Proof of the Constitution and the Configuration of D-Glucose. Journal of Chemical Education, Vol. 46, p. 55, 1969.
Hudson: Emil Fischer's Discovery of the Configuration of Glucose. Journal of Chemical Education, Vol. 18, p. 353, 1941.
Hussey and Scherer: Rayon—Today and Tomorrow. Journal of Chemical Education, Vol. 7, p. 2543, 1930.
McCord and Getchell: Cotton. Chemical & Engineering News, November 14, 1960, p. 106.
Nye: Wooden Models of Asymmetric Structures. Journal of Chemical Education, Vol. 46, p. 175, 1969.
Thomas: The Production of Chemical Cellulose from Wood. Journal of Chemical Education, Vol. 35, p. 493, 1958.
Vennoz: Construction and Uses of an Inexpensive Polarimeter. Journal of Chemical Education, Vol. 46, p. 459, 1969.

CHAPTER 21

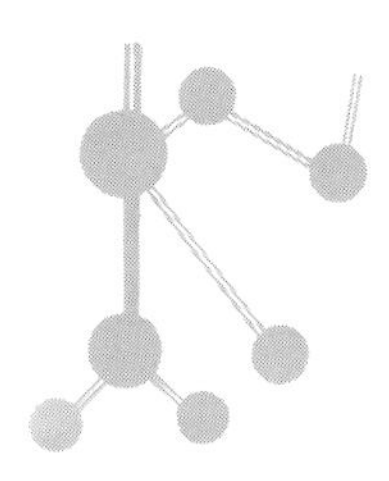

LIPIDS

The common fats and oils compose a class of organic compounds of biological importance. The general term for this type of material is lipid. **Lipids** are characterized by the presence of fatty acids or their derivatives and by their solubility in fat solvents such as acetone, alcohol, ether, and chloroform. Lipids are essential constituents of practically all plant and animal cells. In the human body they are concentrated in cell membranes and in brain and nervous tissue. Repeated extraction of specimens of body tissue with hot fat solvents will invariably yield a mixture of lipid components.

Chemically, lipids are composed of five main elements: carbon, hydrogen, oxygen, and occasionally nitrogen and phosphorus. They may be classified according to the scheme that is followed in this chapter:

Fats—esters of fatty acids with glycerol.
Phosphatides—compounds that contain phosphorus, fatty acids, glycerol, and a nitrogenous compound.
Sphingolipids—compounds that contain a fatty acid, phosphoric acid, choline, and an amino alcohol, sphingosine.
Glycolipids—composed of a carbohydrate, a fatty acid, and an amino alcohol.
Steroids—high molecular weight cyclic alcohols.
Fat soluble vitamins—vitamins A, D, E, and K.
Waxes—esters of fatty acids with alcohols other than glycerol.

FATTY ACIDS

Since all fats are esters of fatty acids and glycerol, it may be well to consider the composition and properties of these substances before discussing lipids in general. Fatty acids, although not lipids themselves, are sometimes classified as derived lipids, since they are constituents of all the above types of lipids except sterols. The fatty acids that occur in nature almost always have an even

TABLE 21–1 SOME IMPORTANT FATTY ACIDS OCCURRING IN NATURAL FATS

Name	Formula	Carbon Atoms	Position of Double Bonds	Occurrence
Saturated				
Butyric	C_3H_7COOH	4		Butter fat
Caproic	$C_5H_{11}COOH$	6		Butter fat
Caprylic	$C_7H_{15}COOH$	8		Coconut oil
Capric	$C_9H_{19}COOH$	10		Palm kernel oil
Lauric	$C_{11}H_{23}COOH$	12		Coconut oil
Myristic	$C_{13}H_{27}COOH$	14		Nutmeg oil
Palmitic	$C_{15}H_{31}COOH$	16		Animal and vegetable fats
Stearic	$C_{17}H_{35}COOH$	18		Animal and vegetable fats
Arachidic	$C_{19}H_{39}COOH$	20		Peanut oil
Unsaturated				
Palmitoleic (1=)°	$C_{15}H_{29}COOH$	16	Δ9†	Butter fat
Oleic (1 =)	$C_{17}H_{33}COOH$	18	Δ9	Olive oil
Linoleic (2=)	$C_{17}H_{31}COOH$	18	Δ9, 12	Linseed oil
Linolenic (3 =)	$C_{17}H_{29}COOH$	18	Δ9, 12, 15	Linseed oil
Arachidonic (4 =)	$C_{19}H_{31}COOH$	20	Δ5, 8, 11, 14	Lecithin

° Number of double bonds.
† Δ9 indicates a double bond between carbon 9 and 10, Δ12 between carbon 12 and 13, and so forth.

number of carbon atoms in their molecules. They are usually straight chain organic acids that may be saturated or unsaturated. Some of the important fatty acids that occur in natural fats are listed in Table 21–1.

In the series of saturated fatty acids, those up to and including capric acid are liquid at room temperature. The most important saturated fatty acids are **palmitic** and **stearic acids.** They are components of the majority of the common animal and vegetable fats.

Unsaturated fatty acids are characteristic constituents of oils. **Oleic acid,** which contains one double bond, is the most common unsaturated fatty acid.

FATS

From a chemical standpoint fats are esters of fatty acids and glycerol. Since glycerol contains three hydroxyl groups, it will form a triple ester with fatty acids. The combination of 3 molecules of fatty acid with 1 molecule of glycerol may be illustrated as shown in the accompanying diagram.

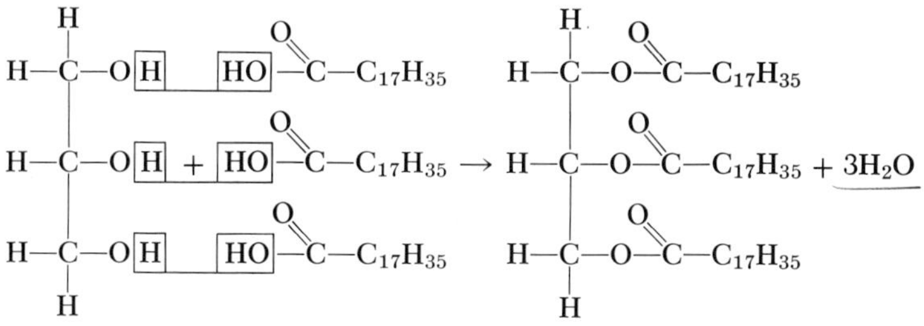

$$
\begin{array}{lcccl}
\text{H} & & & & \text{H} \\
\text{H—C—O}\boxed{\text{H}} & & \boxed{\text{HO}}\text{—C(=O)—}C_{17}H_{35} & & \text{H—C—O—C(=O)—}C_{17}H_{35} \\
\text{H—C—O}\boxed{\text{H}} & + & \boxed{\text{HO}}\text{—C(=O)—}C_{17}H_{35} & \rightarrow & \text{H—C—O—C(=O)—}C_{17}H_{35} + 3H_2O \\
\text{H—C—O}\boxed{\text{H}} & & \boxed{\text{HO}}\text{—C(=O)—}C_{17}H_{35} & & \text{H—C—O—C(=O)—}C_{17}H_{35} \\
\text{H} & & & & \text{H} \\
\text{Glycerol} & & \text{3 molecules of stearic acid} & & \text{Tristearin, a fat}
\end{array}
$$

Tristearin is called a **simple glyceride** because all the fatty acids in the fat molecule are the same. Other examples of simple glycerides would be tripalmitin and triolein. In most naturally occurring fats, different fatty acids are found in the same molecule. These are called **mixed glycerides** and may contain both saturated and unsaturated fatty acids.

Both fats and oils are esters of fatty acids and glycerol. The main difference between fats and oils is the relatively high content of unsaturated fatty acids in the oils. A fat that contains short chain saturated fatty acids may also exist as a liquid at room temperature.

Most of the common animal fats are glycerides that contain saturated and unsaturated fatty acids. Since the saturated fatty acids predominate, these fats are solid at room temperature. Beef fat, mutton fat, lard, and butter are important examples of animal fats. Butter fat is readily distinguished from other animal fats because of its relatively high content of short chain fatty acids.

Glycerides that are found in vegetables usually exist as oils rather than fats. Vegetable oils such as olive oil, corn oil, cottonseed oil, and linseed oil are characterized by their high content of oleic, linoleic, and linolenic acids. Coconut oil, like butter fat, contains a relatively large percentage of short chain fatty acids.

Reactions of Fats. ***Glycerol Portion.*** When glycerol or a liquid containing glycerol is heated with a dehydrating agent, **acrolein** is formed. Acrolein has a very pungent odor and is sometimes formed by the decomposition of glycerol in the fat of frying meats.

$$\underset{\text{Glycerol}}{\begin{array}{l} CH_2OH \\ | \\ CHOH \\ | \\ CH_2OH \end{array}} + \xrightarrow{KHSO_4} \underset{\text{Acrolein}}{\begin{array}{l} H-C{=}O \\ \quad | \\ \quad CH \\ \quad \| \\ \quad CH_2 \end{array}} + 2H_2O$$

The formation of acrolein is often used as a test for fats, since all fats yield glycerol when they are heated.

Hydrolysis. Fats may be hydrolyzed to form free fatty acids and glycerol by the action of acid, superheated steam, or the enzyme lipase. In hydrolysis of a fat, the 3 water molecules (that were split out when the 3 fatty acid molecules combined with 1 glycerol molecule in an ester linkage to make the fat molecule) are replaced with the resultant splitting of the fat into glycerol and fatty acids. Commercially, fats are a cheap source of glycerol for use in the manufacture of high explosives and pharmaceuticals.

$$\underset{\text{Tripalmitin}}{\begin{array}{l} CH_2-O-\overset{O}{\overset{\|}{C}}-C_{15}H_{31} \\ | \\ CH-O-\overset{O}{\overset{\|}{C}}-C_{15}H_{31} \\ | \\ CH_2-O-\overset{O}{\overset{\|}{C}}-C_{15}H_{31} \end{array}} + \Delta \longrightarrow \underset{\text{Glycerol}}{\begin{array}{l} CH_2OH \\ | \\ CHOH \\ | \\ CH_2OH \end{array}} + \underset{\text{Palmitic acid}}{3C_{15}H_{31}COOH}$$

Hydrolysis by an alkali is called **saponification,** and produces glycerol and salts of the fatty acids that are called soaps. In the laboratory, fats are usually saponified by an alcoholic solution of an alkali. The fats are more soluble in hot alcohol and the reaction is therefore more rapid.

Rancidity. Many fats develop an unpleasant odor and taste when they are allowed to stand in contact with air at room temperature. The two common types of rancidity are **hydrolytic** and **oxidative.** Hydrolytic changes in fats are the result of the action of enzymes or microorganisms producing free fatty acids. If these acids are of the short chain variety, as is butyric acid, the fats develop a rancid odor and taste. This type of rancidity is common in butter.

The most common type of rancidity is the oxidative type. The unsaturated fatty acids in fats undergo oxidation at the double bonds. The combination with oxygen results in the formation of peroxides, volatile aldehydes, ketones, and acids.

Heat, light, moisture, and air are factors that accelerate oxidative rancidity. Modern packaging has helped considerably in this connection, although a more important contribution has been the development of **"antioxidants."** These compounds are usually phenolic in structure (quinones, tocopherol, and catechols) and inhibit the oxidation of fats. The majority of the vegetable shortenings on the market as well as certain brands of lard are protected from rancidity by the addition of antioxidants.

Hydrogenation. It has already been stated that the main difference between oils and fats is the number of unsaturated fatty acids in the molecule. Vegetable oils may be converted into solid fats by the addition of hydrogen to the double bonds of the unsaturated fatty acids. As an illustration, oleic acid may be hydrogenated to produce stearic acid:

$$\underset{\text{Oleic acid}}{CH_3(CH_2)_7CH{=}CH(CH_2)_7COOH} \rightarrow \underset{\text{Stearic acid}}{C_{17}H_{35}COOH}$$

This process has important applications in industry and is used to prepare lard substitutes and shortening from vegetable oils, such as cottonseed oil. The extent of hydrogenation is controlled to produce a fat of the desired consistency, since complete saturation would result in a brittle, tallow-like product. The resultant fat, commercial examples of which are "Crisco" and "Spry," contains approximately 20 to 25 per cent of saturated fatty acids, 65 to 75 per cent of oleic acid, and 5 to 10 per cent of linoleic acid.

SOAPS. The process of saponification, or alkaline hydrolysis, of a fat produces soap as one of the end products (Fig. 21–1). **Soaps** may be defined as metallic salts of fatty acids. The saponification of a fat may be represented as follows:

$$\underset{\text{Tristearin}}{\begin{array}{l} CH_2{-}O{-}\overset{O}{\overset{\|}{C}}{-}C_{17}H_{35} \\ | \\ CH{-}O{-}\overset{O}{\overset{\|}{C}}{-}C_{17}H_{35} \\ | \\ CH_2{-}O{-}\overset{O}{\overset{\|}{C}}{-}C_{17}H_{35} \end{array}} + 3NaOH \longrightarrow \underset{\text{Glycerol}}{\begin{array}{l} CH_2OH \\ | \\ CHOH \\ | \\ CH_2OH \end{array}} + \underset{\substack{\text{Sodium stearate} \\ \text{(soap)}}}{3C_{17}H_{35}\overset{O}{\overset{\|}{C}}{-}ONa}$$

Figure 21-1 Ribbons of soap which are eventually converted into flakes or bars. (The Procter & Gamble Co.)

Sodium salts of fatty acids are known as **hard soaps,** whereas potassium salts form **soft soaps.** The ordinary cake soaps used in the home are sodium soaps. Yellow laundry soap contains resin, which increases the solubility of soap and its lathering properties and has some detergent action. White laundry soap in the form of bars, soap chips, or powdered soap contains sodium silicate and a water-softening agent such as sodium carbonate or sodium phosphate. When sodium soaps are added to hard water, the calcium and magnesium salts present replace sodium to form insoluble calcium and magnesium soaps. The familiar soap curd formed in hard water is due to these **insoluble soaps.**

Detergents. These compounds are a mixture of the sodium salts of the sulfuric acid esters of lauryl and cetyl alcohols. They may be used in hard water because they do not form insoluble compounds with calcium and magnesium.

Extensive research on new detergents and emulsifying agents has resulted in the development of several hundred products possessing almost any desired property. At present potent enzymes (see Chap. 25) capable of digesting protein material in stains are added to the new detergent products.

Analysis of Lipids. For many years the analysis of lipid mixtures has been most difficult. Determination of the **saponification number** yielded a rough measure of molecular weight and the **iodine number** was used for the content of unsaturated fatty acids. Since these determinations have been replaced by more sensitive methods, their details will not be discussed. Thin-layer chromatography and gas-liquid chromatography are presently the methods of choice for the analysis of lipids.

Thin-layer chromatography is carried out on a thin, uniform layer of silica gel spread on a glass plate and activated by heating in an oven (100° to 250°C). Samples of lipid material in the proper solvent are spotted along one edge of the plate with micropipettes. After evaporation of the solvent, the plates are placed vertically in a covered glass tank which contains a layer of suitable solvent on the bottom. Within a few minutes the lipids are separated by the solvent rising through the thin layer carrying the spots to different locations on the silica gel by a combination of adsorption on the gel and varying distribution in the solvent system. The plates are removed, dried, and sprayed with various detection agents to visualize the lipid components. This technique is very sensitive and can be made quantitative by removing the spots and measuring the concentration of the component by gas-liquid chromatography.

Gas-liquid chromatography is another powerful tool of the lipid chemist. Any substance that is volatile or can be made into a volatile derivative, for example fatty acids being converted into their methyl ester, can be separated and analyzed by this technique. The volatile substance is injected into a long column which contains a nonvolatile liquid on a finely divided inert solid. The column is heated and the volatile material is carried through the tube by an inert gas such as helium. Separation depends on the difference in vapor pressure and the partition coefficients of the components in the nonvolatile liquid. As the fractionated components reach the end of the column, they pass over a detection device that is extremely sensitive to differences in organic material carried by the gas, and it records the changes in the gas flow as peaks on a recorder chart. By the use of helium gas alone as the control, and known lipid components as standards to determine the position and area under the peaks, quantitative analysis of lipids can be achieved.

PHOSPHATIDES

The phosphatides are found in all animal and vegetable cells. They are composed of glycerol, fatty acids, phosphoric acid, and a nitrogen compound or inositol. More specifically, they are esters of **phosphatidic acid** with choline, ethanolamine, serine, or inositol.

Phosphatidyl Choline, or Lecithins. The lecithins are esters of phosphatidic acid and choline.

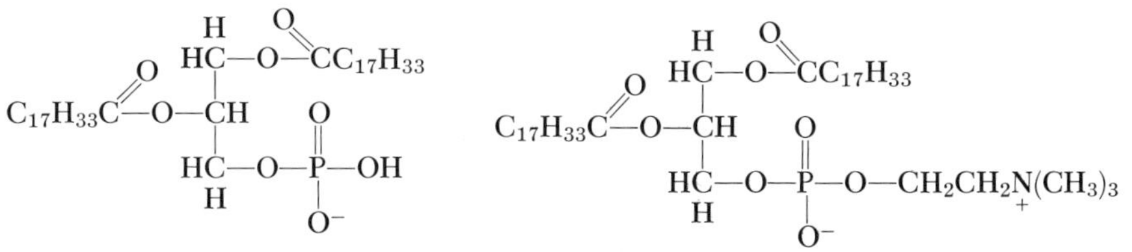

```
                H  O
                HC—O—CC17H33
          O     |
C17H33C—O—CH    O
                |    ||
                HC—O—P—OH
                H    |
                     O⁻
```

L-α-Phosphatidic acid

```
                H  O
                HC—O—CC17H33
          O     |
C17H33C—O—CH    O
                |    ||
                HC—O—P—O—CH2CH2N(CH3)3
                H    |          +
                     O⁻
```

α-Lecithin

The formula is written with the fatty acid on the left side of the central, or β-carbon, to indicate optical activity and an asymmetric carbon atom. Naturally occurring phosphatides have the L form and may contain at least five different fatty acids. In addition, the formula indicates that lecithin exists in the dissociated state since phosphoric acid is a fairly strong acid and choline is a strong base. **Choline** is a quaternary ammonium compound whose basicity in aqueous solution is similar to that of KOH. It is therefore usually represented in the ionic form:

$$(CH_3)_3\overset{+}{N}CH_2CH_2OH + OH^-$$

Choline

The lecithins are constituents of brain, nervous tissue, and egg yolk. From a physiological standpoint they are important in the transportation of fats from one tissue to another and are essential components of the protoplasm of all body cells. In industry lecithin is obtained from soybeans and finds wide application as an emulsifying agent.

If the oleic acid on the central carbon atom of lecithin is removed by hydrolysis, the resulting compound is called **lysolecithin.** Disintegration of the red blood cells, or hemolysis, is caused by intravenous injection of lysolecithin. The venom of snakes such as the cobra contains an enzyme capable of converting lecithins into lysolecithins, which accounts for the fatal effects of the bite of these snakes. A few insects and spiders produce toxic effects by the same mechanism.

Phosphatidyl Ethanolamine, or Cephalins. The cephalins are found in brain tissue and are essentially mixtures of phosphatidyl ethanolamine and phosphatidyl serine. Phosphatidyl ethanolamine may be written as follows:

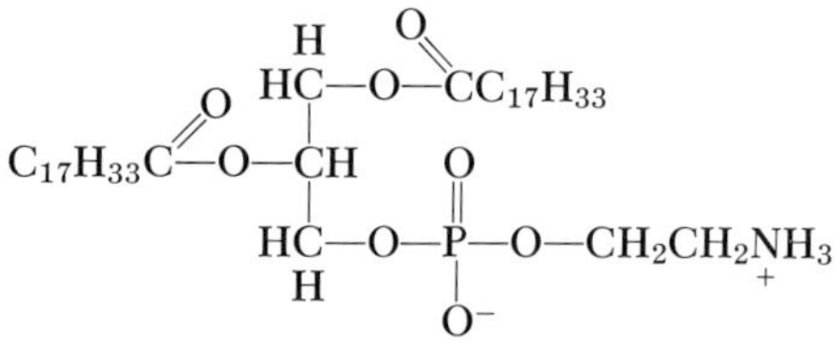

$$\begin{array}{l} \qquad\qquad\qquad\quad H_2C-O-\overset{O}{\overset{\|}{C}}C_{17}H_{33} \\ C_{17}H_{33}\overset{O}{\overset{\|}{C}}-O-CH \qquad O \\ \qquad\qquad\qquad\quad |\qquad\qquad\quad \| \\ \qquad\qquad\qquad\quad H_2C-O-P-O-CH_2CH_2\overset{+}{N}H_3 \\ \qquad\qquad\qquad\qquad\qquad\quad | \\ \qquad\qquad\qquad\qquad\qquad\quad O^- \end{array}$$

Phosphatidyl ethanolamine

The cephalins are involved in the blood-clotting process and are therefore essential constituents of the body.

SPHINGOLIPIDS

Sphingomyelins. The sphingomyelins differ chemically from the lecithins or cephalins. They are composed of a fatty acid, phosphoric acid, choline, and a complex amino alcohol, sphingosine.

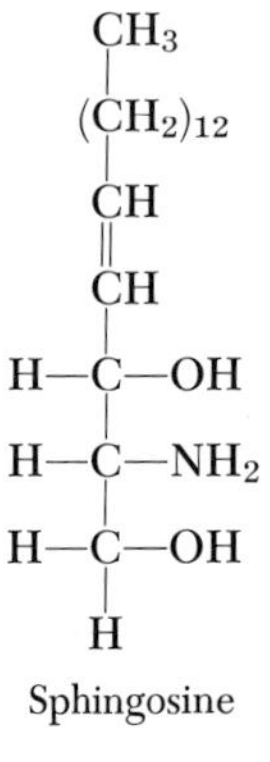

$$\begin{array}{c} CH_3 \\ | \\ (CH_2)_{12} \\ | \\ CH \\ \| \\ CH \\ | \\ H-C-OH \\ | \\ H-C-NH_2 \\ | \\ H-C-OH \\ | \\ H \end{array}$$

Sphingosine

$$\begin{array}{l} CH_3 \\ | \\ (CH_2)_{12} \\ | \\ CH \\ \| \\ CH \\ | \\ H-C-OH \quad O \\ |\qquad\qquad\quad \| \\ H-C-NH-CC_{17}H_{33} \\ |\qquad\qquad O \\ |\qquad\qquad \| \\ H_2C-O-P-O-CH_2CH_2\overset{+}{N}(CH_3)_3 \\ \qquad\qquad | \\ \qquad\qquad O^- \end{array}$$

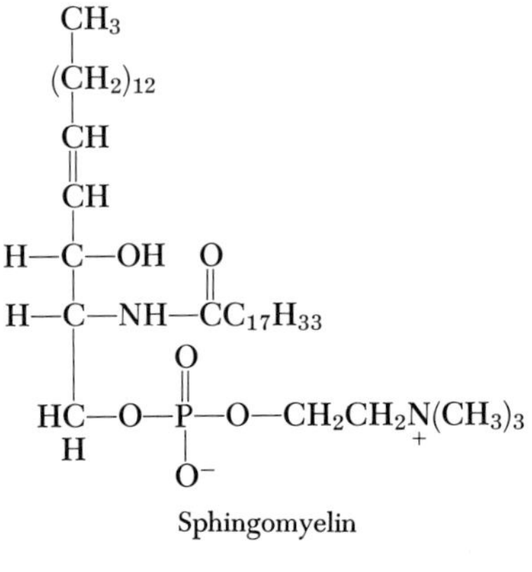

Sphingomyelin

The sphingomyelins are found in large amounts in brain and nervous tissue and are essential constituents of the protoplasm of cells.

GLYCOLIPIDS

The glycolipids are compound lipids that contain a carbohydrate.

Cerebrosides. These lipids are often called **cerebrosides** because they are found in brain and nervous tissue. They are composed of a carbohydrate, sphingosine, and a fatty acid. There are four different cerebrosides, each containing a different fatty acid. The carbohydrate in these lipids is usually galactose, although glucose is sometimes present. A formula for a typical cerebroside may be written as follows:

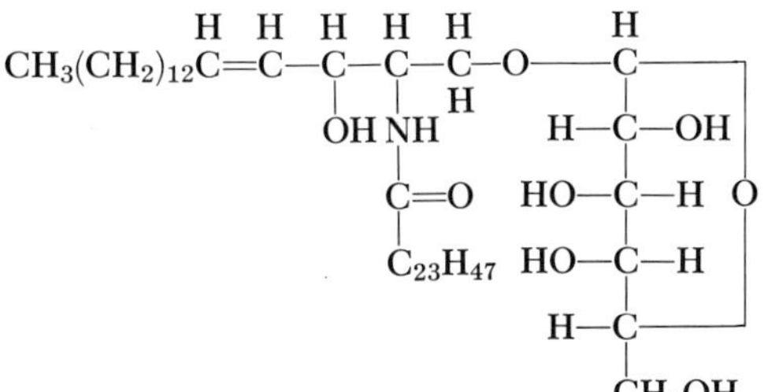

```
              H  H  H  H  H              H
CH3(CH2)12C=C—C—C—C—O———C————┐
                 |   |   H              |               |
               OH NH          H—C—OH       |
                     |                      |               |
                   C=O    HO—C—H   O
                     |                      |               |
                   C23H47  HO—C—H       |
                                            |               |
                                       H—C————┘
                                            |
                                         CH2OH
```

Cerebroside (Kerasin)

STEROIDS

The steroids are derivatives of cyclic alcohols of high molecular weight that occur in all living cells. The lipid material from tissue that is not saponifiable by alkaline hydrolysis contains the steroids. The parent hydrocarbon compound for all the steroids is the sterol nucleus. This structure is an integral part of the cholesterol molecule which may be used to illustrate the lettering system for the rings and the number system for the carbon atoms.

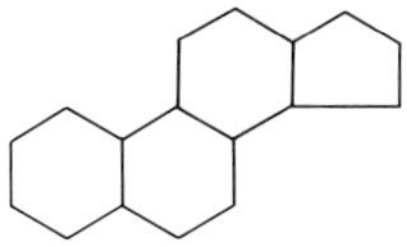

Sterol nucleus

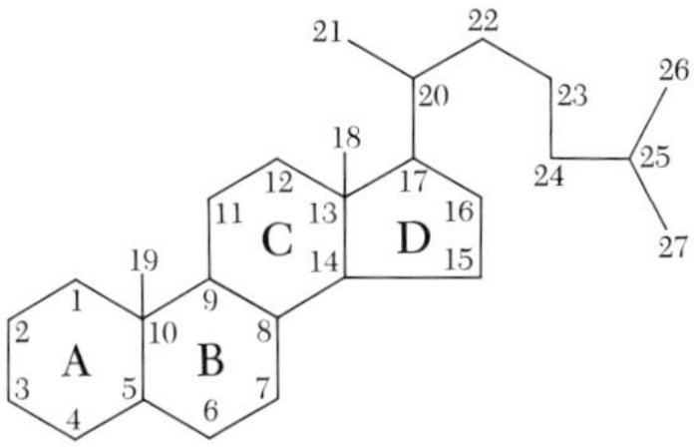

Cholesterol structure designation

The most common sterol is **cholesterol,** which is found in brain and nervous tissue and in gallstones. The structure of cholesterol is shown as follows (each ring is completely saturated and where there is a double bond in the ring it is specifically designated):

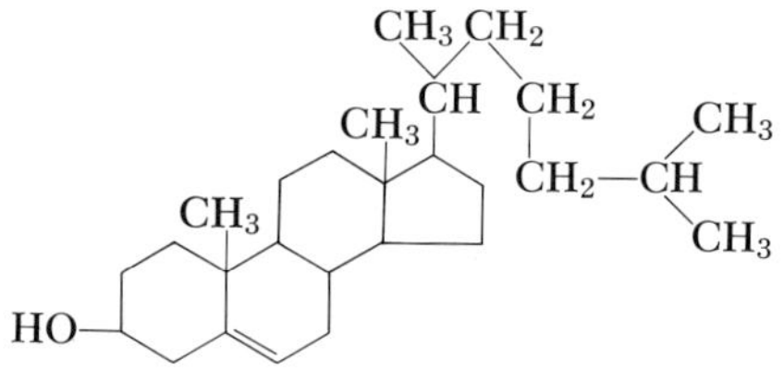

Cholesterol

Cholesterol also reacts with acetic anhydride and sulfuric acid in a dry chloroform solution to yield a green color. This is called the **Liebermann-Burchard reaction** and is the basis for both qualitative detection and quantitative methods for cholesterol. Cholesterol is the precursor of bile acids, sex hormones, hormones of the adrenal cortex, and vitamin D, which will be discussed in the following sections.

BILE SALTS

The bile salts are natural emulsifying agents found in the bile, a digestive fluid formed by the liver. Cholesterol and bile pigments are also important constituents of the bile. Bile is stored in the gall bladder and released at intervals to assist in the digestion and absorption of fats. Glycocholic acid represents the combination of the bile acid, cholic acid, and glycine to form the major bile salt in human bile.

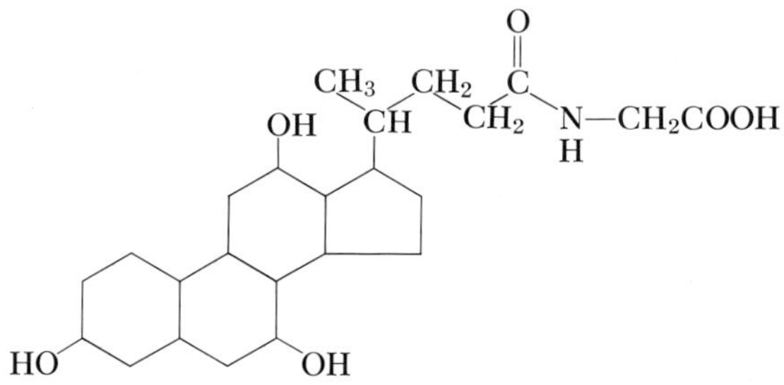

Glycocholic Acid

Glycine is a simple amino acid that will be discussed in the next chapter.

HORMONES OF THE ADRENAL CORTEX

The cortex of the adrenal gland produces a group of hormones with important physiological functions. If the gland exhibits decreased function, as in Addison's disease, electrolyte and water balance are abnormal, carbohydrate and protein metabolisms are adversely affected, and the patient is more sensitive to cold and stress. Typical steroid hormones of the gland are represented on the top of page 311.

Corticosterone was the original name of the first adrenal cortical hormone, which accounts for the naming of other hormones as derivatives of this compound. Three major types of adrenal cortical hormones illustrate the relation of structure to physiological activity.

1. Compounds containing an oxygen on the C-11 position (C—OH, or

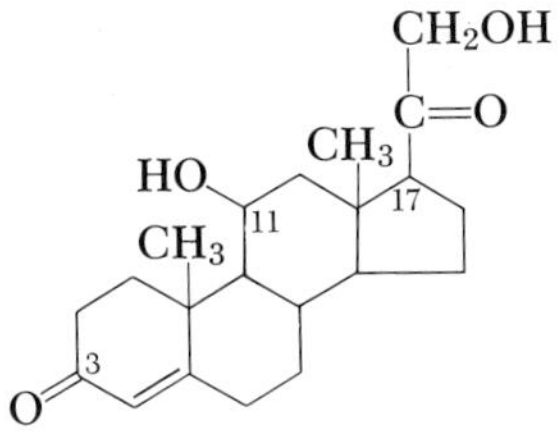

Corticosterone

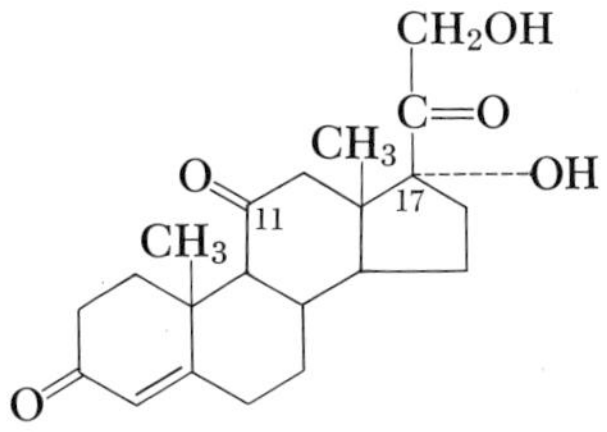

11-Dehydro-17-hydroxycorticosterone
(compound E, cortisone)

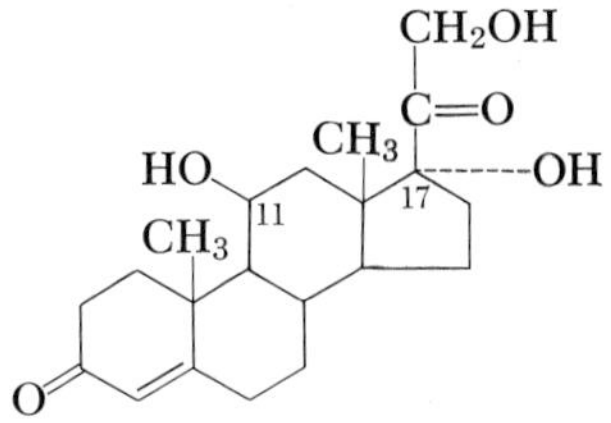

17-Hydroxycorticosterone
(hydrocortisone, cortisol)

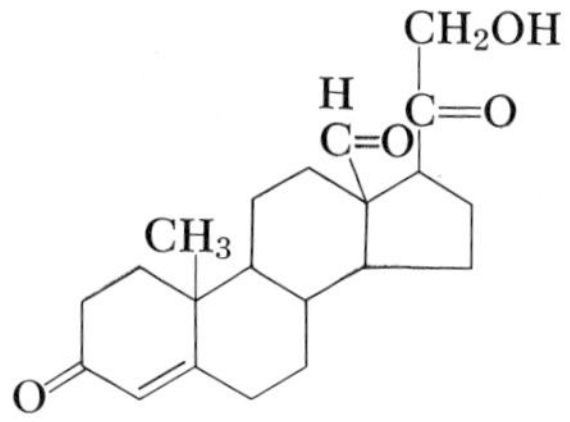

Aldosterone
(aldehyde form)

C=O) exhibit greatest activity in carbohydrate and protein metabolism. Examples are **corticosterone, cortisone,** and **cortisol.**

2. Hormones without an oxygen on the C-11 position have their greatest effect on electrolyte and water metabolism. Examples are **11-deoxycorticosterone** and **11-deoxycortisol.**

3. **Aldosterone** is the only compound without a methyl group at C-18. It is replaced by an aldehyde group that can exist in the aldehyde form or in a hemiacetal form. Aldosterone has a very potent effect on electrolytes and is called a **mineralocorticoid.** In higher doses it also acts on carbohydrate and protein metabolism.

Corticosterone, cortisol, and aldosterone are the major hormones found in the blood, with **cortisol** exerting the greatest effect on carbohydrate and protein metabolism, and aldosterone on the body fluid electrolytes.

When first tried clinically, **cortisone** stimulated considerable excitement in the treatment of rheumatoid arthritis. However, it was subsequently found that the original symptoms would reappear after a period of treatment and that unwanted side effects resulted from the use of this steroid. The pharmaceutical industry prepared and tried many closely related compounds to increase the potency and decrease the side effects of the drug. The 9-fluoro-16-methyl derivative of **prednisolone,** a steroid closely related to cortisol, is 100 to 250 times as potent as cortisone in the treatment of rheumatoid arthritis. The small therapeutic doses that can be applied greatly reduce all of the undesirable side effects.

FEMALE SEX HORMONES

The female sex hormones are steroid in structure and are formed in the ovaries. They are responsible for the development of the secondary sexual characteristics that occur at puberty. These hormones regulate the **estrus cycle,** or **menstrual cycle,** and function in pregnancy (Fig. 21–2).

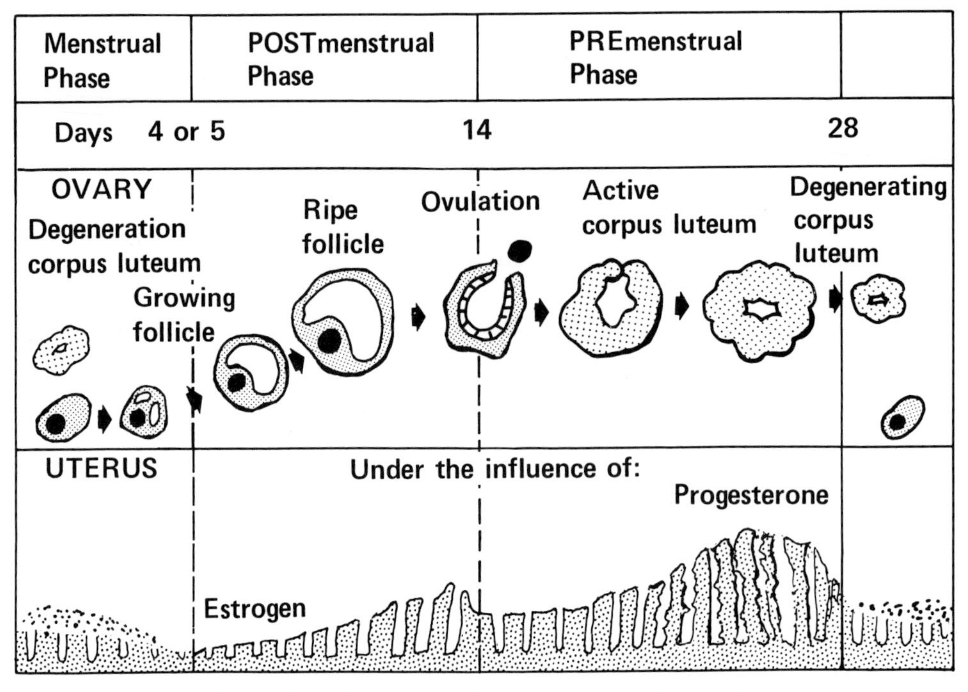

Figure 21-2 The sequence of events in the menstrual cycle.

Hormones of the Follicle. The liquid within the follicle contains at least two hormones, known as **estrone** and **estradiol.** Estrone (theelin) was the first hormone to be isolated from the follicular liquid, but estradiol (dihydrotheelin) is more potent than estrone and may be the principal hormone.

CH_3 O HO

Estrone

CH_3 OH HO

Estradiol

These two compounds are excreted in the urine in increased amounts during pregnancy.

The Hormone of the Corpus Luteum. The hormone produced by the corpus luteum is called **progesterone.** In the body progesterone is converted into **pregnanediol** by reduction before it is excreted in the urine. These two compounds are similar to the estrogens in chemical structure.

CH_3 C=O CH_3 CH_3 O

Progesterone

CH_3 HCOH CH_3 CH_3 HO

Pregnanediol

The main function of progesterone is the preparation of the uterine endometrium for implantation of the fertilized ovum. If pregnancy occurs, this hormone is responsible for the retention of the embryo in the uterus and for the development of the mammary glands prior to lactation.

ORAL CONTRACEPTIVES

The Pill. If the methyl group at position 19 in progesterone (see p. 312) is replaced with a hydrogen atom, the potency of the hormone is increased several fold. After a study of several related compounds, it was found that the further replacement of the side chain at position 17 with a hydroxy group and the addition of an ethinyl group at this position resulted in a hormone with a high degree of potency even when administered orally. Clinically, it was demonstrated that this hormone, norethindrone, prevented ovulation when administered on days 5 to 25 of the menstrual cycle.

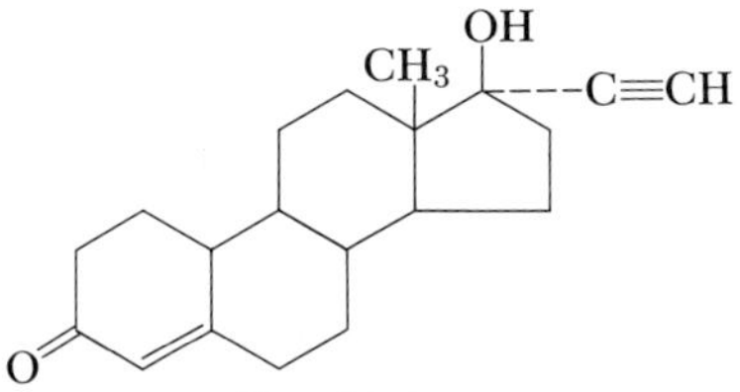

Norethindrone
(19 nor, 17α-ethinyltestosterone)

On withdrawal of the drug, normal menstruation occurred. As may be expected from a study of the events in the menstrual cycle (Fig. 21–2), continued use of this hormone may result in a decrease of estrogen production. For this reason, small amounts of estrogen are included in preparations for oral contraception.

Other compounds similar in structure to norethindrone have been developed and used with estrogens as oral contraceptive agents. Millions of women have used these preparations successfully for several years with a low incidence of side effects. Since there may be some danger in the continuous ingestion of potent hormones that affect the normal menstrual cycle, the present trend is to reduce the doses, for example from 10 mg to 1 mg, to a point where the hormones are still effective contraceptive agents and exhibit fewer side effects.

MALE SEX HORMONES

The male sex hormones are produced by the testes, which are two oval glands located in the scrotum of the male. Small glands in the testes form spermatozoa, which are capable of fertilizing a mature ovum. Between the cells that manufacture spermatozoa are the **interstitial cells,** which produce a hormone called **testosterone.** This hormone is probably converted into other compounds such as **androsterone** before being excreted in the urine. **Dehydro-**

androsterone has also been isolated from male urine but is much less active than the other two hormones. The male sex hormones, or **androgens,** have structures similar to the estrogens.

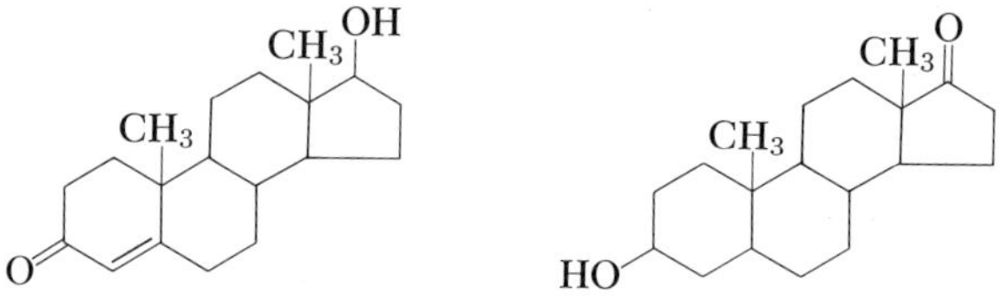

Testosterone

Androsterone

The main function of the androgens in man is the development of masculine sexual characteristics, such as deepening of the voice, the growth of a beard, and the distribution of body hair at puberty. They also control the function of the glands of reproduction (seminal vesicles, prostate, and Cowper's gland).

FAT SOLUBLE VITAMINS

After vitamin C deficiency was related to **scurvy** and vitamin B to **beriberi,** it was noted in experiments prior to 1920 that certain animal fats such as butter and cod liver oil were capable of promoting growth in young rats which were fed a purified diet. These fat soluble vitamins were first collectively called vitamin A, but now include vitamins A, D, E, and K.

VITAMIN A. Vitamin A is closely related to the carotenoid pigments, alpha, beta, and gamma carotene and cryptoxanthin, which are polyunsaturated hydrocarbons. The carotene pigment, beta carotene, is an active precursor of the vitamin.

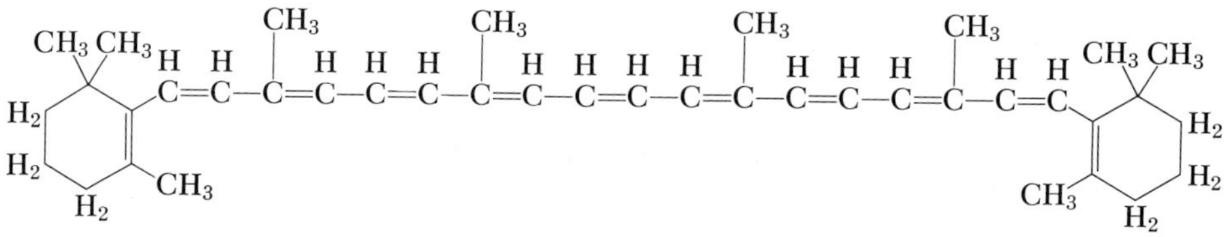

β Carotene

Vitamin A represents half the beta carotene molecule with the ends oxidized to primary alcohol groups.

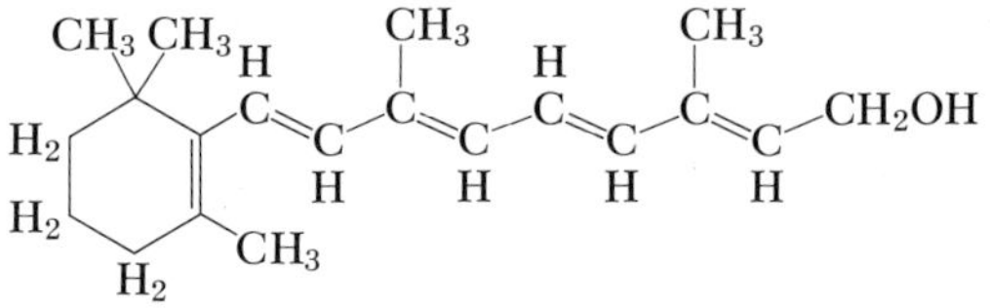

Vitamin A

The carotene pigments and cryptoxanthin can be converted into vitamin A in the animal body. The vitamin is soluble in fat and fat solvents and is a liquid at room temperature.

A diet deficient in vitamin A will not support growth, and the deficiency adversely affects the epithelial cells of the mucous membrane of the eye, the respiratory tract, and the genitourinary tract. The process in which the mucous membrane hardens and dries up is known as **keratinization.** The eye is first to show the effect of a deficiency and one of the first symptoms of the lack of the vitamin is **night blindness.** Later the eyes develop a disease called **xerophthalmia.** This disease is characterized by inflamed eyes and eyelids. The eyes ultimately become infected, and when this infection involves the cornea and lens, sight is permanently lost. A continued deficiency of vitamin A results in extensive infection in the respiratory tract, the digestive tract, and the urinary tract. Vitamin A deficiency also causes sterility, since it affects the lining of the genital tract. It is therefore necessary for normal reproduction and lactation.

The body has the ability to store vitamin A in the liver when it is present in the food in excess of the body requirements. Infants obtain a store of the vitamin in the first milk (colostrum) of the mother, which is ten to one hundred times as rich in vitamin A as ordinary milk.

Vitamin D. ***Chemical Nature.*** Several compounds with vitamin D activity exist, although only two of them commonly occur in antirachitic drugs and foods. These two compounds are produced by the irradiation of ergosterol and 7-dehydrocholesterol with ultraviolet light. Ergosterol is a sterol that occurs in yeast and molds, whereas 7-dehydrocholesterol is found in the skin of animals. Irradiated ergosterol is called **calciferol,** or vitamin D_2; irradiated 7-dehydrocholesterol is called vitamin D_3.

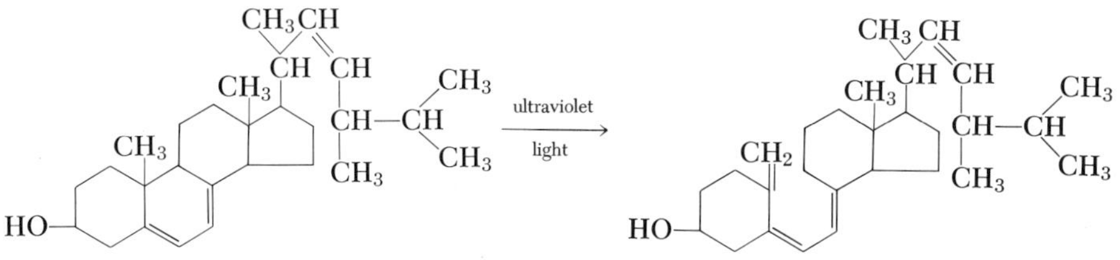

ultraviolet light →

Ergosterol

Vitamin D_2 (calciferol)

The lack of vitamin D in the diet of infants and children results in an abnormal formation of the bones, a disease called **rickets.** Calcium, phosphorus, and vitamin D are all involved in the formation of bones and teeth. Bowed legs, a "rachitic rosary" of the ribs, an abnormal formation of the ribs known as "pigeon breast," and poor tooth development are common signs of vitamin D deficiency in small children. Rickets does not occur in adults after bone formation is complete, although the condition of **osteomalacia** may occur in women after several pregnancies. In osteomalacia, the bones soften and abnormalities of the bony structure may occur.

The main function of vitamin D in the body is to increase the utilization of calcium and phosphorus in the normal formation of bones and teeth. The exact mode of action of the vitamin is not known, although it increases calcium and phosphorus absorption from the intestine, stimulates the activity of the enzyme phosphatase, and is essential for normal growth.

Vitamin E. Vitamin E is chemically related to a group of compounds called **tocopherols.** Alpha-, beta-, and gamma-tocopherol have vitamin E activity, but alpha-tocopherol is the most potent.

```
            CH3
            |
            C     O     CH3          CH3                CH3                CH3
          //  \  /  \  /             |                  |                  |
   CH3—C        C     C—(CH2)3—CH—(CH2)3—CH—(CH2)3—CH—CH3
        |       ||    |
   HO—C         C     CH2
         \\   /   \  /
            C       CH2
            |
            CH3
```

Alpha-tocopherol

Vitamin E is stable to heat but is destroyed by oxidizing agents and ultraviolet light. Oxidative rancidity of fats rapidly destroys the potency of the vitamin.

As yet evidence is not conclusive that vitamin E is required in human nutrition. When animals such as rabbits and rats are maintained on a vitamin E deficient diet, muscular dystrophy, creatinuria, and anemia develop in rabbits, and changes in reproductive organs and function occur in rats.

The tocopherols are excellent antioxidants and prevent the oxidation of several substances in the body, including unsaturated fatty acids and vitamin A. As an antioxidant, vitamin E may protect mitochondrial systems in the cell from irreversible oxidation by lipid peroxides.

Vitamin K. Vitamin K is a derivative of 1,4-naphthoquinone, as is illustrated in the formula for vitamin K:

```
         H       O
         |       ||
         C       C
       //  \   /   \
   HC        C       C—CH3                                                   CH3
   |         ||      ||                                                      |
   HC        C       C—CH2—CH=C—(CH2)3—CH—(CH2)3—CH—(CH2)3—CH
      \\   /   \   /            |            |            |            |
         C       C              CH3          CH3          CH3          CH3
         |       ||
         H       O
```

Vitamin K

The 2-methyl-1,4-naphthoquinones and naphthohydroquinones possess vitamin activity. Vitamin K is fat soluble and therefore is soluble in ordinary fat solvents. It is stable to heat, but is destroyed by alkalis, acids, oxidizing agents, and light.

A diet lacking in vitamin K will cause an increase in the clotting time of blood. This condition produces hemorrhages under the skin and in the muscle tissue. The abnormality in the clotting mechanism is due to a reduction in the formation of **prothrombin,** one of the factors in the normal process.

The vitamin is essential in the synthesis of prothrombin by the liver, but the exact mechanism of synthesis is as yet unknown. It is also thought to be involved in the metabolic reactions and electron transport systems in the mitochondria of cells. The role of prothrombin in the clotting of blood is shown in the following equations:

$$\text{Prothrombin} \xrightarrow[\text{Ca}^{++}]{\text{thromboplastin}} \text{Thrombin}$$

$$\text{Fibrinogen} \xrightarrow{\text{thrombin}} \text{Fibrin clot}$$

Although the clotting process is represented as two simple equations, it involves a large series of factors and reactions in which prothrombin is converted to the enzyme thrombin, which in turn converts plasma **fibrinogen** to the **fibrin clot.**

WAXES

Waxes are simple lipids that are esters of fatty acids and high molecular weight alcohols. Fatty acids such as myristic, palmitic, and carnaubic are combined with alcohols that contain from 12 to 30 carbon atoms. Common, naturally occurring waxes are **beeswax, lanolin, spermaceti,** and **carnauba wax.** Beeswax is found in the structural part of the honeycomb. Lanolin, from wool, is the most important wax from a medical standpoint, since it is widely used as a base for many ointments, salves, and creams. Spermaceti, obtained from the sperm whale, is used in cosmetics, some pharmaceutical products, and in the manufacture of candles. Carnauba wax is obtained from the carnauba palm and is widely used in floor waxes and in automobile and furniture polishes.

Paraffin wax, petrolatum, and lubricating oils are not to be confused with the simple lipids discussed above, because they are merely mixtures of hydrocarbons.

Questions

1. Name and write the formulas for the three most commonly occurring fatty acids.

2. Illustrate the formation of an ester linkage between glycerol and three molecules of butyric acid. How would you name the product?

3. Are most commonly occurring fats composed of simple glycerides or mixed glycerides? Explain.

4. Explain how you would test for the presence of glycerol in the laboratory and why this test can be used as a general test for fats.

5. What type of rancidity occurs in common shortenings? How is this prevented? Explain.

6. Write an equation illustrating the process of saponification. Name all compounds in the equation.

7. How do insoluble soaps, soft soaps, and hard soaps differ from each other?

8. What analytical procedure would you employ to separate a mixture of free fatty acids, phosphatides, and neutral fats into its components? Explain.

9. Explain and illustrate chemically why lecithin is more accurately called phosphatidyl choline.

10. Define and chemically illustrate the structure of a steroid.

11. What is the major structural difference between aldosterone and the other steroid hormones of the adrenal cortex?

12. Write the structure for estrone and indicate how it differs from the male sex hormones.

13. How are oral contraceptives, the pill, related to female sex hormones? Explain.

14. Illustrate the chemical change that occurs when ergosterol is irradiated with ultraviolet light.

15. Describe the characteristic similarities and differences between the structures of the fat-soluble vitamins.

Suggested Reading

Beyler: Some Recent Advances in the Field of Steroids. Journal of Chemical Education, Vol. 37, p. 497, 1960.
Cover: The Identification of Vegetable Oils. A Gas Chromatographic Experiment. Journal of Chemical Education, Vol. 45, p. 120, 1968.
Dowling: Night Blindness. Scientific American, Vol. 215, No. 4, p. 78, 1966.
Frank: Carotenoids. Scientific American, Vol. 194, No. 1, p. 80, 1956.
Kushner and Hoffman: Synthetic Detergents. Scientific American, Vol. 185, No. 4, p. 26, 1951.
Samuels: A Simple Sprayer for Quantitative Thin-Layer Chromatography. Journal of Chemical Education, Vol. 45, p. 438, 1968.
Sarett: The Hormones. Journal of Chemical Education, Vol. 37, p. 184, 1960.
Snell: Soap and Glycerol. Journal of Chemical Education, Vol. 19, p. 172, 1942.

CHAPTER 22

PROTEINS

The chemistry of proteins is more complex than that of carbohydrates and lipids. Proteins were recognized as essential nitrogen-containing constituents of protoplasm by Mulder in 1839. He named them **proteins,** from a Greek word meaning "of prime importance." Proteins are fundamental constituents of all cells and tissues in the body and are present in all fluids except the bile and urine. They are also essential constituents of the diet required for the synthesis of body tissue, enzymes, certain hormones, and protein components of the blood.

Proteins are made by plant cells by a process starting with photosynthesis from carbon dioxide, water, nitrates, sulfates and phosphates. The complicated process of synthesis has not as yet been completely elucidated. Animals can synthesize only a limited amount of protein from inorganic sources and are mainly dependent on plants or other animals for their source of dietary protein. Proteins are used in the body for growth of new tissue, for maintenance of existing tissue, and as a source of energy. When used for energy, they are broken down by oxidation to form simple substances such as water, carbon dioxide, sulfates, phosphates, and simple nitrogen compounds that are excreted from the body. These same products are formed in decaying plant and animal matter. The simple nitrogen compounds such as amino acids and urea are converted into ammonia, nitrites, and nitrates. The growing plants then use these inorganic compounds to form new proteins, and the cycle is completed (Fig. 22–1).

ELEMENTARY COMPOSITION

The five elements that are present in most naturally occurring proteins are **carbon, hydrogen, oxygen, nitrogen,** and **sulfur.** There is a wide variation in the amount of sulfur in proteins. Gelatin, for example, contains about 0.2 per cent, in contrast to 3.4 per cent in insulin.

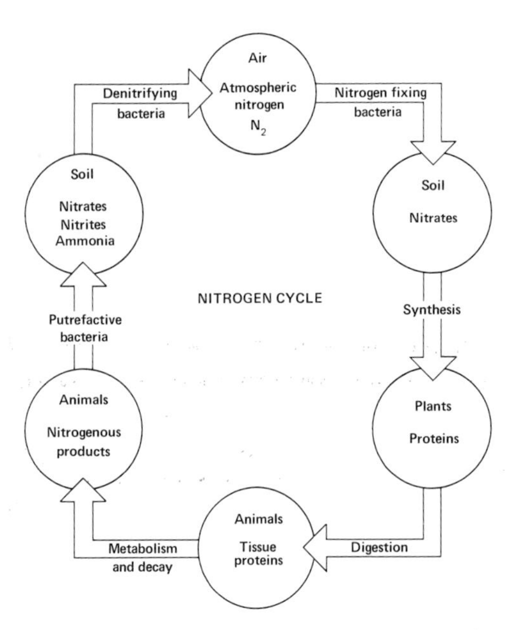

Figure 22-1 A simple diagram showing the events occurring in the nitrogen cycle.

Other elements, such as phosphorus, iodine, and iron, may be essential constituents of certain specialized proteins. Casein, the main protein of milk, contains phosphorus, an element of utmost importance in the diet of infants and children. Iodine is a basic constituent of the protein in the thyroid gland and is present in sponges and coral. Hemoglobin of the blood, which is necessary for the process of respiration, is an iron-containing protein. Most proteins show little variation in their elementary composition; the average content of the five main elements is as follows:

Element	*Average Per Cent*
Carbon	53
Hydrogen	7
Oxygen	23
Nitrogen	16
Sulfur	1

The relatively high content of nitrogen differentiates proteins from fats and carbohydrates.

MOLECULAR WEIGHT

Protein molecules are very large, as indicated by the approximate formula for oxyhemoglobin:

$$C_{2932}H_{4724}N_{828}S_8Fe_4O_{840}$$

The molecular weight of oxyhemoglobin would thus be about 68,000. The common protein egg albumin has a molecular weight of about 34,500. In general, protein molecules have weights that vary from 34,500 to 50,000,000. Their extremely large size can readily be appreciated when they are compared with the molecular weight of a fat such as tripalmitin, which is 807, of glucose, which is 180, or of an inorganic compound such as ammonia, which is 17.

HYDROLYSIS

In addition to their large size, protein molecules are also very complicated. Like any complex molecule, they may be broken down by hydrolysis into smaller molecules whose structure is more easily determined. Common reagents used for the hydrolysis of proteins are acids (HCl and H_2SO_4), bases (NaOH), and enzymes (proteases). The simple molecules that are formed by the complete hydrolysis of a protein are called **amino acids.**

AMINO ACIDS

Before considering the properties and reactions of proteins, it may be well to study the individual amino acids. An amino acid is essentially an organic acid that contains an amino group. If a hydrogen is replaced by an amino group on the carbon atom that is next to the carboxyl group in acetic acid, CH_3COOH, the simple amino acid **glycine** will be formed.

$$\underset{\displaystyle NH_2}{\overset{}{CH_2}}\!-\!COOH$$

The chemical name for this amino acid would be aminoacetic acid. The carbon atom next to the carboxyl group is called the alpha (α) carbon, the next beta (β), the next one gamma (γ), the next delta (δ), and the fifth from the carboxyl group is called the epsilon (ε) carbon atom. Since all the amino acids have an amino group attached to the alpha carbon atom, they are known as **alpha amino acids.**

The amino acids are divided into groups according to their chemical structure. Examples of each group are given in the following classification. The common name for the amino acid is followed by the chemical name in parentheses, after which the abbreviation used in sequence and structure models is given. Example:

Glycine (aminoacetic acid) Gly

I. Aliphatic amino acids
   A. With 1 amino and 1 carboxyl group:

Glycine (aminoacetic acid) Gly

$$\underset{\displaystyle NH_2}{CH_2}COOH$$

Alanine (α-aminopropionic acid) Ala

$$CH_3\!-\!\underset{\displaystyle NH_2}{CH}\!-\!COOH$$

Valine (α-aminoisovaleric acid) Val

$$\begin{array}{l} CH_3-\underset{\displaystyle CH_3}{\underset{|}{CH}}-\underset{\displaystyle NH_2}{\underset{|}{CH}}-COOH \end{array}$$

Leucine (α-aminoisocarproic acid) Leu

$$CH_3-\underset{\displaystyle CH_3}{\underset{|}{CH}}-CH_2-\underset{\displaystyle NH_2}{\underset{|}{CH}}-COOH$$

B. With 1 amino, 1 carboxyl, and 1 hydroxyl group:

Serine (α-amino-β-hydroxypropionic acid) Ser

$$\underset{\displaystyle OH}{\underset{|}{CH_2}}-\underset{\displaystyle NH_2}{\underset{|}{CH}}-COOH$$

Threonine (α-amino-β-hydroxybutyric acid) Thr

$$CH_3-\underset{\displaystyle OH}{\underset{|}{CH}}-\underset{\displaystyle NH_2}{\underset{|}{CH}}-COOH$$

C. With 1 amino and 2 carboxyl groups:

Aspartic acid (α-aminosuccinic acid) Asp

$$\begin{array}{c} COOH \\ | \\ H-C-NH_2 \\ | \\ CH_2 \\ | \\ COOH \end{array}$$

Glutamic acid (α-aminoglutaric acid) Glu

$$\begin{array}{c} COOH \\ | \\ H-C-NH_2 \\ | \\ CH_2 \\ | \\ CH_2 \\ | \\ COOH \end{array}$$

D. With 2 amino and 1 carboxyl group:

Lysine (α, ε-diaminocaproic acid) Lys

$$\underset{\displaystyle NH_2}{\underset{|}{CH_2}}-CH_2-CH_2-CH_2-\underset{\displaystyle NH_2}{\underset{|}{CH}}-COOH$$

Arginine (α-amino-δ-guanidovaleric acid) Arg

$$H_2N-\underset{\displaystyle NH}{\underset{\|}{C}}-NH-CH_2-CH_2-CH_2-\underset{\displaystyle NH_2}{\underset{|}{CH}}-COOH$$

E. Sulfur-containing amino acids:

Cysteine (α-amino-β-thiopropionic acid) Cys

$$\begin{array}{c} SH \\ | \\ CH_2 \\ | \\ H-C-NH_2 \\ | \\ COOH \end{array}$$

Cystine [di(α-amino-β-thiopropionic acid)] Cys-Cys

$$\begin{array}{ccc} S & \text{———} & S \\ | & & | \\ CH_2 & & CH_2 \\ | & & | \\ H-C-NH_2 & & H-C-NH_2 \\ | & & | \\ COOH & & COOH \end{array}$$

Methionine (α-amino-γ-methylthiobutyric acid) Met

$$CH_3-S-CH_2-CH_2-\underset{\displaystyle NH_2}{\underset{|}{CH}}-COOH$$

II. Aromatic amino acids

Tyrosine (α-amino-β-parahydroxyphenyl-propionic acid) Tyr

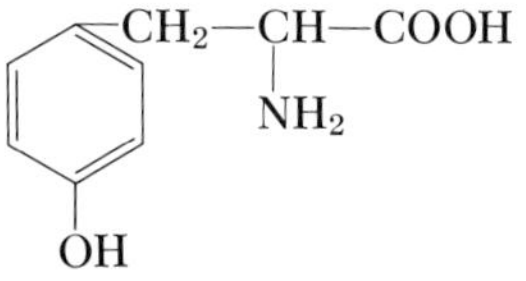

$$HO-C_6H_4-CH_2-CH(NH_2)-COOH$$

Phenylalanine (α-amino-β-phenyl-propionic acid) Phe

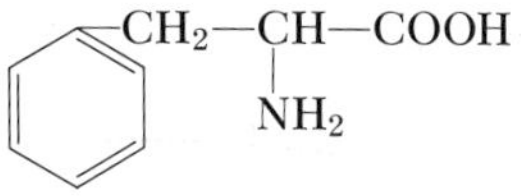

$$C_6H_5-CH_2-CH(NH_2)-COOH$$

III. Heterocyclic amino acids

Tryptophan (α-amino-β-indolpropionic acid) Try

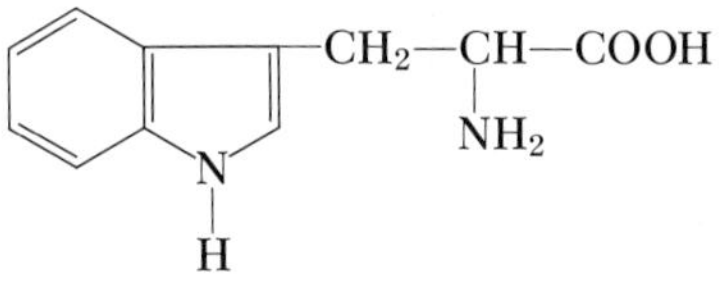

$$\text{Indolyl}-CH_2-CH(NH_2)-COOH$$

Proline (pyrrolidine-2-carboxylic acid) Pro

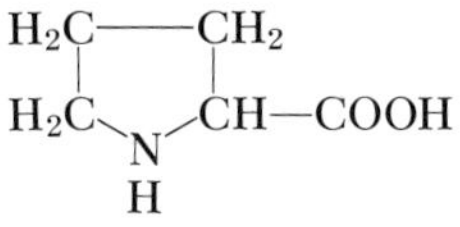

$$\text{(ring: } H_2C-CH_2,\ H_2C-N(H)-CH-COOH)$$

Histidine (α-amino-β-imidazolylpropionic acid) His

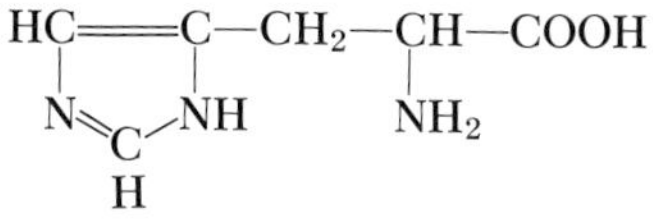

$$HC{=}C(-NH-CH{=}N-)-CH_2-CH(NH_2)-COOH$$

Optical Activity of Amino Acids. All amino acids except glycine contain an asymmetric carbon atom in their formulas. For this reason they may exist in the D or L form. Using alanine as an example of a simple amino acid, we may compare the D and L forms of those of glyceraldehyde and lactic acid:

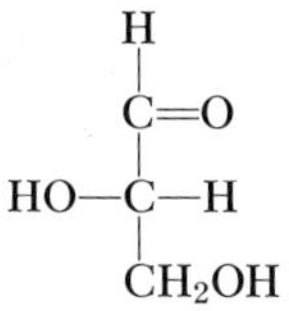

L-Glyceraldehyde	D-Glyceraldehyde	L-Lactic acid
H–C=O; HO–C–H; CH_2OH	H–C=O; H–C–OH; CH_2OH	O=C–OH; HO–C–H; CH_3

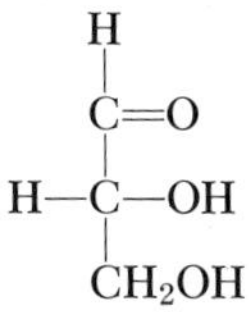

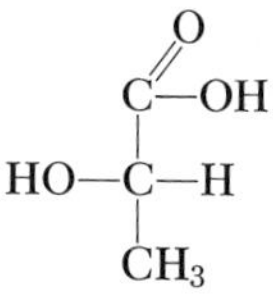

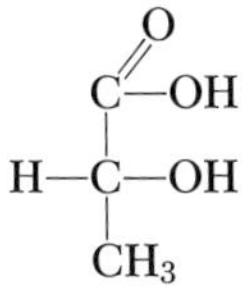

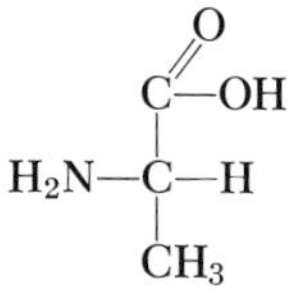

D-Lactic acid	L-Alanine	D-Alanine
O=C–OH; H–C–OH; CH_3	O=C–OH; H_2N–C–H; CH_3	O=C–OH; H–C–NH_2; CH_3

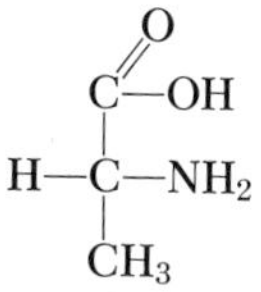

Naturally occurring amino acids from plant and animal sources have the L configuration and would be designated L(+) or L(−), depending on their rotation of plane polarized light. D-alanine and D-glutamic acid have been obtained from microorganisms, especially from their cell walls.

Amphoteric Properties of Amino Acids. Amino acids behave both as weak acids and as weak bases, since they contain at least one carboxyl and one amino group. Substances that ionize as both acids and bases in aqueous solution are called **amphoteric.** An example would be glycine, in which both the acidic and basic groups are ionized in solution to form dipolar ions or **zwitterions.**

$$\begin{array}{l} CH_2\text{—}COO^- \\ | \\ NH_3^+ \end{array}$$

The glycine molecule is electrically neutral, since it contains an equal number of positive and negative ions. Since proteins are composed of amino acids, they are amphoteric substances and are able to neutralize both acids and bases. This property of proteins is responsible for their buffering action in blood and other fluids.

Reactions of Amino Acids. The fact that amino acids can ionize as both weak acids and weak bases and contain amino groups and carboxyl groups suggests a very reactive molecule. Many of the common reactions of organic chemistry may be applied to amino acids.

Reaction with Nitrous Acid. This is the basis of the Van Slyke method for the determination of free primary amino groups.

$$\underset{\displaystyle NH_2}{R\text{—}\underset{|}{C}HCOOH} + HONO \longrightarrow \underset{\displaystyle OH}{R\text{—}\underset{|}{C}HCOOH} + N_2 + H_2O$$

The nitrogen gas that is liberated in the reaction is collected and its volume measured. One half of this nitrogen comes from the amino acid and is used as a measure of the free amino nitrogen.

Reaction with 1-Fluoro-2,4-dinitrobenzene (FDNB). This compound, also called **Sanger's reagent,** reacts with the free amino group of an amino acid to form a yellow colored dinitrophenylamino acid, or DNP-amino acid.

$$R\text{—}\overset{\displaystyle H}{\underset{\displaystyle COOH}{C}}\text{—}NH_2 + F\text{—}C_6H_3(NO_2)_2 \longrightarrow R\text{—}\overset{\displaystyle H}{\underset{\displaystyle COOH}{C}}\text{—}NH\text{—}C_6H_3(NO_2)_2 + HF$$

Dinitrophenylamino acid, or DNP-amino acid

This reaction will be found to be very important in the determination of protein structure, since the reagent reacts with the free amino group of the terminal amino acid in a protein and thus identifies the end amino acid in the structure.

Reaction with Ninhydrin. Amino acids react with ninhydrin (triketohydrindene hydrate) to form CO_2, NH_3, and an aldehyde. The NH_3 that is formed in the reaction combines with a molecule of reduced and a molecule of oxidized ninhydrin and forms a blue-colored compound. This compound may be measured colorimetrically for the quantitative determination of amino acids.

$$\text{Oxidized ninhydrin} + NH_3 + \text{Reduced ninhydrin} \rightarrow \text{Blue-colored compound} + 3H_2O$$

Oxidized ninhydrin　　Reduced ninhydrin　　Blue-colored compound

Color Reactions of Specific Amino Acids. Certain amino acids, whether in the free form as in protein hydrolysates or combined in proteins, give specific color reactions that aid in their detection and determination. The **Millon test** depends on the formation of a red-colored mercury complex with tyrosine, whether free or in proteins, whereas tryptophan reacts with glyoxylic acid to produce a violet color in the **Hopkins-Cole test.** In the **Sakaguchi reaction,** the guanidino group in arginine forms a red color with α-naphthol and sodium hypochlorite, and cysteine and proteins that contain free sulfhydryl groups yield a red color with sodium nitroprusside. Both cystine and cysteine, free and in proteins, form a black precipitate of PbS in the **unoxidized sulfur test.**

POLYPEPTIDES

To understand the structure of the polypeptide molecule the manner in which the amino acids are joined together must be known. Several complicated theories have been proposed to explain polypeptide structure. The most reasonable theory, suggested by Emil Fisher, is that amino acids are joined by the peptide linkage.

The **peptide linkage** may be defined as the linkage between the carboxyl group of one amino acid and the amino group of another amino acid, with the splitting out of a molecule of water. This type of linkage may be illustrated by the union of a molecule of alanine and a molecule of glycine:

$$\underset{\text{Alanine}}{CH_3-\underset{NH_2}{\underset{|}{CH}}-\overset{O}{\overset{/\!/}{C}}-OH} + \underset{\text{Glycine}}{H-\underset{H}{\underset{|}{N}}-CH_2-COOH}$$

$$\downarrow \text{synthesis}$$

$$\underset{\text{Alanylglycine}}{CH_3-\underset{NH_2}{\underset{|}{CH}}-\overset{O}{\overset{/\!/}{C}}-\underset{H}{\underset{|}{N}}-CH_2-COOH} + H_2O$$

The compound alanylglycine, which results from this linkage, is called a **dipeptide.** The union of three amino acids would result in a **tripeptide,** and the combination of several amino acids by the peptide linkage would be called a **polypeptide.** Since each amino acid has lost a water molecule when it joins to two other amino acids in a polypeptide, the remaining compound is called an **amino acid residue.** Proteins may be considered as complex polypeptides. A polypeptide chain illustrating the primary structure of a protein may be represented as follows:

```
            O       R1             O       R3             O
            ‖       |              ‖       |              ‖
            C      CH      NH      C      CH      NH      C
----C     /  \NH /   \C  /   \CH /  \NH /   \C  /   \CH2 /  \NH----
                      ‖        |               ‖        |
                      O        R2              O        R4
```

The R groups are the side chains of the specific amino acids in the polypeptide chain.

The Biuret Test. When a few drops of very dilute copper sulfate solution are added to a strongly alkaline solution of a peptide or protein, a violet color is produced. This is a general test for proteins and is given by peptides that contain two or more peptide linkages. **Biuret** is formed by heating urea and has a structure similar to the peptide structure of proteins:

$$NH_2-\overset{\overset{O}{\|}}{C}-NH-\overset{\overset{O}{\|}}{C}-NH_2$$

Biuret

DETERMINATION OF AMINO ACIDS IN MIXTURES

Prior to 1950, it was thought that unraveling the combinations of the 23 different amino acids which form proteins of molecular weight from 34,500 to 50,000,000 was inconceivable. Proteins were subjected to hydrolysis and several amino acids could be separated and determined in the mixture. As mentioned earlier, the methyl esters of the carboxyl groups of the amino acids could be steam distilled to gain a measure of separation. Some amino acids form insoluble salts with silver ions, picric acid, HCl, tungstic acid and other compounds, and therefore could be identified. The specific color reactions described earlier helped identify arginine, tyrosine, tryptophan, and cysteine. Also certain bacteria were used in the microbiological determination of one amino acid that was required for the growth of that bacteria in a media containing a mixture of amino acids. All of these methods resulted in the separation and determination of amino acids in protein hydrolysates, but were of little value in the determination of the order or sequence of amino acids in a protein molecule. Amino acid sequence determination was made possible by the development of the techniques of chromatography.

Paper Chromatography. Several types of chromatographic techniques have been developed for the analysis of mixtures of molecules such as amino acids. **Paper chromatography** is relatively simple and produces excellent separation, detection, and quantitation of the individual amino acids. The technique of ascending paper chromatography is often used. When a strip of filter paper is held vertically in a closed glass cylinder with its lower end dipped in a mixture of water and an organic solvent such as butyl alcohol, phenol, or collidine, the mixture of water and organic solvent moves up the paper. If a solution containing a mixture of amino acids is added as a small spot just above the solvent level, the individual amino acids will be affected by the water phase and the organic phase as they move up the paper. A solvent partition will occur, and each amino acid will be carried to a particular location on the paper. This location depends on many factors, including the pH, the temperature, the concentration of the solvents, and the time of chromatography. After drying the paper, it is sprayed with a solution of ninhydrin, which yields a blue to purple color with each amino acid. By comparison with known amino acids, separation of the amino acids from a hydrolysate of a protein or polypeptide fragment may be achieved by the proper choice of conditions and solvents. This separation is often improved by a second chromatographic run during with different solvents are used and the dried paper from the first run is turned 90 degrees from the direction of the first migration. The results of a two-dimensional paper chromatography separation are shown in Figure 22–2.

A recent extension of the technique of paper chromatography involves thin-layer chromatography (see Chapter 21, p. 306). Instead of a strip of paper, the amino acids or other molecules to be separated are carried by the solvent mixture through a thin layer of cellulose powder, silica gel, and other adsorbents located on a glass plate, a plastic film, or a specially processed paper backing. The separation is much more rapid and a wide variety of reagents and dyes may be used as detection sprays. The speed of migration, sensitivity, and versatility of thin-layer chromatography make this technique a valuable addition to the tools of research in biochemistry.

Ion Exchange Chromatography. Columns of starch, cellulose powder, and alumina gels have been used to separate amino acids, but it has been difficult to isolate amino acids with similar properties from each other. More successful separations and quantitative determinations of amino acids in mixtures are achieved with **ion exchange chromatography.** Ion exchange resins are insoluble synthetic resins containing acidic or basic groups, such as $—SO_3H$ or —OH. A sulfonated polystyrene resin may be used as a cation exchange resin by the

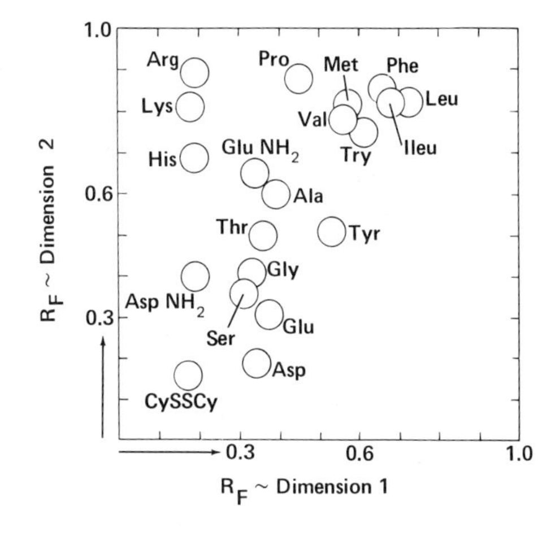

Figure 22–2 A schematic representation of a two-dimensional paper chromatogram. The solvent for dimension 1 is *n*-butanol-acetic acid-water (250:60:250 vol. per vol.) and for dimension 2, phenol-water-ammonia (120:20:0.3 per cent). Each solvent front moves with an R_F equal to 1 in each dimension. (From White et al.: *Principles of Biochemistry,* 3rd Ed. New York, McGraw-Hill, 1964, p. 117.)

addition of Na ions to produce $—SO_3Na$ groups on the surface of the resin. Basic amino acids react with a cation exchange resin as follows:

$$\text{Resin—}SO_3^-Na^+ + \underset{Cl^-}{NH_3^+}\text{—R} \rightarrow \text{Resin—}SO_3^-NH_3\text{—R} + Na^+Cl^-$$

The resin is placed in a column or long glass tube, and the mixture of amino acids, which are dissolved in a small volume of buffer, is placed on top of the column. The column is washed with a buffer solution, and, as the amino acids pass down the column, the basic amino acids react with the $—SO_3Na$ groups, replacing Na^+, and are slowed in their passage. Glutamic and aspartic acids are least affected by the column and come off in the first buffer. They are followed by the neutral amino acids and finally the basic amino acids.

A more efficient removal of the different amino acids from the column is achieved by gradually increasing the pH and concentration of the eluting buffer to force the basic amino acids off the column. By proper choice of resin, buffers, length of column, temperature, and elution rates, and collecting the eluted amino acids in fraction collectors that are coupled to automatic analyzing instruments, it is possible to obtain a quantitative amino acid analysis of a protein hydrolysate in a few hours. The **elution pattern** of representative amino acids from a cation exchange resin is illustrated in Figure 22–3.

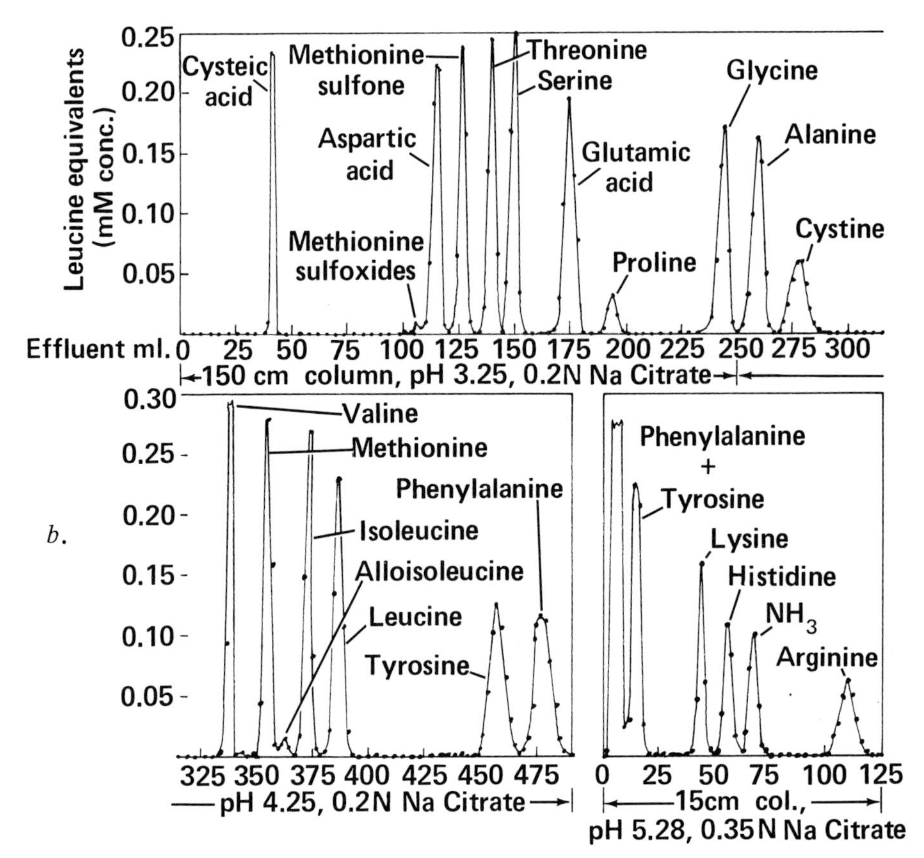

Figure 22–3 Chromatographic fractionation of a synthetic mixture of amino acids on columns of Amberlite IR-120. (From Moore et al.: Anal. Chem., *30:* 1186, 1958.)

SANGER'S ANALYSIS OF THE INSULIN MOLECULE

Sanger's efforts to establish the structure of insulin were made possible by the advances in the methods of protein chemistry, especially the technique of chromatography. In addition, he found that a dinitrophenyl (DNP) group could be attached to free amino groups to form a yellow compound, as was discussed earlier. This DNP group remained attached to the amino acid residue even after hydrolysis was used to split the peptides, therefore making it possible to identify the terminal residue. Employing the DNP method, he first established that each insulin molecule contained two amino acid residues with free amino groups. He concluded, therefore, that insulin consists of two chains. The two chains were held together by the disulfide —S—S— bonds of cystine residues, which could be broken by mild oxidation. The two intact chains were obtained, and it was proved that one contained 21 amino acids whereas the other contained 30 (Fig. 22–4). Each chain was hydrolyzed with acid into smaller pieces and the amino acids identified by chromatography.

Amino Acid Sequence in Peptides and Proteins. After Sanger laid the foundation for the attack on the amino acid sequence of a protein molecule, other workers studied peptides and proteins. Vincent du Vigneaud established the exact structure and sequence in two peptide hormones, **oxytocin** and **vasopressin,** that are elaborated by the posterior lobe of the pituitary gland. Each peptide contained eight amino acids, with the disulfide bridge of cystine across four of the amino acids.

Cys—Tyr—Ileu—Glu—Asp—Cys—Pro—Leu—Gly
NH$_2$ NH$_2$ NH$_2$

Oxytocin

Cys—Tyr—Phe—Glu—Asp—Cys—Pro—Arg—Gly
NH$_2$ NH$_2$ NH$_2$

Vasopressin

The presence of two different amino acids in such a small peptide results in very different physiological activity. Du Vigneaud also succeeded in synthesizing these two molecules from amino acids and demonstrated the similar hormone activity of the synthetic peptides.

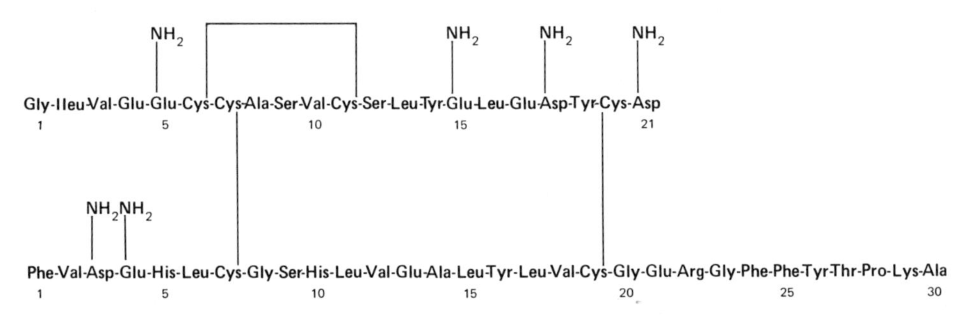

Figure 22–4 Amino acid sequence in beef insulin.

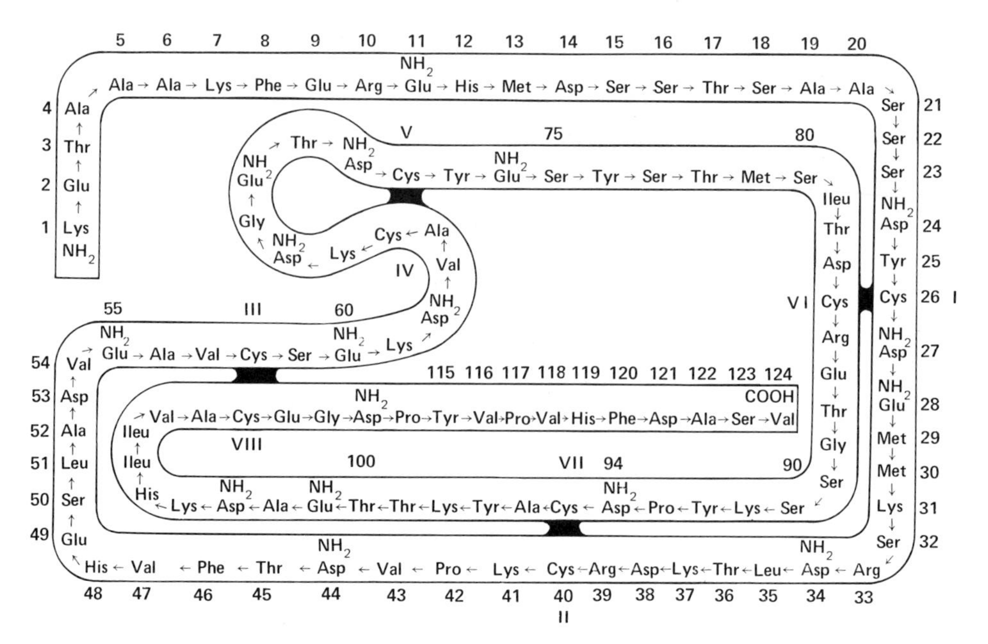

Figure 22-5 The complete amino acid sequence of enzyme ribonuclease. Standard three-letter abbreviations are used to indicate individual amino acid residues. (After Smyth et al.: J. Biol. Chem., *238:* 227, 1963.)

The amino acid sequence of the **adrenocorticotropic hormone, ACTH,** containing 39 amino acids, has been established. Larger protein molecules, such as the enzyme **ribonuclease** (Fig. 22–5), with 124 amino acids, the α and β polypeptide chains of **hemoglobin** (141 and 146 amino acid residues, respectively), and the **tobacco mosaic virus** protein with 158 amino acid residues, have also been characterized.

STRUCTURE OF PROTEINS

The chemical, physical, and biological properties of specific proteins depend on the structure of the molecule as it exists in the native state. Proteins range in complexity from a simple polypeptide, such as vasopressin, with biological activity, to a globular protein such as myoglobin, whose molecule involves cross linkages, helix formation, and folding and conformational forces.

Primary Structure. The amino acid sequence determinations have established the exact structure of the polypeptide chain in simple proteins. The peptide linkage joining amino acids to produce a polypeptide is considered the **primary structure** of a protein.

Secondary Structure. If only peptide bonds were involved in protein structure, the molecules would consist of long polypeptide chains coiled in random shapes. Most **native proteins,** however, are either fibrous or globular in nature, and consist of polypeptide chains joined together or held in definite folded shapes by hydrogen bonds. This influence of hydrogen bonding on the protein molecule is often called the **secondary structure** of the protein. Although **hydrogen bonds** may be formed between several groups on the peptide chains, the most common bonding occurs between the carbonyl and imide groups of the peptide chain backbone, as shown:

```
      RCH            HCR
        C=O----HN
    HN                C=O
      HCR         RCH
  O=C                 NH
        NH----O=C
      RCH            HCR
        C=O----HN
```

This phenomenon explains the joining together of parallel polypeptide chains, which is evident in several fibrous proteins. A further structure that involves hydrogen bonds and accounts for the stereochemistry and the proper bond lengths and angles in the protein molecule is the **α-helix** proposed by Pauling (Fig. 22–6). The α-helix structure consists of a chain of amino acid units wound into a spiral which is held together by hydrogen bonds between a carbonyl group and the imido group of the third amino acid residue further along the chain (Fig. 22–7). Each amino acid residue is 1.5 Å from the next amino acid residue, and the helix makes a complete turn for each 3.6 residues. The helix may be coiled in a right-handed or left-handed direction, but the right-handed helix is the most stable.

Tertiary Structure. The polypeptide chains of globular proteins are more extensively folded or coiled than those of fibrous proteins. This results from the activity of several types of bonds that hold the structure in a more complex and rigid shape. These bonds are responsible for the **tertiary structure**

Figure 22-6 Linus Pauling (1901–) Professor of Chemistry, California Institute of Technology. Winner of the 1954 Nobel Prize for Chemistry and the 1962 Nobel Prize for Peace.

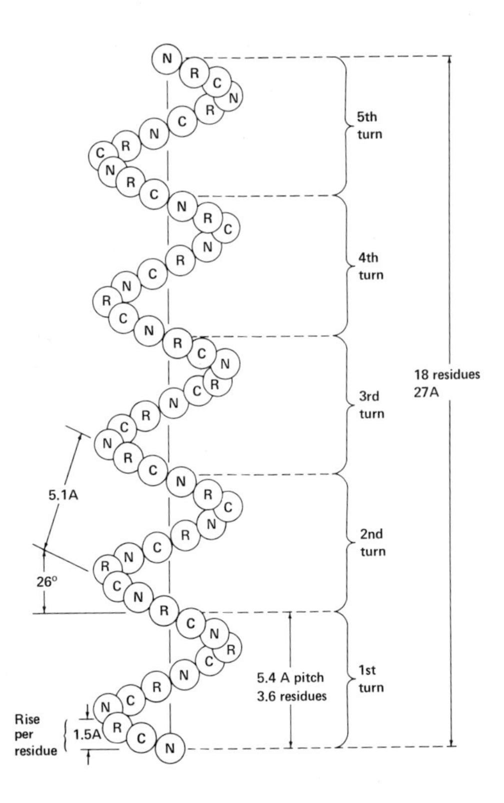

Figure 22-7 Representation of a polypeptide chain as an α-helical configuration. Here A stands for the angstrom unit. [From Pauling and Corey: *Proc. Intern. Wool Textile Research Conf.*, *B*, 249 (1955), as redrawn in Anfinsen: *The Molecular Basis of Evolution*, John Wiley and Sons, New York, 1959, p. 101.]

of proteins, and they exert stronger forces than hydrogen bonds in holding together polypeptide chains or folds of individual chains. A strong **covalent bond** is formed between two cysteine residues, resulting in the disulfide bond. **Salt linkages,** or **ionic bonds,** may be formed between the basic amino acid residues of lysine and arginine and the dicarboxylic amino acids such as aspartic and glutamic. Also, there are many examples of **hydrophobic bonding** that result from the close proximity of aromatic groups or of like aliphatic groups on amino acid residues. Examples of these bonds may be seen in Figure 22–8.

Several types of physical measurements, including sedimentation in a high intensity centrifugal field, osmotic pressure, light scattering, optical rotation, and x-ray analysis, have been used to obtain information concerning the tertiary structure of proteins.

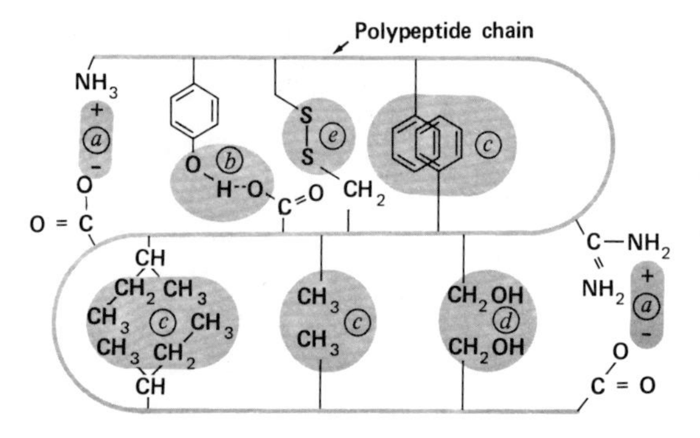

Figure 22-8 Some types of noncovalent bonds which stabilize protein structure: (*a*) Electrostatic interaction; (*b*) hydrogen bonding between tyrosine residues and carboxyl groups on side chains; (*c*) hydrophobic interaction of nonpolar side chains caused by the mutual repulsion of solvent; (*d*) dipole-dipole interaction; (*e*) disulfide linkage, a covalent bond. (Adapted from Anfinsen: *The Molecular Basis of Evolution*, John Wiley and Sons, New York, 1959, p. 102.)

Sedimentation of Proteins. Measurement of the rate of sedimentation of proteins in a centrifugal field of high intensity was made possible by the development of the **ultracentrifuge** by Svedberg in the 1920's. When a solution containing a single protein is centrifuged, the protein molecules will form a single boundary moving through the solvent away from the center of rotation. The size, shape, and especially the molecular weight of a protein determine its sedimentation rate. The **sedimentation constant, s,** obtained from these studies, is characteristic of a protein molecule.

Molecular Weight by Osmotic Pressure. The osmotic pressure of a protein solution is the pressure required to prevent the diffusion of solvent through a semipermeable membrane into a solution of the protein dissolved in the solvent. An osmometer may be constructed by placing the protein solution in a sack of semipermeable membrane, immersing the sack in the solvent, and measuring the pressure developed in the sack by the rise of the protein solution in a capillary tube. Although many factors must be carefully controlled, the molecular weight of proteins may be determined by these measurements.

X-ray Diffraction Analysis. A three-dimensional picture of the protein molecule, **myoglobin,** has been developed from x-ray diffraction data by Kendrew and his co-workers. Myoglobin has a molecular weight of 17,000 and consists of a single polypeptide chain with 153 amino acid residues. It crystallizes readily from muscle extracts of the sperm whale. When x-rays strike an atom, they are diffracted (reflected) in proportion to the number of extranuclear electrons in the atom. The heavier atoms with a high atomic number, therefore, produce more diffraction than lighter atoms. A crystal, such as a protein crystal of myoglobin, when bombarded by a beam of monochromatic x-rays, yields a photographic pattern of the electron density of the atoms in the molecule. From a large series of electron density photographs in different planes, a three-dimensional picture of myoglobin was constructed. The model of the myoglobin that resulted from these studies is shown in Figure 22–9. The structural resolution achieved by Kendrew indicated that the major portion of the polypeptide chain was in the form of the right-handed helix proposed by Pauling.

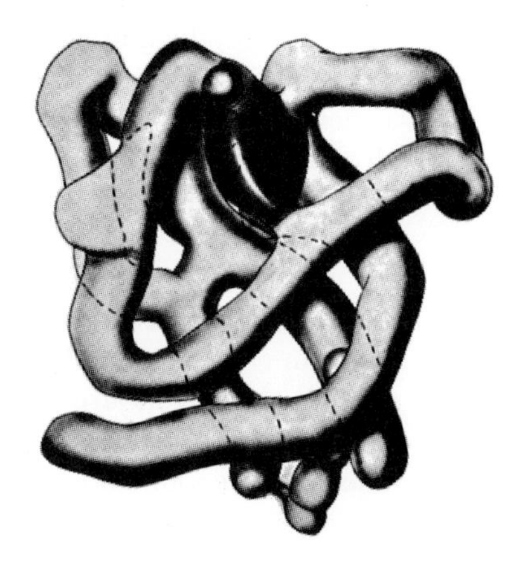

Figure 22–9 Model of the myoglobin molecule, derived from the 6 Å Fourier synthesis. The heme group is a dark gray disk (center top). (From Kendrew: Science, *139:* 1261, 1963.)

Hemoglobin, the respiratory protein of the red blood cell, has been studied in a similar fashion by Perutz and his group, and has been found to consist of four polypeptide chains (two α chains and two β chains) and four heme groups. Heme is composed of four heterocyclic pyrrole rings joined by one iron (Fe^{+2}) atom through the four nitrogen atoms. One of the nitrogen atoms of a histidine residue in the protein globin is joined to the iron in heme, as is the molecular oxygen which heme binds reversibly in the process of respiration. The three-dimensional shape of the hemoglobin α chain and β chain are similar to that of the myoglobin molecule.

QUATERNARY STRUCTURE. This level of protein structure involves the polymerization, or degree of aggregation, of protein units. The hemoglobin molecule is a good example of subunit structure in proteins. **Native hemoglobin** has a molecular weight of 68,000 in a neutral solution. If the solution is diluted, made acid or 4M with urea, the molecular weight changes to 34,000. This dissociation is due to the four polypeptide chains which exist in hemoglobin as two pairs of α and β chains. The enzyme **phosphorylase a,** which will be discussed under carbohydrate metabolism, contains four subunits which are inactive as catalysts until they are joined as a tetramer. **Insulin** is another example of a protein hormone containing subunits, and there are several proteins that are split into subunits when their disulfide bonds are converted to sulfhydryl groups.

CLASSIFICATION OF PROTEINS

Proteins are most often classified on the basis of their chemical composition or solubility properties. Of the three main types, **simple proteins** are classified by solubility properties, **conjugated proteins** by the non-protein groups, and **derived proteins** by the method of alteration.

Simple proteins such as **protamines** in the form of salmine and sturine from fish sperm, **histones** in the form of the globin in hemoglobin, and **albumins** in the form of egg albumin and serum albumin are all soluble in water, and protamines and histones contain a high proportion of basic amino acids. The **globulins** as lactoglobulin in milk are insoluble in water but soluble in dilute salt solutions; the **glutelins** as glutenin in wheat are insoluble in water and dilute salt solutions, but are soluble in dilute acid or alkaline solutions; **prolamines** as zein in corn and gliadin in wheat are soluble in 70 to 80 per cent ethyl alcohol; whereas the **albuminoids** as keratin in hair, horn, and feathers are insoluble in all the solvents mentioned above and can be dissolved only by hydrolysis.

Conjugated proteins include **nucleoproteins,** which consist of a basic protein such as histones or protamines combined with nucleic acid. They are found in cell nuclei and mitochondria. **Phosphoproteins** as casein in milk and vitellin in egg yolk are proteins linked to phosphoric acid; **glycoproteins** are composed of a protein and a carbohydrate and occur in mucin in saliva and mucoids in tendon and cartilage, whereas **chromoproteins** such as hemoglobin and cytochromes consist of a protein combined with a colored compound.

Lipoproteins are proteins combined with lipids such as fatty acids, fats, and lecithin, and are found in serum, brain, and nervous tissue.

Derived proteins are an indefinite type of protein produced, for example, by partial hydrolysis, denaturation, and heat, and are represented by proteoses, peptones, metaproteins, and coagulated proteins.

DETERMINATION OF PROTEINS

It is often desirable to know the protein content of various foods and biological material. The analysis of the protein content of such material is based on its nitrogen content. Since the average nitrogen content of proteins is 16 per cent, the protein content of a substance may be obtained by multiplying its nitrogen value by the factor $100/16 = 6.25$. For example, if a certain food contained 2 per cent nitrogen on analysis, its protein content would equal 2 times 6.25, or 12.5 per cent.

Since the determination of nitrogen requires considerable equipment and is time consuming, several colorimetric methods for the quantitative estimation of protein have been developed. Methods based on the biuret test and the ninhydrin test are commonly used when the total protein content of many specimens is required. The aromatic amino acids in proteins absorb ultraviolet light at a wave length of 280 nm. The measurement of light absorption at 280 nm is a convenient method for determining the amount of protein in a solution or in an eluate from a chromatograph column.

DENATURATION OF PROTEINS

Denaturation of a protein molecule causes changes in the structure that result in marked alterations of the physical properties of the protein. When in solution, proteins are readily denatured by standing in acids or alkalies, shaking, heating, reducing agents, detergents, organic solvents, and exposure to x-rays and light. Some of the effects of **denaturation** are loss of biological activity, decreased solubility at the isoelectric point, increased susceptibility to hydrolysis by proteolytic enzymes, and increased reactivity of groups that had been masked by the folding of chains in the native protein. Examples of the last mentioned are the uncovered SH groups of cysteine and the OH groups of tyrosine.

The cleavage of several hydrogen bonds and of several possible disulfide bonds often results in the loss of biological activity by denaturation. In some proteins, the native configuration is so stable that denaturation changes are reversible. Hemoglobin, for example, can be reversibly denatured.

PRECIPITATION OF PROTEINS

One of the most important characteristics of proteins is the ease with which they are precipitated by certain reagents. Many of the normal functions in the body are essentially precipitation reactions; for example, the clotting of blood or the precipitation of casein by rennin during digestion. Since animal tissues are chiefly protein in nature, reagents that precipitate protein will have a marked toxic effect if introduced into the body. Bacteria, which are mainly protein, are effectively destroyed when treated with suitable precipitants. Many of the common poisons and disinfectants act in this way. The following paragraphs contain a brief summary of the most common methods of protein precipitation.

By Heat Coagulation. When most protein solutions are heated, the protein becomes insoluble and precipitates, forming coagulated protein. Many protein foods coagulate when they are cooked. Tissue proteins and bacterial proteins are readily coagulated by heat. Routine examinations of urine specimens for protein are made by heating the urine in a test tube to coagulate any protein that might be present.

By Alcohol. Alcohol coagulates all proteins except the prolamines. A 70 per cent solution of alcohol is commonly used to sterilize the skin, since it effectively penetrates the bacteria. A 95 per cent solution of alcohol is not effective because it merely coagulates the surface of the bacteria and does not destroy them.

By Inorganic Acids. Proteins are precipitated from their solutions by acids such as hydrochloric, sulfuric, and nitric acid. Casein, for example, is precipitated from milk as a curd when acted on by the hydrochloric acid of the gastric juice.

By Salts of Heavy Metals. Salts of heavy metals, such as mercuric chloride and silver nitrate, precipitate proteins. Since proteins behave as zwitterions, they will ionize as negative charges in neutral or alkaline solutions. The reaction with silver ions may be illustrated as follows:

$$\underset{\text{Protein}}{\text{R—}\underset{\displaystyle NH_2}{\underset{|}{\text{CH}}}\text{—COO}^-} + Ag^+ \rightarrow \underset{\text{Silver proteinate}}{\text{R—}\underset{\displaystyle NH_2}{\underset{|}{\text{CH}}}\text{—COOAg}}$$

These salts are used for their disinfecting action and are toxic when taken internally. A protein solution such as egg white or milk, when given as an antidote in cases of poisoning with heavy metals, combines with the metallic salts. The precipitate that is formed must be removed by the use of an emetic before the protein is digested and the heavy metal is set free to act on the tissue protein. A silver salt such as Argyrol is used in nose and throat infections, and silver nitrate is used to cauterize wounds and to prevent gonorrheal infection in the eyes of newborn babies.

By Alkaloidal Reagents. Tannic, picric, and tungstic acids are common alkaloidal reagents that will precipitate proteins from solution. When in acid solution the protein as a zwitterion ionizes as a positive charge. It will therefore react with picric acid as shown:

$$\underset{\text{Protein}}{\underset{\displaystyle NH_3^+}{\text{R—}\overset{}{\text{CH}}\text{—COOH}}} + \text{picric acid} \rightarrow \underset{\text{Protein picrate}}{\underset{\displaystyle NH_3\text{—picrate}}{\text{R—CH—COOH}}}$$

Tannic and picric acids are sometimes used in the treatment of burns. When a solution of either of these acids is sprayed on extensively burned areas, it precipitates the protein to form a protective coating; this excludes air from the burn and prevents the loss of water. In an emergency, strong tea may be used as a source of tannic acid for the treatment of severe burns. Many other therapeutic agents have been used in the treatment of burns, the most recent being penicillin. Nevertheless, considerable quantities of tannic and picric acid preparations are still employed for this purpose.

By Salting Out. Most proteins are insoluble in a saturated solution of a salt such as ammonium sulfate. When it is desirable to isolate a protein from a solution without appreciably altering its chemical nature or properties, the protein may be precipitated by saturating the solution with $(NH_4)_2SO_4$. After filtration, the excess $(NH_4)_2SO_4$ is usually removed by dialysis. This salting out process finds wide application in the isolation of biologically active proteins.

Questions

1. List the five main elements present in proteins with their average content. Using this list, how would you differentiate proteins from carbohydrates and fats?

2. Name three important agents for the hydrolysis of proteins. What products would be formed on hydrolysis?

3. Explain the basis of the name "amino acid" from the standpoint of organic chemistry.

4. Write the formula for L-threonine. Explain why this is an "alpha" amino acid.

5. Illustrate with equations two reactions that involve the amino group of an amino acid.

6. What advantage would the use of ion exchange resin columns have over paper chromatography in the separation of amino acid mixtures? Explain.

7. Briefly explain the application of chromatography and Sanger's reagent to the analysis of the insulin molecule.

8. How does the primary structure differ from the tertiary structure of a protein? Explain.

9. Describe two methods that can be used to determine the amount of protein in a solution.

10. Why is a protein solution used as an antidote in cases of poisoning with heavy metals? Explain.

11. Why are preparations containing tannic or picric acids sometimes used in the treatment of burns?

12. Illustrate with equations the precipitation of proteins with tannic acid and with silver salts.

Suggested Reading

Dewhurst: Student Experiments on the Gel Filtration of Proteins. Journal of Chemical Education, Vol. 46, p. 864, 1969.
Doty: Proteins. Scientific American, Vol. 197, No. 3, p. 173, 1957.
Elmore: Peptides and Proteins. Cambridge University Press, New York, 1968.
Kendrew: Myoglobin and the Structure of Proteins. Science, Vol. 139, p. 1259, 1963.
Li: The ACTH Molecule. Scientific American, Vol. 209, No. 1, p. 46, 1963.
Merrifield: The Automatic Synthesis of Proteins. Scientific American, Vol. 218, No. 3, p. 56, 1968.
Perutz: The Hemoglobin Molecule. Scientific American, Vol. 211, No. 5, p. 64, 1964.
Sande: Conformation of Peptides. Speculations Based on Molecular Models. Journal of Chemical Education, Vol. 45, p. 587, 1968.
Stein and Moore: The Structure of Proteins. Scientific American, Vol. 204, No. 2, p. 81, 1961.

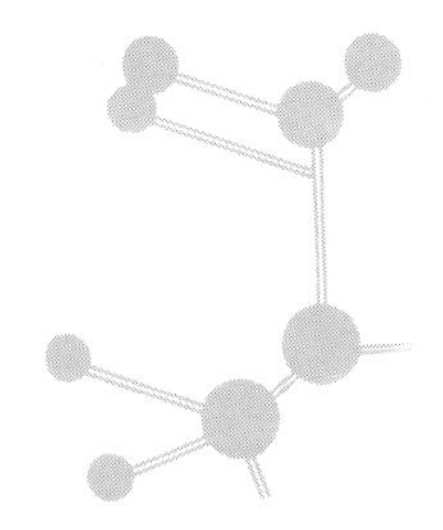

CHAPTER 23

NUCLEIC ACIDS

Nucleic acids were first isolated from cell nuclei and were named after their source. They were thought to be fairly simple groups that were conjugated with proteins to form nucleoproteins. These proteins are characterized by their content of **basic amino acids** such as **arginine** and **lysine.** We know now that nucleic acids are polymers of large molecular weight with nucleotides as the repeating unit. They are present in the nucleus and cytoplasm of all living cells and are responsible for the transmission of genetic information and the synthesis of proteins by the cell. The progressive hydrolysis of a nucleoprotein may be represented as follows:

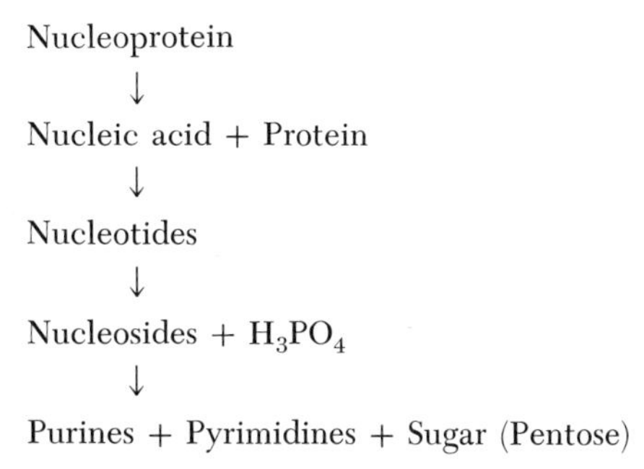

The components of nucleic acids are thus **phosphoric acid,** a **sugar,** and a **mixture of purines and pyrimidines.** To obtain an understanding of the structure of the complex molecules we shall first consider the separate hydrolysis products.

THE PENTOSE SUGARS

For many years the two major sources of nucleic acids were yeast and thymus tissue. On hydrolysis the yeast nucleic acid yielded β-D-ribose, whereas thymus nucleic acid contained β-2-deoxy-D-ribose. Both pentose sugars occur in the furanose form.

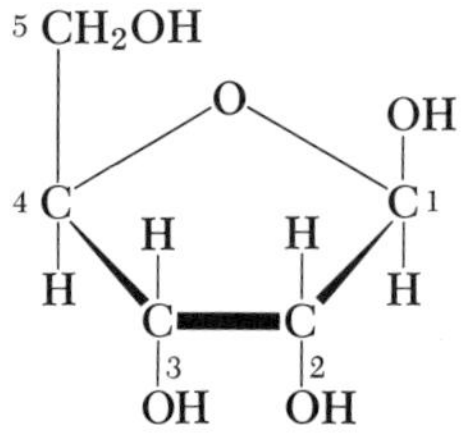

β-D-Ribose 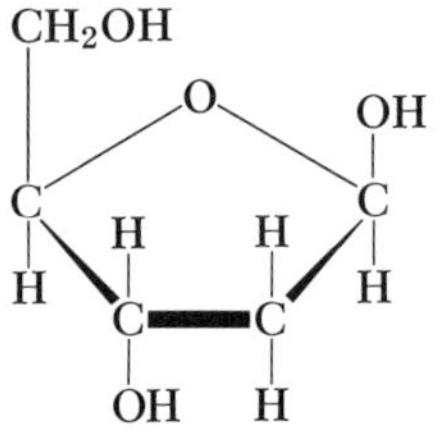 β-2-deoxy-D-Ribose

THE PYRIMIDINE AND PURINE BASES

In Chapter 18 we discussed the heterocyclic rings, including the pyrimidine nucleus. The pyrimidine bases found in nucleic acids include cytosine, uracil, and thymine, which are represented as follows:

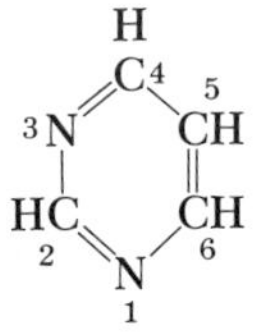

Pyrimidine	Cytosine	Uracil	Thymine
	(2-oxy-4-amino pyrimidine)	(2,4-dioxy pyrimidine)	(5-methyl uracil)

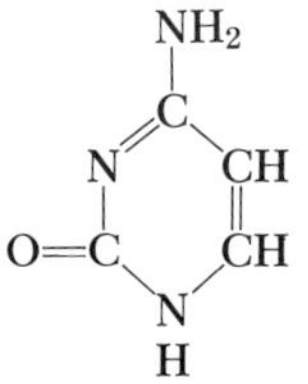

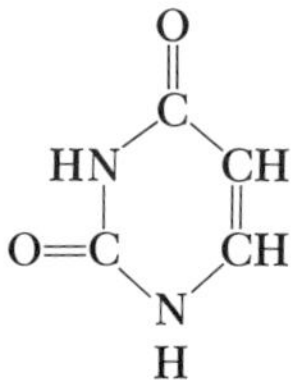

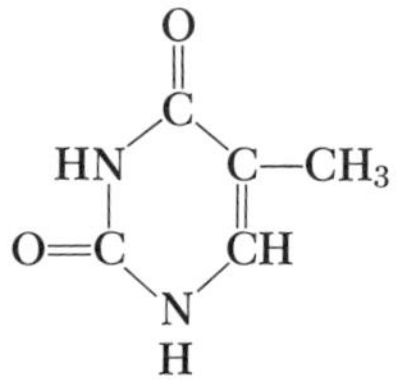

The purine bases found in both yeast and thymus nucleic acids are adenine and guanine.

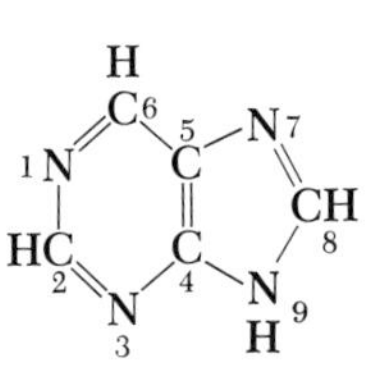

Purine	Adenine	Guanine
	(6-amino purine)	(2-amino-6-oxy purine)

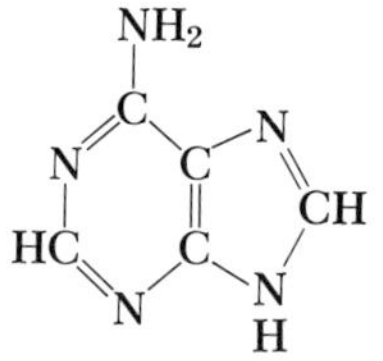

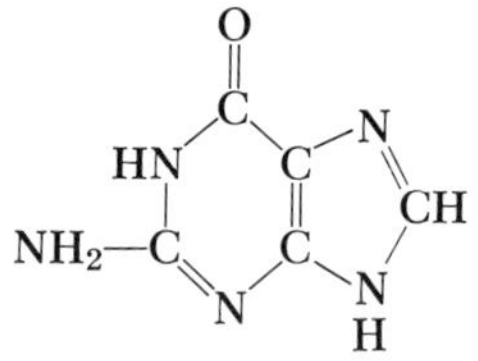

NUCLEOSIDES

When a purine or pyrimidine base is combined with β-D-ribose or β-2-deoxy-D-ribose, the resultant molecule is called a **nucleoside.** The linkage of the two components is from the nitrogen of the base (position 1 in pyrimidines, 9 in purines) to carbon 1 of the pentose sugars. Important examples of nucleosides are cytidine and adenosine:

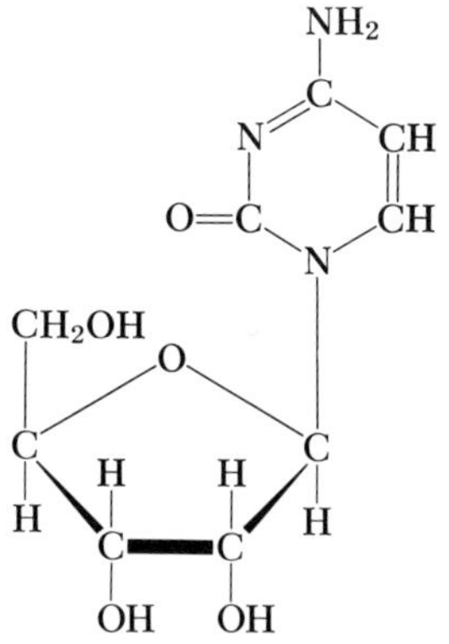

Cytidine

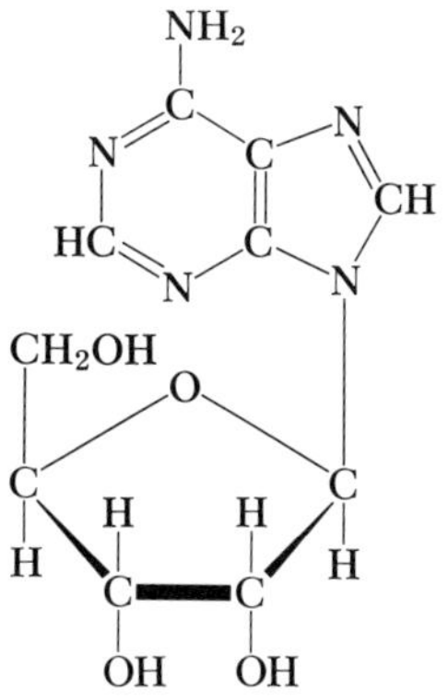

Adenosine

NUCLEOTIDES

When a phosphoric acid is attached to a hydroxyl group of the pentose sugar in the nucleoside by an ester linkage, the result is a **nucleotide.** The esterification may occur on the 2′, 3′, or 5′ hydroxyl of ribose and the 3′ or 5′ hydroxyl of deoxyribose. In the nucleotides, the carbon atoms of ribose or deoxyribose are designated by prime numbers to distinguish them from the atoms in the purine or pyrimidine bases. Yeast nucleic acid contains four mononucleotides: adenylic acid, guanylic acid, cytidylic acid, and uridylic acid. All these compounds include D-ribose in their structure and are named as acids because of the ionizable hydrogens of the phosphate group. Thymus nucleic acid yields nucleotides on hydrolysis that contain β-2-deoxy-D-ribose and thymidylic acid instead of uridylic acid. These structures may be represented by adenylic acid and uridylic acid, two compounds of prime importance in muscle metabolism and carbohydrate metabolism:

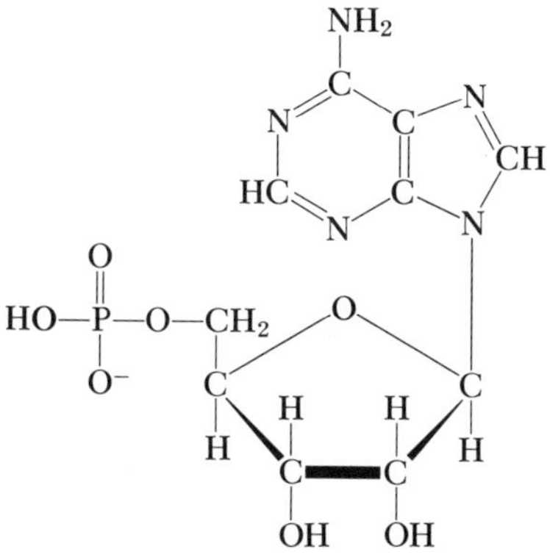

Adenylic acid (AA)
(adenosine-5′-monophosphate, AMP)

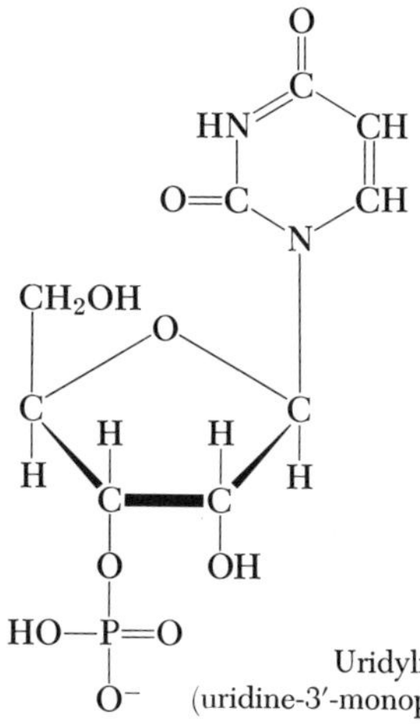

Uridylic acid
(uridine-3′-monophosphate, UMP)

In addition to the nucleotides that are integral components of yeast and thymus nucleic acids, several nucleotides and their derivatives occur free in the tissues and are essential constituents of tissue metabolism. For example, adenylic acid (AA), also designated adenosine monophosphate (AMP), is found in muscle tissue. Adenylic acid may also exist as the diphosphate (adenosine diphosphate, ADP) or as the triphosphate (adenosine triphosphate, ATP). These two compounds are sources of high energy phosphate bonds and are involved in many metabolic reactions.

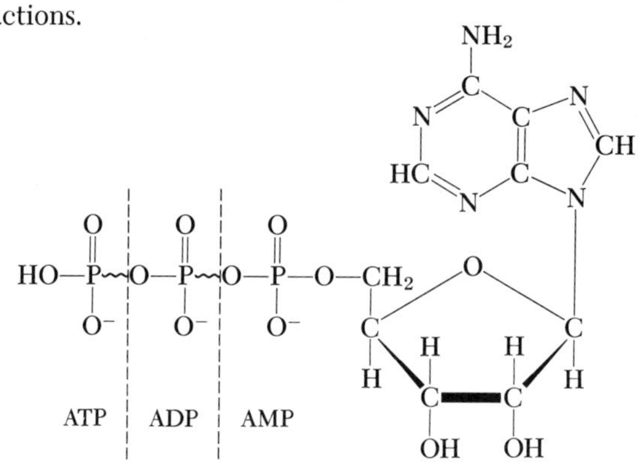

ATP | ADP | AMP

Adenosine triphosphate

Nucleotides also combine with vitamins to form coenzymes which will be discussed in Chapter 25.

NUCLEIC ACIDS

The nucleic acids of yeast and thymus tissue have already been considered. It may be recalled that β-D-ribose was found in yeast and β-2-deoxy-D-ribose in thymus nucleic acids. Since there are other sources of nucleic acids, they have been generally classified into **ribose nucleic acid, RNA,** and **deoxyribose nucleic acid, DNA.** By complete hydrolysis the two types have been found to contain the following components:

RNA	DNA
Adenine	Adenine
Guanine	Guanine
Cytosine	Cytosine
Uracil	Thymine
D-ribose	D-2-deoxyribose
Phosphoric Acid	Phosphoric Acid

THE STRUCTURE OF DNA. The structure of DNA has been shown to consist of chains of nucleotides linked together with phosphate groups that connect carbon atom 3 of one sugar molecule to the number 5 carbon of the next sugar. The 3,5 linkage from one deoxyribose molecule to another is characteristic in DNA and is the predominant linkage in RNA. The molecular weights of DNA preparations from various sources exhibit a range from 6 to 16 million and illustrate the large size of the polymer. It is more difficult to determine the sequences of nucleotides in a DNA molecule than the amino acids in a protein because there are only four common nucleotide units. A series of hydrolytic experiments utilizing weak acids, bases, and enzymes has established a 1:1 ratio

Figure 23-1 *a* James Watson (1928–) Professor of Biology, Harvard University. Shared the Nobel Prize for Medicine and Physiology in 1962 with Crick and Wilkins. Double-helical structure of DNA and role of RNA in protein synthesis. *b* Francis Crick (1916–) member British Medical Research Council, Laboratory on Molecular Biology. Shared the Nobel Prize for Medicine and Physiology in 1962 with Watson and Wilkins. Proposed model for double-helical structure of DNA.

between cytosine and guanine nucleotides and a similar 1:1 ratio between adenine and thymine nucleotides.

By physical-chemical methods, such as x-ray diffraction, the DNA molecule has been shown to consist of two chains of nucleotides coiled in a double helical structure, forming a molecule that is very long by contrast to its diameter. Watson and Crick (Fig. 23–1) suggested that the pair of nucleotide chains run in opposite directions around the helix so that an adenine of one chain is bonded to a thymine of the other by hydrogen bonds, and a guanine of one chain is joined by hydrogen bonds to a cytosine of the other as shown in Figure 23–2. Thymine of the first chain is connected by two hydrogen bonds to adenine of the second, and cytosine of the first chain is joined to guanine of the second chain by three hydrogen bonds. The chains consist of deoxyribose nucleotides joined together by phosphate groups with the bases projecting perpendicularly from the chain into the center of the helix. For every adenine or guanine projecting into the central portion, a thymine or cytosine molecule must project toward it from a second chain. Because of the base pairing, the two chains are not identical, and they do not run in the same direction with respect to the linkages between deoxyribose and the base. The chains in the DNA structure are therefore considered **antiparallel.** A portion of the helix emphasizing the hydrogen bonding and the antiparallel chain structure is shown in Figure 23–3.

The Structure of RNA. The RNA molecules in a cell are found mainly in the cytoplasm and about 60 per cent of the RNA is found in the ribosomes. Of the three common types of RNA in a cell, *ribosomal RNA* (r-RNA) has the largest molecular size. It is an integral part of the ribosomes or polysomes and is involved in the synthesis of proteins in the cytoplasm. *Messenger RNA*

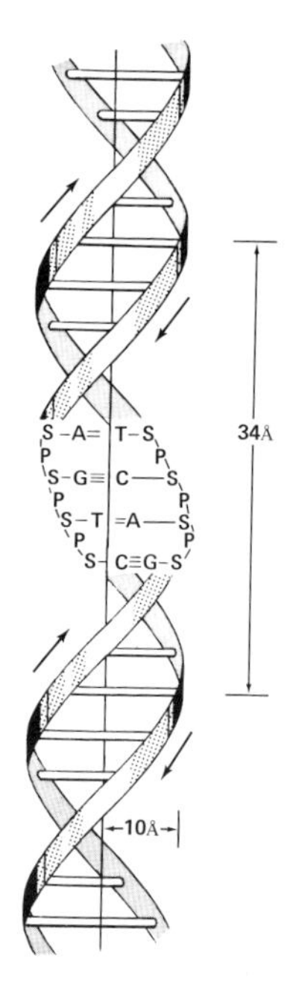

Figure 23-2 Double helix of DNA. Here P means phosphate diester, S means deoxyribose, A=T is the adenine-thymine pairing, and G≡C is the guanosine-cytosine pairing. (Adapted from Conn and Stumpf: *Outlines of Biochemistry*, 2nd Ed. New York, Wiley, 1963, p. 108.)

(m-RNA) is a second type of RNA that forms a chain or strand of molecules that join several ribosomes in a template for the synthesis of a specific protein. The m-RNA molecules are complementary to a segment of the DNA in the nucleus and thus carry the genetic message from the DNA to the protein synthesizing sites. The smallest RNA molecules are called *transfer RNA* (t-RNA) and have molecular weights from 25,000 to 40,000. Their function is to bind

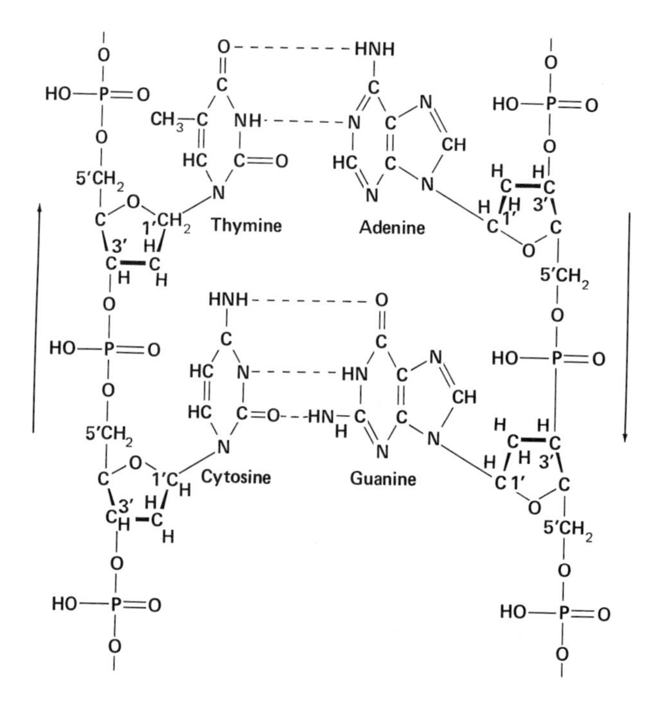

Figure 23-3 Hydrogen bonding with antiparallel chains. (Adapted from Conn and Stumpf: *Outlines of Biochemistry*, 2nd Ed. New York, Wiley, 1963.)

specific amino acids and transport them to a specific site on the messenger RNA which incorporates them in the growing polypeptide chain.

There is very little information available concerning the exact structure of r-RNA or m-RNA, although they are apparently single-stranded molecules. The structures for some of the smaller t-RNA molecules have been proposed. They are thought to be composed of single strands of nucleotides bent into cloverleaf type structures to give the maximum number of hydrogen bonded pairs. The closed loop of the chain at one end of the cloverleaf contains a sequence of three bases that serve as an anticodon for a specific amino acid (see Chap. 28). The role of these RNA molecules in protein synthesis will be discussed in greater detail in Chapter 28.

THE BIOLOGICAL IMPORTANCE OF THE NUCLEIC ACIDS

Originally RNA was associated only with yeast and was thought to be restricted to plant sources. DNA from thymus tissue represented the nucleic acids of animal tissues. As methods for their determination have been developed, both RNA and DNA have been found in practically all types of cells. DNA appears to be restricted to the nucleus, most specifically to the chromosomes, whereas RNA occurs both in the cytoplasm and nucleus of a cell.

From the standpoint of genetics it is highly significant that the amount of DNA in each cell nucleus is constant for a given species. It has also been found that the number of genes in a cell nucleus is exactly equal to the number of DNA molecules. Only one-half the DNA in somatic cell nuclei is found in the sperm cell nuclei. Since the sperm nuclei contain only one set of chromosomes, as opposed to the two sets in somatic cell nuclei, it is apparent that the DNA content of the cell nucleus is closely related to number of chromosome sets present. In cells containing more than two sets of chromosomes, there is a corresponding increase in the amount of DNA. From all the available evidence it appears that DNA is the integral part of chromosomes primarily concerned with the genetic process, and that it carries the code of genetic information for metabolic function and control within each cell. The overall function of DNA and its relation to RNA and protein synthesis are outlined in Figure 23–4.

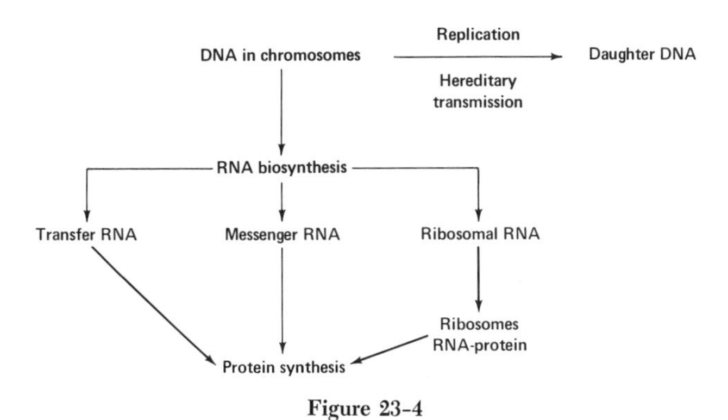

Figure 23–4

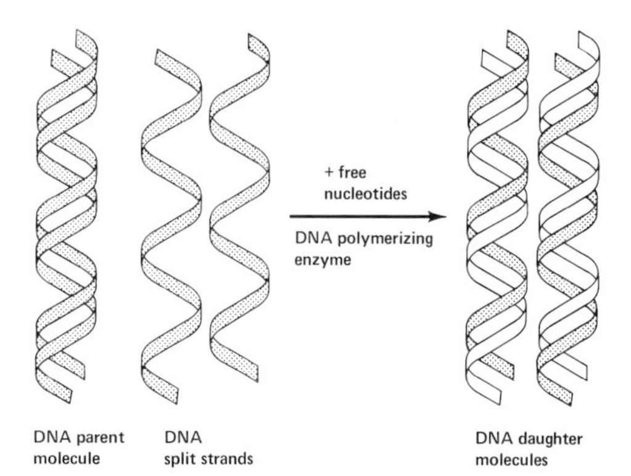

Figure 23-5 Replication of DNA chains as suggested by Watson and Crick. White strands are newly synthesized DNA.

The method of DNA replication was proposed by Watson and Crick based on their model of DNA. They suggested that the paired molecules of a DNA helix first separate from each other, as shown in Figure 23-5. Deoxyribonucleotides that were in the uncombined form in the cell then become attached by hydrogen bonding to the nucleotide bases of the separated strands of DNA. The base pairing follows the patterns of DNA with adenine and thymine, and also guanine and cytosine deoxyribonucleotides being joined by hydrogen bonds. The nucleotide chains are then linked together by a polymerizing catalyst or enzyme, and a new DNA helix is formed that is identical to the parent molecule.

In reduplication of cell nuclei, which is necessary in cell division, the double helix may unravel and each of the original chains may serve as a template for the synthesis of another chain. From our discussion of the structure of DNA it will be recalled that adenine can combine only with thymine, and guanine only with cytosine. Experimental evidence favors this hypothesis. In the laboratory it has been demonstrated that pure DNA preparations from a particular species of bacteria or bacteriophage, when added to another species of bacteria, will serve as a template to direct the recipient cells to develop the characteristics of the donor bacteria. It is quite possible that the multiplication of viruses within cells may occur by the same process. For example, type 1 poliomyelitis virus, which has been crystallized, contains an RNA that serves as a template to infect cells with this type of virus. Present knowledge suggests that RNA functions mainly in the cytoplasm of a cell as a template for the synthesis of specific cellular proteins. A close relation exists between DNA of the nucleus and RNA of the cytoplasm, since one chain of DNA and one chain of RNA could twist around each other to form a double helix and thus influence the RNA template.

Questions

1. Write formulas to illustrate each of the four types of components obtained by complete hydrolysis of a nucleic acid.

2. Name and write the formula for a typical nucleoside.

3. What is the difference between ATP and ADP; ADP and AMP? Illustrate these differences with a composite formula.

4. What are the two major types of nucleic acids? List the composition of each type.

5. How are the nucleotides linked together chemically when they form nucleic acids?

6. Prepare a sketch of the DNA molecule showing the double helix and the hydrogen bonding between the bases.

7. What is meant by the antiparallel chain structure of DNA?

8. Briefly describe the function and importance of DNA in the cell.

9. Illustrate the process of replication of DNA.

10. Briefly describe the three major types of RNA molecules in a cell.

11. How does the structure of RNA differ from that of DNA? Explain.

Suggested Reading

Crick: The Genetic Code. Scientific American, Vol. 215, No. 4, p. 55, 1966.
Deamer: ATP Synthesis. The Current Controversy. Journal of Chemical Education, Vol. 46, p. 198, 1969.
Fraenkel-Conrat and Stanley: The Chemistry of Life 2. Implications of Recent Studies of a Simple Virus. Chemical & Engineering News, May 15, 1961, p. 136.
Hanawalt and Haynes: The Repair of DNA. Scientific American, Vol. 216, No. 2, p. 36, 1967.
Holley: The Nucleotide Sequence of a Nucleic Acid. Scientific American, Vol. 214, No. 2, p. 30, 1966.
Kornberg: The Synthesis of DNA. Scientific American, Vol. 219, No. 4, p. 64, 1968.
Luria: The Recognition of DNA in Bacteria. Scientific American, Vol. 222, No. 1, p. 88, 1970.
Mirsky: The Discovery of DNA. Scientific American, Vol. 218, No. 6, p. 78, 1968.
Tomasz: Cellular Factors in Genetic Transformation. Scientific American, Vol. 220, No. 1, p. 38, 1969.
Watson: Double Helix. Atheneum, New York, 1968.
Yanofsky: Gene Structure and Protein Structure. Scientific American, Vol. 216, No. 5, p. 80, 1967.

CHAPTER 24

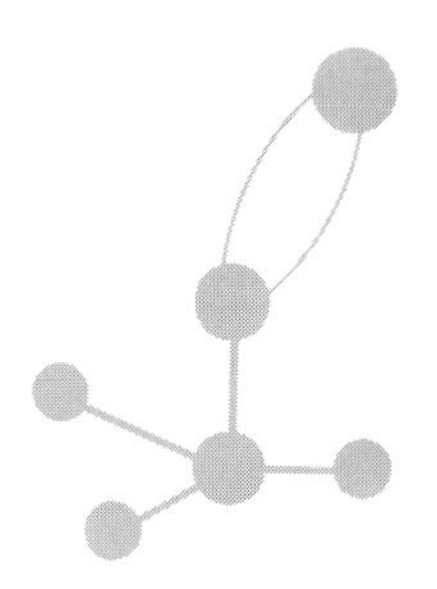

BIOCHEMISTRY OF THE CELL AND HIGH ENERGY COMPOUNDS

Within a span of 25 years biochemistry has progressed from nutritional studies on the whole animal, to tissue slices of important organs, to tissue homogenates, and currently to a study of the cell. Considerable research is directed toward an understanding of the chemical reactions that occur in the cell and their relation to cellular function and structure. The use of labeled compounds and cytochemical techniques have assisted the biochemist in the location of the cellular site for specific reactions, especially those involving enzymes. The **light microscope** has been invaluable in the study of staining reactions and the rough morphology of the cell. With the advent of the **electron microscope** the fine structure of the cell was revealed, and a whole new area of biochemical research was made available. A more complete understanding of the relation of structure to function is now within the grasp of biochemists.

Cells differ in size, appearance, and structure, depending on their function, but a typical animal cell has the features illustrated in Figure 24–1. Structures common to all cells include the cell membrane, nucleus, mitochondria, endoplasmic reticulum, ribosomes, and Golgi apparatus. A typical plant cell, Figure 24–2, also includes the same structures. Fortunately for cytological studies, the electron microscope can be focused on each subcellular component to reveal its structural details.

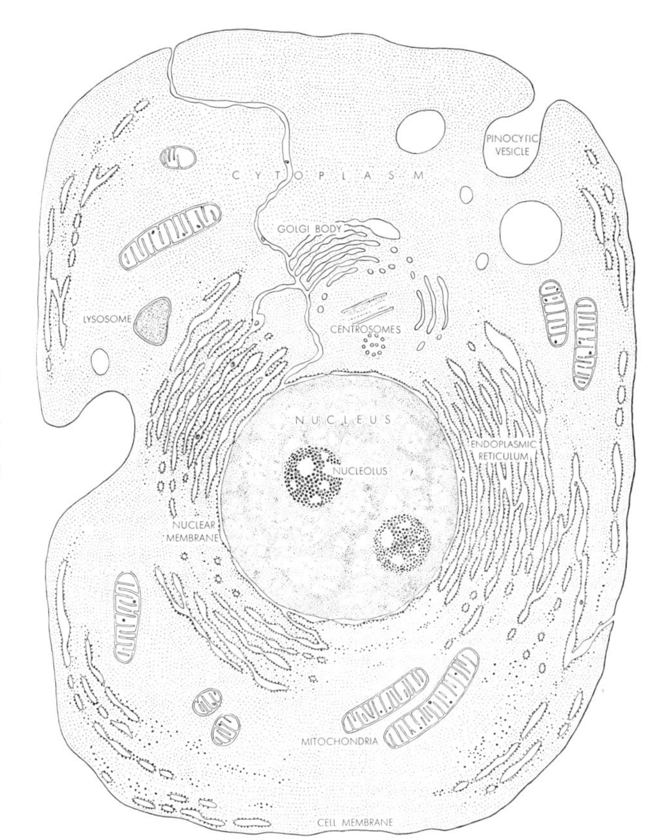

Figure 24-1 A diagram of a typical cell based on an electron micrograph. (From the Living Cell, by Jean Brachet. Copyright © 1961 by Scientific American, Inc. All rights reserved.)

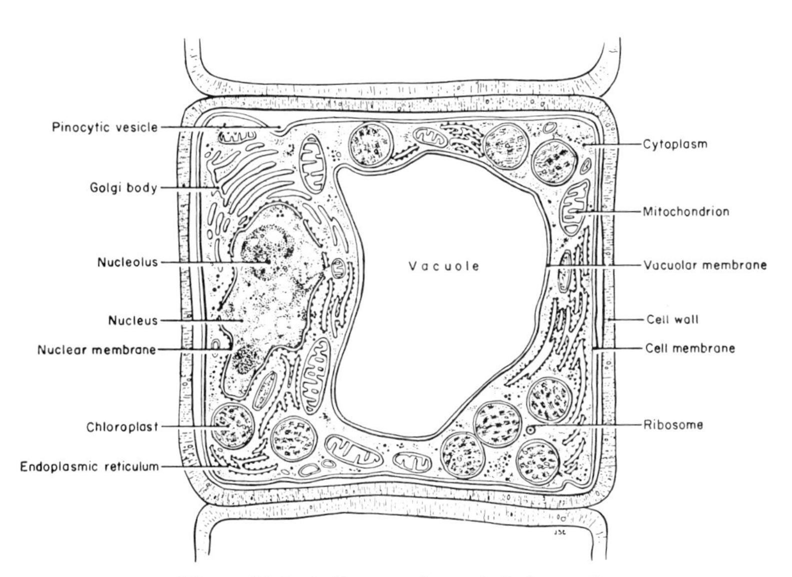

Figure 24-2 A diagram of a typical plant cell.

SUBCELLULAR COMPONENTS

Cell Membrane. All of the subcellular components of a cell are contained within a definite cell wall or membrane. This membrane plays a vital role in the passage of nutrient and waste material into and out of the cell. In addition to the cell membrane, plant cells have rigid walls that surround and protect the membrane. The cell walls consist of cellulose and other polysaccharides. The **cell membrane** is composed of lipids and protein arranged in such a fashion that water-soluble and lipid-soluble substances can pass through the membrane. The permeability of living membranes has never been adequately explained. Although many cells are bathed in a fluid rich in sodium and chloride ions and low in potassium ions, the cell contents are rich in potassium ions and low in sodium and chloride ions. The different rates of absorption of monosaccharides and amino acids from the small intestine emphasize the importance of the membrane in selective permeability toward small ions and molecules.

Nucleus. The nucleus is roughly spherical in shape and is surrounded by a double layered membrane that is more porous than the cell membrane. In many cells the outer membrane is connected with the nuclear membrane by one or more channels through the cytoplasm. In addition there is usually a connection between the endoplasmic reticulum and the double-layered nuclear membrane. It has long been recognized that the **nucleus** serves as a site for the transmission and regulation of hereditary characteristics of the cell. This control is an essential feature of the **chromosomes** that are composed of **deoxyribose nucleic acid** (DNA), and basic protein. The nucleus also contains one or more small, dense, round bodies called nucleoli. These bodies contain DNA and **ribose nucleic acid** (RNA), and appear to be involved in the synthesis of RNA and proteins.

Mitochondria. These subcellular particles are shaped like an elongated oval 2 to 7μ in length and 1 to 3μ in diameter. The walls of the mitochondria are double layered membranes with projections that extend inward toward the center of the particle, increasing the surface of the membrane inside the mitochondrion. These projections inside the mitochondria are called **crista.** The liquid in the matrix of the particle contains protein, neutral fat, phospholipids, and nucleic acids. In contrast to the nucleus, the nucleic acids in the mitochondria are mostly RNA with only small amounts of DNA. The **mitochondria** have been called the *powerhouses of the cell,* since they are the site of major oxidative processes and oxidative phosphorylation which result in the formation of ATP.

Chloroplasts. Plant cells contain highly pigmented particles 3 to 6μ in diameter called **chloroplasts** (Figure 24–2). These particles contain the green pigment chlorophyll and play a major role in photosynthesis. The structure and function of chloroplasts parallel those of mitochondria.

Endoplasmic Reticulum and Microsomes. The **endoplasmic reticulum** is composed of a network of interconnected, thin, membrane-like tubules or vesicles. In some areas of the cytoplasm the membranes are covered with dark round bodies about 0.015μ in diameter called ribosomes. These areas are known as rough endoplasmic reticulum contrasted to the smooth reticulum which does not have ribosomes adsorbed on its surface. The endoplasmic reticulum and accompanying ribosomes are also called microsomes or the microsomal region of a cell. The specific particles, the **ribosomes,** are the site for the synthesis of proteins within the cell.

THE GOLGI APPARATUS. This is also called the Golgi body or complex and consists of an orderly array of flattened sacs with smooth membranes associated with small vacuoles of varying size. The **Golgi apparatus** is often connected to the cell membrane by a channel and serves as a way station in the transport of substances produced in other subcellular particles. In liver cells, for example, the Golgi apparatus is usually located close to the small bile canals and is involved in the transport and excretion of substances such as bilirubin glucuronide into the bile.

LYSOSOMES. These particles are spherical in shape with an average diameter of 0.4μ. They contain several soluble hydrolytic enzymes (hydrolases) that exhibit an optimum pH in the acid range. The lysosomal membrane is lipoprotein in nature and prevents the enzymes from escaping into the cellular cytoplasm. The membrane also prevents the substrates for the enzymes from entering the cell. When the cell is injured and the membrane is broken, the released enzymes cause cellular breakdown. In autolysis of tissue, whether normal (as involution of the thymus gland at puberty), pathological, or postmortem, the lysosomal enzymes destroy cellular tissue. In fact, one of the main functions suggested for these particles is to help clear tissues of dead cells. The processes of phagocytosis and pinocytosis involve the engulfment of foreign material into vesicles or vacuoles and the digestion of this material. These particles may be converted into lysosomes to assist in the hydrolysis of phagocytosed material.

VACUOLES AND VESICLES. These particles are roughly spherical in shape and vary in size from 0.1 to 0.7μ in diameter. They are often found close to the Golgi apparatus and to channels involved in the entrance and excretion of material to and from the cell. Vacuoles may serve as temporary storage sacs, or as bodies involved in the removal of foreign material from the cell.

CYTOPLASM. The **cytoplasm** is the general protoplasmic mass in which the definite subcellular components described above are embedded. At present all of the essential compounds and macromolecules in the cell not associated with definite particles are thought to exist in the cytoplasm. Many soluble enzymes are found in the cytoplasm, particularly those associated with the conversion of glucose to pyruvic or lactic acids. Considerable research remains to be done on the components of the cell and the cytoplasm with respect to enzyme and coenzyme distribution and their role in various metabolic reactions.

TABLE 24-1 BIOCHEMICAL FUNCTIONS OF CELLULAR COMPONENTS

Cellular Components	*Biochemical Composition*	*Biochemical Function*
Cell membrane	Lipoprotein	"Active" transport, ion transport
Nuclei	DNA, basic proteins	DNA, RNA regulation of metabolism
Mitochondria	Lipoprotein, phospholipid, RNA, flavoprotein	Krebs cycle, oxidative phosphorylation, fatty acid oxidation
Endoplasmic reticulum, ribosomes	RNA, protein	Protein synthesis
Lysosomes	Enzymes, protein	Hydrolysis of foreign matter
Cytoplasm	Protein, enzymes	Glycolysis
Golgi apparatus	RNA, protein	Intracellular transfer and cellular excretion

BIOCHEMICAL FUNCTION OF CELLULAR COMPONENTS

Many investigators have concentrated their research activities on the biochemical reactions that occur in a specific subcellular particle. The mitochondria and ribosomes particularly have been the subject of several research studies. It is obviously not possible at present to reconstruct the exact biochemical functions of the intact cell, but a combination of cytochemical techniques and research on reactions within the separated particles provide a greater understanding of the overall process. Some of the biochemical functions that have been associated with cellular components are listed in Table 24-1.

HIGH-ENERGY CELLULAR COMPOUNDS

To understand more fully the reactions catalyzed by enzymes in the metabolic processes of the body, we must consider the energy relationships that are involved. The energy released from one reaction within a cell may be used almost simultaneously in another reaction that is essential in cellular economy. Energy produced in the cell may also be used as heat, for mechanical work as in muscular contraction, or as an electric impulse in nerve transmission. Many of the reactions in metabolism produce chemical energy which is stored in **high-energy compounds.** These high-energy compounds are used to drive essential reactions in the metabolic cycles of carbohydrate, lipid, and protein metabolism.

Early investigations on the nature of muscular contraction revealed that the presence of the high-energy compound creatine phosphate has a driving force in muscle reactions. Studies on the oxidation of glucose and especially the metabolic cycles of carbohydrate oxidation emphasized the role of **adenosine triphosphate, ATP,** and this energy-rich compound has become the key in linking endergonic processes to those that are exergonic.

High-energy compounds are often complex phosphate esters that yield large amounts of free energy on hydrolysis. A more detailed consideration of the energy released by the stepwise hydrolysis of ATP will illustrate the high energy concept.

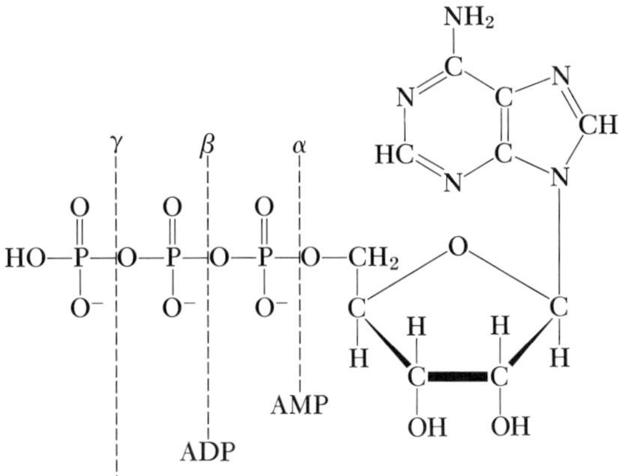

γ β α

HO—P(=O)(O⁻)—O—P(=O)(O⁻)—O—P(=O)(O⁻)—O—CH_2—(ribose)—adenine

AMP

ADP

Adenosine triphosphate, ATP

$$\text{ATP} \xrightarrow{\text{hydrolysis}} \text{ADP} + H_3PO_4 \qquad -8000 \text{ cal}$$

$$\text{ADP} \xrightarrow{\text{hydrolysis}} \text{AMP} + H_3PO_4 \qquad -6500 \text{ cal}$$

$$\text{AMP} \xrightarrow{\text{hydrolysis}} \text{Adenosine} + H_3PO_4 \qquad -2200 \text{ cal}$$

Several explanations have been proposed for the release of energy on the hydrolysis of high-energy compounds. These include the fact that these compounds are unstable in acid and alkaline solutions and are readily hydrolyzed. A major reason for the release of energy involves the *type of bond structure* in these compounds. The β and γ bonds in ATP are anhydride linkages that involve a large amount of repulsion energy between the phosphates, which is released on hydrolysis.

Other phosphorus-containing, high-energy compounds include:

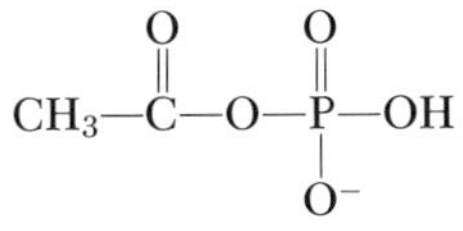

CH_3—C(=O)—O—P(=O)(O⁻)—OH

Acetyl phosphate
−10,000 cal

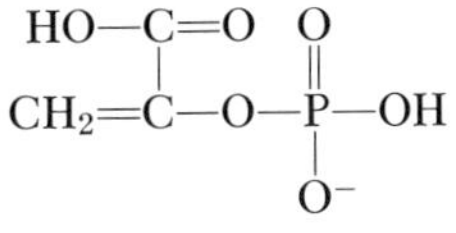

HO—C=O
CH_2=C—O—P(=O)(O⁻)—OH

Phosphoenolpyruvic acid
−12,000 cal

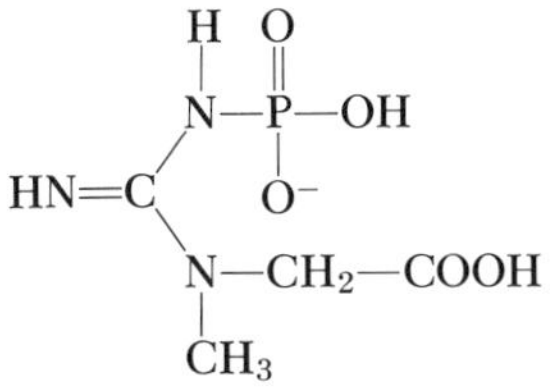

HN=C(—NH—P(=O)(O⁻)—OH)(—N(CH_3)—CH_2—COOH)

Creatine phosphate
−10,000 cal

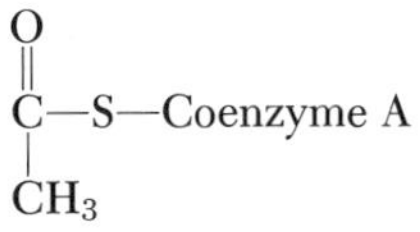

CH_3—C(=O)—S—Coenzyme A

Acetyl coenzyme A
−8200 cal

The top two compounds in the previous illustrations have anhydride linkages between a phosphate and either a carbonyl or acid enol group. Creatine phosphate, the high-energy compound in muscle, has a direct linkage between phosphate and nitrogen, whereas the acyl mercaptide linkage in acetyl coenzyme A is also characteristic of an energy-rich compound. In every instance the high-energy compound is readily hydrolyzed to products that undergo spontaneous reactions. These reactions result in forms that are thermodynamically more stable.

The five energy-rich compounds described in the preceding section illustrate five types of these compounds.

1. Pyrophosphates such as ATP
2. Acyl phosphates such as acetyl phosphate
3. Enolic phosphates such as phosphoenolpyruvic acid
4. Guanidinium phosphates such as creatine phosphate
5. Acyl thioesters such as acetyl coenzyme A

Simple phosphate esters, such as glucose-6-phosphate and 3-phosphoglyceric acid, are not considered as high-energy compounds, and yield less energy on hydrolysis.

Glucose-6-phosphate
−3300 cal

3-Phosphoglyceric acid
−3300 cal

In the discussion of metabolism that follows, there will be many examples of the use of high-energy compounds in the storage of energy, the transmission of energy, and the coupling of energy obtained from foodstuffs to the utilization of that energy for cellular reactions.

THE FORMATION OF ATP

Since adenosine triphosphate has been marked as a key compound in the storage of chemical energy and in the coupling of exergonic reactions to endergonic reactions in the cell, it is a major driving force in the metabolic reactions in the tissue. Although ATP can be formed by light energy in the process of photosynthesis, which will be discussed in Chapter 26, the present discussion will consider its formation in the cytoplasm in the absence of oxygen (substrate level phosphorylation) and in the **mitochondria** by the process of oxidative phosphorylation.

Substrate Level ATP. In the anaerobic (absence of oxygen) scheme of carbohydrate metabolism (Embden-Meyerhof pathway) glucose is phosphorylated and is eventually broken down to the 3-carbon phosphorylated derivative.

In the following two reactions ADP is converted into the energy-rich compound ATP with the assistance of catalysts, called **enzymes.** These reactions can take place in the absence of O_2 and in the cytoplasm, and are termed **substrate level phosphorylations.**

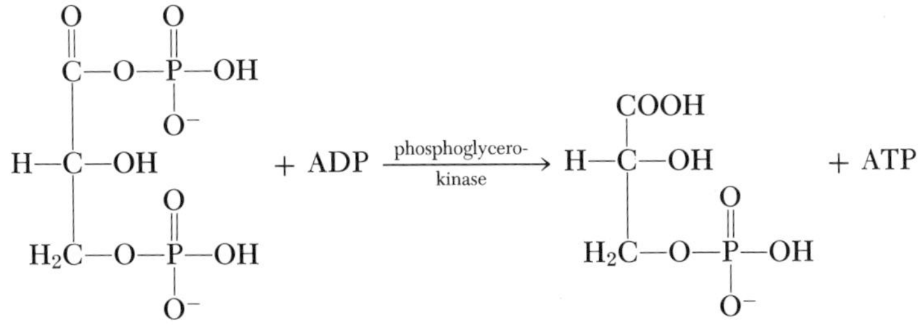

$$\begin{array}{c} O=C-O-P(=O)(OH)-O^- \\ | \\ H-C-OH \\ | \\ H_2C-O-P(=O)(OH)-O^- \end{array} + ADP \xrightarrow[\text{kinase}]{\text{phosphoglycero-}} \begin{array}{c} COOH \\ | \\ H-C-OH \\ | \\ H_2C-O-P(=O)(OH)-O^- \end{array} + ATP$$

1,3-Diphosphoglyceric acid → 3-Phosphoglyceric acid

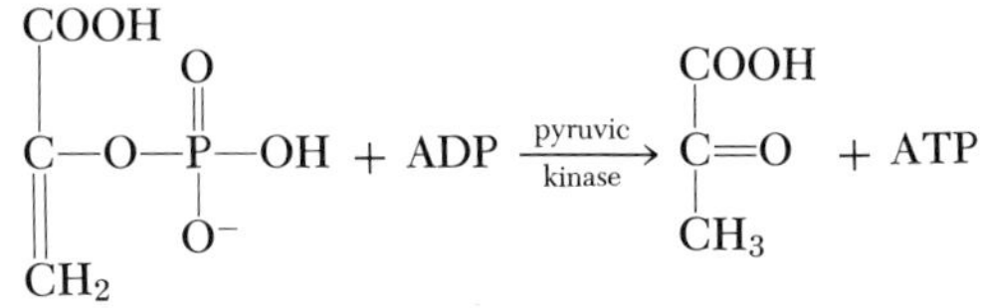

$$\begin{array}{c} COOH \\ | \\ C-O-P(=O)(OH)-O^- \\ \| \\ CH_2 \end{array} + ADP \xrightarrow[\text{kinase}]{\text{pyruvic}} \begin{array}{c} COOH \\ | \\ C=O \\ | \\ CH_3 \end{array} + ATP$$

Phosphoenolpyruvic acid → Pyruvic acid

Oxidative Phosphorylation ATP. One of the major aerobic or oxidative schemes of carbohydrate metabolism (Krebs cycle) involves the reaction of intermediate compounds with the production of several moles of ATP. An **electron transport system** in the mitochondria of the cell actively transports electrons from a reduced metabolite to oxygen with the assistance of enzymes and coenzymes, as shown in Figure 24–3. P_i is inorganic phosphate. NAD is nicotinamide adenine dinucleotide and FAD is flavin adenine dinucleotide, both coenzymes that will be described in the following chapter. Other intermediate compounds in Figure 24–3 between the reduced metabolite and oxygen, besides NAD and FAD, are coenzyme Q and the cytochromes. The overall reactions involve first

$$NAD + Metabolite \cdot H_2 \rightarrow Metabolite + NADH_2$$

then

$$NADH_2 + 3\ ADP + 3\ P_i + \tfrac{1}{2}O_2 \rightarrow NAD + 3\ ATP + H_2O$$

Although it is theoretically possible to obtain four moles of ATP in the complete cycle, only three are formed experimentally. A reduced metabolite that is linked

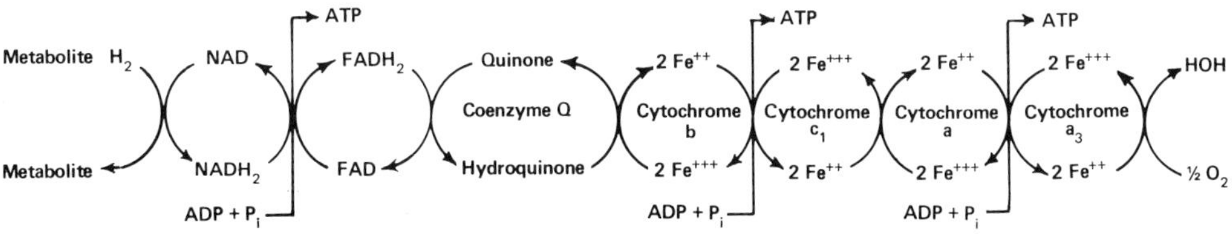

Metabolite H_2 / Metabolite — NAD / $NADH_2$ — $FADH_2$ / FAD — Quinone / Coenzyme Q / Hydroquinone — $2\ Fe^{++}$ / Cytochrome b / $2\ Fe^{+++}$ — $2\ Fe^{+++}$ / Cytochrome c_1 / $2\ Fe^{++}$ — $2\ Fe^{++}$ / Cytochrome a / $2\ Fe^{+++}$ — $2\ Fe^{+++}$ / Cytochrome a_3 / $2\ Fe^{++}$ — HOH / $\tfrac{1}{2}\ O_2$

ADP + P_i → ATP (three sites)

Figure 24–3 Oxidative phosphorylation and the electron transport system.

to FAD rather than NAD will form only two moles of ATP in the part of the cycle from FAD to oxygen.

NAD, **nicotinamide adenine dinucleotide** (structure shown in the next chapter), is a dinucleotide composed of AMP linked to nicotinamide-ribose-phosphate. The nicotinamide portion of the molecule is involved in the oxidation and reduction reactions in oxidative phosphorylation.

FAD, **flavin adenine dinucleotide** (structure shown in next chapter), is a dinucleotide composed of a flavin-ribose-phosphate linked to AMP. The reduction of FAD in the electron transport system involves the flavin portion of the molecule.

Coenzyme Q is a lipid soluble quinone, sometimes called ubiquinone-10 for the ten isoprene units found in the side chain (the number may vary from 0 to 10). This coenzyme is readily reduced to the hydroquinone form during the transport of hydrogen.

The **cytochromes** are oxidation-reduction pigments that consist of iron-porphyrin complexes known as **heme,** which is also an integral part of hemoglobin, the respiratory pigment of the red blood cells. The heme in cytochrome c, for example, is attached to a protein molecule by coordination with two basic amino acid residues, and by thioether linkages formed by the addition of a sulfhydryl group from each of two molecules of cysteine in the protein molecule. Cytochrome c is an electron carrier in the oxidative phosphorylation cycle, in which the iron atom of heme is changed from Fe^{+++} to Fe^{++} as shown:

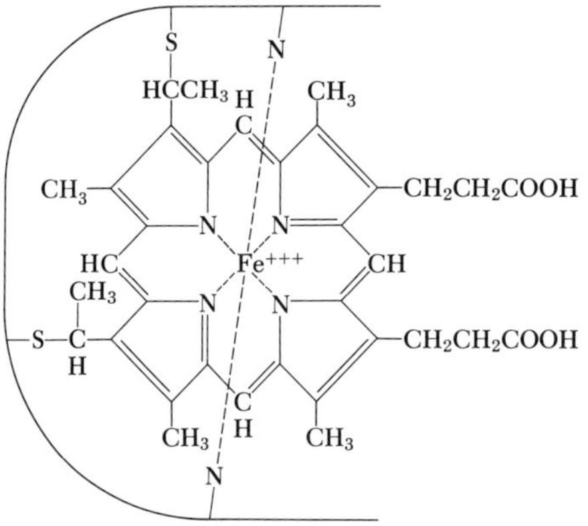

Cytochrome c (oxidized)

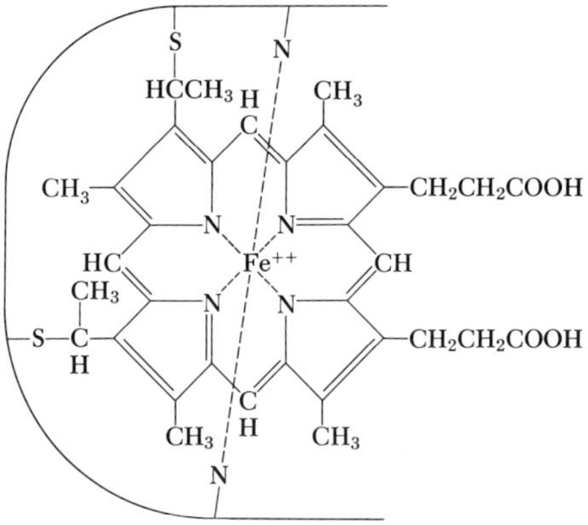

Cytochrome c (reduced)

The exact nature and detailed functional mechanism of the oxidative phosphorylation cycle is as yet not completely understood. Recently Green and his co-workers have isolated a large **electron transport particle** from the mitochondria of cells. They then separated the particle into four complexes, and, after a study of their composition and function, proposed a relationship between the complexes and the electron transport system (see p. 357).

Biochemical energy in the form of ATP is an essential driving force in many metabolic reactions in the cells and tissues. As has been described, several complex reactions are involved in the synthesis of this vital compound, and

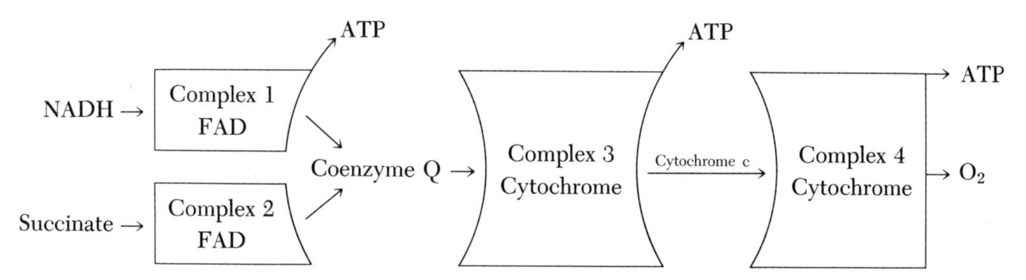

it should be emphasized that three moles of ATP are formed when the electrons from $NADH_2$ are transported through the system to oxygen. Also two moles of ATP are formed when electrons from $FADH_2$ are transported to oxygen. These relationships will assist in the understanding of the energy balance in the metabolic cycles.

Questions

1. What major instrumental development enabled the biochemist to study subcellular components?

2. Briefly describe three functions of a cell membrane.

3. Biochemists have intensively studied the mitochondria for several years. Why should they be so interested in this subcellular particle?

4. Compare and contrast the Golgi apparatus and the endoplasmic reticulum.

5. What is the nature and function of the lysosomes?

6. How do the enzymes in the lysosomes differ from those in the cytoplasm?

7. Briefly explain why phosphoenolpyruvic acid is a high-energy compound.

8. If 3-phosphoglyceric acid were converted to 1,3-diphosphoglyceric acid, what would happen to the ATP and what energy change would you expect?

9. Why is the oxidative phosphorylation mechanism also called the electron transport system? Explain.

10. What common reaction is involved with NAD, FAD, and coenzyme Q during the course of oxidative phosphorylation?

Suggested Reading

Baserga and Kisieleski: Autobiographies of Cells. Scientific American, Vol. 209, No. 2, p. 103, 1963.
Changeux: The Control of Biochemical Reactions. Scientific American, Vol. 212, No. 4, p. 36, 1965.
Green: The Mitochondrion. Scientific American, Vol. 210, No. 1, p. 63, 1964.
Heidt: The Path of Oxygen from Water to Molecular Oxygen. Journal of Chemical Education, Vol. 43, p. 623, 1966.
Kirschbaum: Biological Oxidations and Energy Conservation. Journal of Chemical Education, Vol. 45, p. 28, 1968.
Neutra and Leblond: The Golgi Apparatus. Scientific American, Vol. 220, No. 2, p. 100, 1969.
Nomura: Ribosomes. Scientific American, Vol. 221, No. 4, p. 28, 1969.
Racker: The Membrane of the Mitochondrion. Scientific American, Vol. 218, No. 2, p. 32, 1968.
Rich: Polyribosomes. Scientific American, Vol. 209, No. 6, p. 44, 1963.

CHAPTER 25

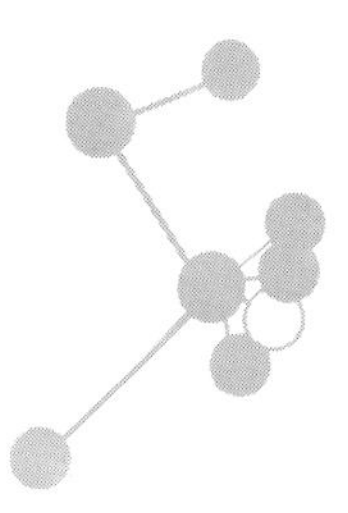

ENZYMES

A rapidly developing field of biochemical research involves the study of enzyme-controlled reactions in the cells of animals, plants, and microorganisms. It has been estimated that over one fourth of the biochemists in this country are directly engaged in some form of enzyme research. Many of the reactions that occur in living organisms are not only accelerated by enzymes but would not occur to any appreciable extent at body temperature. The magnitude of the problem facing the enzyme chemist is indicated by the estimate of as many as 1000 separate enzymes in a single cell.

THE CHEMICAL NATURE OF ENZYMES

Enzymes have always been considered as catalysts and are often compared to inorganic catalytic agents such as platinum and nickel. These inorganic agents are often used in conjunction with high temperatures, high pressures, and favorable chemical conditions. Few of these conditions occur when an enzyme reacts in body tissue, at body temperature, and at the pH of body fluids. Enzymes were originally defined as catalysts, organic in nature, formed by living cells, but independent of the presence of the cells in their action. A more current definition would state that *enzymes are proteins, formed by a living cell, which catalyze a reaction by lowering the activation energy so the rate of reaction is compatible with the conditions in the cell.*

By now it is generally accepted that all enzymes are protein in nature. The purification, crystallization, and inactivation procedures exactly parallel those for pure proteins. For example, excessive heat, alcohol, salts of heavy metals, and inorganic acids will cause coagulation or precipitation of the protein material and thus inactivate an enzyme.

The naming of enzymes has become increasingly complex as many new specific enzymes have been described. Originally they were named according to their source or according to the method of separation when they were discovered. As the family of enzymes grew, they were named in a more orderly fashion by adding the ending **-ase** to the root of the name of the substrate. An enzyme's **substrate** is the compound or type of substance upon which it acts. For example, sucrase catalyzes the hydrolysis of sucrose, lipase is an enzyme that hydrolyzes lipids, and urease is the enzyme that splits urea. This system also was used to name types of enzymes such as proteases, oxidases, and hydrolases. The discovery of so many enzyme mechanisms in the past few years has resulted in a mass of complex substrates and enzyme nomenclature. The problem was assigned to a Commission on Enzymes of the International Union of Biochemistry, whose members studied the system of nomenclature for six years before publishing a report in 1961. They were not in favor of eliminating all the names in common usage, but recommended the use of two names for an enzyme. One was the trivial name, either the one in common use or a simple name describing the activity of the enzyme. The other was constructed by the addition of the ending **-ase** to an accurate chemical name for the substrate. They also devised a numbering system to cover all enzymes, including those yet undiscovered. Each enzyme number consists of four digits, which clearly indicate the main type of enzyme, the nature of the chemical group affected by its action, the type of bond that is split, and the type of acceptor that is involved in an oxidation or reduction reaction. For the purposes of our discussion we shall use the trivial names or add **-ase** to the substrate.

Purification and Potency of Enzymes. Crude extracts of enzymes from tissues or cells may be obtained by grinding the tissue in metal grinders or between ground glass surfaces, by alternate freezing and thawing, by exposure to ultra-sound, or by the process of autolysis. The enzymes may be separated from the extracts by protein precipitation, adsorption on ion exchange resins, electrophoresis (adsorption assisted by the passage of an electrical current), and extraction with various solvents. The first enzyme to be obtained in crystalline form was urease, which was crystallized by Sumner in 1926. Since then about 75 other enzymes have been crystallized.

As enzyme preparations are purified and crystallized they naturally increase their potency of action. The activity of an enzyme, or more specifically one unit of an enzyme, is defined as that amount which will catalyze the transformation of 1 micromole of substrate per minute. A **micromole** is 1 millionth of a gram molecular weight. The potency, or **specific activity,** is expressed as units of enzyme per milligram of protein; the **molecular activity** is defined as units per micromole of enzyme at optimal substrate concentration. To compare the relative activity of different enzymes the **turnover number** is used. This is defined as the number of moles of substrate transformed per mole of enzyme per minute at a definite temperature. Turnover numbers vary from about 10,000 to 5,000,000, with the enzyme catalase exhibiting the highest activity.

PROPERTIES OF ENZYMES

Specificity of Action. Perhaps the major difference between inorganic catalysts, such as platinum and nickel, and enzymes is the specificity of action of the latter. Platinum, for example, will act as a catalyst for several reactions. Enzymes may exhibit different types of specificities. Lipase catalyzes the breaking of the ester linkage between glycerol and fatty acids in lipids, but it will not affect proteins or carbohydrates. Urease exhibits absolute specificity in action in that it catalyzes the splitting of a single compound, urea. Other enzymes exhibit stereo chemical specificity; D-amino acid oxidase is specific for D-amino acids and will not affect the natural L-amino acids. Arginase catalyzes the hydrolysis of L-arginine to urea and ornithine, but will not act on the D-isomer. In general, the specificity of action accounts for the large numbers of enzymes found in cells and tissues, and for the fact that enzymes are involved in all the metabolic reactions that occur in the cell.

ENZYME ACTIVITY

The activity of an enzyme is affected by many factors. Most important are the concentrations of the substrate and the enzyme, and the temperature and the pH of the reaction. In addition, the rate of enzyme reaction is affected by the nature of the end products, the presence of inhibitors, and light. The activity may be measured by following the chemical change that is catalyzed by the enzyme. The substrate is incubated with the enzyme under favorable conditions, and samples are withdrawn at short intervals for analysis of the end products or analysis of the decrease of substrate concentration. The enzyme lipase, for example, catalyzes the hydrolysis of fat molecules to fatty acids and glycerol. A simple method of measuring the activity of lipase would involve a determination of the rate of appearance of fatty acid molecules.

Effect of Substrate. It is generally agreed that before an enzyme can exert its catalytic activity it first combines with its substrate to form a complex. Although a larger portion of the enzyme molecule is probably essential for its catalytic action, there is a definite area on the surface of the enzyme where the substrate is combined. The particular chemical grouping at this spot is called the *active center* of the enzyme. There is evidence for the presence of more than one active center on the surface of most enzymes. Michaelis and Menten in 1913 first expressed the concept of the **enzyme substrate complex** as a transition state in enzyme reactions.

$$\underset{\text{Enzyme}}{E} + \underset{\text{Substrate}}{S} \underset{K_2}{\overset{K_1}{\rightleftharpoons}} \underset{\text{Enzyme-substrate complex}}{ES} \xrightarrow{K_3} E + \underset{\text{Products}}{P}$$

The formation of the ES complex permits the overall reaction to proceed at a lower energy of activation. A constant, K_m, known as the Michaelis constant, is related to the three velocity constants.

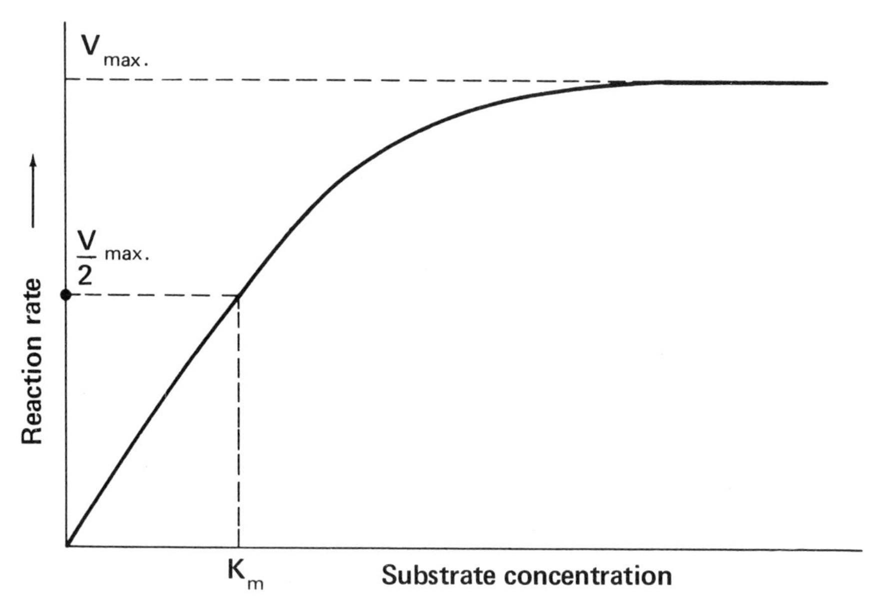

Figure 25-1 Effect of substrate concentration on the reaction rate when the enzyme concentration is held constant.

As seen from Figure 25-1, K_m by definition equals the substrate concentration [S] at one half the maximal velocity, $V_{max.}$, where $V_{max.}$ equals the maximum velocity at the saturation concentration of the substrate. Under proper conditions of temperature and pH, the Michaelis constant, K_m, is approximately equal to the dissociation constant of the ES complex. Conversely, the reciprocal of K_m, or $1/K_m$, is a measure of the *affinity of an enzyme for its substrate.*

Effect of Enzyme. When a purified enzyme is used, the rate of reaction is proportional to the concentration of the enzyme over a fairly wide range (Fig. 25-2). The substrate concentration must be kept constant and remain in excess of that required to combine with the enzyme.

Effect of pH. The hydrogen ion concentration, or pH, of the reaction mixture exerts a definite influence on the rate of enzyme activity. If a curve

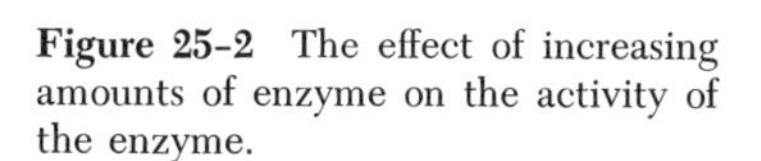
Figure 25-2 The effect of increasing amounts of enzyme on the activity of the enzyme.

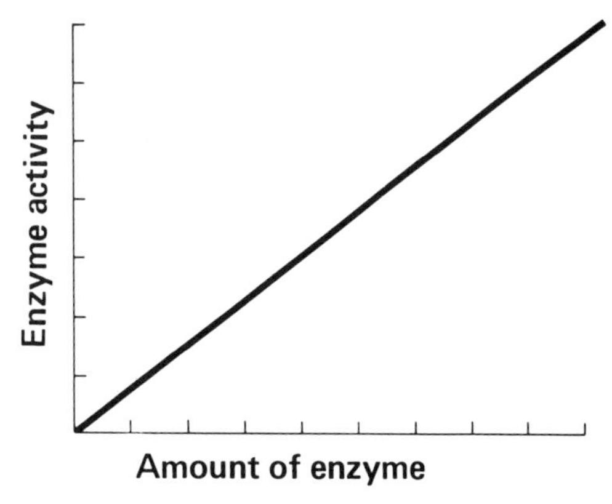

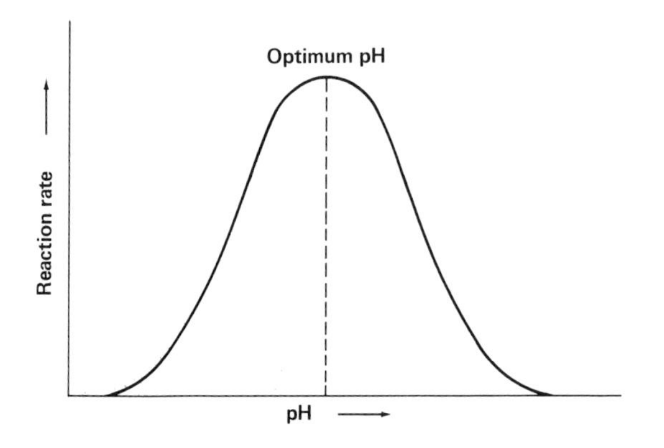

Figure 25-3 The effect of pH on enzyme activity.

is plotted comparing changes in pH with the rate of enzyme activity, it takes the form of an inverted U or V (Fig. 25–3). The maximum rate occurs at the **optimum pH,** with a rapid decrease of activity on either side of this pH value. The optimum pH of an enzyme may be related to a certain electric charge on the surface or to optimum conditions for the binding of the enzyme to its substrate. Most enzymes exhibit an optimum pH value close to 7, although pepsin is most active at pH 1.6 and trypsin at pH 8.2. Pepsin has no activity in an alkaline solution, whereas trypsin is inactive in an acid solution.

Effect of Temperature. The speed of most chemical reactions is increased two or three times for each 10°C rise in temperature. This is also true for reactions in which an enzyme is the catalyst, although the temperature range is fairly narrow. The activity range for most enzymes occurs between 10° and 50°C; the **optimum temperature** for enzymes in the body is around 37°C. The increased rate of activity observed at 50°C or above is short-lived, because the increased temperature first denatures and then coagulates the enzyme protein, thereby destroying its activity. The optimum temperature of an enzyme is therefore dependent on a balance between the rise in activity with increased temperature and the denaturation or inactivation by heat (Fig. 25–4). For any 10° rise in temperature the change in rate of enzyme activity is known as the Q_{10} value, or temperature coefficient. The Q_{10} value for most enzymes varies from 1.5 to 3.0.

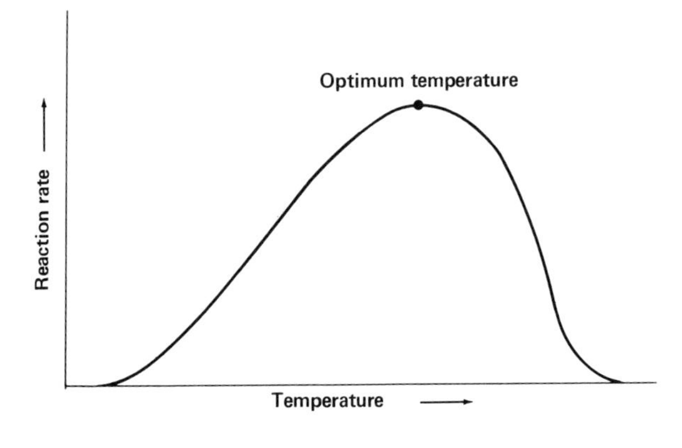

Figure 25-4 The effect of temperature on enzyme activity.

Effect of End Products. The end products of an enzyme reaction have a definite effect on the rate of activity of the enzyme. If they are allowed to increase in concentration without removal, they will slow the reaction. Some end products, when acid or alkaline in nature, may affect the pH of the mixture and thus decrease the rate of reaction. The effect of the end products on the activity of the enzyme is sometimes expressed as a chemical feedback system, with inhibition or decrease in rate called **negative feedback.** The activity of enzymes in the cell may be controlled to some extent by this chemical feedback system.

Several enzymes contain sulfhydryl groups (SH) which are associated with their active centers. Oxidizing agents change these groups to disulfide linkages and cause inactivation of the enzyme, whereas reducing agents restore the SH groups and activate the enzymes.

ACTIVATION OF ENZYMES

In the body many enzymes are secreted in an inactive form to prevent their action on the very glands and tissues that produce them. Also, during the process of purification an enzyme may become inactive. They may be activated by several agents: a change of pH, the addition of inorganic ions, or the addition of organic compounds.

The requirements for enzyme activity are further complicated by the fact that several enzymes require the presence of a metal ion for their activity. Representative metals required by enzymes are zinc, magnesium, iron, cobalt, and copper. Carbonic anhydrase, an enzyme that catalyzes the formation of carbonic acid from CO_2 and H_2O, requires zinc and is inactivated when this metal is removed.

PROENZYMES

A proenzyme is the precursor of the active enzyme in the body. For example, pepsinogen is the proenzyme of pepsin and trypsinogen is the inactive form of trypsin. When pepsinogen is secreted into the stomach, it is converted into pepsin by hydrogen ions of the hydrochloric acid. The pepsin then activates more pepsinogen to form more of the active enzyme. Trypsinogen is secreted by the pancreas and is activated in the intestine by enterokinase.

COENZYMES

In early studies on the enzymes of yeast it was observed that dialysis of a solution of the enzymes inactivated them. Dialysis consisted of placing the enzyme solution inside a sack made of dialyzing membrane, such as cellophane, and suspending the sack in water or in a buffer solution. When the dialyzed material was added to the enzymes, they again exhibited activity. The cofactor in the dialysate was called a **coenzyme.** Since that time several coenzymes have been discovered, and they have been found to consist of small organic molecules. If the organic molecule, or non-protein portion, is readily separated from the enzyme, it is called a coenzyme. If it is firmly attached to the protein portion of the enzyme, it is called a **prosthetic group.** Most enzymes may therefore be considered as conjugated proteins composed of an inactive protein molecule called the **apoenzyme** combined with the prosthetic group or coenzyme. The complete, conjugated, active molecule is called a **holoenzyme.**

As our knowledge of intermediary metabolism increases, it becomes apparent that *vitamins or derivatives of vitamins serve as coenzymes or prosthetic groups* in enzymatic reactions involving oxidation, reduction, and decarboxylation (removal of carboxyl groups). The water-soluble vitamin B complex contains several vitamins that exhibit the properties of coenzymes.

VITAMIN B_1. Vitamin B_1, or thiamine, contains a pyrimidine ring and another heterocyclic ring containing sulfur called thiazole.

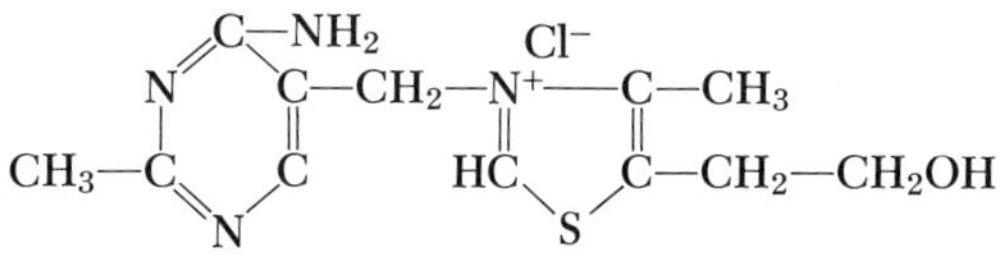

```
           C—NH2           Cl⁻
         //    \
        N       C—CH2—N⁺———C—CH3
        |       ||     ||    ||
  CH3—C        C      HC     C—CH2—CH2OH
        \\    /          \  /
          N               S
```

Thiamine chloride

A deficiency of the vitamin in the diet results in a disease called **polyneuritis** in animals and **beriberi** in man. The peripheral nerves of the body are involved, with muscle cramps, numbness of the extremities, pain along the nerves, and eventually atrophy of muscles, edema, and circulatory disturbances occurring in the body.

Thiamine occurs free in cereal grains, but occurs as the coenzyme, *thiamine pyrophosphate,* in yeast and meat.

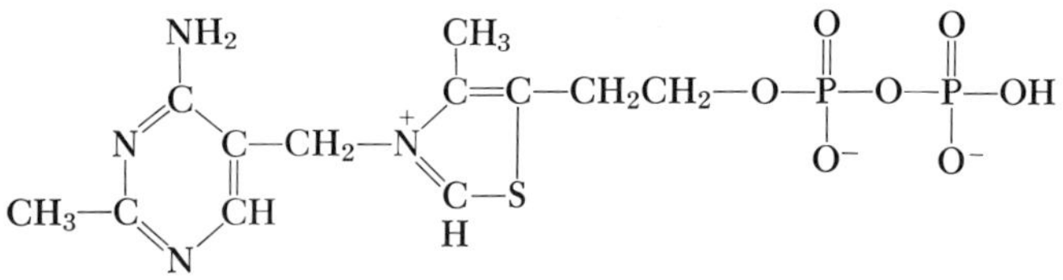

```
          NH2            CH3                      O      O
          |              |                        ||     ||
          C              C=C—CH2CH2—O—P—O—P—OH
        //  \           +     |                   |      |
       N      C—CH2—N         |                   O⁻     O⁻
       |      ||         \\   |
  CH3—C      CH           C—S
       \\   /             H
         N
```

Thiamine pyrophosphate (cocarboxylase) (TPP)

Cocarboxylase functions in the oxidative decarboxylation of pyruvic acid to form acetaldehyde and carbon dioxide. The thiazole ring is the active site of this function.

RIBOFLAVIN. Riboflavin, or vitamin B_2, is composed of a pentose alcohol, ribitol, and a pigment, flavin. A deficiency of vitamin B_2 in the diet of animals such as the rat, dog, and chicken causes lack of growth, loss of hair, and cataracts of the eyes. Lack of the vitamin in the human affects vision and causes inflammation of the cornea, and sores and cracks in the corners of the mouth.

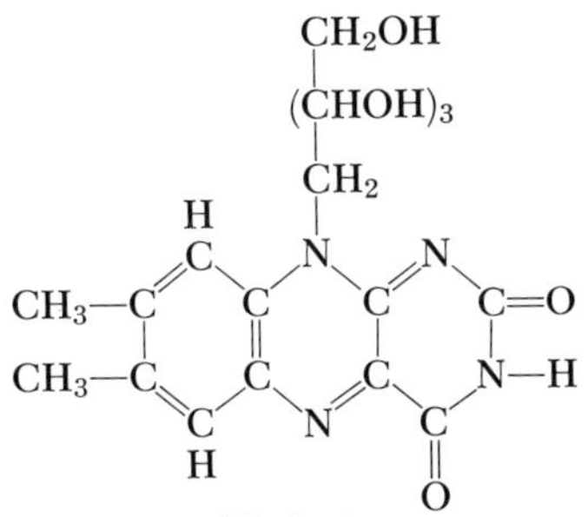

Riboflavin

The vitamin functions as a coenzyme; in fact, it occurs in foods as a component of two flavin coenzymes, FMN and FAD. The structures of **flavin mononucleotide, FMN,** and **flavin adenine dinucleotide, FAD,** are represented as follows:

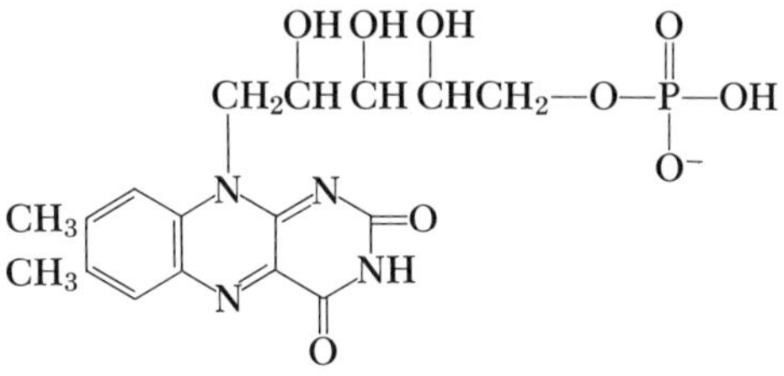

Flavin mononucleotide (FMN)

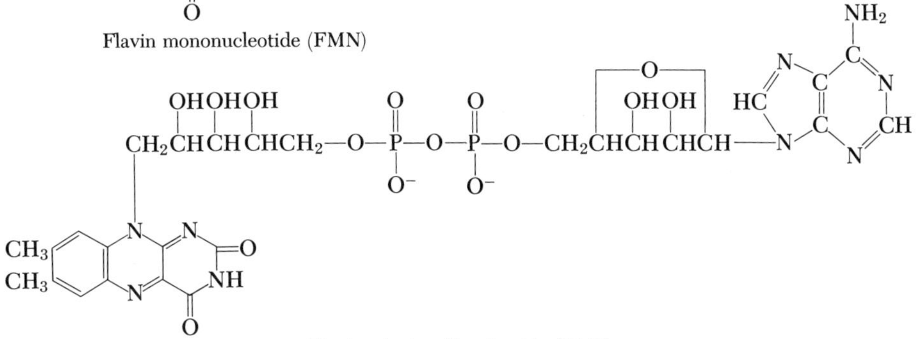

Flavin adenine dinucleotide (FAD)

Both FMN and FAD serve as coenzymes for a group of enzymes which catalyze oxidation-reduction reactions. Glutathione reductase, succinic dehydrogenase, and D-amino acid oxidase are examples of these enzymes. In the preceding chapter the mechanism for the reduction of FAD to $FADH_2$ by $NADH_2$ in the oxidative phosphorylation process was described. The flavin portion of the molecule is the active site for the oxidation-reduction reactions.

Nicotinic Acid and Nicotinamide. These two compounds have comparatively simple structures and as vitamins are called **niacin.**

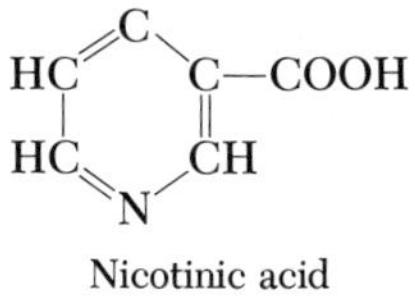

Nicotinic acid

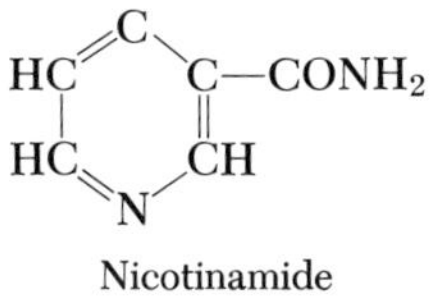

Nicotinamide

A deficiency of niacin in the diet results in **pellagra** in man and blacktongue in dogs. Pellagra is a disease characterized by skin lesions that develop on parts of the body that are exposed to sunlight. A sore and swollen tongue, loss of appetite, diarrhea, and nervous and mental disorders are typical symptoms of the disease.

Niacin is an essential component of two important coenzymes, **nicotinamide-adenine dinucleotide, NAD,** and **nicotinamide-adenine dinucleotide phosphate, NADP.**

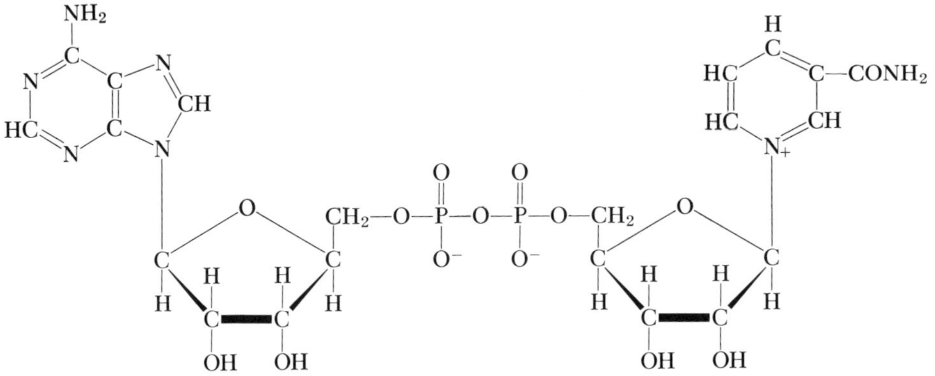

Nicotinamide-adenine dinucleotide (NAD)

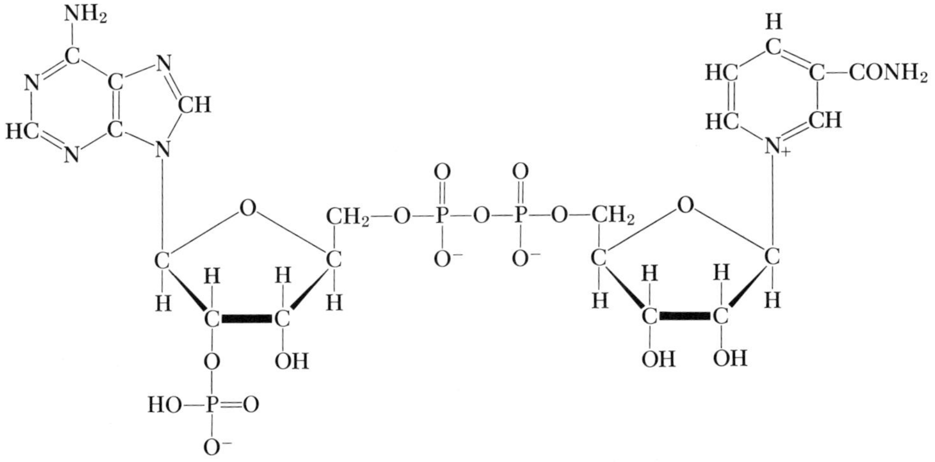

Nicotinamide-adenine dinucleotide phosphate (NADP)

The nicotinamide portion of NAD and NADP is involved in the mechanism of the oxidation-reduction reactions with which these coenzymes are involved.

Both NAD and NADP are coenzymes for dehydrogenases, which are enzymes that catalyze oxidation-reduction reactions. Lactic dehydrogenase, for example, catalyzes the oxidation of lactic acid to form pyruvic acid, with NAD serving as a coenzyme and being reduced to $NADH_2$ in the reaction. Alcohol dehydrogenase and glucose-6-phosphate dehydrogenase also require NAD as a coenzyme. Many enzymes of clinical diagnostic significance, such as lactic dehydrogenase, may be determined quantitatively in body fluids by the change in form of the coenzyme that occurs in the reaction.

$$\text{NAD} \rightleftharpoons \text{NADH}_2$$

Zero absorption of light at 340 nm — Strongly absorbs light at 340 nm

By measuring the change of absorbance of the solution at a wavelength of 340 nm in a spectrophotometer, the concentration of **dehydrogenase enzyme** responsible for the change can be determined.

Pyridoxine. The original name for this vitamin was **vitamin B_6,** which is a general name for **pyridoxine** and two closely related compounds, **pyridoxal** and **pyridoxamine.** These compounds, like nicotinic acid, are pyridine derivatives.

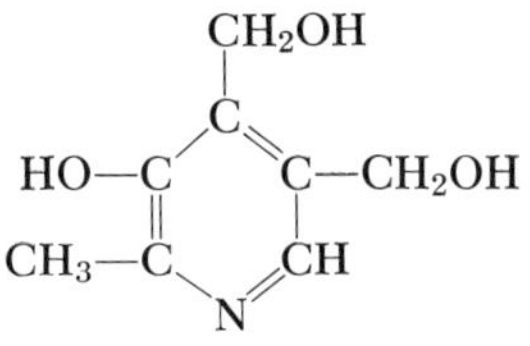

CH_2OH | HO—C, C—CH_2OH, CH_3—C, CH, N (pyridine ring)

Pyridoxine

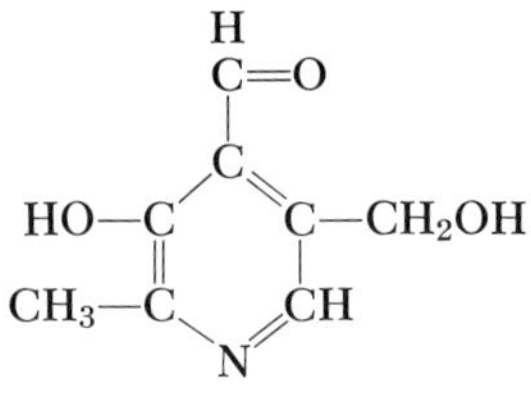

H—C=O | HO—C, C—CH_2OH, CH_3—C, CH, N (pyridine ring)

Pyridoxal

CH_2NH_2 | HO—C, C—CH_2OH, CH_3—C, CH, N (pyridine ring)

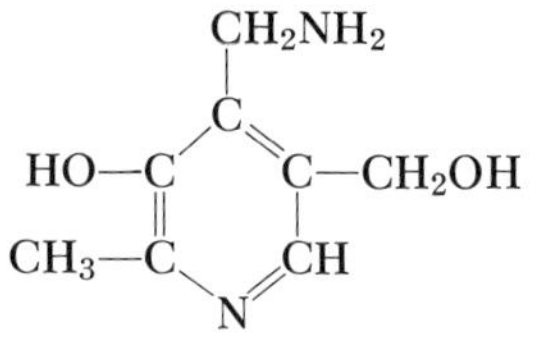

Pyridoxamine

A deficiency of vitamin B_6 in the diet of young rats results in a dermatitis called **acrodynia,** which is characterized by swelling and edema of the ears, nose, and paws. Pigs, cows, dogs, and monkeys exhibit central nervous system disturbances on a pyridoxine-deficient diet.

The phosphate derivatives of pyridoxal and pyridoxamine occur in vitamin B_6 sources and serve as the coenzyme forms of the vitamin.

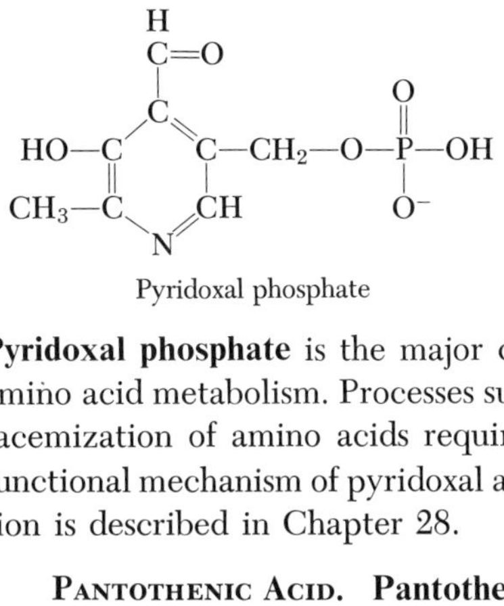

H—C=O | HO—C, C—CH_2—O—P(=O)(—OH)—O^-, CH_3—C, CH, N (pyridine ring)

Pyridoxal phosphate

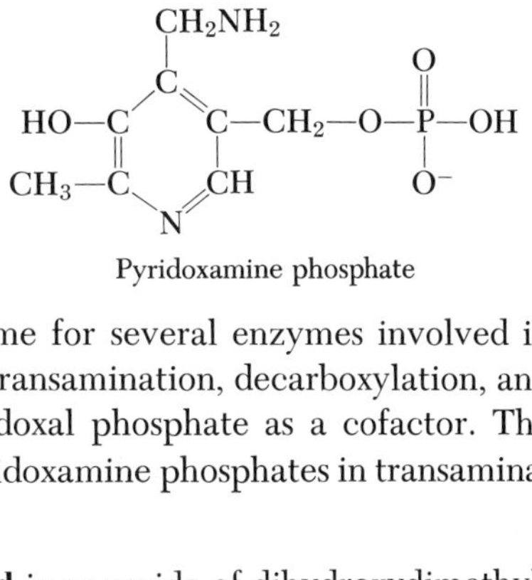

CH_2NH_2 | HO—C, C—CH_2—O—P(=O)(—OH)—O^-, CH_3—C, CH, N (pyridine ring)

Pyridoxamine phosphate

Pyridoxal phosphate is the major coenzyme for several enzymes involved in amino acid metabolism. Processes such as transamination, decarboxylation, and racemization of amino acids require pyridoxal phosphate as a cofactor. The functional mechanism of pyridoxal and pyridoxamine phosphates in transamination is described in Chapter 28.

Pantothenic Acid. Pantothenic acid is an amide of dihydroxydimethylbutyric acid and alanine.

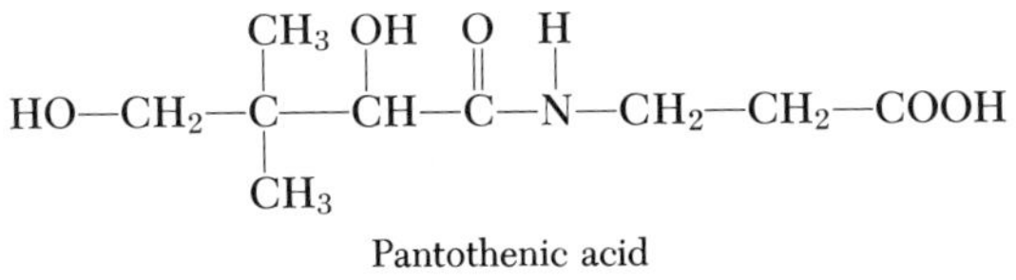

$$\text{HO—CH}_2\text{—C(CH}_3)_2\text{—CH(OH)—C(=O)—N(H)—CH}_2\text{—CH}_2\text{—COOH}$$

Pantothenic acid

Many animals show deficiency symptoms on diets lacking pantothenic acid; for example, the rat fails to grow, and exhibits a dermatitis, graying of hair, and adrenal cortical failure. In recent dietary research on pantothenic acid deficiency in man, such symptoms as emotional instability, gastrointestinal tract discomfort, and a burning sensation in the hands and feet have been observed. The coenzyme form of this vitamin is known as **coenzyme A.**

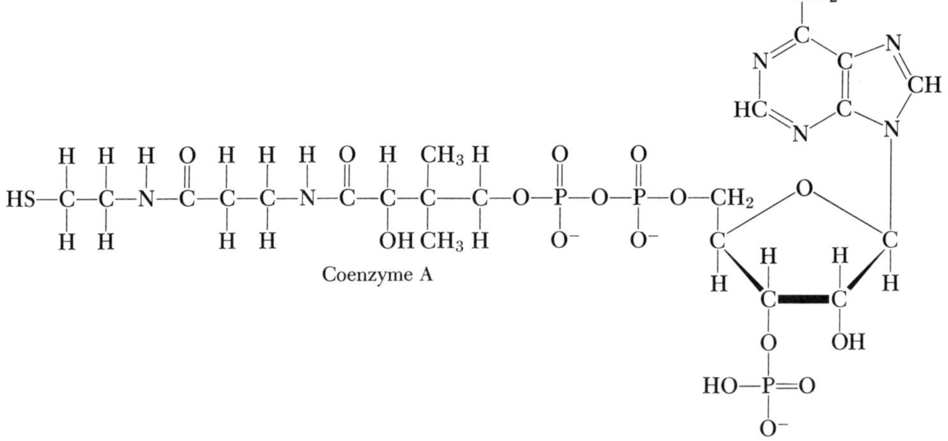

Coenzyme A

The functional group of the coenzyme is the —SH group, resulting in the abbreviated form CoASH. In biological systems it functions mainly as **acetyl CoA,** and is involved in acetylation reactions, synthesis of fats, synthesis of steroids, and other metabolic reactions that will be discussed in subsequent chapters. The formation of acetyl CoA involves a reaction of the functional —SH group with a lipoic acid complex. The mechanism of this reaction is shown later in the section on lipoic acid. Acetyl CoA is also important as a source of acetate for the Krebs cycle.

Vitamin B_{12}. Vitamin B_{12} has a complex chemical structure that is centered about an atom of cobalt bound to the four nitrogen atoms of a tetrapyrrole, to a nucleotide, and to a cyanide group. It is called **cyano-cobalamin,** and is represented on the top of page 369.

Vitamin B_{12}, is useful in the treatment of the anemias that develop in humans and animals. **Pernicious anemia** in particular responds most readily to treatment with the vitamin. In addition to increasing the hemoglobin and the red cell count, vitamin B_{12} administration also produces a remission of the neurological symptoms of anemia. The coenzyme form of the vitamin occurs in nature and is known as coenzyme B_{12}. It is an unstable compound in which the CN or OH group attached to the cobalt atom in vitamin B_{12} is replaced by the nucleoside, adenosine, as shown:

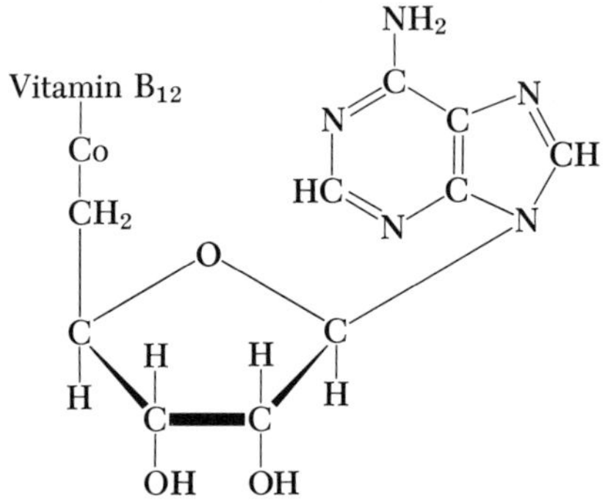

Coenzyme B_{12}

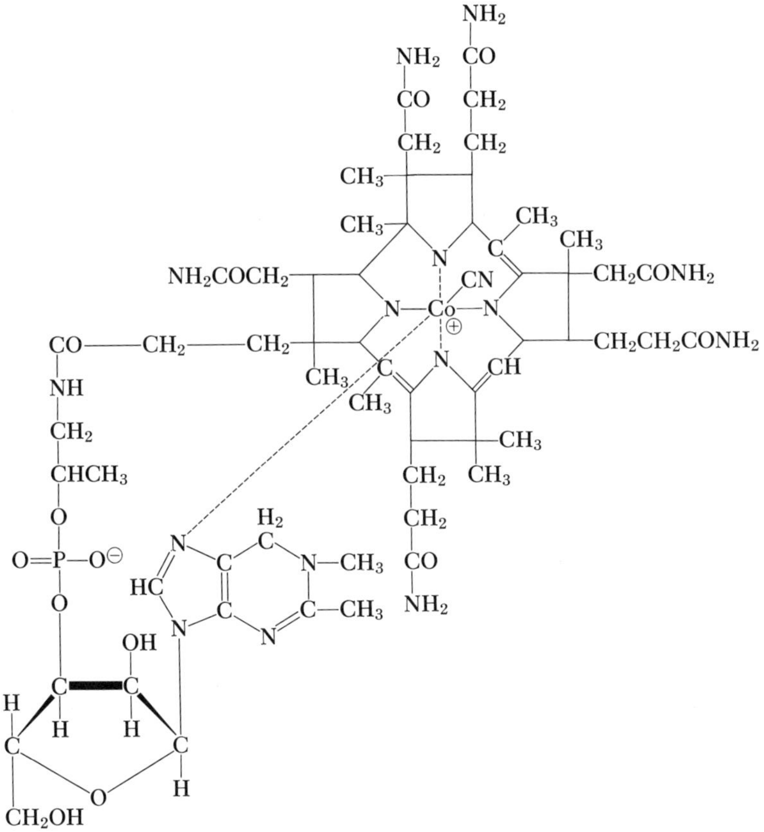

Vitamin B_{12} (cyanocobalamin)

The coenzyme is readily converted into either cyano- or hydroxycobalamin in the presence of cyanide or light. Coenzyme B_{12} functions in several important reactions in metabolism.

ENZYME INHIBITORS

The activity of an enzyme may be inhibited by an increase in temperature, a change in pH, and the addition of a variety of protein precipitants. More specific inhibition can be achieved by the addition of an oxidizing agent to attack SH groups, or inhibitors such as iodoacetamide and *p*-chloromercuribenzoate that react with SH groups. Cyanide forms compounds with metals essential for enzyme action, whereas fluoride combines with magnesium and inhibits enzymes that require Mg. Cyanide, for example, may remove a metal such as copper that is essential for the activity of the enzyme.

$$\underset{\text{Active enzyme (holoenzyme)}}{\text{Protein-Cu}} + 2CN^- \rightleftarrows Cu(CN)_2 + \underset{\text{Inactive enzyme (apoenzyme)}}{\text{Protein}}$$

Sodium azide and monoiodoacetate are also potent inhibitors. This type of compound usually combines with a group at the active site of the enzyme and cannot be displaced by additional substrate. These inhibitors are called **non-competitive inhibitors,** since their degree of inhibition is not related to the concentration of the substrate.

Compounds that directly compete with the substrate for the active site on the enzyme surface in the formation of the enzyme-substrate complex are called **competitive inhibitors.** An example of competitive inhibition would be the action of sulfanilamide on the utilization of *p*-aminobenzoic acid in the body. The similarity of these two compounds may readily confuse the enzyme involved in the utilization of this B vitamin in the synthesis of tetrahydrofolic acid, the active coenzyme of folic acid.

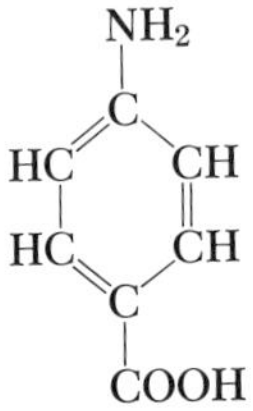

NH_2 — C(=CH–CH=C–CH=CH ring: HC, CH, HC, CH) — COOH

p-Aminobenzoic acid

NH_2 — C(ring: HC, CH, HC, CH) — SO_2NH_2

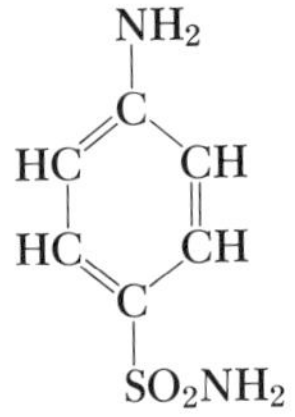

Sulfanilamide

The action of drugs in the body may depend on specific inhibitory effects on a particular enzyme in the tissues. The highly toxic nerve poison diisopropylfluorophosphate inhibits acetylcholine esterase, an enzyme essential for normal nerve function, by forming an enzyme inhibitor compound by attachment to a hydroxyl group on a serine residue in the enzyme.

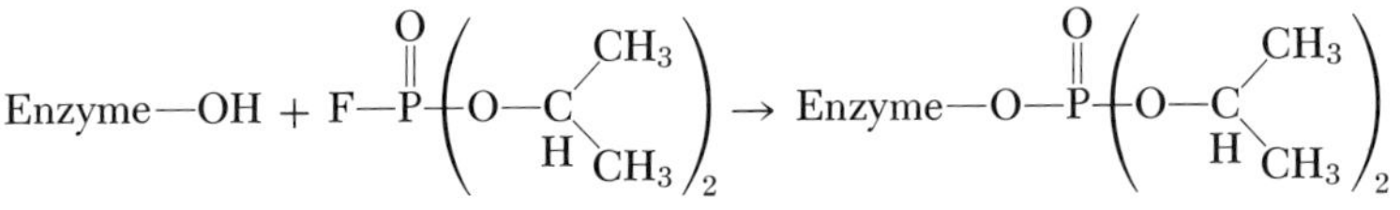

$$\text{Enzyme—OH} + \text{F—}\overset{\text{O}}{\overset{\|}{\text{P}}}\left(\text{O—}\underset{\text{H}}{\text{C}}(CH_3)_2\right)_2 \rightarrow \text{Enzyme—O—}\overset{\text{O}}{\overset{\|}{\text{P}}}\left(\text{O—}\underset{\text{H}}{\text{C}}(CH_3)_2\right)_2$$

Acetylcholine esterase — Diisopropylfluorophosphate (DFP) — Enzyme inhibitor compound

Antibiotic drugs may act by inhibiting enzyme and coenzyme reactions in microorganisms. Penicillin, for example, adversely affects cell wall construction in bacteria. A similar mechanism may be involved in the action of insecticides and herbicides.

Inhibitors of enzyme action in the body are called **antienzymes.** The tapeworm is a classic example of a protein-rich organism that is not digested in the intestine of the host. Substances that inhibit the activity of pepsin and trypsin have been isolated from the tapeworm. A trypsin inhibitor has been found in the secretion of the pancreas and in milk made from fresh soya beans. This substance exhibits properties similar to those of enzymes, and its activity is destroyed by heat. It may be formed by the pancreas to control the production of trypsin.

CLASSIFICATION OF ENZYMES

As new enzymes are discovered and further enzyme mechanisms are elaborated the problem of classification becomes more complex. Perhaps the most satisfactory basis for classification depends on the type of reaction influenced by the enzyme. On this basis, the Commission on Enzymes of the International Union of Biochemistry proposed that enzymes be classified in six main divisions. Although new names are introduced for some of the groups of enzymes that catalyze specific chemical reactions, we will retain the familiar names of enzymes that are included as examples within the six groups.

1. **Oxidoreductases**—Enzymes in this group are involved in physiological oxidation processes in the body. They include the common oxidation-reduction enzymes, examples of which are the dehydrogenases, oxidases, peroxidases, and hydrases.
2. **Transferases**—This group of enzymes catalyzes the transfer of essential chemical groups from one compound to another. Important examples are transaminases, transmethylases, and transacylases.
3. **Hydrolases**—This large group of enzymes is involved in the hydrolysis of compounds by the introduction of water molecules. They include the digestive enzymes such as amylase, sucrase, lipase, and the proteases.
4. **Lyases**—These enzymes catalyze the removal of chemical groups from compounds nonhydrolytically. Decarboxylases that catalyze the removal of CO_2 from organic molecules and aldolase, an enzyme involved in glycolysis, are examples of lyases.
5. **Isomerases**—This group of enzymes catalyzes different types of isomerization. The names of the enzymes indicate their activity. For example, *cis-trans* isomerases, racemases, intramolecular transferases, and epimerases are included in this group.
6. **Ligases**—These enzymes catalyze the linking together of two molecules with the breaking of a pyrophosphate bond of ATP or a similar triphosphate. An example would be an amino acid, RNA ligase, which is involved in protein synthesis in the cell.

Questions

1. How would you explain the definition of an enzyme?

2. What is a substrate? Give an example of a substrate and enzyme using the trivial name for the enzyme. Give an example of an enzyme and its substrate based on modern nomenclature.

3. What role does the formation of an enzyme-substrate complex play in the action of an enzyme? Explain.

4. Draw a graph representing the change in activity of an enzyme as the amount of its substrate is increased from zero to a maximum concentration. Explain the shape of the curve obtained.

5. How would you define the Michaelis constant, K_m, of an enzyme? Why is the K_m value important in enzyme reactions?

6. What is meant by (1) optimum pH? (2) optimum temperature of enzyme reactions?

7. How is the phenomenon of negative feedback related to the end products of an enzyme reaction? Could negative feedback control be of value in the body? Explain.

8. Name the coenzymes that are involved in oxidation-reduction reactions in the body.

9. Discuss the mechanism of function of any one coenzyme, including chemical structures in your discussion.

10. Cyanide is a very potent poison. Explain how cyanide may exert its toxic properties.
11. List the major classes of enzymes and give an example of each type.

Suggested Reading

Classification and Nomenclature of Enzymes. Science, Vol. 137, p. 405, 1962.
Enzyme Nomenclature. Recommendations 1964 of the International Union of Biochemistry. New York, Elsevier Publishing Co., 1965.
Neurath: Protein-Digesting Enzymes. Scientific American, Vol. 211, No. 6, p. 68, 1964.
Phillips: Three-Dimensional Structure of an Enzyme Molecule. Scientific American, Vol. 215, No. 5, p. 78, 1966.
Shaw: The Kinetics of Enzyme Catalyzed Reactions. Journal of Chemical Education, Vol. 34, p. 22, 1957.
Sumner: The Story of Urease. Journal of Chemical Education, Vol. 14, p. 255, 1937.
Wroblewski: Enzymes in Medical Diagnosis. Scientific American, Vol. 205, No. 2, p. 99, 1961.

CHAPTER 26

CARBOHYDRATE METABOLISM

In the preceding chapters on biochemistry we have considered the chemistry of carbohydrates, lipids, proteins, and nucleic acids. These substances are taken into the body in food and ordinarily cannot be utilized directly in the form in which they are ingested. Before the food can be absorbed and utilized by the body, it must be broken down into small, relatively simple molecules. The process by which complex food material is changed into simple molecules is called **digestion.** Digestion involves the hydrolysis of carbohydrates into monosaccharides, fats into glycerol and fatty acids, and proteins into amino acids, by the action of the hydrolases or hydrolytic enzymes.

DIGESTION

SALIVARY DIGESTION. Food taken into the mouth is broken into smaller pieces by chewing and is mixed with saliva, which is the first of the digestive fluids. Saliva contains **mucin,** a glycoprotein that makes the saliva slippery, and **ptyalin,** an enzyme that catalyzes the hydrolysis of starch to maltose. Since this enzyme has little time to act on starches in the mouth, its main activity takes place in the stomach before it is inactivated by the acid gastric contents. The most important functions of saliva are to moisten and lubricate the food for swallowing and to initiate the digestion of starch to dextrins and maltose.

GASTRIC DIGESTION. When food is swallowed it passes through the esophagus into the stomach. During the process of digestion the food is mixed with **gastric juice,** which is secreted by many small tubular glands located in the walls of the stomach. Gastric juice is a pale yellow, strongly acid solution containing the enzymes **pepsin** and **rennin.** Pepsin is the principal enzyme in gastric juice and it catalyzes the hydrolysis of large protein molecules into smaller more soluble molecules of *proteoses* and *peptones*. The optimum pH of pepsin is 1.5 to 2.0; thus, it is ideally suited for the digestion of protein in normal stomach contents with a pH of 1.6 to 1.8. Rennin converts casein of milk into a soluble protein. The mixing action of the stomach musculature and the process of digestion produce a liquid mixture called chyme which passes through the pyloric opening into the intestine.

INTESTINAL DIGESTION. The acid chyme is neutralized by the alkalinity of the three digestive fluids, **pancreatic juice, intestinal juice,** and **bile,** in the first part of the small intestine known as the duodenum. There are enzymes in the pancreatic juice that are capable of digesting proteins, fats, and carbohydrates. The pancreatic proteases are **trypsin, chymotrypsin,** and **carboxypolypeptidase,** whereas the pancreatic lipase is called **steapsin.** The enzyme **amylopsin** in pancreatic juice is an amylase similar to ptyalin in the saliva. This enzyme splits starch into maltose. The most important enzymes present in the intestinal juice are **aminopolypeptidase** and **dipeptidase,** and the three disaccharide-splitting enzymes, **sucrase, lactase,** and **maltase.** Cane sugar is the main source of dietary sucrose; milk contains lactose, and maltose comes from the partial digestion of starch by ptyalin and amylopsin. Sucrase, lactase, and maltase split these disaccharides into their constituent monosaccharides, thus completing the digestion of carbohydrates. In the normal adult, little or no digestion of fat occurs as the food passes through the mouth and the stomach. When fat enters the duodenum, the gastrointestinal tract hormone **cholecystokinin** is secreted and is carried by the blood to the gallbladder, where it stimulates that organ to empty its bile into the small intestine. Bile acids and bile salts are good detergents and emulsify fats for digestion by **pancreatic lipase.** Another hormone that is secreted when the chyme enters the duodenum is **secretin.** This hormone enters the circulation and stimulates the pancreas to release pancreatic juice into the intestine. The enzyme pancreatic lipase is activated by the bile salts and splits fat into fatty acids, glycerol, soaps, mono-, and diglycerides. The enzymes trypsin, chymotrypsin, carboxypolypeptidase, aminopolypeptidase, and dipeptidase in the small intestine act on native proteins, proteoses and peptones, and polypeptides. These molecules are gradually split into *amino acids,* which are the end products of protein digestion.

ABSORPTION

The monosaccharides glucose, fructose, and galactose are absorbed directly into the bloodstream through the capillary blood vessels of the **villi.** The villi are finger-like projections on the inner surface of the small intestine that greatly increase the effective absorbing surface. There are approximately five million villi in the human small intestine, and each villus is richly supplied with both lymph and blood vessels. Considerable evidence exists to indicate that the intestinal mucosa possesses the property of **selective absorption,** which is not possessed by a nonliving membrane. The exact mechanism by which hexose sugars are absorbed by the intestine against a concentration gradient has not been established, although it probably involves a phosphorylation process. The three monosaccharides are absorbed at different rates. Galactose is absorbed more rapidly than glucose, which is absorbed more rapidly than fructose. In the absorption process, as the end products of fat digestion pass through the intestinal mucosa they are reconverted into triglycerides, which then enter the lymph circulation. Bile salts are essential in absorption, both because of their effect on the solubility of fatty acids and because of direct involvement in the absorption process. Phospholipids are split into their component structures by digestive enzymes and are also resynthesized in the intestinal mucosa during absorption. Short-chain fatty acids may be directly absorbed into the blood and carried to the liver. Amino acids are absorbed through the intestinal mucosa directly into the blood stream by an active process that requires energy and enzymes. Each amino acid, like the monosaccharides, has a different rate of absorption, and there may be a different mechanism for types of amino acids, such as acidic, basic, neutral, L forms, and D forms.

THE BLOOD SUGAR

After the monosaccharides are absorbed into the blood stream, they are carried by the portal circulation to the liver. Fructose and galactose are phosphorylated by liver enzymes and are either converted into glucose or follow similar metabolic pathways. The metabolism of carbohydrates, therefore, is essentially the metabolism of glucose.

The concentration of glucose in the general circulation is normally 70 to 90 mg per 100 cc of blood. This is known as the **normal fasting level** of blood sugar. After a meal containing carbohydrates, the glucose content of the blood increases, causing a temporary condition of **hyperglycemia.** In cases of severe exercise or prolonged starvation, the blood sugar value may fall below the normal fasting level, resulting in the state of **hypoglycemia.** After an ordinary meal the glucose in the blood reaches hyperglycemic levels; this may be returned to normal by the following processes:

1. Storage
   (a) as glycogen
   (b) as fat
2. Oxidation to produce energy
3. Excretion by the kidneys

The operation of these factors in counteracting hyperglycemia is illustrated in Figure 26–1. The space between the vertical lines may be compared to a thermometer, with values expressed as milligrams of glucose per 100 cc. of blood. During active absorption of carbohydrates from the intestine the blood sugar level rises, causing a temporary hyperglycemia. In an effort to bring the glucose concentration back to normal, the liver may remove glucose from the blood stream, converting it into glycogen for storage. The muscles will also take glucose from the circulation to convert it to glycogen or to oxidize it to produce energy. If the blood sugar level continues to rise, the glucose may be converted into fat and stored in the fat depots. These four processes usually control the hyperglycemia; but if large amounts of carbohydrates are eaten and the blood sugar level exceeds an average of 160 mg of glucose per 100

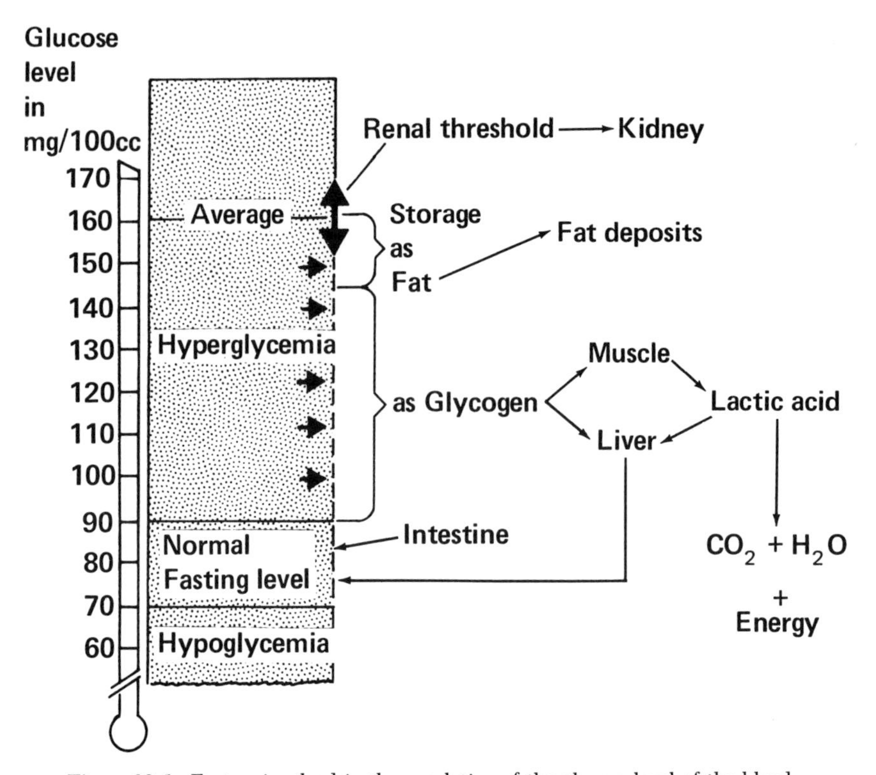

Figure 26-1 Factors involved in the regulation of the glucose level of the blood.

cc, the excess is excreted by the kidneys. The blood sugar level at which the kidney starts excreting glucose is known as the **renal threshold** and has a value from 150 to 170 mg per 100 cc.

In addition to the above factors, there are more specific reactions of the liver and the hormones to bring about regulation of the level of the glucose in the blood. The liver, for example, functions both in the removal of sugar from the blood and in the addition of sugar to the blood. During periods of hyperglycemia the liver stops pouring sugar into the blood stream and starts to store it as liver glycogen. During fasting the liver supplies glucose to the blood by breaking down its glycogen and by forming glucose from other food material such as amino acids or glycerol. The liver is assisted in this control process by several hormones.

HORMONES AND THE BLOOD SUGAR LEVEL

The properties and action of enzymes have already been discussed. In metabolism there are many related chemical reactions under the influence of enzymes. Another important group of regulating agents is the **hormones.** The hormones are formed mainly in the endocrine glands, which are also called ductless glands since their secretions diffuse directly into the blood stream. Enzyme action is more specific than hormone action, and the factors involved in the action of a hormone appear to be related to the action of other hormones.

In a normal individual, major cellular processes depend on an endocrine or **hormone balance,** and a disturbance in this balance results in metabolic abnormalities. In the regulation of a body process, a hormone probably has control over several specific enzyme-catalyzed reactions.

INSULIN. Although it was demonstrated as early as 1889 that removal of the pancreas of an animal would result in diabetes mellitus, it was not until 1922 that Banting and Best developed a method for obtaining active extracts of the pancreas. Within a short time insulin became available in sufficient quantities for the treatment of diabetes. It was first crystallized in 1926. More recently, as a result of the brilliant work of Sanger and his co-workers, a molecule consisting of two chains of amino acids with a molecular weight of 6000 has been described. The native molecule is thought to consist of four chains with a molecular weight of 12,000. Since it is a protein, it is not effective when taken by mouth, because the proteolytic enzymes of the gastrointestinal tract cause its hydrolysis and destroy its activity. Insulin is usually injected into the muscles when administered to a diabetic.

Insulin lowers the blood sugar level by increasing the conversion of glucose into liver and muscle glycogen, by regulating the proper oxidation of glucose by the tissues, and by preventing the breakdown of liver glycogen to yield glucose. In muscle and adipose tissue, insulin acts by increasing the rate of transport of glucose across membranes into the cells. Also, there is considerable evidence that in liver tissue insulin acts by controlling the phosphorylation of glucose to form glucose-6-phosphate, which is the first step in the formation of glycogen. In the absence of an adequate supply of insulin the transformation of extracellular glucose to intracellular glucose-6-phosphate is retarded.

Diabetes Mellitus. If the pancreas fails to produce sufficient insulin, the condition of **diabetes mellitus** results. The failure of the storage mechanisms in the absence of insulin causes a marked increase in the blood sugar level. Glucose is ordinarily excreted in the urine because the renal threshold is exceeded. The impairment of carbohydrate oxidation causes the formation of an excess of ketone bodies. Some of these ketone bodies are acid in nature, and the severe acidosis that results from the lack of insulin causes **diabetic coma,** which is sometimes fatal to a diabetic patient. When the correct dosage of insulin is injected, carbohydrate metabolism is properly regulated and the above symptoms do not appear.

GLUCAGON. Glucagon is a hormone that is produced by the α-cells of the pancreas. It is a polypeptide of known amino acid sequence with a molecular weight of about 3500. Glucagon causes a rise in the blood sugar level by increasing the activity of the enzyme liver phosphorylase, which is involved in the conversion of liver glycogen to free glucose. The activation of phosphorylase depends on the presence of the compound cyclic-3′,5′-adenosine monophosphate (AMP), whose formation is increased by the action of glucagon (see p. 380).

EPINEPHRINE. This hormone is produced by the central portion, or medulla, of the adrenal glands. Epinephrine is antagonistic to the action of insulin in that it causes glycogenolysis in the liver with the liberation of glucose. It stimulates an enzyme to produce cyclic-3′,5′-AMP from ATP and is also involved in the activation of phosphorylase. In addition to hyperglycemia, it also increases blood lactic acid by converting muscle glycogen to lactic acid. Continued secretion of epinephrine occurs under the influence of strong emotions such as fear or anger. This mechanism is often used as an emergency function to provide instant glucose for muscular work. The hyperglycemia that results often exceeds the renal threshold, and glucose is excreted in the urine.

Adrenal Cortical Hormones. Hormones such as **cortisone** and **cortisol** are produced by the outer layer or cortex of the adrenal gland. These hormones, especially those with an oxygen on position 11, have an effect on carbohydrate metabolism. In general they stimulate the production of glucose in the liver by increasing gluconeogenesis from amino acids. The cortical hormones are therefore antagonistic to insulin.

Anterior Pituitary Hormones. Of the many hormones secreted by the anterior lobe of the pituitary gland, the growth hormone, ACTH and the diabetogenic hormone affect the blood sugar level. The **growth hormone** causes the liver to increase its formation of glucose, but at the same time it stimulates the formation of insulin by the pancreas. Its action is complex and not completely understood. **ACTH,** the adrenocorticotropic hormone, stimulates the function of the hormones of the adrenal cortex and their action on the blood sugar level. The **diabetogenic hormone,** when injected into an animal, causes permanent diabetes and exhaustion of the islet tissue of the pancreas.

Although the overall control of the blood sugar level depends on the action of the liver and a balanced action of several hormones, it is readily apparent that insulin plays a major role in the normal process and is an important factor in the control of diabetes mellitus.

GLYCOGEN

As may be recalled from Chapter 20, glycogen is a polysaccharide with a branched structure composed of linear chains of glucose units joined by α-1,4 linkages and with α-1,6 linkages at the branch points. During absorption of the carbohydrates, the excess glucose is stored as glycogen in the liver. Normally this organ contains about 100g of glycogen, but it may store as much as 400g. The glycogen in the liver is readily converted into glucose and serves as a reservoir from which glucose may be drawn if the blood sugar level falls below normal. The formation of glycogen from glucose is called **glycogenesis,** whereas the conversion of glycogen to glucose is known as **glycogenolysis.** The muscles also store glucose as glycogen, but muscle glycogen is not as readily converted into glucose as is liver glycogen.

Glycogenesis. The process of glycogenesis is not just a simple conversion of glucose to glycogen. As we have learned previously, insulin is involved in the action of glucokinase in the phosphorylation of glucose to glucose-

6-phosphate. The glucose-6-phosphate is then converted to glucose-1-phosphate with the aid of the enzyme phosphoglucomutase. The glucose-1-phosphate then reacts with uridine triphosphate (UTP) to form an active nucleotide, uridine diphosphate glucose (UDPG). In the presence of a branching enzyme and the enzyme UDPG-glycogen-transglucosylase, the activated glucose molecules of UDPG are joined in glucosidic linkages to form glycogen. These reactions may be represented as follows:

$$\text{Glucose} \xrightarrow[\text{ATP} \curvearrowright \text{ADP}]{\text{(insulin)}\ \text{glucokinase}} \text{Glucose-6-phosphate}$$

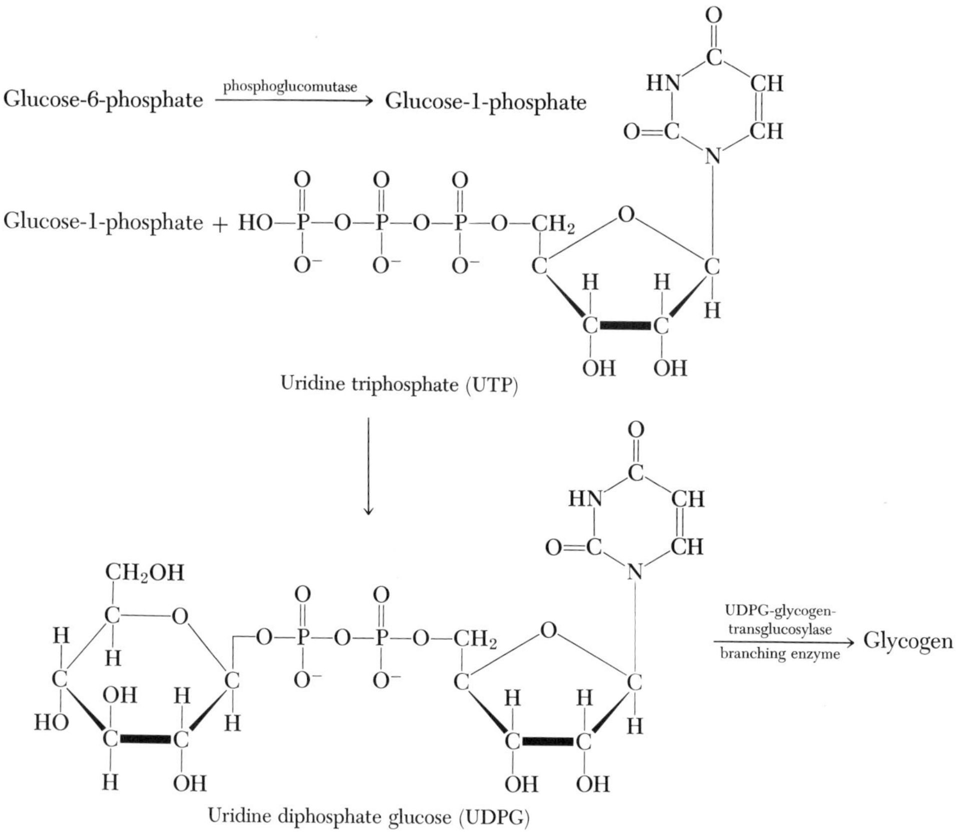

$$\text{Glucose-6-phosphate} \xrightarrow{\text{phosphoglucomutase}} \text{Glucose-1-phosphate}$$

$$\text{Glucose-1-phosphate} + \text{HO—P(=O)(O}^-\text{)—O—P(=O)(O}^-\text{)—O—P(=O)(O}^-\text{)—O—CH}_2\text{—(ribose)—uracil}$$

Uridine triphosphate (UTP)

↓

$$\text{Glucose—O—P(=O)(O}^-\text{)—O—P(=O)(O}^-\text{)—O—CH}_2\text{—(ribose)—uracil} \xrightarrow[\text{branching enzyme}]{\text{UDPG-glycogen-transglucosylase}} \text{Glycogen}$$

Uridine diphosphate glucose (UDPG)

Glycogenolysis. The process of glycogenolysis liberates glucose into the blood stream to maintain the blood sugar level during fasting and to supply energy for muscular contraction. In the liver the reaction is initiated by the action of the enzyme phosphorylase, which splits the 1,4 glucosidic linkages in glycogen. The enzyme phosphorylase exists in two forms: an active form, **phosphorylase a,** and an inactive form, **phosphorylase b.** Phosphorylase b is converted to the active form of the enzyme by ATP in the presence of Mg^{++} and phosphorylase b kinase, as shown:

$$2\ \text{Phosphorylase b} + 4\ \text{ATP} \xrightarrow[\text{cyclic-3}',5'\text{-AMP}]{\text{kinase, Mg}^{++}} \text{Phosphorylase a} + 4\ \text{ADP}$$

The phosphorylase b kinase is activated by cyclic-3′,5′-AMP, a derivative of adenylic acid.

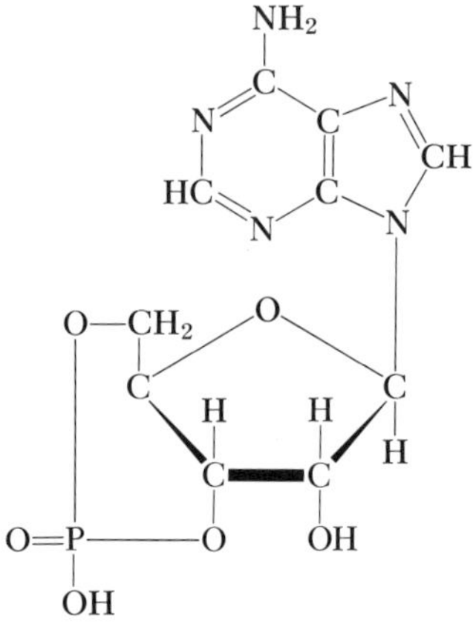

Cyclic-3′,5′-AMP

As stated in an earlier chapter, both epinephrine and glucagon influence the activation of phosphorylase by increasing the formation of cyclic-3′,5′-AMP. Other enzymes assist the breakdown to glucose-1-phosphate, which is subjected to the reversed action of phosphoglucomutase to yield glucose-6-phosphate. A specific enzyme in the liver, glucose-6-phosphatase, acts on glucose-6-phosphate to produce glucose. This enzyme is not present in muscle; therefore, muscle glycogen cannot serve as a source of blood glucose. These reactions may be illustrated as follows:

$$\text{Glycogen} \xrightarrow[\text{debranching enzyme}]{\text{phosphorylase}} \text{Glucose-1-phosphate}$$

$$\text{Glucose-1-phosphate} \xrightarrow{\text{phosphoglucomutase}} \text{Glucose-6-phosphate}$$

$$\text{Glucose-6-phosphate} \xrightarrow[\text{in liver}]{\text{glucose-6-phosphatase}} \text{Glucose}$$

GLYCOLYSIS

The ready availability of muscle preparations and their use in the development of physiology led to an early study of the biochemical changes associated with muscular contraction. It was observed that when a muscle contracts in an anaerobic medium, glycogen disappears and pyruvic and lactic acids are formed. In the presence of oxygen, or under aerobic conditions, the glycogen is re-formed, and the pyruvic and lactic acids disappear. Further studies demonstrated that one fifth of the lactic acid formed during glycolysis is oxidized to CO_2 and water, whereas the remaining four fifths is converted to glycogen.

Substances other than the carbohydrates in food and the lactic acid from muscular contraction may be converted into glycogen. These glycogenic compounds are formed by the process of **gluconeogenesis,** which is the conversion of non-carbohydrate precursors into glucose. Examples of these precursors are the glycogenic amino acids, the glycerol portion of fat, and any of the metabolic breakdown products of glucose, such as pyruvic acid, which may form glucose by reversible reactions in metabolism. The reactions discussed in the above section can be summarized in the **lactic acid cycle.**

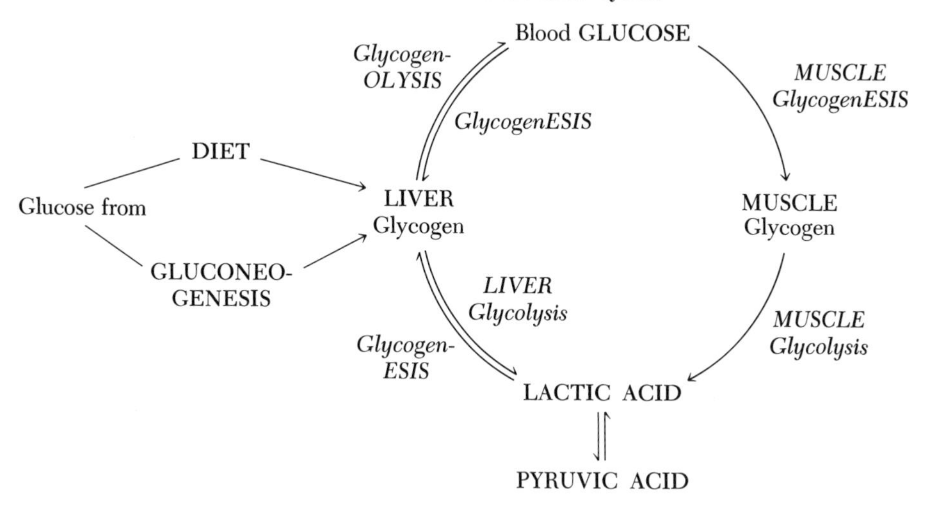

The Lactic Acid Cycle

OXIDATION OF CARBOHYDRATES

Glucose is ultimately oxidized in the body to form CO_2 and H_2O with the liberation of energy. **Glucose-6-phosphate** is a principal compound in the metabolism of glucose. As discussed earlier, it may be formed by the phosphorylation of glucose under the control of insulin. Once it is formed, it may be converted to glycogen or to free glucose, or it may be metabolized by several mechanisms or pathways. The two major pathways of glucose-6-phosphate metabolism are the **anaerobic,** or **Embden-Meyerhof, pathway** followed by the **aerobic,** or **Krebs, cycle.** The largest proportion of energy available from the oxidation of the glucose molecule is liberated from the Krebs cycle, but the Embden-Meyerhof pathway is essential in the formation of pyruvic acid used in the Krebs cycle.

Anaerobic, or Embden-Meyerhof, Pathway of Glycolysis. The first stage of the Embden-Meyerhof pathway involves the formation of glucose-6-phosphate from muscle or liver glycogen or from glucose. The glucose-6-phosphate is converted to fructose-6-phosphate by the enzyme isomerase, and this compound forms fructose-1,6-diphosphate under the influence of enzymes and ATP. The diphosphate is then split into two triose monophosphates, which are converted to 1,3-diphosphoglyceric acid. This compound undergoes a series of transformations through 3-phosphoglyceric to 2-phosphoglyceric acid to phosphoenolpyruvic acid to finally form pyruvic acid. Under anaerobic conditions the pyruvic acid is further reduced to lactic acid. The Embden-Meyerhof series of reactions may be represented as shown on p. 382.

The requirement for and liberation of ATP in this anaerobic pathway is emphasized by the shaded areas. One mole of ATP is required for the phos-

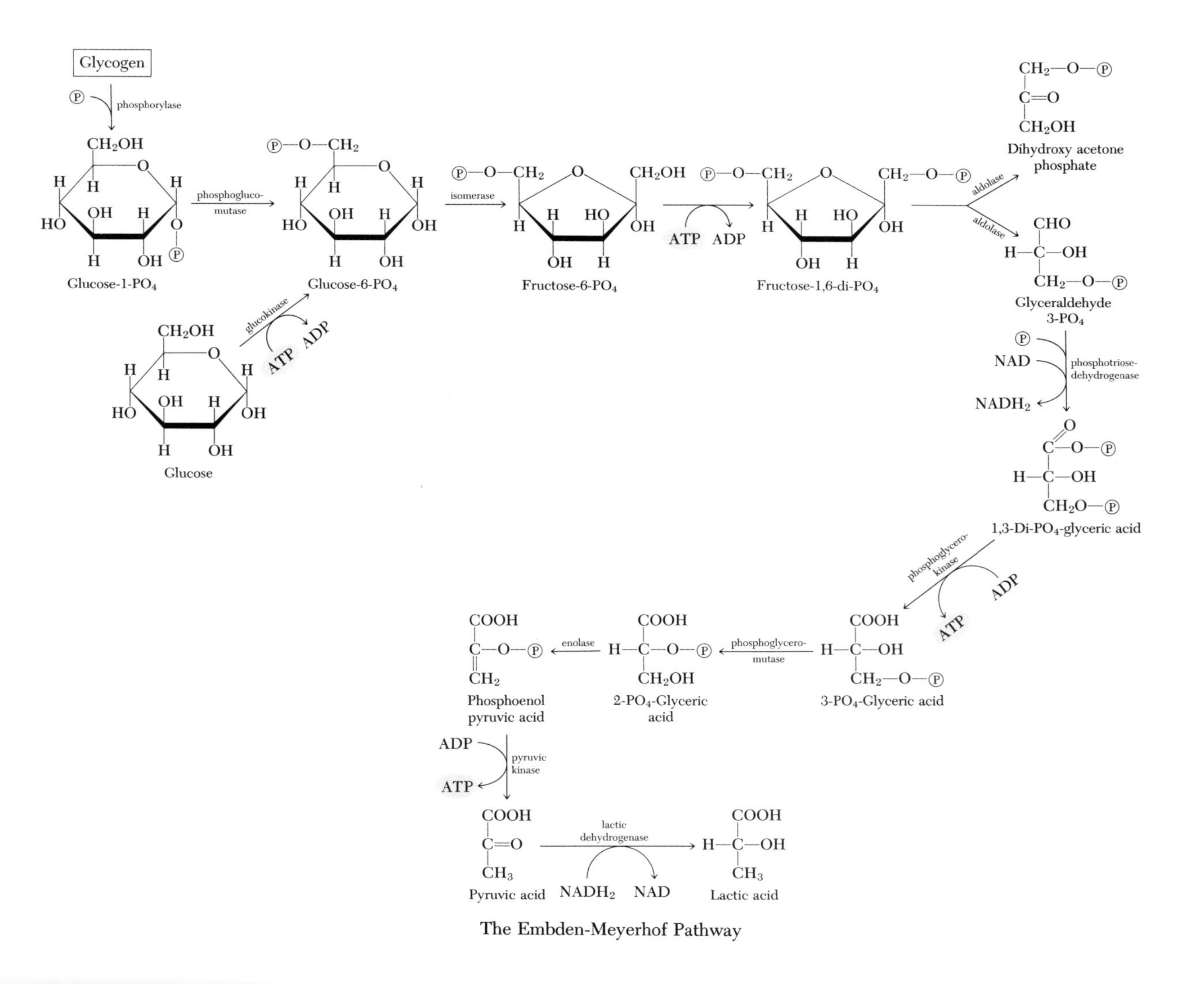

The Embden-Meyerhof Pathway

phorylation of glucose and one more for the conversion of fructose-6-phosphate to fructose-1,6-diphosphate. The reaction of 1,3-diphosphoglyceric acid to form 3-phosphoglyceric acid liberates 1 mole of ATP per triose molecule or 2 moles of ATP per glucose molecule. The conversion of phosphoenolpyruvic acid to pyruvic acid also yields 2 moles of ATP per glucose molecule. For each mole of glucose broken down in the Embden-Meyerhof pathway, therefore, *2 moles of ATP are consumed and 4 moles are liberated, with a net gain of 2 moles of ATP.*

Aerobic, or Krebs, Cycle. The pyruvic acid and the one fifth of the lactic acid that is not converted to liver glycogen are eventually oxidized with the formation of CO_2 and energy. The lactic acid may be oxidized to re-form pyruvic acid, which is then converted to acetyl coenzyme A by pyruvic oxidase in the presence of Mg^{+2}, NAD, lipoic acid, thiamine pyrophosphate (TPP), and coenzyme A. The acetyl coenzyme A combines with oxalacetic acid to form citric acid. The citric or tricarboxylic acid cycle then follows a series of reactions in which a molecule of pyruvic acid disappears as CO_2 and H_2O, with the liberation of energy. The **citric acid,** or **Krebs, cycle** may be represented as shown in the diagram on page 384).

The overall reaction for the conversion of pyruvic acid to carbon dioxide and water may be written as:

$$C_3H_4O_3 + \frac{5}{2}O_2 + 15ADP + 15P_i \rightarrow 3CO_2 + 2H_2O + 15ATP$$

The moles of ATP formed and CO_2 liberated in one turn of the Krebs cycle are shown in shaded areas. It may be recalled from a consideration of the electron transport mechanism in oxidative phosphorylation, page 355, that the oxidation of $NADH_2$ or $NADPH_2$ (through $NADH_2$) by way of the cytochrome system yields 3 moles of ATP per mole of $NADH_2$. Starting with $FADH_2$, the system yields 2 moles of ATP per mole of $FADH_2$.

When one molecule of glucose is completely oxidized it liberates 686,000 calories. Each molecule of glucose subjected to the Embden-Meyerhof pathway liberates 8 moles of ATP (6 moles from the $NADH_2$ formed in the conversion of glyceraldehyde-3-phosphate to 1,3-diphosphoglyceric acid, and 2 moles net yield of ATP formed directly). Since each mole of glucose forms 2 moles of pyruvic acid, the Krebs cycle will yield 2×15 or 30 moles of ATP per molecule of glucose. *A total of 38 moles of ATP are therefore formed by the oxidation of a molecule of glucose.* Since each mole of ATP will yield approximately 8000 calories on hydrolysis, the 38 moles are equivalent to 304,000 calories. This series of reactions is therefore capable of storing about 44 per cent of the available calories in the form of the high energy compound ATP, to be used in muscular work and for other energy requirements.

ALTERNATE PATHWAYS OF CARBOHYDRATE OXIDATION

Pathways other than the Embden-Meyerhof and Krebs cycles have been proposed for the oxidation of carbohydrates. The most generally accepted alternate pathway is called the **phosphogluconate oxidative pathway,** the

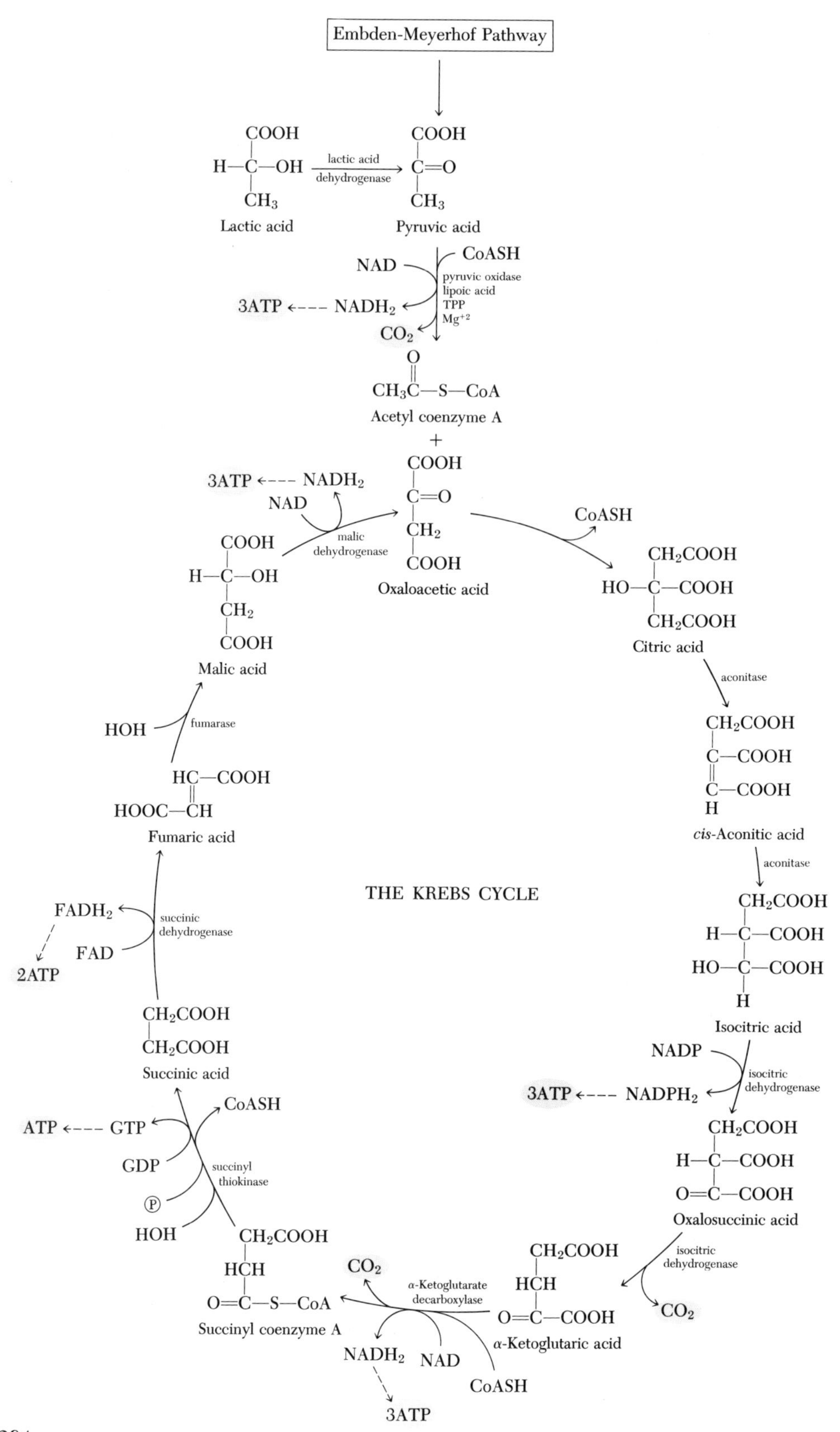
Embden-Meyerhof Pathway
COOH
H—C—OH
CH3
Lactic acid
lactic acid
dehydrogenase
COOH
C=O
CH3
Pyruvic acid
NAD
CoASH
pyruvic oxidase
lipoic acid
TPP
Mg+2
3ATP ←--- NADH2
CO2
O
CH3C—S—CoA
Acetyl coenzyme A
+
COOH
C=O
CH2
COOH
Oxaloacetic acid
3ATP ←--- NADH2
NAD
malic
dehydrogenase
CoASH
CH2COOH
HO—C—COOH
CH2COOH
Citric acid
aconitase
CH2COOH
C—COOH
C—COOH
H
cis-Aconitic acid
aconitase
CH2COOH
H—C—COOH
HO—C—COOH
H
Isocitric acid
NADP
isocitric
dehydrogenase
3ATP ←--- NADPH2
CH2COOH
H—C—COOH
O=C—COOH
Oxalosuccinic acid
isocitric
dehydrogenase
CO2
CH2COOH
HCH
O=C—COOH
α-Ketoglutaric acid
CO2
α-Ketoglutarate
decarboxylase
NADH2
NAD
CoASH
3ATP
CH2COOH
HCH
O=C—S—CoA
Succinyl coenzyme A
CoASH
ATP ←--- GTP
GDP
succinyl
thiokinase
P
HOH
CH2COOH
CH2COOH
Succinic acid
FADH2
succinic
dehydrogenase
FAD
2ATP
HC—COOH
HOOC—CH
Fumaric acid
HOH
fumarase
COOH
H—C—OH
CH2
COOH
Malic acid
THE KREBS CYCLE

hexose monophosphate shunt, or the **pentose phosphate pathway.** This alternate pathway involves the direct oxidation of glucose-6-phosphate to 6-phosphogluconic acid with the coenzyme NADP forming $NADPH_2$. Ribose-5-phosphate, fructose-6-phosphate, and glyceraldehyde-3-phosphate are formed in this pathway. The ribose may be used in the synthesis of nucleotides, whereas fructose and glyceraldehyde can be used in the Embden-Meyerhof pathway. Also, the pathway provides for the formation of $NADPH_2$, which is essential in the synthesis of compounds such as fatty acids and steroids. The phosphogluconate oxidative pathway may be represented as follows:

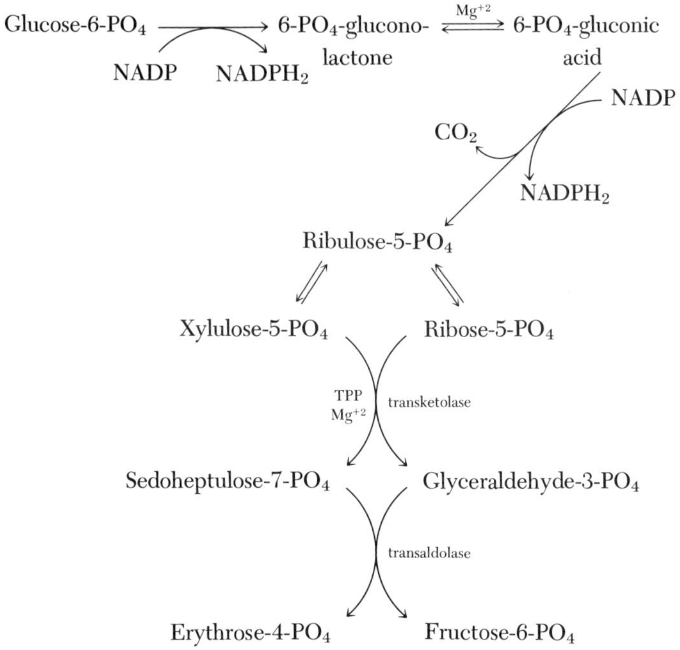

Glucose-6-PO_4 → (NADP → $NADPH_2$) → 6-PO_4-glucono-lactone ⇌ (Mg^{+2}) 6-PO_4-gluconic acid

6-PO_4-gluconic acid → (NADP → $NADPH_2$; → CO_2) → Ribulose-5-PO_4

Ribulose-5-PO_4 ⇌ Xylulose-5-PO_4

Ribulose-5-PO_4 ⇌ Ribose-5-PO_4

Xylulose-5-PO_4 + Ribose-5-PO_4 → (TPP, Mg^{+2}; transketolase) → Sedoheptulose-7-PO_4 + Glyceraldehyde-3-PO_4

Sedoheptulose-7-PO_4 + Glyceraldehyde-3-PO_4 → (transaldolase) → Erythrose-4-PO_4 + Fructose-6-PO_4

MUSCLE CONTRACTION

Contraction of muscle fibers under anaerobic conditions leads to the formation of lactic acid and eventually to muscle fatigue. Muscle glycogen disappears during this process. In the presence of oxygen, the muscle regains its glycogen, loses lactic acid, and recovers its ability to contract. The high-energy compound **creatine phosphate** also changes form during contraction and subsequent recovery. The process of glycolysis supplies ATP, and the creatine phosphate and ATP join forces in muscular contraction.

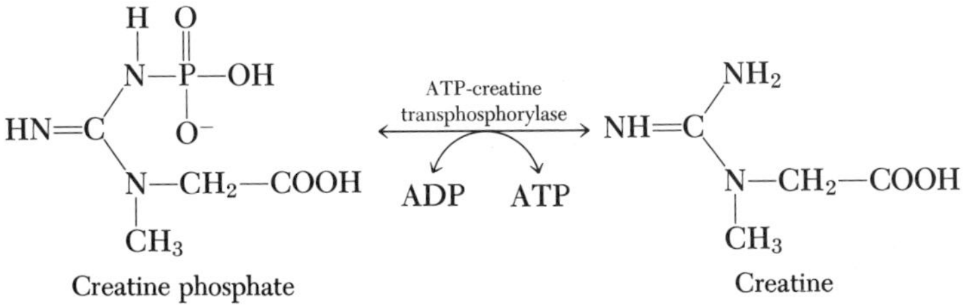

$$HN{=}C(NH{-}P(=O)(OH)O^-)(N(CH_3){-}CH_2{-}COOH) + ADP \xrightleftharpoons{\text{ATP-creatine transphosphorylase}} NH{=}C(NH_2)(N(CH_3){-}CH_2{-}COOH) + ATP$$

Creatine phosphate ⇌ Creatine

The straight and curved arrows with heads on each end denote reversible reactions, and you may recall that a direct linkage between nitrogen and phosphorus, as in creatine phosphate, denotes a high-energy compound. The reaction is readily reversible, and the muscle continues to contract as long as creatine phosphate is present.

During recovery, when more ATP is formed from glycolysis, the creatine phosphate is regenerated. The ATP is the direct source of energy for muscular work, and the function of creatine phosphate and glycolysis is to supply the ATP. Since there is only a small amount of ATP in the muscle at any instant, the supply of ATP needed for muscular work is obtained from creatine phosphate, the process of glycolysis, and the recovery of muscle glycogen. Thus muscular contraction is dependent on the cooperative action of several systems in carbohydrate metabolism.

PHOTOSYNTHESIS

Carbohydrates are formed in the cells of plants from carbon dioxide and water. In the presence of sunlight and chlorophyll, the green pigment of leaves, these two compounds react to form pentoses, trioses, fructose, and more complex sugars. **Chlorophyll** is a protoporphyrin derivative containing magnesium that is located in the chloroplasts of green leaves.

Originally the reaction between carbon dioxide and water to form carbohydrates was represented as follows:

$$CO_2 + H_2O \xrightarrow[\text{chlorophyll}]{\text{sunlight}} \underset{\text{Simple sugar}}{C_6H_{12}O_6} + 6\ O_2$$

This process by which plants convert the energy of sunlight to form food material is called **photosynthesis.** Although photosynthesis is represented as a simple chemical reaction, it is more complex and includes several intermediates of the phosphogluconate and Embden-Meyerhof pathways.

The use of isotopes and radioactive tracers has greatly assisted the research workers in this field. When a green leaf is grown in an atmosphere of $^{14}CO_2$, the radioactive carbon appears very rapidly in a three-carbon atom and a five-carbon atom intermediate, and later in glucose and starch.

The reaction involving the conversion of light energy into chemical energy is called the **light reaction.** This transformation of energy occurs during photosynthetic phosphorylation and takes place in the chloroplasts of plants. The essential reaction that occurs in **photophosphorylation** can be represented as follows:

$$ADP + P_i \xrightarrow{\text{light energy}} ATP$$

The photochemical process is initiated by the absorption of light by chlorophyll, which produces an excited-state molecule in which several electrons are raised from their normal energy level to a higher level in the double bond structure of chlorophyll. These excited electrons flow from chlorophyll to an iron-containing protein, **ferredoxin,** and bring about the reduction of NADP to form $NADPH_2$, which is used in the CO_2 fixation reactions of photosynthesis. Some of the excited electrons flow from ferredoxin through flavin pigments to a quinone structure called **plastoquinone,** then to **cytochrome pigments,** and then back to chlorophyll and their normal energy level. During this cycle some of the energy is given up by coupling in the reaction of ADP with P_i to form ATP.

The incorporation of carbon into carbohydrates has been called the **dark reaction,** since it is not dependent on light energy. This fixation of carbon dioxide during photosynthesis takes place in a cycle of reactions. A pentose phosphate, **ribulose-1,5-diphosphate,** is the key compound that combines with carbon dioxide, with the formation of two molecules of 3-phosphoglyceric acid. The phosphoglyceric acid is then reduced to 3-phosphoglyceraldehyde by a dehydrogenase enzyme with NADP as a coenzyme. This triose phosphate is condensed to fructose phosphate and finally to glucose by the action of **aldolase.** The reactions occurring in this cycle that convert carbon dioxide to glucose are similar to those in the phosphogluconate oxidative pathway (see p. 385).

Questions

1. Describe the digestion of starch in the gastrointestinal tract.

2. Briefly describe the digestion and absorption of dietary fat.

3. Outline the process of digestion and absorption of proteins in the diet.

4. What is the normal fasting level of blood glucose? What values of blood glucose would be considered hypoglycemic? hyperglycemic?

5. Discuss the factors involved in counteracting the normal hyperglycemia that occurs after a meal.

6. Explain the function of the liver in the control of the normal blood sugar level.

7. List the main hormones that are involved in the control of the blood sugar level. If the insulin production of the body drops below normal, what happens to the blood sugar? Explain.

8. Describe the process of glycogenesis.

9. What is cyclic-3′,5′-AMP? What role does it play in glycogenesis?

10. How do the reactions of the lactic acid cycle explain the fate of the lactic acid formed by the process of glycolysis? Explain.

11. How many moles of ATP per glucose molecule are required to run the Embden-Meyerhof pathway, and how many moles are produced?

12. One turn of the Krebs cycle will yield how many moles of ATP? Why the large number compared to the Embden-Meyerhof pathway?

13. What are the major products of the phosphogluconate oxidative pathway? How are they used in other pathways and cycles?

14. What is the relation between ATP and creatine phosphate in muscular contraction?

15. What is photosynthesis? Explain the difference between the so-called "light reaction" and the "dark reaction."

16. Explain how any three compounds formed in the CO_2 fixation scheme in photosynthesis could enter the reactions of the Embden-Meyerhof or Krebs cycle.

Suggested Reading

Arnon: The Role of Light in Photosynthesis. Scientific American, Vol. 203, No. 5, p. 104, 1960.
Bergen: Tracer Isotopes in Biochemistry. Journal of Chemical Education, Vol. 29, p. 84, 1952.
Calvin and Bassham: The Photosynthesis of Carbon Compounds. New York, W. A. Benjamin Inc., 1962.
Heidt: The Path of Oxygen from Water to Molecular Oxygen. Journal of Chemical Education, Vol. 43, p. 623, 1966.
Horecker: Pathways of Carbohydrate Metabolism and Their Physiological Significance. Journal of Chemical Education, Vol. 42, p. 244, 1965.
Lehninger: Energy Transformation in the Cell. Scientific American, Vol. 202, No. 5, p. 102, 1960.
Levine: The Mechanism of Photosynthesis. Scientific American, Vol. 221, No. 6, p. 58, 1969.
Park: Advances in Photosynthesis. Journal of Chemical Education, Vol. 39, p. 424, 1962.

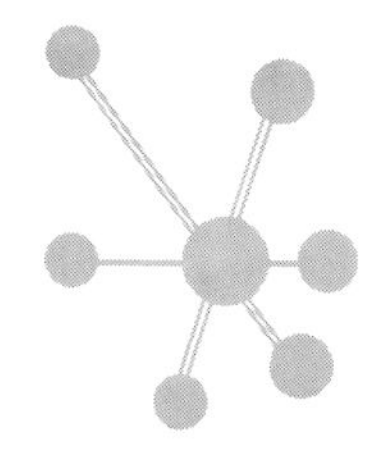

CHAPTER 27

LIPID METABOLISM

In the preceding chapter it was pointed out that the energy for many of the body activities is derived from the metabolism of carbohydrates. Fat stored in the fat depots of the body also represents a rich source of energy, especially since the caloric value of fats is more than twice that of carbohydrate or protein. The major stores of foodstuff in the body are therefore carbohydrate or fat in nature, and the body obtains most of its required energy from the oxidation of carbohydrates or fats.

Although the major energy source is fat, the metabolism of the lipids also involves phospholipids, glycolipids, and sterols. These latter substances are not stored in the fat depots but are essential constituents of tissues that play a role in fat transport and in many cellular metabolic reactions. These lipid derivatives also function as components of cell membranes, nerve tissue, membranes of subcellular particles such as microsomes and mitochondria, and chloroplasts in green leaves.

Unsaturated fatty acids such as **linoleic** and **linolenic** are essential components of cellular lipids that must be obtained in the diet, since they cannot be synthesized by the body. Cholesterol can be readily synthesized by the tissues and is currently a topic of considerable controversy, since there may be a relation between dietary cholesterol, blood cholesterol levels, and atherosclerosis. The digestion of dietary lipids and the absorption of the end products were discussed in Chapter 26.

BLOOD LIPIDS

The blood lipids to a certain extent parallel the behavior of the blood sugar. Their concentration in the blood increases after a meal and the level is returned to normal by processes of storage, oxidation, and excretion.

The lipids of the blood are constantly changing in concentration as lipids are added by absorption from the intestine, by synthesis, and by removal from the fat depots; they are removed by storage in the fat depots, oxidation to produce energy, synthesis to produce tissue components, and excretion into the intestine. The **normal fasting level** of blood lipids is usually measured in the plasma. Average values for young adults are as follows:

	mg/100 ml
Total lipids	510
Triglycerides	150
Phospholipids	200
Total cholesterol	160

The triglycerides, phospholipids, and cholesterol in the plasma are combined with protein as lipoprotein complexes. These **lipoproteins** are bound to the α- and β-globulin fractions of the plasma proteins and are transported in this form. A small amount of **nonesterified fatty acids (NEFA)** is always present in the blood and is bound to the albumin fraction of the plasma for transportation. These free fatty acids are thought to be the most active form of the lipids involved in metabolism. Their concentration is affected by the mobilization of fat from fat depots and by the action of several hormones.

FAT STORAGE

Fats may be removed from the blood stream by storage in the various fat depots. When fat is stored under the skin, it is usually called **adipose tissue.** However, considerable quantities of fat may be stored around such organs as the kidneys, heart, lungs, and spleen. This type of depot fat acts as a support for these organs and helps to protect them from injury. Recent studies employing the electron microscope reveal two major types of storage fat. One type is composed almost entirely of fat globules and has the characteristics of a storage depot. The second type contains many cells and a more extensive blood circulation, and is metabolically active, converting glycogen to fat and releasing fatty acids to other tissues as energy sources.

Obesity. Obesity is the condition in which excessive amounts of fat are stored in the fat depots. In a small percentage of cases, obesity is due to a disorder of certain endocrine glands, but as a general rule it results from eating more food than the body requires. Most of the food consumed by an adult is used to produce energy, and food in excess of that necessary to fulfill the energy requirements of the body is stored as fat. Thin people generally are more active than fat people and are able to eat larger amounts of food without putting on weight.

Many people apparently eat all they want and yet maintain a fairly constant weight over long periods of time. This weight control may be due to the appetite, which is abnormally increased in people that are gaining weight and decreased in those who are losing weight. Recent investigations, however, have cast doubt on the simple explanation of overeating as the only factor responsible for obesity. Apparently there are some individuals who can maintain obesity or increase their weight on a low calorie diet. If they attempt to lose weight by decreasing the intake of food, their rate of metabolism decreases and they require fewer calories to maintain their activity. This combination of *endocrine balance, rate of metabolism,* and *difference in requirement for calories* is as yet not completely understood.

THE SYNTHESIS OF TISSUE LIPIDS

Lipids such as **phospholipids, glycolipids,** and **sterols** are essential constituents of cells, protoplasm, and tissues in various parts of the body. They are also involved in specialized functions, i.e., blood clotting mechanisms and in transportation of lipids in the blood. The adipose tissue that is stored around organs of the body does not contain the same proportion of saturated or unsaturated fatty acids as the food fat and therefore must also be synthesized. The most important organ in the body concerned with lipid synthesis is the liver. It is able to synthesize phospholipids and cholesterol and to modify all blood fats by lengthening or shortening, and saturating or unsaturating, the fatty acid chains.

Lecithin is used in transporting fats to the various tissues and may be involved in the oxidation of fats. Another essential phosopholipid is **cephalin,** which is a vital factor in the clotting of blood. Special fats and oils in the body such as milk fat, various sterols, the natural oil of the scalp, and the wax of the ear are examples of lipids synthesized from the fats of the food.

OXIDATION OF FATTY ACIDS

Fatty acids that arise from the breakdown of any lipid, but especially from fats, are oxidized completely to form CO_2, water, and energy. The glycerol portion of fats is converted into glycerol aldehyde and enters the chain of reactions followed by the three carbon derivatives from carbohydrate metabolism.

The oxidation of fatty acids occurs in a series of reactions that require several enzymes and cofactors, with the production of acetyl coenzyme A. The acetyl CoA molecules then enter the Krebs cycle to form CO_2, H_2O, and energy. Early research by Knoop in 1904 established the fact that fatty acids were oxidized on the beta-carbon atom with the subsequent splitting off of two carbon fragments. In his **theory of beta-oxidation** he stated that acetic acid was split off in each stage of the process that reduced an 18-carbon fatty acid to a two-carbon acid.

In the past few years the detailed reactions, with their enzymes and cofactors, have been worked out, and Knoop's theory has been confirmed. Instead of acetic acid, the key compound in the reactions is acetyl CoA. Five reactions are involved in the conversion of a long chain fatty acid into a CoA derivative with two less carbon atoms and a molecule of acetyl CoA. These reactions are outlined in the scheme that is shown in the accompanying diagram.

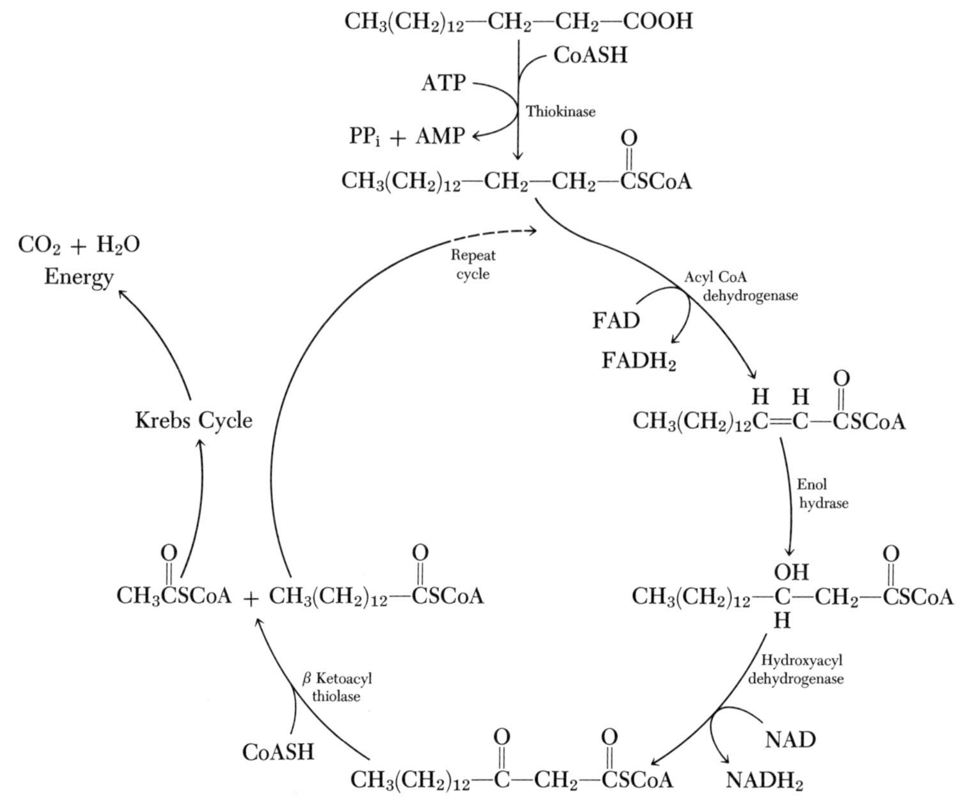

The first reaction initiates the series and involves the activation of a fatty acid molecule by conversion into a coenzyme A derivative. A dehydrogenase enzyme, with FAD as the coenzyme, desaturates the fatty acid; then a hydration is catalyzed by an enol hydrase. The hydroxyl group on the β-carbon atom is oxidized by a dehydrogenase with NAD as a coenzyme. The oxidized derivative plus coenzyme A is split into a fatty acid molecule with 2 less carbons, and acetyl CoA is formed. The acetyl CoA enters the Krebs cycle to form CO_2 and H_2O, plus energy. The new fatty-acid coenzyme A derivative re-enters the cycle and again loses an acetyl CoA molecule. *Palmitic acid would require seven turns of the cycle to form* 8 *acetyl CoA moles.*

During the oxidation of palmitic acid, 7 $FADH_2$ and 7 $NADH_2$ moles would be formed. When these molecules enter the electron transport chain, they would form ATP as shown:

$$
\begin{array}{lrl}
7\ FADH_2 \rightarrow & 14\ ATP & \\
7\ NADH_2 \rightarrow & \underline{21\ ATP} & \\
 & 35\ ATP & \\
 & \underline{-\ 1\ ATP} & \text{used in first reaction} \\
 & 34\ \text{Net ATP} &
\end{array}
$$

In the seven turns of the cycle, 8 acetyl CoA moles are formed; then, when oxidized in the Krebs cycle,

$$8\ \text{Acetyl CoA} + 16\ O_2 \xrightarrow[\text{cycle}]{\text{Krebs}} 16\ CO_2 + 8\ H_2O + 8\ \text{CoASH}$$

In the oxidative phosphorylation process, 3 moles of ATP are formed for each oxygen atom, or $\frac{1}{2}O_2$, consumed. Therefore,

$$16\ O_2 = 32\ O \times 3 = 96\ ATP$$

The sum of 34 + 96 = 130 ATP for the complete oxidation of palmitic acid in the above scheme. The total combustion of palmitic acid yields 2,338,000 calories, and when compared to cellular oxidation,

$$\frac{130 \times 8000\ \text{cal./mole} \times 100}{2{,}338{,}000} = 48\%$$

This represents a very efficient conservation of energy in the form of ATP molecules when palmitic acid is completely oxidized by the tissues. The previous discussion emphasizes the statement that food fat is an effective source of available energy. Also, a contributing factor to this efficient utilization is the fact that all the enzymes utilized in the β-oxidation scheme, the Krebs cycle, oxidative phosphorylation, and electron transport are found in the **mitochondria** of the cell.

SYNTHESIS OF FATTY ACIDS

The β-oxidation pathway in the mitochondria can be reversed to form fatty acid molecules, but this accounts for only a small percentage of the fatty acids synthesized in the tissues. The **cytoplasm** of the cell is the major site, and acetyl coenzyme A is the starting material, for the synthesis. Acetyl coenzyme A is carboxylated to form **malonyl coenzyme A** under the influence of acetyl CoA carboxylase in the presence of ATP and the vitamin biotin. Malonyl CoA then forms a complex with an enzyme system called fatty acid synthetase to catalyze the addition of two carbons of the malonic acid to a small organic acid CoA derivative and this forms a β-keto enzyme complex. The β-keto group is converted to a hydroxyl group by NADPH, and then a water molecule is split out to form an α,β-unsaturated derivative. The unsaturated compound plus NADPH forms a 2-carbon-longer fatty acid enzyme derivative, from which the enzyme is split off before adding another mole of coenzyme A to start through the cycle again. Several of these reactions are similar to the reverse of the oxidative cycle, and by adding 2-carbon fragments in each turn of the cycle eventually form palmitic acid. The overall reaction may be represented as follows:

$$CH_3-\overset{\overset{\large O}{\|}}{C}-SCoA + 7\ \underset{\text{Malonyl CoA}}{\overset{\overset{\large COO^-}{|}}{CH_2}-\overset{\overset{\large O}{\|}}{C}-SCoA} + 14\ NADPH$$

$$\downarrow$$

$$CH_3(CH_2CH_2)_7COOH + 8\ HSCoA + 7\ CO_2 + 14\ NADP$$

SYNTHESIS OF TRIGLYCERIDES

Triglycerides are synthesized in the tissues from glycerol and fatty acids in activated forms. The active form of glycerol is **glycerophosphate,** which reacts with two fatty acid CoA derivatives to form a **diglyceride,** which then reacts with another mole of fatty acid CoA to form a **triglyceride.** An outline of the process can be illustrated by the following scheme:

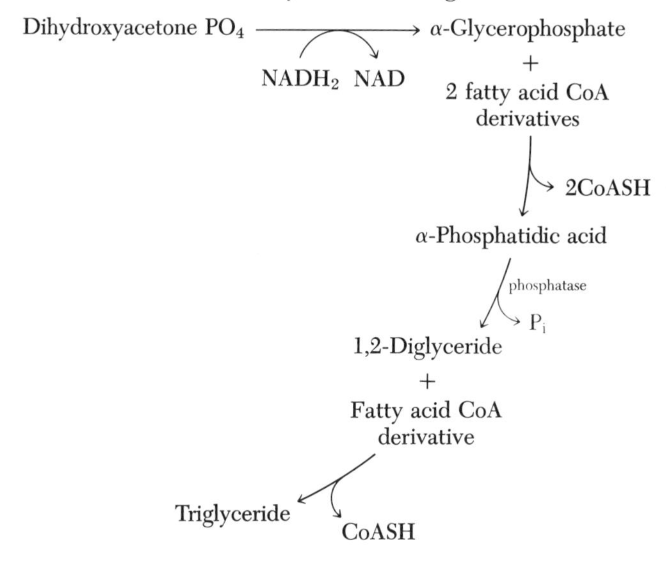

FORMATION OF KETONE BODIES

The ketone, or acetone, bodies consist of **acetoacetic acid, β-hydroxybutyric acid,** and **acetone.** In a normal individual they are present in the blood in small amounts, but large amounts are present in the blood and urine during starvation and in the condition of diabetes mellitus. In general, any condition that results in a restriction of carbohydrate metabolism, with a subsequent increase in fat metabolism to supply the energy requirements of the body, will produce an increased formation of ketone bodies. This condition is called **ketosis.**

The precursor of the ketone bodies is acetoacetic acid which is formed in the liver from acetoacetyl CoA, a normal intermediate in the beta-oxidation of fatty acids. It may also be formed by the condensation of two molecules of acetyl CoA. Both methods of formation can be represented in the normal reversible reaction as follows:

$$\underset{\text{Acetyl CoA}}{2\ CH_3\overset{\overset{O}{\|}}{C}SCoA} \xrightleftharpoons{\text{thiolase}} \underset{\text{Acetoacetyl CoA}}{CH_3\overset{\overset{O}{\|}}{C}CH_2\overset{\overset{O}{\|}}{C}SCoA} + CoASH$$

The liver contains a deacylase enzyme which readily converts acetoacetyl CoA to the free acid.

$$CH_3\overset{O}{\overset{\|}{C}}CH_2\overset{O}{\overset{\|}{C}}SCoA + H_2O \xrightarrow{\text{deacylase}} CH_3\overset{O}{\overset{\|}{C}}CH_2COOH + CoASH$$

Acetoacetyl CoA — Acetoacetic acid

The other ketone bodies are formed from acetoacetic acid; acetone by decarboxylation and β-hydroxybutyric acid by the action of a specific enzyme, as shown in the accompanying diagram.

$$CH_3\overset{O}{\overset{\|}{C}}CH_2COOH$$

$$\xrightarrow{-CO_2} CH_3\overset{O}{\overset{\|}{C}}CH_3 \quad \text{Acetone}$$

$$\xrightarrow[NADH_2 \rightarrow NAD]{\beta\text{-hydroxybutyric dehydrogenase}} CH_3\overset{H}{\underset{OH}{C}}\text{—}CH_2COOH \quad \beta\text{-Hydroxybutyric acid}$$

PHOSPHOLIPID METABOLISM

Knowledge concerning the metabolism of the phospholipids is incomplete, although they are known to serve many important functions in the body. Because their molecules are more strongly dissociated than any of the other lipids, they tend to be more soluble in water, to lower surface tension at oil-water interfaces, and to be involved in the electron transport system in the tissues. They would have a tendency to concentrate at cell membranes, and are probably involved in the transport mechanisms for carrying fatty acids and lipids across the intestinal barrier and from the liver and fat depots to other body tissues. Phospholipids are essential components of the blood clotting mechanism, and sphingomyelin is one of the principal components of the myelin sheath of nerves.

STEROL METABOLISM

The metabolism of sterols is mainly concerned with cholesterol and its derivatives. The **synthesis of cholesterol** and its relation to the other steroids of the body has been the subject of considerable research. Using either stable or radioactive isotopes, it has been shown that cholesterol can be synthesized from two-carbon compounds such as acetyl CoA. It can also be synthesized from acetoacetyl CoA and other intermediates. Although the synthesis of cholesterol occurs in many tissues in the body, the liver is the main site of cholesterol formation.

Cholesterol is a key compound in the synthesis of essential steroids such as bile acids, sex hormones, adrenal cortical hormones, and vitamin D. Not only is cholesterol converted to bile acids by the liver, but it is also excreted as such in the bile. The concentration of cholesterol in the blood is apparently dependent on the dietary intake of sterols and neutral fats, and the synthesis of cholesterol by the liver. The normal level in the blood gradually increases with age and ranges from 150 to 200 mg/100 ml. Blood cholesterol levels are often determined in patients to assess their cholesterol status. Many methods have been devised for this determination, and many of them are modifications of the Liebermann-Burchard reaction described in Chapter 21. If the cholesterol level in the blood is maintained at an abnormally high concentration, such as 200 to 300 mg/100 ml, deposition of cholesterol plaques may occur in the aorta and lesser arteries. This condition, known as **atherosclerosis** or **arteriosclerosis,** is seen in older persons and often results in circulatory or heart failure. Considerable research effort is being directed at this problem in an attempt to reduce the cholesterol level in the blood of these patients and thus alleviate the symptoms of the disease.

CORRELATION OF CARBOHYDRATE AND FAT METABOLISM

From a nutritional standpoint it has long been apparent that carbohydrate can be converted into fat in the body. When glucose tagged with ^{14}C was fed to animals, the fatty acids of liver and other tissue fat were found to be labeled with ^{14}C. The conversion of fat to carbohydrate has long been open to question. The glycerol portion of fat is closely related to the three-carbon intermediates of carbohydrate metabolism, but it has been more difficult to demonstrate a direct relation between fatty acids and glucose. Since the role of acetyl CoA has been established in both carbohydrate and fat metabolism, it is apparent that the acetyl CoA from fatty acid oxidation can enter the Krebs cycle in the same fashion as this compound formed from pyruvic acid. More recently it has been shown that ^{14}C labeled fatty acids are converted to ^{14}C labeled glucose in a diabetic animal.

Questions

1. List the important lipids in a normal individual's blood and their approximate concentration.

2. Discuss some of the advantages and disadvantages of a generous supply of adipose tissue.

3. Briefly explain the major difference between Knoop's scheme for the oxidation of fatty acids and the modern scheme.

4. How do you account for the large number of moles of ATP formed in the oxidation of fatty acids?

5. How does the scheme for the synthesis of fatty acids differ from that for the oxidation of fatty acids? How is it similar?

6. Outline the scheme for the formation of acetoacetic acid in the liver.

7. Show how the other ketone bodies are formed from acetoacetic acid.

8. What is ketosis? Explain the cause of this condition in the body.

9. Name four important compounds in the body that are synthesized from cholesterol. From what compound is cholesterol synthesized?

10. Why is atherosclerosis receiving so much attention in our society?

Suggested Reading

Bergen: Tracer Isotopes in Biochemistry. Journal of Chemical Education, Vol. 29, p. 84, 1952.
Gibson: The Biosynthesis of Fatty Acids. Journal of Chemical Education, Vol. 42, p. 236, 1965.
Green: Metabolism of Fats. Scientific American, Vol. 190, No. 1, p. 32, 1954.
Green: The Synthesis of Fat. Scientific American, Vol. 202, No. 2, p. 46, 1960.
Spain: Atherosclerosis. Scientific American, Vol. 215, No. 2, p. 48, 1966.

CHAPTER 28

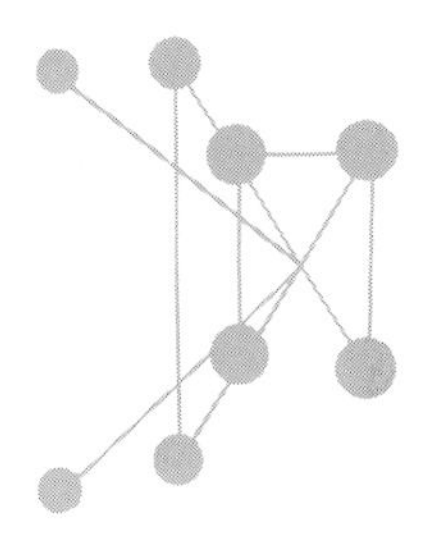

PROTEIN METABOLISM

The metabolism of proteins is concerned with a large variety of complex molecules. Tissue proteins of various species of animals, plants, and microorganisms all have specific structures and compositions. Protein enzymes and hormones, plasma proteins, and the protein of hemoglobin and of various nucleoproteins represent other types of proteins. **Anabolism,** or the synthesis of new proteins for growth and development, involves the building of different amino acids into the proper sequences and spatial arrangements to produce specific protein molecules. The process of **catabolism** of proteins to produce energy involves many general metabolic reactions and many that are specific for the metabolism of each of the twenty different amino acids. The digestion of dietary protein and the absorption of amino acids were discussed in Chapter 26.

THE AMINO ACID POOL

After absorption, the amino acids are carried by the portal circulation to the liver and subsequently to all the tissues of the body. The liver removes some amino acids for its requirements and adds to the blood those it has synthesized. Other tissues add amino acids to the blood as their proteins undergo catabolism, or breakdown. In contrast to carbohydrate and fat metabolism, there are no storage depots for proteins or amino acids. The increased concentration of amino acids that occurs from the process of absorption, synthesis, or catabolism represents a temporary pool of amino acids which may be used for metabolic purposes. This pool of amino acids is available to all tissues and may be synthesized into new tissue proteins, blood proteins, hormones, enzymes, or nonprotein nitrogenous substances such as creatine and glutathione. The relationship that exists between the **amino acid pool** and protein metabolism in general may be represented as shown in the accompanying diagram.

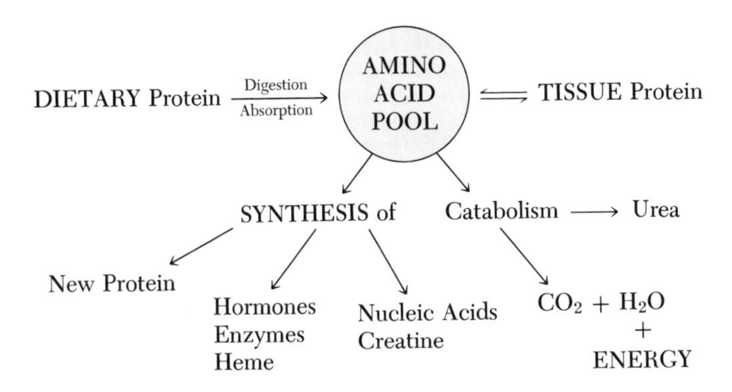

THE DYNAMIC STATE OF BODY PROTEIN

Until the late 1930's it was believed that the body proteins of the adult human were stable molecules and that the majority of the amino acids from the diet were catabolized to produce energy. A small proportion was thought to be used for maintenance and repair of the existing tissue proteins. When isotopes became available, Schoenheimer and his associates demonstrated that tissue proteins exist in a *dynamic state of equilibrium*. When the nitrogen of an amino acid was labeled with N^{15} and incorporated in the diet of an animal, about 50 per cent of the N^{15} was found in the tissues of the animal, and a greater percentage was found in the nitrogen of amino acids other than that specifically fed. This indicated that the amino acids of tissue proteins were constantly changing places with those in the amino acid pool, and that the body proteins were extremely labile molecules.

More recent research using isotopically labeled amino acids has shown that tissue proteins vary considerably in their rate of turnover of amino acid molecules. The **turnover rate** represents the amount of protein synthesized or degraded per unit time, and the **turnover time** is usually expressed as the half-life of a protein in the tissues. Liver and plasma proteins have turnover times (a half-life) of 6 days, in contrast to 180 days for muscle protein and 1000 days for some collagen proteins. Muscle and connective tissue proteins appear to have a very prolonged turnover compared to liver and plasma proteins, which are rapidly synthesized from the amino acids in the amino acid pool. The concept of the dynamic state of body protein requires modification in view of the individuality of specific proteins.

THE SYNTHESIS OF PROTEIN

The process of synthesis of protein is always occurring in the body, especially in those tissues with a rapid turnover rate. A growing child or animal is continually building new tissue and therefore makes the greatest demand on the amino acid pool. The individual amino acids required for protein synthesis are apparently sorted out by the body and used to construct specific protein molecules. Although the tissues, particularly the liver, are able to synthesize some amino acids, others must be present in the diet to assure a complete and proper assortment for synthetic purposes.

ESSENTIAL AMINO ACIDS. The amino acids that cannot be synthesized by the body and must therefore be supplied by the dietary protein are called **essential amino acids.** If an essential amino acid is lacking in the diet, the body is unable to synthesize tissue protein. If this condition occurs for any length of time, a negative nitrogen balance will exist, and there will be a loss of weight, a lowered level of serum protein, and a marked edema. Extensive feeding experiments on laboratory rats have established the following amino acids as essential for growth:

Histidine	Isoleucine
Methionine	Leucine
Arginine	Lysine
Tryptophan	Valine
Threonine	Phenylalanine

From the studies of Rose and his associates at the University of Illinois on the amino acid requirements of man, it has been suggested that all of the above ten except histidine and possibly arginine are essential to maintain nitrogen balance. An individual is in **nitrogen balance** when the nitrogen excreted equals the nitrogen intake in a given period of time. A growing child or a patient recovering from a prolonged illness is in **positive nitrogen balance.** Starvation, a wasting disease, or a diet lacking sufficient amounts of essential amino acids can result in a **negative nitrogen balance.**

Many common dietary proteins are deficient in one or more of these essential amino acids. Gelatin, for example, lacks tryptophan and is therefore an **incomplete protein.** Zein and gliadin, the prolamines of corn and wheat respectively, are deficient in lysine, and zein is also low in tryptophan. Although an incomplete protein will not support growth when it is the only protein in the diet, we seldom confine ourselves to the consumption of a single protein. In an ordinary mixed diet the essential amino acids are best supplied by proteins of animal origin, such as meat, eggs, milk, cheese, and fish. A daily intake of protein of 1 to 1.5 grams per kilogram of body weight is recommended as adequate for body needs.

MECHANISM OF PROTEIN SYNTHESIS. For many years the synthesis of proteins in the body was explained as a reversal of the action of hydrolyzing enzymes, causing the formation of peptide bonds instead of the splitting of the bonds. The mechanism by which the tissues assured the proper assortment of amino acids in a newly synthesized protein was not explained. Although complete details of the synthesis of tissue proteins with the proper concentration and sequence of amino acids have not been settled, knowledge is growing rapidly in this field. In fact, a very active field of research at the present time involves the mechanism of protein synthesis, the sequence of amino acids in the protein being synthesized, and the nature of the genetic code responsible for this sequence.

Protein synthesis is initiated by the activation of amino acids. This process occurs by the combination of the amino acid with ATP and an enzyme specific for the amino acid, with the splitting off of two molecules of phosphoric acid.

$$\mathrm{R{-}\underset{NH_2}{\overset{H}{C}}{-}COOH} + \mathrm{ATP} + \xrightarrow{\text{Specific enzyme}} \overbrace{\mathrm{R{-}\underset{NH_2}{\overset{H}{C}}{-}\overset{O}{\overset{\|}{C}}{-}O{-}\underset{O^-}{\overset{O}{\overset{\|}{P}}}{-}O{-}Ribose\text{-}Adenine}}^{\text{Enzyme}} + \mathrm{PP_i}$$

Amino acyl adenylate-enzyme complex

The second step of the process involves the transfer of the activated amino acid to a **soluble RNA molecule.** The soluble molecules, s-RNA, are small (mol. wt. 30,000) nucleic acid molecules with a terminal grouping of cytidylic-cytidylic-adenylic acid represented as s-RNA-C-C-A. The transfer of the activated amino acid is under the influence of the same enzyme that produced the activation and CTP, or cytidine triphosphate.

$$\mathrm{s\text{-}RNA{-}C{-}C{-}A} + \overbrace{\mathrm{R{-}\underset{NH_2}{\overset{H}{C}}{-}\overset{O}{\overset{\|}{C}}{-}O{-}\underset{O^-}{\overset{O}{\overset{\|}{P}}}{-}O{-}Ribose\text{-}Adenine}}^{\text{Enzyme}} \xrightarrow{\text{CTP}}$$

$$\mathrm{s\text{-}RNA{-}C{-}C{-}A{-}O{-}\overset{O}{\overset{\|}{C}}{-}\underset{NH_2}{\overset{H}{C}}{-}R} + \mathrm{AMP} + \text{Enzyme}$$

s-RNA bound amino acid

The third step involves the transfer of s-RNA-bound amino acids to the ribosomes of the cellular cytoplasm. Ribosomes are nucleoprotein particles composed of basic proteins and **ribosomal RNA** (r-RNA). In the cytoplasm ribosomes are joined by strands of messenger RNA to form polyribosomes or polysomes, which consist of 4 to 6 ribosomes. The transfer reactions require ATP, GTP, soluble transfer enzymes, and reduced glutathione.

Another type of RNA that is closely associated with protein synthesis and the ribosomes is **messenger RNA** (m-RNA). It is synthesized by RNA polymerase under the direction of the DNA in the nucleus, and it carries information to the ribosomes to direct the sequence of alignment of s-RNA-bound amino acids. The m-RNA chains hold ribosomes together in polysomes, and the s-RNA-bound amino acids apparently attach to a site on m-RNA. Apparently ribosomes possess at least two binding sites for s-RNA bound amino acids and a third site for unbound s-RNA. It appears that one molecule of amino acid bound s-RNA is readily attached to the messenger RNA-ribosome complex. The binding of a second amino acid bound s-RNA requires GTP and a soluble transfer enzyme and involves moving the first amino acid s-RNA over one position on the messenger RNA-ribosome complex. Peptide bond formation takes place by attachment of the first amino acid to the free amino group of the second amino acid bound to s-RNA. The third amino acid s-RNA molecule then attaches to the messenger RNA ribosome complex moving the second dipeptide s-RNA over another position. The separated s-RNA on the third position then splits off of the messenger-RNA-ribosome to join a newly activated amino acid, while

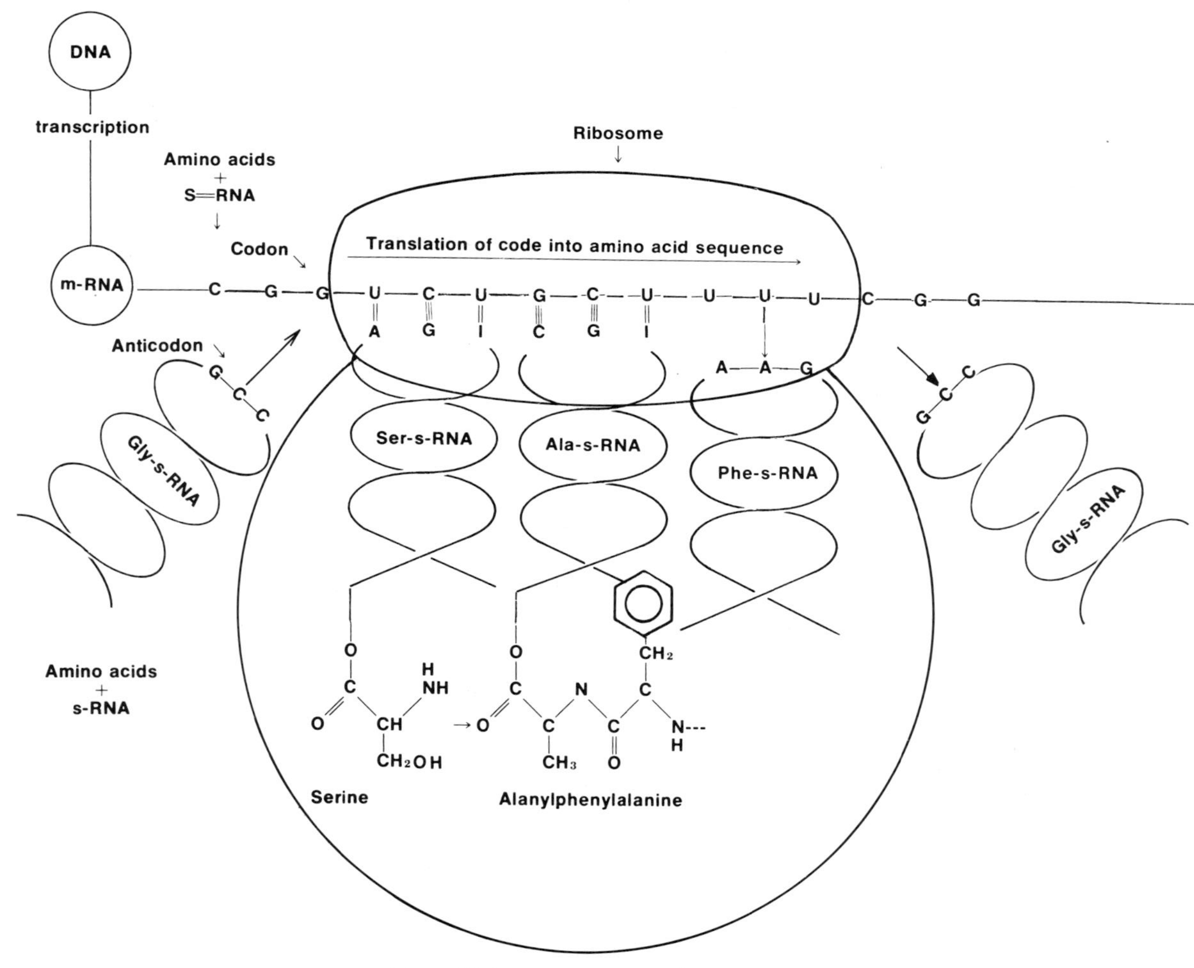

Figure 28-1 Representative scheme of protein synthesis.

the new amino acid forms an additional peptide linkage to produce a tripeptide. This procedure is repeated until a polypeptide of sufficient size is formed and then splits off to form a specific protein molecule. This first process of overall information transfer from DNA to m-RNA is called **transcription** (transcribing the message). The specific programing of the amino acids on the polysomes or m-RNA molecules to synthesize a protein containing a definite sequence of amino acids is called **translation** (translating the code). As the polypeptide increases in size, it eventually splits off from its attachment to the m-RNA on the polysomes to form a protein chain.

The nature of the **genetic code** carried by the nuclear DNA molecules to direct the sequence of amino acids in the synthesis of a specific protein is the subject of current research. The 20 amino acids must be coded by the use of the four nucleotide residues of DNA represented by A, C, U, and G, or **adenine, cytosine, uracil,** and **guanine,** in the nucleotides. Research carried out by Nirenberg has established a basis for understanding the nature of the genetic code involved in protein synthesis. Apparently the sequence of bases on the m-RNA directs the specific synthesis of the amino acid sequence of a protein. The unit that codes for a single amino acid, called a **codon,** is composed of a group of three adjacent nucleotide residues on the m-RNA chain. The next three nucleotide residues on the chain then code for the next amino acid. The base sequence on the s-RNA that corresponds to the codon for an individual amino acid is called an anticodon. The overall scheme of protein synthesis as described above is represented in Figure 28–1. Since each sequence is three residues long, a total of 4^3 or 64 different combinations of residues would be available for coding. The first codon to be demonstrated by Nirenberg's research

TABLE 28–1

Amino Acids	*Codons or Code Words for m-RNA*
Alanine	CCG, CUG, CAG, GCU
Arginine	GUC, CGC, AGA, CGC
Asparagine	CUA, UAA, CAA
Aspartic acid	GUA, GCA
Cysteine	UGU, GGU, UGC
Glutamic acid	GAA, GAG, AUG
Glutamine	ACA, UAC, CAG
Glycine	CGG, UGG, AGG
Histidine	CAC, AUC
Isoleucine	UAU, AAU, CAU
Leucine	CUC, UGU, UAU
Lysine	AAA, AAG, AUA
Methionine	UGA
Phenylalanine	UUU, UUC
Proline	CCC, CAC, CUC
Serine	UCG, CCU, UCU
Threonine	ACA, UCA, CAC
Tryptophan	UGG
Tyrosine	UAU, ACU
Valine	GUG, UUG, GUA

was **polyuridylic acid,** or UUU, which coded for **phenylalanine.** The code is **nonoverlapping** in that it requires the action of a specific group of three residues intact on the m-RNA chain. Also, the code is said to be **degenerate** in that more than one codon may be employed by m-RNA to insert a specific amino acid into the peptide chain. Codons proposed for specific amino acids are listed in Table 28–1. The exact order of the bases in the codon is not known.

The factors involved in the complete synthesis of a large, complex protein molecule still excite the imagination. The synthesis of a simple protein such as insulin whose sequence and cross-linkage is known is not too difficult to visualize. Questions that remain unanswered are the mechanisms that establish the spatial arrangements of the polypeptide chains, the various crosslinkages, and the physiological activity of enzyme, hormone, and virus molecules.

METABOLIC REACTIONS OF AMINO ACIDS

The amino acids in the metabolic pool that are not immediately used for synthesis can undergo several metabolic reactions. They may follow the path of catabolism through deamination, urea formation, and energy production, or they may assist in the synthesis of new amino acids by the process of reamination and transamination. **Deamination, reamination, transamination,** and **urea formation** are processes common to all amino acids and are therefore very important to protein metabolism.

Deamination. A general reaction of catabolism is the splitting off of the amino group of an amino acid, with the formation of ammonia and a keto acid. This process is called **oxidative deamination** and is catalyzed by enzymes found in liver and kidney tissue called **amino acid oxidases.** These enzymes are generally flavoprotein enzymes containing either flavin adenine dinucleotide, FAD, or flavin mononucleotide, FMN. The enzyme dehydrogenates the amino acid to form an imino acid, which is hydrolyzed to a keto acid and ammonia. The process may be illustrated with a type formula for an amino acid.

$$\underset{\text{Amino acid}}{\mathrm{R{-}\underset{\displaystyle NH_2}{\underset{|}{C}H}{-}COOH}} + \text{Amino acid oxidase} \xrightarrow[\text{FAD or FMN}]{\text{protein-}} \underset{\text{Imino acid}}{\mathrm{R{-}\underset{\displaystyle NH}{\underset{\|}{C}}{-}COOH}} + \text{Protein-}FADH_2 \text{ or } FMNH_2$$

$$\underset{\text{Imino acid}}{\mathrm{R{-}\underset{\displaystyle NH}{\underset{\|}{C}}{-}COOH}} \xrightarrow[H_2O]{\text{hydrolysis}} \underset{\text{Keto acid}}{\mathrm{R{-}\underset{\displaystyle O}{\underset{\|}{C}}{-}COOH}} + NH_3$$

The fate of the keto acid depends on the amino acid from which it is derived. In general the catabolism of each amino acid must be studied separately. Glycine, for example, is the simplest amino acid, yet it can be transformed metabolically to formate, acetate, ethanolamine, serine, aspartic acid, fatty acids, ribose, purines, pyrimidines, and protoporphyrin. This amino acid may therefore play a role in carbohydrate, lipid, protein, nucleic acid, and hemoglobin metabolism, and it admirably illustrates the interrelationships that exist among the different types of metabolism in the body. Other amino acids undergo complex metabolic reactions that are beyond the scope of this book. In general, amino acids are classed **glycogenic** or **ketogenic,** indicating their capacity to form glucose or glycogen and to follow the path of carbohydrates, or to enter the metabolic reactions of lipids and to form ketone bodies.

Transamination. The process of deamination results in the formation of many keto acids that are capable of accepting an amino group to form a new amino acid. That this process of **reamination** occurs was readily apparent from the isotope-labeling experiments of Schoenheimer. He observed a ready exchange of amino groups of dietary amino acids and tissue amino acids. A major mechanism for the conversion of keto acids to amino acids in the body is known as **transamination.** The original transamination reactions involved glutamic and aspartic acids. Glutamic acid, for example, could react with oxalacetic acid in the presence of a transaminase to form a new keto acid, α-ketoglutaric, and the new amino acid, aspartic acid.

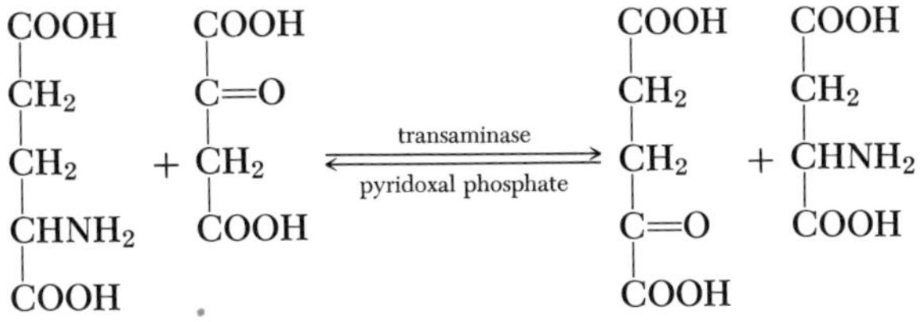

$$\underset{\text{Glutamic acid}}{\begin{array}{c}COOH\\|\\CH_2\\|\\CH_2\\|\\CHNH_2\\|\\COOH\end{array}} + \underset{\text{Oxalacetic acid}}{\begin{array}{c}COOH\\|\\C{=}O\\|\\CH_2\\|\\COOH\end{array}} \underset{\text{pyridoxal phosphate}}{\overset{\text{transaminase}}{\rightleftharpoons}} \underset{\alpha\text{-Ketoglutaric acid}}{\begin{array}{c}COOH\\|\\CH_2\\|\\CH_2\\|\\C{=}O\\|\\COOH\end{array}} + \underset{\text{Aspartic acid}}{\begin{array}{c}COOH\\|\\CH_2\\|\\CHNH_2\\|\\COOH\end{array}}$$

The coenzyme **pyridoxal phosphate** is required in the reaction and **pyridoxamine phosphate** is formed.

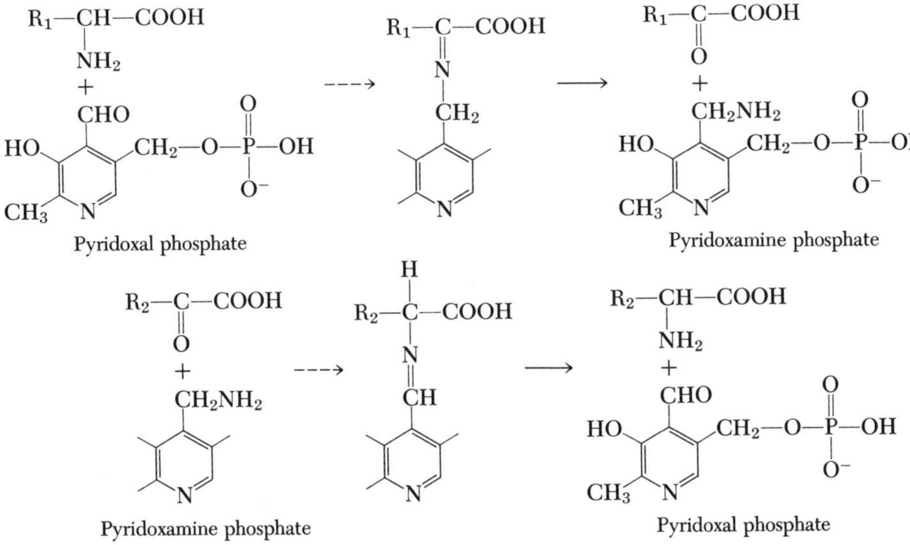

R_1—CH(—NH_2)—COOH + Pyridoxal phosphate (HO, CH_3, CHO, CH_2—O—P(=O)(—OH)—O^- pyridine ring) ---→ R_1—C(=N—CH_2—pyridine ring)—COOH ⟶ R_1—C(=O)—COOH + Pyridoxamine phosphate (HO, CH_3, CH_2NH_2, CH_2—O—P(=O)(—OH)—O^- pyridine ring)

Pyridoxal phosphate

Pyridoxamine phosphate

R_2—C(=O)—COOH + CH_2NH_2 (pyridine ring) ---→ R_2—C(H)(—N=CH—pyridine ring)—COOH ⟶ R_2—CH(—NH_2)—COOH + Pyridoxal phosphate (HO, CH_3, CHO, CH_2—O—P(=O)(—OH)—O^- pyridine ring)

Pyridoxamine phosphate

Pyridoxal phosphate

The enzyme that catalyzes the reaction in the serum is called SGOT, serum glutamic oxalacetic transaminase, and a sharp rise in its concentration in the serum is indicative of myocardial infarction, a heart condition involving the cardiac muscle. There are many specific transaminases that serve as catalysts in the transfer of amino groups from amino acids to a variety of keto acids. Transamination reactions serve as important links joining carbohydrate, fat, and protein metabolism. A keto acid from any source can be used for the synthesis of an amino acid to be incorporated in tissue protein.

Formation of Urea. The ammonia, carbon dioxide, and water that result from the deamination and oxidation of the amino acids are combined to form urea. Urea formation takes place in the liver by a fairly complicated series of reactions, first described by Krebs and his coworkers. The ammonia and carbon dioxide combine with the amino acid ornithine to form another amino acid, citrulline. Another molecule of ammonia then combines with the citrulline to form the amino acid arginine, which is then hydrolyzed by means of the enzyme arginase, present in the liver, to form urea and ornithine. The ornithine may then enter the beginning of the cycle and combine with more ammonia and carbon dioxide from protein catabolism.

In recent years the detailed mechanism of the cycle has been worked out. Apparently ornithine does not react directly with CO_2 and NH_3 to form citrulline, but reacts with a compound called **carbamyl phosphate.** This compound is synthesized from ATP, CO_2, and NH_3 in the presence of the specific enzyme carbamyl phosphate synthetase and the cofactors N-acetylglutamate and Mg^{+2}.

$$CO_2 + NH_2 + 2ATP \xrightarrow[Mg^{+2},\ \text{N-acetylglutamate}]{\text{carbamyl phosphate synthetase}} NH_2{-}\overset{\overset{\displaystyle O}{\|}}{C}{-}O{-}\underset{\underset{\displaystyle O^-}{|}}{\overset{\overset{\displaystyle O}{\|}}{P}}{-}OH + 2ADP + P_i$$

Carbamyl phosphate

The formation of arginine is also not a simple reaction of citrulline and NH_3 but involves a combination with aspartic acid to form argininosuccinic acid, which then splits into arginine and fumaric acid. The currently accepted **urea cycle** can be represented as shown on page 407.

As urea is formed in the liver it is removed by the blood stream, carried to the kidneys, and excreted in the urine. Urea is the main end-product of protein catabolism and accounts for 80 to 90 per cent of the nitrogen that is excreted in the urine.

CREATINE AND CREATININE

Creatine and creatinine are two nitrogen-containing compounds that are usually associated with protein metabolism in the body. **Creatine** is widely distributed in all tissues but is especially abundant in muscle tissue, where it is combined with phosphoric acid as **phosphocreatine,** or **creatine phosphate.** In the contraction of muscles, phosphocreatine apparently plays an important role as a reservoir of high-energy phosphate bonds readily convertible to ATP.

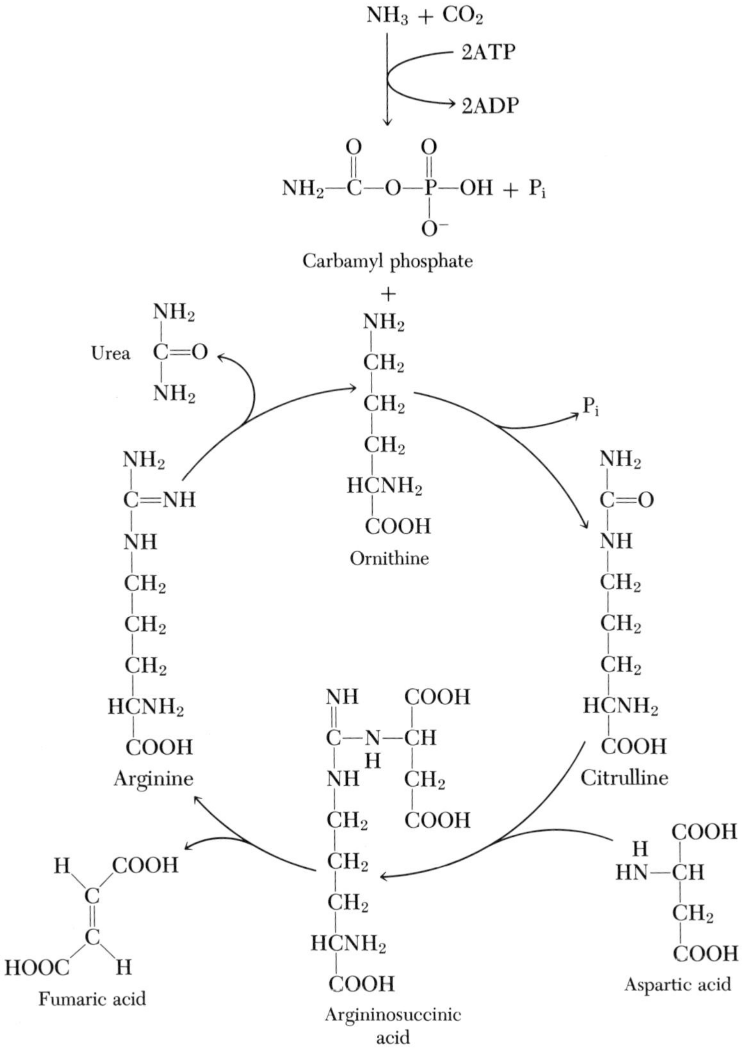

$$NH_3 + CO_2 \xrightarrow[\text{2ADP}]{\text{2ATP}} NH_2{-}\overset{\overset{O}{\|}}{C}{-}O{-}\overset{\overset{O}{\|}}{\underset{\underset{O^-}{|}}{P}}{-}OH + P_i$$

Carbamyl phosphate

+

$NH_2{-}CH_2{-}CH_2{-}CH_2{-}HCNH_2{-}COOH$

Ornithine

P_i

$NH_2{-}C{=}O{-}NH{-}CH_2{-}CH_2{-}CH_2{-}HCNH_2{-}COOH$

Citrulline

$HOOC{-}CH_2{-}CH(COOH){-}NH{-}H$

Aspartic acid

$NH{=}C(NH{-}CH_2{-}CH_2{-}CH_2{-}HCNH_2{-}COOH){-}\underset{H}{N}{-}CH(COOH){-}CH_2{-}COOH$

Argininosuccinic acid

$H(COOH)C{=}C(H)COOH$

Fumaric acid

$NH_2{-}C{=}NH{-}NH{-}CH_2{-}CH_2{-}CH_2{-}HCNH_2{-}COOH$

Arginine

Urea $NH_2{-}C{=}O{-}NH_2$

The energy for the initial stages of muscular contraction probably comes from the hydrolysis of this compound to form creatine and phosphoric acid. These two substances are later combined during the recovery period of the muscle (see page 385). Creatine is synthesized from the amino acids glycine, arginine, and methionine.

Creatinine is also present in the tissues but is found in much larger amounts in the urine. It is formed from either creatine phosphate or creatine and is an end product of creatine metabolism in muscle tissue. The relationship of creatinine to creatine and creatine phosphate is outlined on p. 408.

NUCLEOPROTEIN METABOLISM

In Chapter 23 nucleoproteins were shown to be constituents of nuclear tissue composed of a protein conjugated with nucleic acids. The important

$$HN{=}C(NH_2)-N(CH_3)-CH_2-COOH \xrightarrow{ATP \rightarrow ADP} HN{=}C(NH-PO(OH)O^-)-N(CH_3)-CH_2-COOH$$

Creatine → Creatine phosphate

Creatine $\xrightarrow{-H_2O}$ Creatinine ← Creatine phosphate ($-P_i$)

Creatinine: $HN{=}C{-}NH{-}C{=}O$ ring closed through $N(CH_3){-}CH_2$

nucleic acids DNA and RNA are essential constituents of the cell nucleus, the chromosomes, and viruses and are involved in the synthesis of protein. During the process of digestion the protein is split from the nucleic acids and is broken down to amino acids. The nucleic acids are first attacked by ribonuclease and deoxyribonuclease to form nucleotides that are further hydrolyzed by nucleotidases to form phosphates and nucleosides. The nucleosides are absorbed through the intestinal mucosa and split by nucleosidases of the tissues into D-ribose, deoxyribose, purines, and pyrimidines. In metabolism the amino acids and sugar follow the ordinary process of protein and carbohydrate utilization. The phosphoric acid is used to form other phosphorus compounds in the body or may be excreted in the urine as phosphates.

Purine Metabolism. The purines that are formed by the hydrolysis of nucleosides in the tissues undergo catabolic changes, forming uric acid, which is excreted in the urine. The common nucleosides, adenosine and guanosine, are split into ribose plus adenine, hypoxanthine, guanine, and xanthine, respectively. These purines are not completely broken down to NH_3, CO_2, H_2O, and energy, but are progressively oxidized with the assistance of specific enzymes.

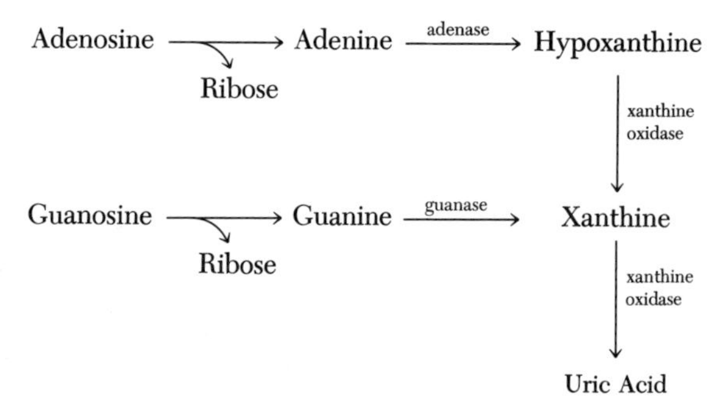

In most mammals other than man and apes the uric acid is converted into **allantoin,** a more soluble substance.

The synthesis of purines has been elaborated in recent years. This is a very

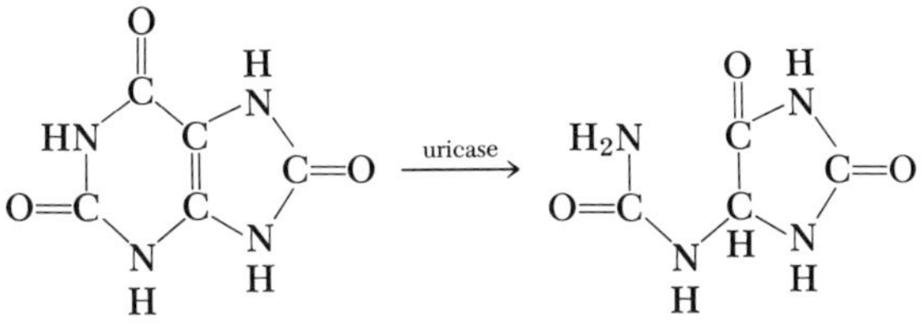

```
        O                                 O H
        ‖      H                          ‖ |
        C      N                          C─N
    HN ╱ ╲ C ╱  ╲            H2N         |    ╲
     |     ‖      C=O  ──uricase──→       |    C=O
  O=C      C ╲  ╱                  O=C    C  ╱
     ╲  N ╱    N                      ╲ N ╱H N
        H      H                         H    H
      Uric acid                        Allantoin
```

complex process involving several steps, specific enzymes, and cofactors. Inosinic acid in the form of a mononucleotide is formed first and serves as an intermediate in the synthesis of adenylic acid and guanylic acid.

Pyrimidine Metabolism. The pyrimidines that result from nucleoside hydrolysis in the tissues can be broken into small molecules in catabolism. Apparently the ring opens and NH_3, CO_2, and amino acids are formed, with the subsequent production and excretion of urea and ammonia.

Pyrimidine synthesis starts with ammonia and carbon dioxide reacting with ATP to form **carbamyl phosphate.** This compound combines with aspartic acid to form **carbamyl aspartate,** which is converted to **orotic acid** and eventually to **uridylic acid.** The mononucleotide of uridylic acid apparently serves as the starting material for the synthesis of other pyrimidine nucleotides.

Synthesis of Nucleic Acids. From the composition of nucleic acids described in Chapter 23, it is apparent that the synthesis would involve the copolymerization of four ribonucleotides for RNA and four deoxyribonucleotides for DNA. Recently Ochoa and his colleagues have isolated enzymes from plant, animal, or bacterial sources that are capable of synthesizing RNA from ribonucleotide mixtures. These enzymes are called **RNA polymerases** and require the presence of the four nucleotide triphosphates, DNA, and Mg^{+2} to synthesize RNA. The DNA molecule has also been synthesized by Kornberg and his coworkers by treating a mixture of deoxyribonucleotides with an enzyme isolated from bacteria. In the presence of all four deoxyribonucleotides, a DNA primer, and Mg^{+2}, the enzyme **DNA polymerase** can synthesize DNA. These nucleic acid molecules are similar to those from natural sources and are used to study the properties, compositions, and reactions of these large molecules.

CORRELATION OF CARBOHYDRATE, LIPID, AND PROTEIN METABOLISM

The correlation between carbohydrate and lipid metabolism has already been discussed. Since the catabolism of amino acids results in keto acids such as pyruvic acid, it can readily be seen that these products could enter the metabolic scheme of the carbohydrates. Furthermore, the glycogenic or glucose-forming amino acids and the ketogenic amino acids could enter the carbohydrate and lipid metabolism schemes. The over-all correlation of the three major types of metabolism is represented in the following diagram:

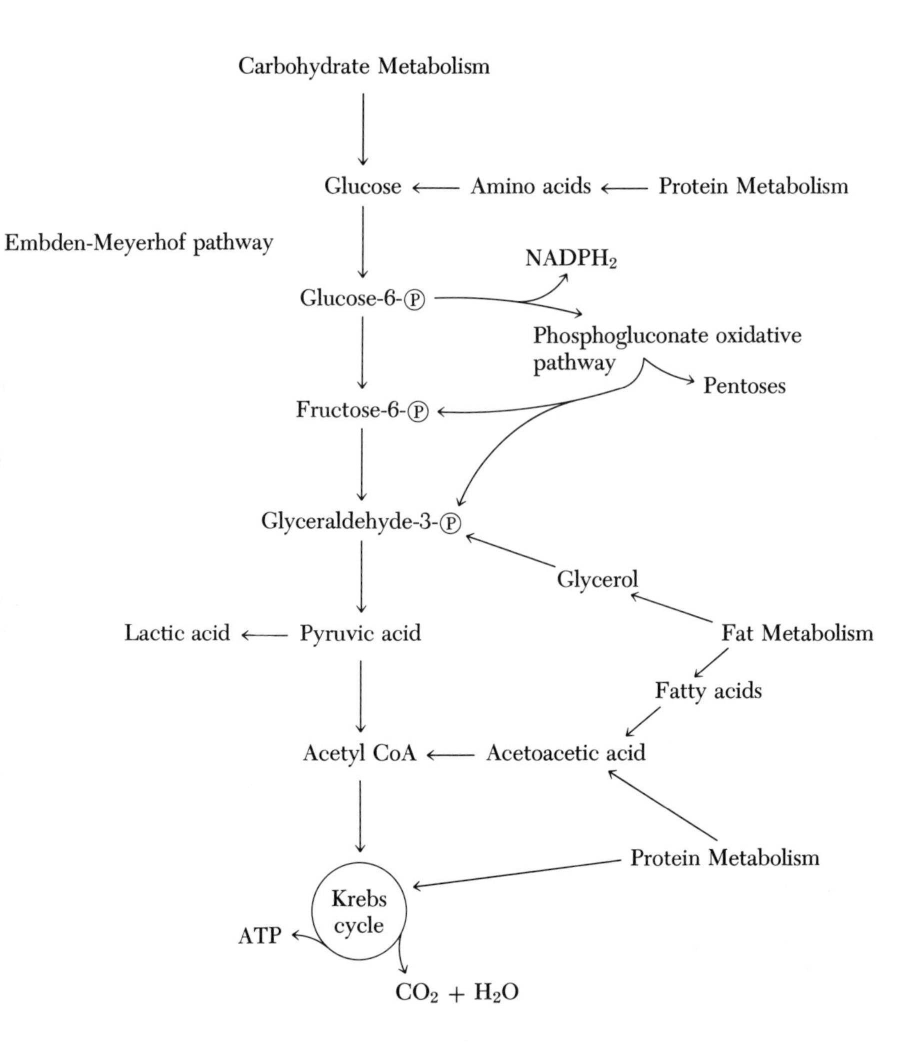

Questions

1. What concept of storage in protein metabolism replaces glycogen stores and fat depots in carbohydrate and lipid metabolism? Explain.

2. What is meant by the dynamic state of tissue proteins in the body?

3. What is an essential amino acid? What would happen to a growing child that was deprived of adequate amounts of these amino acids? Why?

4. In protein synthesis, illustrate the processes of transcription and translation.

5. What is a codon? An anticodon? How do they function in the synthesis of a protein?

6 What are the processes of deamination, reamination, and transamination? Illustrate one process with equations.

7. How does pyridoxal phosphate function as a coenzyme in the process of transamination?

8. Outline the urea cycle using names of compounds rather than chemical structures.

9. What products would result from the complete hydrolysis of RNA? Briefly describe the metabolic fate of each of the products.

10. Explain how the process of transamination can serve as a common link between carbohydrate, fat, and protein metabolism.

Suggested Reading

Bergen: Tracer Isotopes in Biochemistry. Journal of Chemical Education, Vol. 29, p. 84, 1952.
Clark and Marcker: How Proteins Start. Scientific American, Vol. 218, No. 1, p. 36, 1968.
Gorini: Antibiotics and the Genetic Code. Scientific American, Vol. 214, No. 4, p. 102, 1966.
Howe: Amino Acids in Nutrition. Chemical & Engineering News, July 23, 1962, p. 74.
Kornberg: Biologic Synthesis of Deoxyribonucleic Acid. Science, Vol. 131, p. 1503, 1960.
Roth: Ribonucleic Acid and Protein Synthesis. Journal of Chemical Education, Vol. 38, p. 217, 1961.
Scrimshaw and Behar: Protein Malnutrition in Young Children. Science, Vol. 133, p. 2039, 1961.
Yanotsky: Gene Structure and Protein Structure. Scientific American, Vol. 216, No. 5, p. 80, 1967.

CHAPTER 29

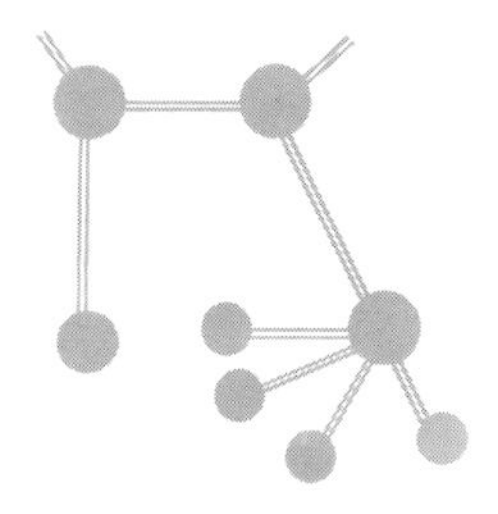

BODY FLUIDS

The complex function of a single cell requires that the intracellular fluid and the extracellular fluid surrounding the cell maintain a fairly constant chemical composition. This may be accomplished by the simple diffusion of cellular material back and forth between the intra- and extracellular fluids. As cells combine to form tissues and organs and finally the entire body, the process of diffusion is not sufficient to maintain the proper fluid balance. A circulatory, or transportation, system capable of carrying nutrient material, enzymes, hormones, and other regulatory substances to the tissues, and waste products to the organs of excretion, is required.

The **blood** is the most active transport system and consists of cellular elements suspended in plasma, the liquid medium. The **tissue fluid,** which surrounds the tissues, and the **lymph,** a slow-moving fluid that is similar to plasma and is carried in a system of vessels called the **lymphatics,** also assist the transportation system of the body. Lymph and tissue fluid are known together as **interstitial fluid** and make up about 15 per cent of the body's weight. Important interchange of material occurs from the blood to the tissue through the medium of the interstitial fluid. Plasma and interstitial fluid are collectively considered as **extracellular fluids.** A 160-pound man would have about 6.5 liters of blood, 3.5 liters of plasma, 10.5 liters of interstitial fluid, and 35 liters of intracellular fluid.

The intracellular fluids are separated from the extracellular fluids by the cellular membranes. Water, electrolytes, nutrient material, and waste products must pass through the membrane to maintain cellular function. The cell membrane is freely permeable to water, O_2, CO_2, urea, and glucose, but exhibits a selective permeability toward electrolytes, such as Na^+, K^+, Cl^-, Ca^{+2}, Mg^{+2}, and HCO_3^-. In experiments during which heavy water, D_2O, was injected into the extracellular fluid, it was found that the body required about 120 minutes to establish equilibrium between the intra- and extracellular fluids.

If water balance is to be maintained in the body, it is obvious that fluid intake and fluid excretion must be equal. The intake and output of a normal adult of average size can be represented as in the following tabulation:

Intake	*ml/Day*	*Output*	*ml/Day*
		Insensible perspiration	
		Lungs	700
		Skin	300
Water in beverages	1200	Sweat	300
Water in food	1500	Feces	200
Water of oxidation	300	Urine	1500
	3000		3000

Some concept of the magnitude of the daily fluid turnover by the body can be gained from the following data:

Daily Volume of Digestive Fluids

	ml/Day
Saliva	1,500
Gastric secretion	2,500
Intestinal secretion	3,000
Pancreatic secretion	700
Bile	500
	8,200
Fluids lost from the body	3,000
Total fluid turnover	11,200
Compared to:	
Plasma volume	3,500
Total extracellular fluid	14,000

In general, if the fluid intake exceeds the output for any length of time, the tissue fluid will increase in volume and **edema** will result. An increased flow of water from the blood vessels into the tissues results in swollen tissues, especially in the legs and feet, and is known as edema. The excessive loss of body fluids that may occur from vomiting, diarrhea, or copious sweating causes **dehydration** of the body.

BLOOD

Some idea of the importance of blood to the body can be gained from a consideration of its major functions:

1. The blood transports nutrient material to the tissues and waste products of metabolism to the organs of excretion.
2. It functions in respiration by carrying oxygen to the tissues and carbon dioxide back to the lungs.
3. It distributes regulatory substances, such as hormones, vitamins, and certain enzymes, to the tissues in which they exert their action.
4. The blood contains white corpuscles, antitoxins, precipitins, and so on, which serve to protect the body against microorganisms.

5. It plays an important role in the maintenance of a fairly constant body temperature.
6. It aids in the maintenance of acid-base balance and water balance.
7. It contains a clotting mechanism that protects against hemorrhage.

Volume. Blood comprises approximately 9 per cent, or one-eleventh to one-twelfth, of the body weight. This means that a person weighing 150 pounds has about six liters of blood. The loss of a small quantity of blood in bleeding or the donation of 500 ml for transfusion or to the Red Cross has no serious effect on the body. The blood volume is rapidly regenerated after such loss, and the missing constituents are replaced in a reasonably short period.

FORMED ELEMENTS

The two major portions of blood are the formed elements and the plasma. When separated by centrifugation, the formed elements occupy from 40 to 45 per cent by volume of the blood. This fraction contains the red blood cells, white blood cells, and thrombocytes (platelets).

SERUM AND PLASMA

When freshly drawn blood is allowed to stand, it clots, and a pale yellow fluid soon separates from the clotted material. This fluid is called **serum** and is blood minus the formed elements and fibrinogen, which is used in the clotting process. If, on the other hand, blood collected in the presence of an anticoagulant is centrifuged, the fluid portion that separates from the cells is called **plasma.** The plasma contains fibrinogen as well as other important proteins. From 55 to 60 per cent of the blood volume is plasma.

Plasma Proteins. The proteins of the plasma are present in a concentration of about 7 per cent. The most important of these are **albumin,** the **globulins,** and **fibrinogen.** The globulins have been separated into several fractions of different molecular size and properties by the technique of **electrophoresis.** More recent techniques, including column, starch gel, cyanogum electrophoresis, and immunoelectrophoresis, are capable of separating plasma into more than 20 different proteins.

The plasma proteins have several functions in the body. One of the most essential is the maintenance of the effective osmotic pressure of the blood, which controls the water balance of the body. The globulins contain immunologically active antibodies against such diseases as diphtheria, influenza, mumps, and measles. These immune principles are especially concentrated in the **gamma globulin** fraction. Fibrinogen is an essential component in the blood clotting process.

Plasma Electrolytes. The electrolytes that are present in the body fluids consist of positively charged ions, or **cations,** and negatively charged ions, or **anions.** They are mainly responsible for the osmotic pressure of the fluids and are involved in maintenance of the acid-base and water balance of the body.

The major cations in the body fluids are Na^+, K^+, Ca^{+2}, and Mg^{+2}, whereas the major anions are HCO_3^-, Cl^-, HPO_4^-, SO_4^{-2}, organic acids, and protein. In a consideration of electrolyte balance, the concentration of these ions is expressed in milliequivalents per liter of body fluid. Milliequivalents per liter (mEq/l) equals the atomic weight expressed in milligrams per liter divided by the charge, or

$$\frac{\text{mg. at. wt./l}}{\text{charge}} = \text{mEq/l}$$

The electrolyte balance of the ions in the plasma is shown in the following tabulation:

Cations	*mEq/l*	*Anions*	*mEq/l*
Na^+	142	HCO_3^-	27
K^+	5	Cl^-	103
Ca^{+2}	5	HPO_4^-	2
Mg^{+2}	3	SO_4^{-2}	1
		Organic acids	6
		Protein	16
Total	155	Total	155

Comparisons of the electrolyte composition of body fluids and changes in disease are often illustrated in vertical bar graphs devised by Gamble. Variations in the electrolyte content of normal plasma, interstitial fluid, and intracellular fluid are shown in Figure 29–1.

In the discussion of water balance we listed the volumes of the daily secretions of digestive fluids. Their electrolyte composition must also be taken into account, especially when they are lost from the body by vomiting or diarrhea.

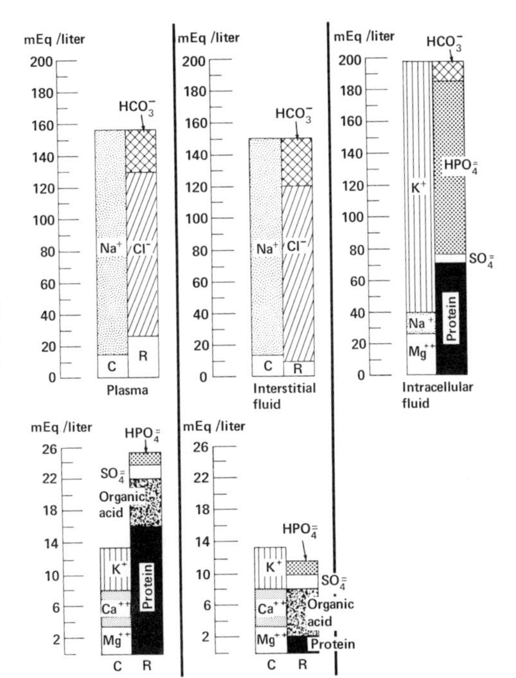

Figure 29–1 Electrolyte composition of normal body fluids. (The expanded scales below show the individual electrolytes in the C and R spaces.)

ACID-BASE BALANCE

The normal metabolic processes of the body result in the continuous production of acids, such as carbonic, sulfuric, phosphoric, lactic, and pyruvic. In cellular oxidations the main acid end product is H_2CO_3 with 10 to 20 moles formed per day, which is equivalent to one to two liters of concentrated HCl. Although some foods yield alkaline end products, the acid type predominates, and the body is faced with the necessity of continually removing the large quantities of acids that are formed within the cells. An added restriction is that these products must be transported to the organs of excretion via the extracellular fluids without a great change in their H^+ concentration. The 7.35 to 7.45 pH range of blood is one of the most rigidly controlled features of the electrolyte structure. The means of accomplishing this is the mechanism of the regulation of acid-base balance, which involves water and electrolyte balance, hemoglobin and blood buffers, and the action of the lungs and the kidneys.

BODY BUFFER SYSTEMS

The ability of extracellular fluids to transport acids from the site of their formation in the cells to the site of their excretion in the lungs and kidneys without an appreciable change in pH depends on the presence of effective buffer systems in these fluids and in the red blood cells. A **buffer** has already been defined as a mixture of a weak acid and its salt that resists changes in pH when small amounts of acid or base are added to the system (Chapter 7). The buffers in the plasma and extracellular fluid include the bicarbonate, phosphate, and plasma protein systems, which are represented as follows:

$$\frac{H_2CO_3}{BHCO_3} \qquad \frac{BH_2PO_4}{B_2HPO_4} \qquad \frac{\text{H Plasma Protein}}{\text{B Plasma Protein}}$$

The B stands for the base or cation.

In the red blood cells both bicarbonate and phosphate buffers are present along with two important hemoglobin buffers, as shown below:

$$\frac{\text{H Hemoglobin}}{\text{B Hemoglobin}} \qquad \frac{\text{H Oxyhemoglobin}}{\text{B Oxyhemoglobin}}$$

The buffers that are most effective in the regulation of acid-base balance are the bicarbonate, the plasma protein, and the hemoglobin buffers. Later in the chapter the important role of hemoglobin and its derivatives in the transportation of CO_2 from the tissues to the lungs without a change in pH will be considered.

The Bicarbonate Buffer. The bicarbonate buffer is by far the most important single buffer in acid-base balance. It is closely related to the constant production of CO_2, H_2CO_3, and $BHCO_3$; to the reactions of hemoglobin and oxyhemoglobin in the red cells; to the respiratory control of the H_2CO_3 concentration; and to the effect of the kidneys on $BHCO_3$ concentration. To illustrate the action of the buffer, reactions for the addition of a small amount of a strong acid and a strong base may be written as follows:

$$HCl + NaHCO_3 \rightarrow NaCl + H_2CO_3 \quad (1)$$
$$NaOH + H_2CO_3 \rightarrow H_2O + NaHCO_3 \quad (2)$$

In reaction (1), the basic member of the buffer pair reacts with the acid to form neutral NaCl and the acid member of the buffer pair. In reaction (2), the acid partner buffers the base to form water and the basic partner.

The buffering capacity of a buffer pair is related to its effectiveness in limiting changes of pH when acid or alkali is added to the system. The **Henderson-Hasselbalch equation** expresses the relation between the pH of the system, the pK of the buffer, and the concentration of each member of the buffer pair. The equation for the bicarbonate buffer is as follows:

$$pH = pK + \log \frac{BHCO_3}{H_2CO_3}$$

The pK is different for each buffer and is determined by measuring the pH of a solution that contains equal concentrations of the two components that make up the buffer pair. For example, if equal concentrations of $NaHCO_3$ and H_2CO_3 are present in a solution the pH equals 6.1, or

$$pH = pK = 6.1 \text{ when } \frac{BHCO_3}{H_2CO_3} = \frac{1}{1}, \text{ since } \log 1 = 0$$

The correct equation for the bicarbonate buffer is

$$pH = 6.1 + \log \frac{BHCO_3}{H_2CO_3}$$

At the pH of blood (pH 7.4) it can be determined from the equation that the concentration of $BHCO_3$ would be 20 times that of the H_2CO_3.

$$pH = 6.1 + \log \frac{20}{1} = 6.1 + 1.3 = 7.4$$

This unequal concentration of the buffer pairs would seem to indicate that the bicarbonate buffer is very ineffective at the pH of the blood. Because of the respiratory control of the H_2CO_3 concentration, however, buffering is remarkably effective, as shall be seen in the discussion that follows.

The 1:20 ratio for H_2CO_3 and $BHCO_3$ in the plasma at pH 7.4 includes the normal values of 1.35 mEq/l for H_2CO_3 and 27 mEq/l for $BHCO_3$. In disease conditions, this ratio may be changed by increases or decreases in H_2CO_3 or $BHCO_3$, producing an acidosis or alkalosis. The most common changes involve the $BHCO_3$ concentration and are described as **metabolic acidosis** or **alkalosis.** Diseases that alter respiratory function affect the H_2CO_3 concentration and produce what is called **respiratory acidosis** or **alkalosis.** These are the four major abnormalities in acid-base balance.

HEMOGLOBIN

Hemoglobin is a conjugated protein composed of the pigment **heme** and the protein **globin,** which is a histone. Heme is a complex molecule containing iron, four pyrrole groups, and hydrocarbon chains, as shown on the following page:

```
                 CH3    CH=CH2

          HC—               =CH
                    N
  CH3—                           —CH3
               N---Fe—N
  CH2—                           —CH=CH2
   |                N  OH
  CH2     HC=               =CH
   |
  COOH

                 CH2    CH3
                  |
                 CH2—COOH
```

Heme

It has been determined that four heme molecules combine with one globin to form hemoglobin, which has a molecular weight of approximately 68,000. The point of attachment of the heme and globin is thought to involve the two propionic acid groups and iron. Since globin is a strongly basic histone protein, the acid groups of heme are probably attached to two basic groups in the globin, whereas the iron is attached to the imidazole nitrogen of histidine. The hemoglobin molecule may be simply represented as follows:

```
Heme       Heme
    \     /
    Globin
    /     \
Heme       Heme
```

Hemoglobin

The normal concentration of hemoglobin in the blood varies from 14 to 16g per 100 cc. This means that a 150-pound person would have a total of approximately 900g of hemoglobin. Since the red blood cells that contain the pigment are constantly being broken down, there is a continuous degradation of hemoglobin into other pigments in the body; for example, **bilirubin,** which is converted into pigments responsible for the characteristic color of bile, urine, and feces.

Hemoglobin is often called the **respiratory pigment** of the blood and has the property of combining with gases to form various derivatives. Most important of these is **oxyhemoglobin,** a combination of oxygen and hemoglobin.

$$\underset{\text{Hemoglobin}}{Hb} + O_2 \rightleftharpoons \underset{\text{Oxyhemoglobin}}{HbO_2}$$

It can be seen from the equation that the reaction is reversible, a very important consideration in respiration since hemoglobin combines with oxygen in the lungs and carries it to the tissues where it gives up its oxygen to form hemoglobin again.

Normal and abnormal hemoglobin molecules may be separated by the electrophoresis of a drop of whole blood on a strip of special filter paper or cellulose acetate. Adult hemoglobin, fetal hemoglobin, and hemoglobin in sickle cell anemia are called hemoglobin a, f, and s, respectively. In addition, there are hemoglobins c, d, e, h, i, and j, which are present in various blood abnormalities. The majority of these hemoglobins differ by the substitution of only one amino acid in the amino acid sequence of their β-chains. It is surprising that this minor change in the molecule accounts for different rates of migration in electrophoresis and variations in physiological properties.

RESPIRATION

By far the most important function of hemoglobin is in respiration; to help carry **oxygen** from the lungs to the tissues and **carbon dioxide** from the tissues to the lungs. The transportation of oxygen by the blood depends on the reversible reaction between hemoglobin and oxygen.

$$Hb + O_2 \rightleftharpoons HbO_2$$

The oxygen capacity of the blood, about 1000 ml, is sufficient for normal tissue requirements. Some conception of the role of hemoglobin may be gained by a comparison of the oxygen capacity of plasma and whole blood. One liter of plasma can carry only 3 ml of oxygen in solution. In the absence of hemoglobin, the body's circulatory system would have to contain over 300 liters of fluid to supply oxygen to the tissues. This would represent a system four to five times our body weight.

In the process of respiration, hemoglobin comes into contact with a relatively rich oxygen atmosphere (partial pressure of 100 torr) in the alveoli of the lungs to form oxyhemoglobin. The oxyhemoglobin is carried by the arterial circulation to the tissues where a low oxygen concentration (partial pressure of 40 torr) and a high carbon dioxide concentration (partial pressure of 60 torr) combine to release the oxygen to the tissues. The carbon dioxide is then carried back to the lungs for excretion and the cycle is repeated.

URINE

In the previous chapters on metabolism several mechanisms that operate to maintain the constituents of the blood within fairly narrow limits of concentration have been considered. The utilization and transformation of nutrient material in the blood was emphasized in the chapters on metabolism. The removal of the waste products of metabolism, such as drugs, toxic substances, excess water, inorganic salts, and excess acid or basic substances, is essential to maintain the normal composition of the blood. The kidneys play a major role in the regular excretion of these substances from the blood and tissue fluids. The kidneys, therefore, are essential for the maintenance of blood and tissue fluid volume, the electrolyte and acid-base balance of the body, and the maintenance of normal osmotic pressure relationships of the blood and body fluids.

Water, carbon dioxide, and other volatile substances are eliminated from the body by the lungs. The skin excretes small amounts of water, inorganic salts, nitrogenous material, and lipids. Some inorganic salts are eliminated by the intestine, and the liver is involved in the excretion of cholesterol, bile salts, and bile pigments. Compared to the kidney, the other organs of excretion play a minor role.

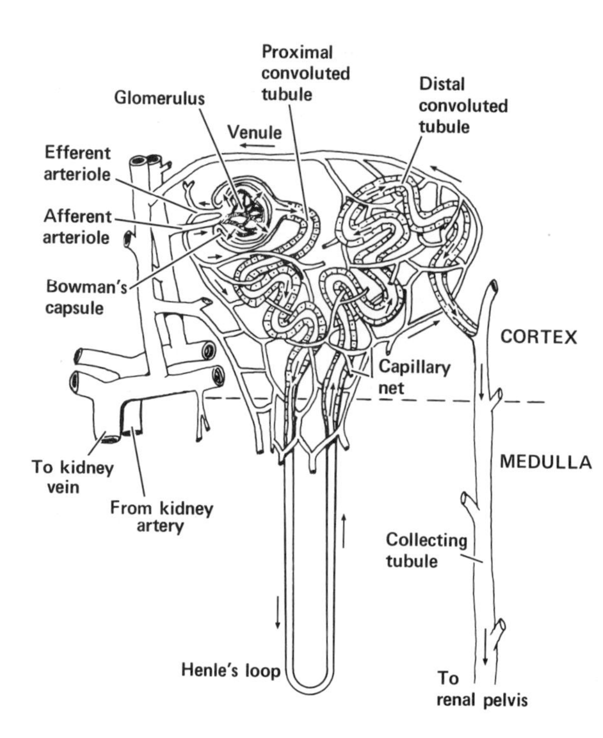

Figure 29-2 Diagram of a single kidney tubule and its blood vessels. (Adapted from Villee: Biology. 5th Ed. Philadelphia, W. B. Saunders Co., 1967.)

THE FORMATION OF URINE. The kidney may be regarded as a filter through which the waste products of metabolism are passed to remove them from the blood. The blood enters the kidney by means of the renal arteries, which break up into smaller branches leading to the small filtration units called **malpighian corpuscles.** Each human kidney contains approximately 1,000,000 of these units. A malpighian corpuscle consists of a mass of capillaries from the renal artery which form the **glomerulus.** The glomerulus is enclosed within a capsule called **Bowman's capsule** which opens into a long tubule. Several of these tubules are connected to larger **collecting tubules** which carry the urine to the bladder. These anatomical structures of the kidney are illustrated in Figure 29–2.

The most generally accepted theory for the formation of urine can be outlined as follows: As the blood passes through the glomerulus, the constituents other than protein filter through the capillary walls and enter the tubules. As this filtrate passes down the tubules, a large proportion of the water and any substances which are of value to the body, such as glucose, certain inorganic salts, and amino acids, are reabsorbed into the blood stream. These substances are called **threshold substances.** Waste products of metabolism, such as urea and uric acid, are not completely reabsorbed, and therefore pass into the collecting tubules for excretion. A few substances, such as creatinine and potassium, are partially removed from the blood through excretion by the tubules in addition to filtration by the glomerulus. The function of the renal glomeruli may be regarded as that of ultrafiltration producing a protein-free filtrate of the plasma, followed by a process of selective reabsorption by the renal tubules. Some concept of the magnitude of this process can be gained from the daily values for filtration and reabsorption. In a normal individual,

170–180 liters of water, 1000g of NaCl, 360g of $NaHCO_3$, and 170g of glucose are filtered through the glomeruli, and 168.5–178.5 liters of water, 988g of NaCl, 360g of $NaHCO_3$, and 170g of glucose are reabsorbed by the tubules in order to excrete about 30g of urea, 12g of NaCl, and other waste products in about 1500 ml of urine.

REGULATORY POWER OF THE KIDNEY

Although the kidney is considered mainly as an organ of excretion, it plays a regulatory role in water, electrolyte, and acid-base balance.

Water Balance. The control of the water content of blood, interstitial fluid, cell fluid, and digestive fluids in the body depends on normally functioning kidneys. Water in excess of body requirements is readily excreted by the kidney. When, however, water is needed to maintain the normal concentration of body fluids, the hypothalamus, in conjunction with the pituitary gland, secretes an antidiuretic hormone called **vasopressin.** This hormone causes an antidiuresis that results in less water being excreted in the urine and more being reabsorbed into the body fluids. Apparently changes in the osmotic pressure of the plasma regulate the secretion of vasopressin, which in effect is the "fine control" of urine volume.

Electrolyte Balance. To a certain extent the excretion or retention of electrolytes such as Na^+, K^+, Cl^-, HCO_3^-, and HPO_4^{-2} depends on the water balance of the body fluids. Electrolytes are excreted or reabsorbed along with the movement of water in the tubules. The steroid hormones of the adrenal cortex, such as **aldosterone** which is called a **mineralocorticoid,** exert more specific control over the excretion of electrolytes by the tubules. When the plasma has too much water, its osmotic pressure decreases, and it is said to be hypotonic. Under these conditions there is an increased secretion of adrenal cortex hormones that increase the retention or reabsorption of electrolytes into the plasma. If the plasma electrolytes were too concentrated and the osmotic pressure increased, the secretion of the cortical hormones would be depressed and the secretion of vasopressin would increase to assist in the excretion of electrolytes and the retention of water.

Acid-Base Balance. In the tubules Na reabsorption, in part at least, is involved in the regulation of acid-base balance through the action of the Na^+—H^+ exchange. This phase of Na^+ excretion is regulated to some extent by the pH of the blood plasma and the capacity of the tubular cells to acidify the urine by H^+ and NH_3 formation.

The glomerular filtrate contains the electrolytes and acids and bases present in the plasma. The chief cation is Na^+, and the chief anions are Cl^-, HCO_3^-, and HPO_4^{-2}, as electrolytes of NaCl, $NaHCO_3$, and Na_2HPO_4. At a pH of 7.4, about 95 per cent of the CO_2 is present as $NaHCO_3$ and about 83 per cent of the PO_4 as Na_2HPO_4. Normally, as the glomerular filtrate passes into the tubules most of the water is reabsorbed, and the greater proportion of the Na^+ is taken up by the tubular cells in exchange for H^+ formed in the tubular cells from H_2CO_3. This Na^+ is returned to the plasma in association with the HCO_3^-

formed from H_2CO_3 in the tubular cells. The formation of H^+ and HCO_3^- from H_2CO_3 in both the proximal and distal tubular cells is catalyzed by the enzyme carbonic anhydrase, as follows:

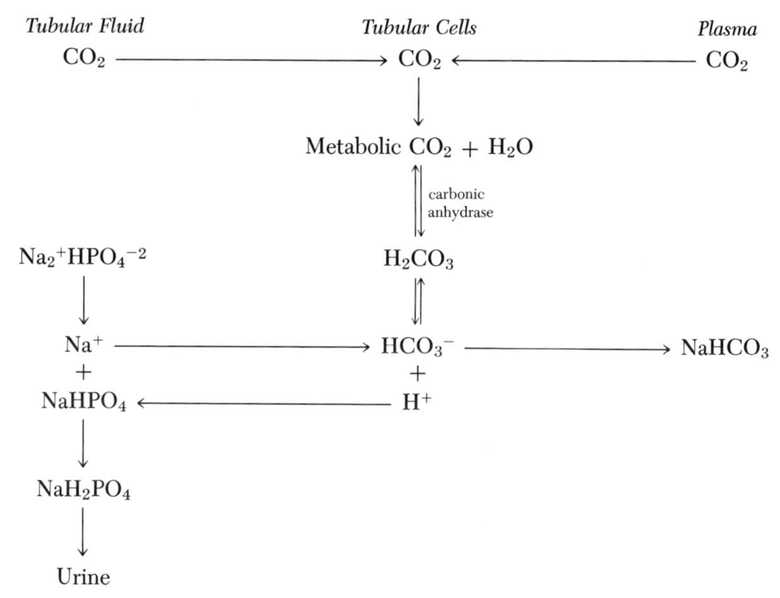

By varying the rate of ventilation, the lungs control the H_2CO_3 concentration in the plasma.

The task of the kidneys is to stabilize the concentration of the bicarbonate. This problem involves two aspects: first, the salvaging of all, or nearly all, of the bicarbonate contained in the glomerular filtrate (equivalent to approximately one pound of $NaHCO_3$ per 24 hours); and second, the neutralization of nonvolatile acids (H_2SO_4 and H_3PO_4). The kidney may conserve base in two ways: by conversion of neutral or basic salts to acid salts for excretion, as shown previously for Na_2HPO_4, and by the synthesis of ammonia.

The mechanism for the synthesis of ammonia to further conserve Na^+ and fixed base may be represented as follows:

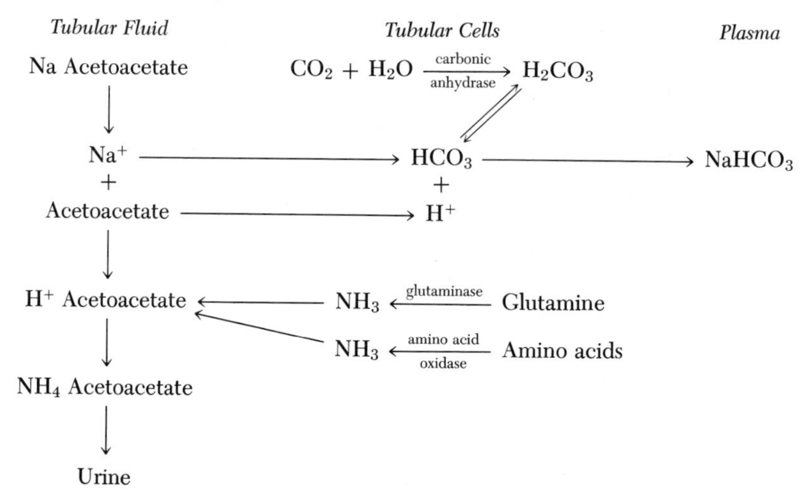

The urinary excretion of the ammonium ion in a person on a normal average diet is 30 to 70 mEq per day. In conditions of acidosis that might result from uncontrolled diabetes mellitus, acetone bodies, such as acetoacetic acid, are formed. The sodium salt of this acid must be excreted by the kidney, and to conserve the Na^+ for plasma buffers H^+ and NH_3 are formed in the tubular cells of the kidney.

Questions

1. Outline the parameters involved in the balance between fluid intake and fluid output in the body.

2. Explain the difference between whole blood, plasma, and serum.

3. Illustrate with a bar graph the electrolyte composition of normal plasma.

4. How is the pK of a buffer related to its buffering capacity? Explain.

5. Calculate the pH of a bicarbonate buffer solution that contains 20mEq./l. of $BHCO_3$ and 2mEq./l. of H_2CO_3.

6. What is the chemical nature of heme, and what is its relation to hemoglobin?

7. Outline the essential functions of the kidney.

8. Describe the essential processes in the formation of the urine.

9. What hormones are involved in the control of water and electrolyte balance by the kidney? Explain how they function.

10. Explain how the process of H^+ formation in the kidney tubules helps conserve Na^+ for the plasma.

11. Why would the kidney tubules increase the synthesis of ammonia in the condition of diabetes mellitus? Explain.

Suggested Reading

Cushny: The Secretion of the Urine. New York, Longmans, Green and Co., Inc., 1926.
Merrill: The Artificial Kidney. Scientific American, Vol. 205, No. 1, p. 56, 1961.
Smith: The Kidney. Scientific American, Vol. 188, No. 1, p. 40, 1953.
Surgenor: Blood. Scientific American, Vol. 190, No. 2, p. 54, 1954.

CHAPTER 30 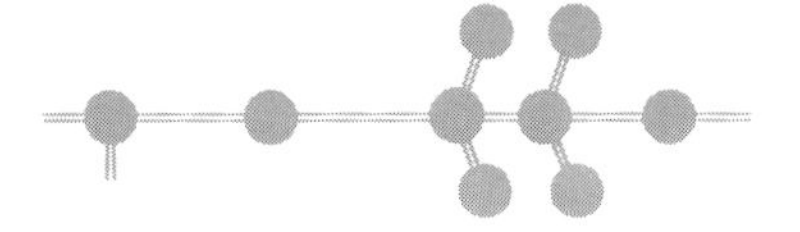

BIOCHEMISTRY OF DRUGS

If we accept a single definition of disease as *dis ease,* headaches, minor aches and pains, malnutrition, dietary deficiencies, metabolic abnormalities, endocrine disturbances, infections, and malignant cancer all qualify as diseases. Volumes have been written on diseases of the cell, tissues and organs of the body, and the therapeutic agents or drugs used to combat the disease process. It is becoming more and more apparent that all disease has a biochemical basis. The biochemistry of normal and abnormal heredity, deficiency diseases, errors of metabolism, and the process of infection is receiving considerable attention, study, and research.

To aid in the understanding of the recent emphasis on health-related research, examples of different types of diseases and their treatment will be discussed. So many people are occasionally inconvenienced with headaches and minor aches and pains that their cause and treatment with analgesic drugs should be of interest. In general, these pains are caused by swelling of tissue, resulting in pressure on peripheral nerves, and also minor inflammation of tissues, accompanied by an increase in body temperature which affects nerve endings. A common analgesic drug that serves as the basis for a multitude of headache, cold, and flu remedies is acetylsalicylic acid or **aspirin.** When combined with **phenacetin** and **caffeine,** the resultant preparation is the common APC tablets. The compound **N-acetyl-p-amino phenol** is a metabolic product of phenacetin and has replaced this drug in several preparations. Aspirin is an analgesic drug in that it reduces inflammation and swelling of tissues, exerts an antipyretic action in reducing fever, and probably exerts a chemical action on the peripheral nerves. Phenacetin and N-acetyl-p-amino phenol exhibit some of the effects of aspirin and are synergistic with respect

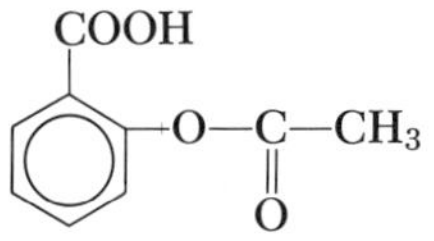

Aspirin 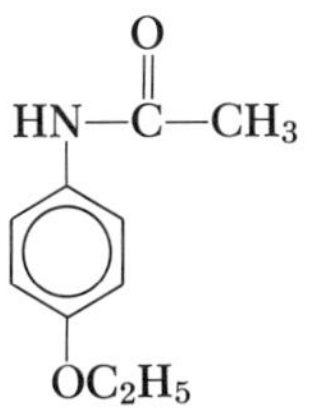   Phenacetin 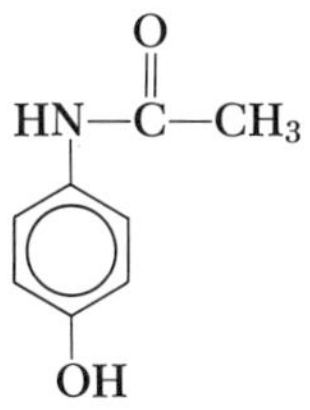   N-acetyl-p amino phenol 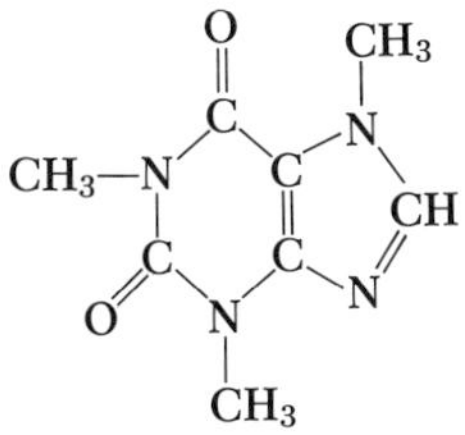   Caffeine

to its action. Caffeine is a diuretic and assists the kidney in excretion of the drug and the circulatory system in the transport of the drugs. The addition of buffering agents to aspirin has been found to speed the absorption of the drug into the blood and tissues and has resulted in products such as Bufferin.

ANTIHISTAMINES

Histamine, which is formed by the decarboxylation of the amino acid histidine (see p. 323), is a powerful pharmacological agent with effects on the vascular system, smooth muscles, and exocrine glands, especially the gastric glands. The administration or release of histamine causes dilatation of capillaries and small blood vessels with a subsequent drop in systemic blood pressure; the dilatation of cerebral vessels results in a histamine headache which may be very severe. Smooth muscles, especially the bronchioles, are stimulated by histamine and may cause respiratory problems in persons suffering from bronchial asthma and other pulmonary diseases. Histamine is a powerful gastric secretogogue and produces a copious secretion of gastric juice of high acidity. It also stimulates nerve endings and causes itching when introduced into the superficial layers of the skin.

Antihistamines are drugs that antagonize the pharmacological actions of histamine and also reduce the intensity of allergic reactions. A portion of the chemical structure of various antihistamines is similar to that in histamine, and these drugs act as competitive antagonists to histamine. Apparently they occupy the receptor sites on the effector cells and exclude histamine from these sites. The common core of the chemical structure in both histamine and antihistamines is a substituted ethylamine (see p. 278). It is believed that it is this portion of the molecule that competes with histamine for the receptors.

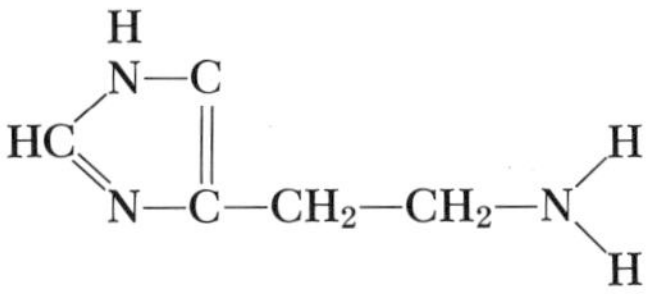

Histamine 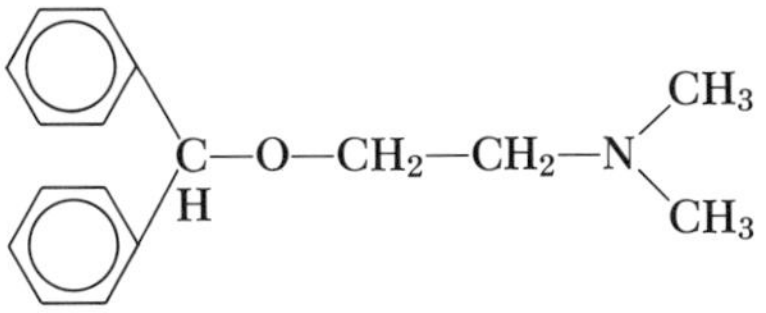   Benadryl (Diphenhydramine)

Therapeutically the antihistamines are most commonly used in the symptomatic treatment of various allergic diseases. Patients with bronchial asthma, hay fever, allergic rhinitis, and chronic rhinitis with superimposed acute colds

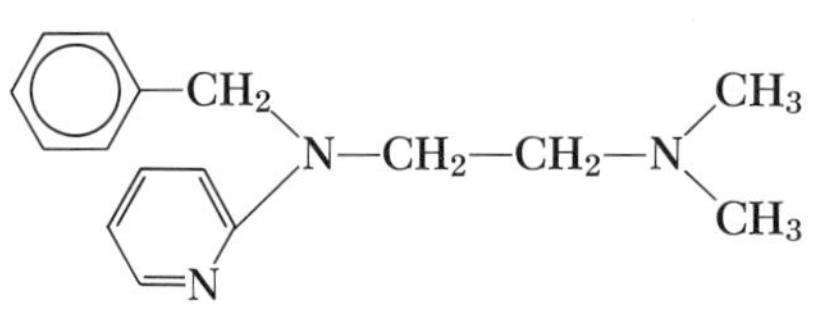

Pyribenzamine (Tripelennamine)

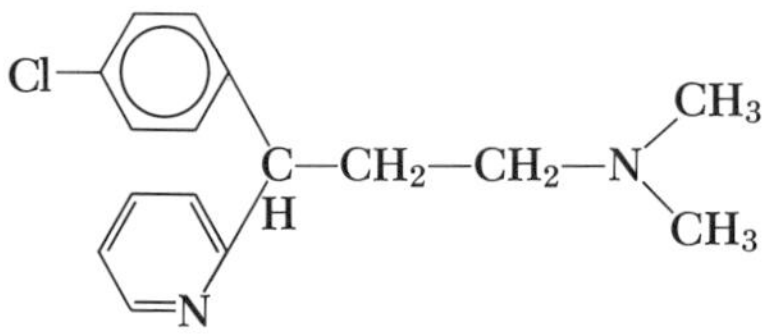

Chlor-trimeton (Chlorpheniramine)

gain considerable relief by the use of antihistamines. Various types of allergic dermatitis, contact dermatitis, insect bites, and poison ivy are benefited by the topical application of antihistamine-containing lotions. Some of these drugs, especially **Dramamine,** which relies on diphenhydramine as the active agent, are very effective against motion sickness. One common side effect of most antihistamines is their tendency to induce sedation, which restricts their daytime use when it is necessary to operate motor vehicles. **Chlor-trimeton,** shown above, is less prone to produce drowsiness than most other preparations. The prominent hypnotic effect of antihistamines related to **Benadryl** and **Pyribenzamine** has resulted in their use in various proprietary remedies for insomnia, such as Sominex and Nytol.

TRANQUILIZERS, SEDATIVES, AND HALLUCINOGENIC DRUGS

Diseases of the brain and nervous system ranging from undue concern and nervousness through inability to sleep, anxiety symptoms, neurotic behavior, and pathology of brain tissue are common conditions in these accelerated times. Drugs such as sedatives, tranquilizers, and psychic energizers are too commonly prescribed and used. Mind-expanding and hallucinogenic drugs are receiving considerable attention, and finally a new drug, L-dopa, shows great promise in the treatment of Parkinson's disease, a condition involving changes in the brain tissue.

Mild tranquilizers or sedatives, such as **meprobamates** and **barbiturates,** are related to urea and pyrimidines. More potent tranquilizers, often used in mental institutions, are reserpine and chlorpromazine. **Reserpine** is an alkaloid obtained from the roots of the *Rauwolfia serpentina*, a plant found mainly in India. Reserpine depletes stores of the biogenic amines, such as catecholamines and serotonin, in the brain and other tissues and thus supresses their action on the central nervous system, resulting in tranquilization. **Chlorpromazine** is an example of the phenothiazine drugs, which contain the complex three-ring structure shown in the formula below. The drug is widely used in the treatment of psychiatric patients and in the treatment of nausea and vomiting.

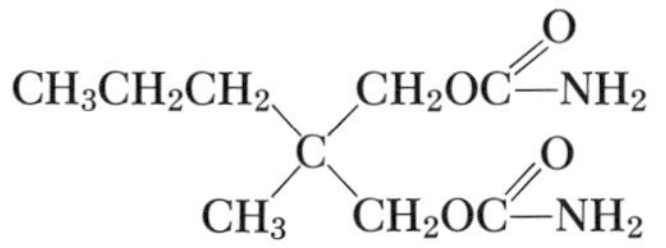

Meprobamate

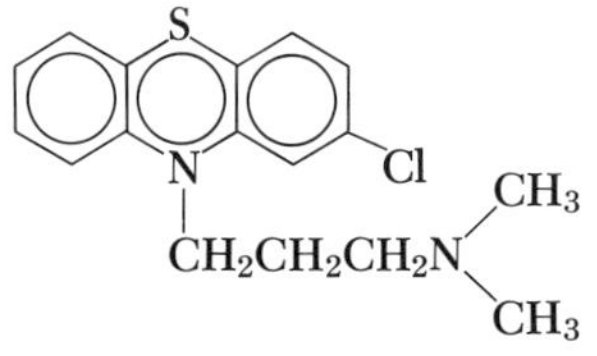

Chlorpromazine

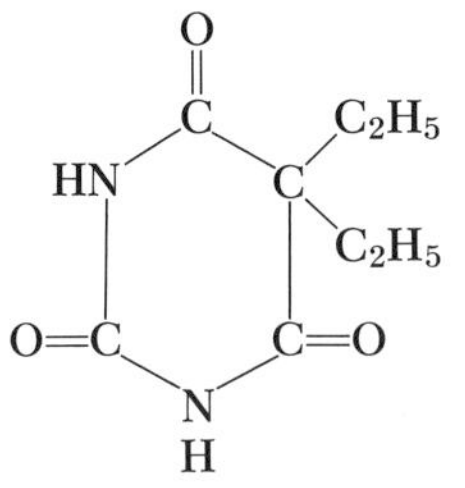

Barbital

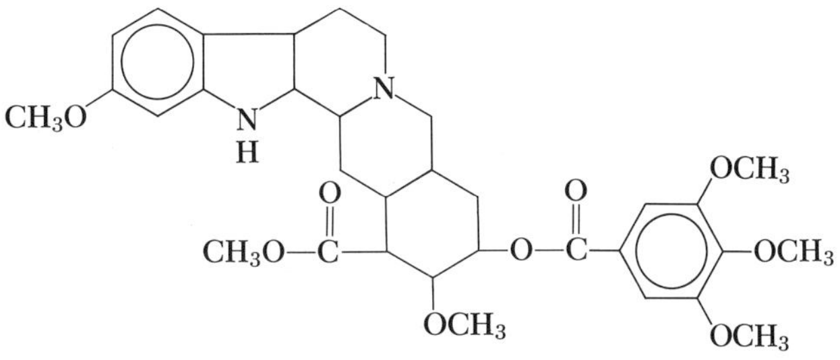

Reserpine

Continued use of mild sedatives or tranquilizers, like barbiturates, leads to a dependency on the drug. Overdosage or combinations of alcoholic beverages and barbiturates may lead to coma and accidental death. The more potent tranquilizers are subjected to strict control by physicians and are not as widely prescribed. When drugs with strong analgesic and sedative properties are required, narcotic drugs, such as **morphine** and **demerol,** are employed. Morphine, first isolated from the opium poppy, has a complex chemical structure containing a phenanthrene nucleus and a piperidine nucleus. **Codeine,** which is also commonly used as a weaker narcotic, is a methyl ester of morphine. **Heroin** is the diethyl ester of morphine and is used by drug addicts because it is more water-soluble and faster-acting than morphine. Demerol was synthesized as a substitute for morphine and is less habit-forming.

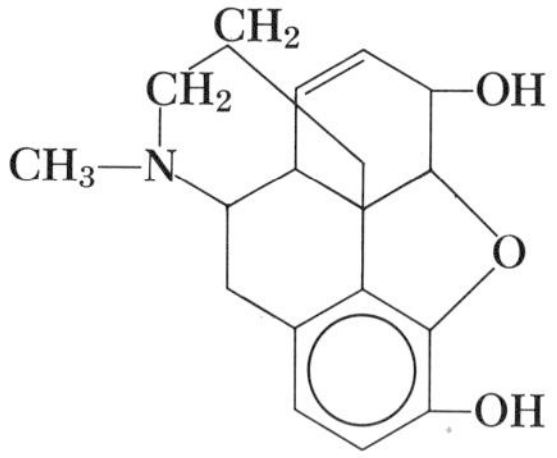

Morphine

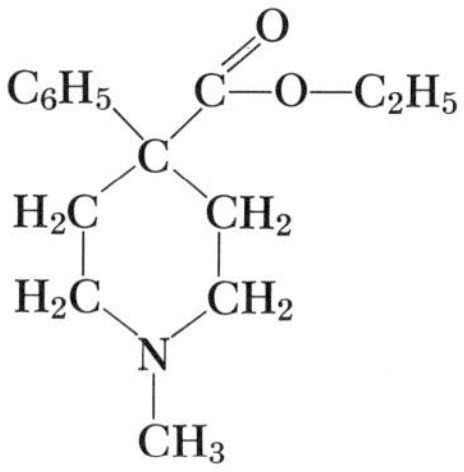

Demerol

Antidepressant or mood-elevating drugs, sometimes called psychic energizers, are being prescribed in increasing quantities. Derivatives of hydrazine directly affect brain and nervous system function by inhibiting the enzyme monoamine oxidase, which apparently results in an antidepressant action. Iproniazid and **isocarboxazid** are examples of potent drugs in this family. The popular antidepressant **Benzedrine** is an amphetamine derivative that is often used as a psychic energizer by truck drivers on long trips and students cramming for exams. Unfortunately, the enthusiastic use of the drug leads to marked nervous irritability and loss of appetite.

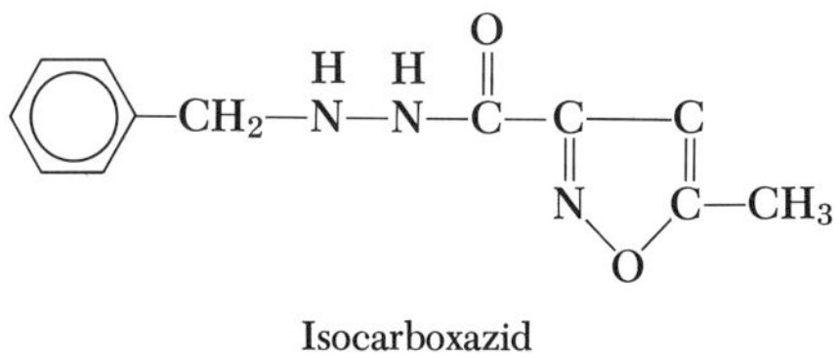

Isocarboxazid

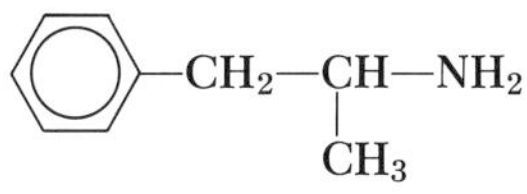

Benzedrine

A long step further in the use of mind-influencing drugs is represented by the **psychedelic** or **hallucinogenic drugs.** It is difficult to characterize the medical effect of these drugs, although many nonmedical experiments are being conducted. In view of the change in personality and mental state of an individual taking this type of drug, the results of ingestion are often unpredictable. **Marijuana** represents a mild type of hallucinogen and is obtained from the *Cannabis sativa* or hemp plant. The most potent marijuana is obtained from the yellow resin produced from the flowers of the ripe plant and is called **hashish.** Chemically the drug is a derivative of an alcohol, **cannabinol,** and the active constituent is believed to be a delta-L form, which has recently been synthesized. **Mescaline** and **lysergic acid diethylamide (LSD)** are examples of more potent hallucinogens.

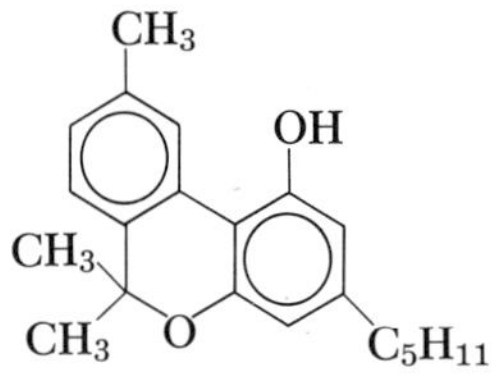

Cannabinol

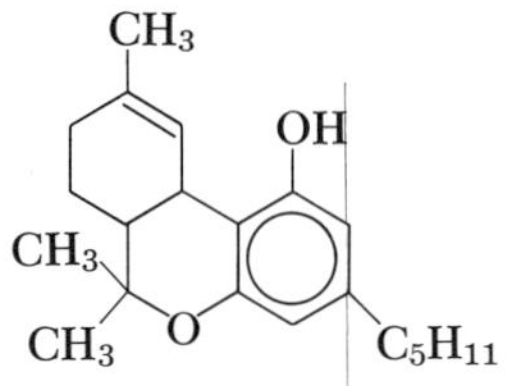

Delta-L-trans-tetrahydrocannabinol

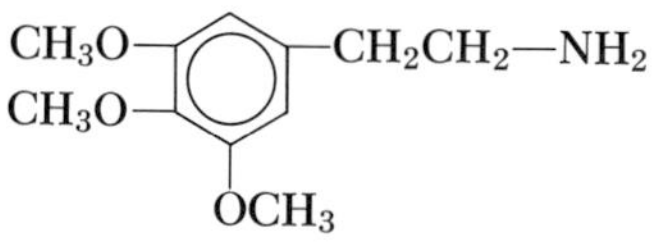

Mescaline

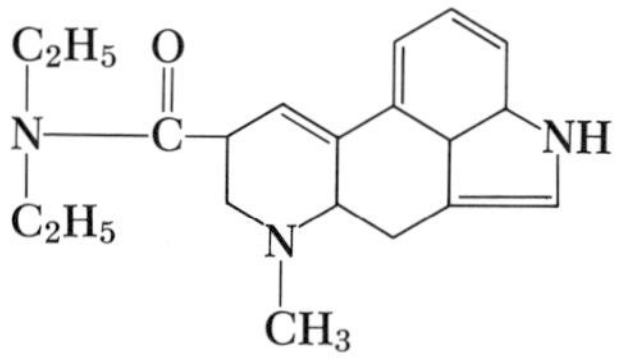

Lysergic acid diethylamide (LSD)

For years, there has been little advance in the treatment of **Parkinson's disease.** This is a disease of the brain affecting the metabolism of dopamine, epinephrine, and norepinephrine known as paralysis agitans, since it involves both a progressive paralytic rigidity and tremors of the extremities. Recently it was found that the dopamine concentration of certain areas of the brain were markedly deficient in chronic patients who had died of the disease. Since dopamine will not penetrate the brain tissue when carried in the blood, the precursor L-dihydroxyphenylalanine, L-dopa, was administered and found to increase the dopamine concentration in the target areas. The use of the drug **L-dopa** promises considerable relief and improvement of symptoms for these patients and is being hailed as a breakthrough in Parkinson's disease therapy.

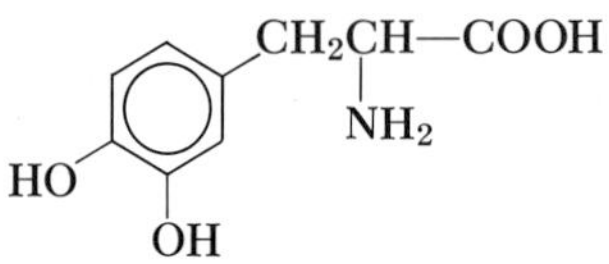

L-Dopa
(L-dihydroxyphenylalanine)

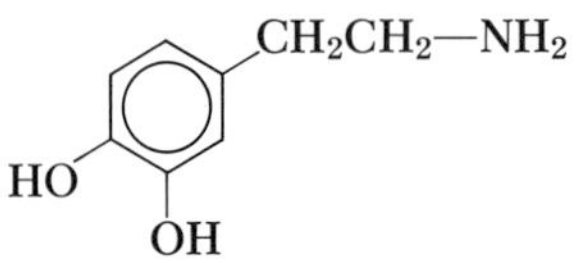

Dopamine

INFECTIOUS DISEASE

The body has to guard against infection by bacteria or viruses from birth to death. The newborn infant is fortified with antibodies against disease upon receiving the gamma globulins in the mother's milk. The layer of skin covering the body, the hydrochloric acid of the gastric juice, the digestive enzymes, and the various phagocytic cells in the circulation, all serve as a first-line defense against infection. Vaccination during childhood stimulates the production of antibodies to certain diseases. Until the mid-thirties, a serious infection or infectious disease was viewed with alarm by physicians and laymen alike. Recovery depended mainly on the body's natural defense and supportive general therapy. The first major group of chemotherapeutic agents were the **sulfa drugs.** These drugs were antibacterial agents that inhibited the synthesis of a compound like folic acid that was essential for the continued growth of the invading bacteria. Sulfanilamide, for example, acts as a competitive inhibitor of the enzyme that is involved in the utilization of p-aminobenzoic acid in the synthesis of tetrahydrofolic acid by the bacteria (see p. 370). **Sulfanilamide** was found

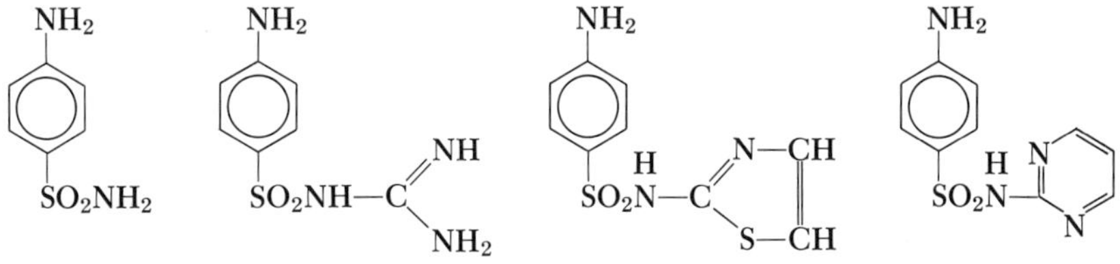

Sulfanilamide Sulfaguanidine Sulfathiazole Sulfadiazine

to be effective in the treatment of streptococcus infections, pneumonia, puerperal fever, gonorrhea, and gas gangrene. The drug is only slightly soluble in water and may damage the kidney by accumulation in that organ during excretion. Other toxic reactions, including methemoglobinemia, resulted in the development of other derivatives, such as **sulfaguanidine, sulfathiazole,** and **sulfadiazine.** A thorough study of the therapeutic properties of each sulfa drug resulted in better treatment and control of various infectious diseases. Sulfadiazine, for example, is less toxic than the other sulfa drugs, yet is one of the most effective in the treatment of pneumonia and staphylococcus infections.

A few years after the development of the sulfa drugs, a new type of antimicrobial agent was accidently discovered by Fleming. He observed that a staphylococcus culture on a bacterial plate did not grow around the periphery of a blue-green mold that had contaminated the culture plate. **Penicillin** was isolated from the secretion of the mold and was termed an antibiotic, since it interfered with the growth of the bacteria. A large number of antibiotic agents have been isolated from similar experiments with other molds. It required nine years of intensive research to synthesize penicillin. Other antibiotics, including **streptomycin, tetracycline,** and **prostaphlin,** have been synthesized, and prostaphlin shows considerable promise as an effective control of infections caused by staphylococcal bacteria. As in the case of the sulfa drugs, a family of antibiotics with specific antimicrobial properties has now been developed.

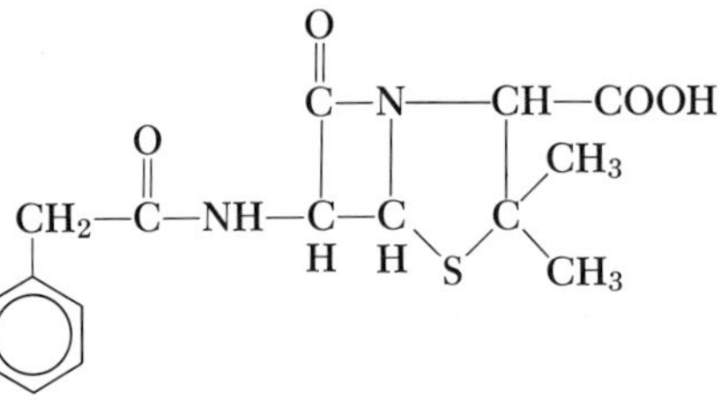

Penicillin G

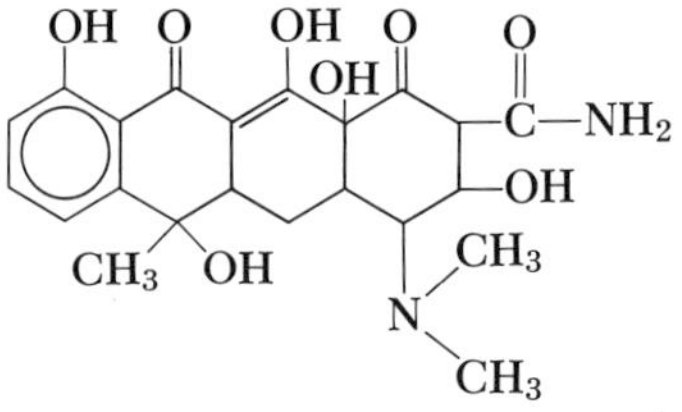

Tetracycline

COLLAGEN DISEASES AND INFLAMMATION

There are several types of collagen diseases associated with inflammatory changes in connective tissue which affect mainly the joints, skin, heart, and muscle. **Lupus erythematosus** and **rheumatoid arthritis** are examples of collagen disease. Arthritis is most common and affects men and especially women in their forties and fifties. For many years, aspirin has been used as a mild anti-inflammatory agent in arthritis and continues as the maintenance therapy. More recently **cortisone** has been found to reduce inflammation of the joints in arthritis and to reverse the course of this and other collagen diseases. The dosage level required to produce these desirable effects also produced several unwanted side effects. Steroid derivatives of cortisone were developed to decrease the incidence of side effects and increase the therapeutic potency of the drugs. The 9-fluoro-16-methyl derivative of **prednisolone,** a steroid closely related to cortisol, is 100 to 250 times as potent as cortisone in the treatment of rheumatoid arthritis. At present, arthritic patients are maintained with aspirin and given small doses of potent steroid drugs whenever an acute inflammatory process flares up in their joints.

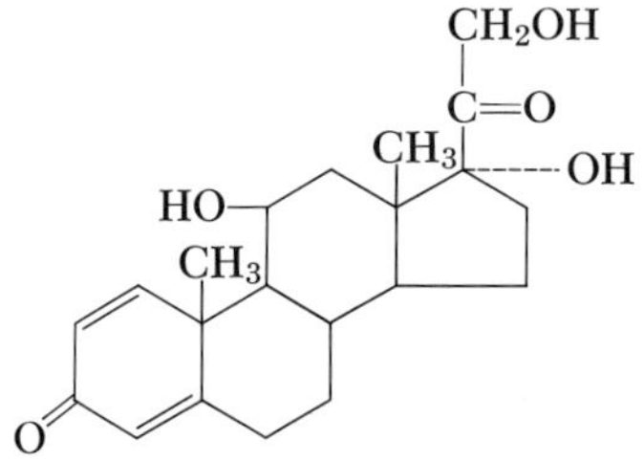

Prednisolone

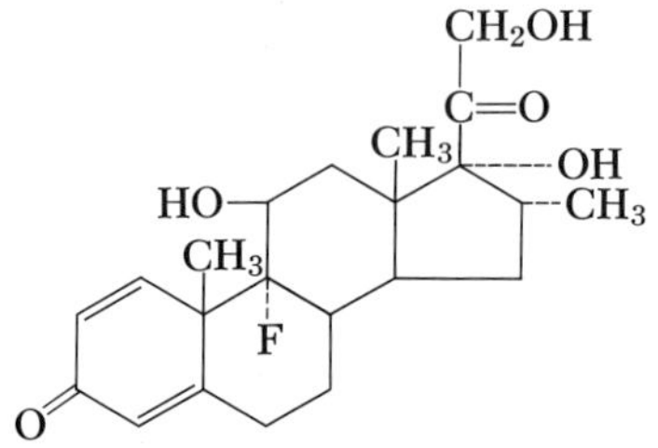

9-Fluoro-16-methyl prednisolone

ANTIFERTILITY DRUGS

Another type of steroid drug related to the sex hormones is the **antifertility drugs.** These drugs are unique in that they are given to inhibit a normal physiological process, whereas the great majority of drugs is used to treat a disease process or to alleviate the symptoms of a disease. As early as 1937 it was shown that the hormone progesterone would inhibit ovulation in rabbits. In the 1950's, several laboratories attempted to develop orally active steroids that possessed the properties of progesterone. **Norethindrone, 17α-ethynyl-19-**

nortestosterone, and **norethynodrel,** the progestational component of Enovid, resulted from these studies. These drugs were related to testosterone and contained a 17α-ethynyl group.

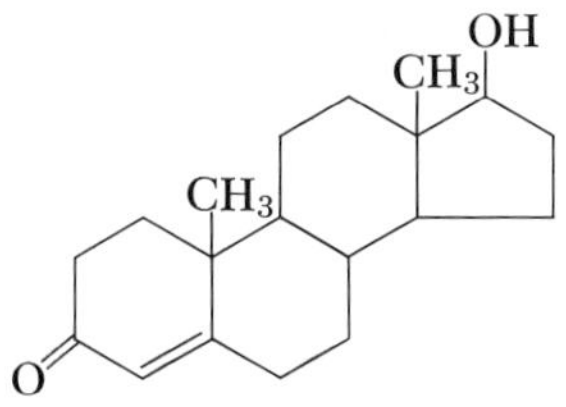

Testosterone

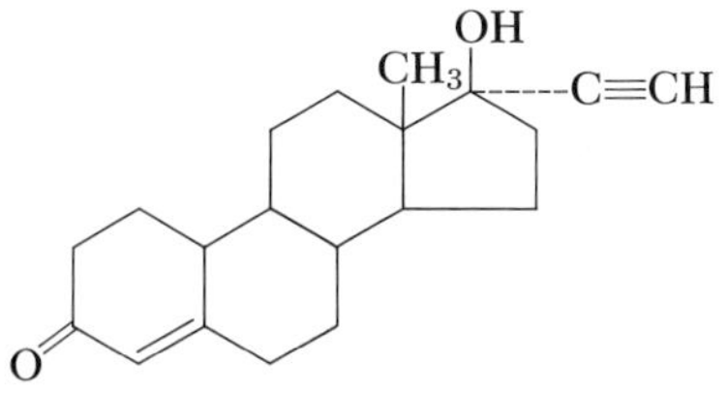

Norethindrone

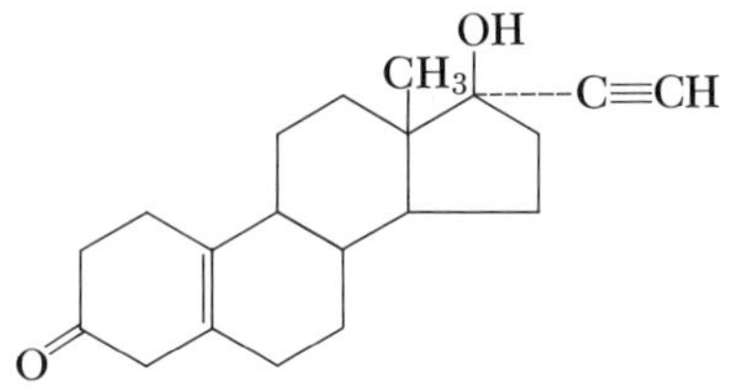

Norethynodrel

Since progesterone is relatively inactive when given orally, derivatives of this compound were studied for oral potency. It was found that 17-acetoxy-progesterone was active orally and that the addition of an α-methyl group to produce **medroxyprogesterone acetate** further enhanced this activity.

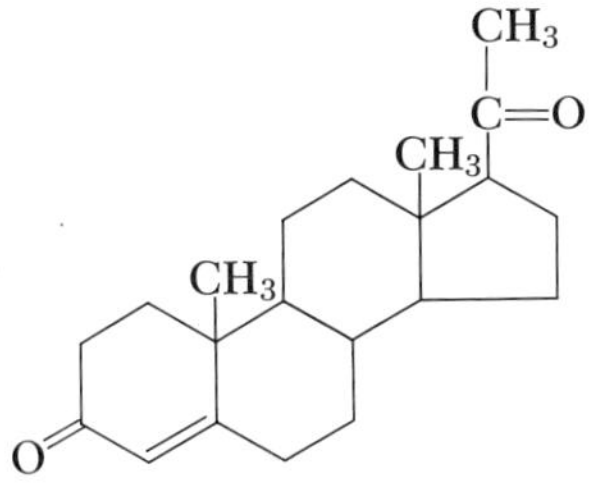

Progesterone

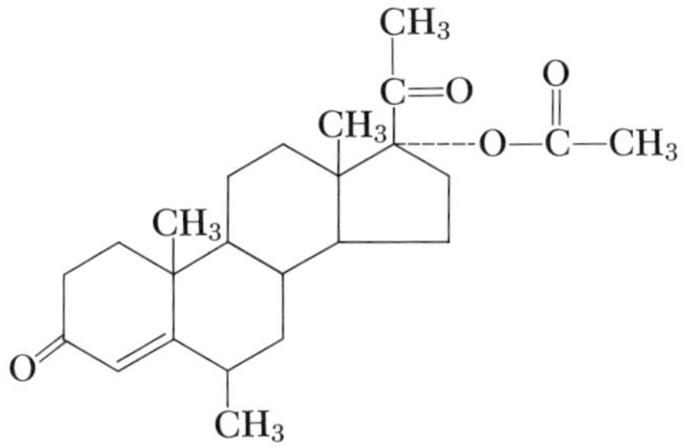

Medroxyprogesterone acetate

In clinical trials in Puerto Rico, Haiti, and the United States, the testosterone and progesterone derivatives described above were found to effectively suppress ovulation in women. Also, it was discovered that the oral drugs suppressed the production of estrogen. A combination of orally effective progesterone and estrogen active drugs are commonly used in "the pill." The original dose was 10 mg, but this is being reduced toward 1 mg to decrease the occurrence of such side effects as nausea, headaches, dizziness, and thrombosis. The drug is usually taken on days five to 25 of the menstrual cycle and then withdrawn to permit normal menstruation. It may be possible in the near future that these drugs may be injected intramuscularly with effective action from one to six months.

From a study of the mode of action of oral contraceptives it was concluded that the primary effect is inhibition of follicular maturation (see p. 312), which prevents the occurrence of ovulation.

DIABETES MELLITUS

Diabetes mellitus, because of its frequency, is probably the most important metabolic disease. The fundamental difficulty in the disease is a relative or complete lack of insulin, which is necessary for normal carbohydrate metabolism. Since the metabolic pathways of carbohydrates, fats, and proteins are known to be closely interwoven, any essential fault in carbohydrate metabolism also involves the metabolism of fat and protein, as well as water, electrolyte, and acid-base balance. There is evidence that diabetes is a hereditary disease and that the genetic tendency toward the disease results in the common condition known as prediabetes, which exists in many relatives of diabetics. The chemistry of **insulin** (p. 329) and its function in diabetes (p. 377) have already been discussed. The beta cells of the islet tissue in the pancreas produce insulin, and any condition resulting in hyperglycemia stimulates the pancreas to secrete greater quantities of this hormone. To control diabetes, insulin must be injected into the muscle tissue daily. In view of the discomfort and inconvenience to the patient, many efforts have been made to prolong the action of insulin and to develop drugs with insulin-like activity that can be taken orally. The combination of insulin with other proteins has resulted in preparations with prolonged activity. **Protamine zinc insulin** is the best example of such a combination. It has been known for several years that salicylates produce a hypoglycemic effect, but the potency was not sufficient to warrant use of the drug as an insulin substitute. In 1942 it was observed that a sulfamidothiazole compound exhibited a potent hypoglycemic effect. Related compounds were tested, and by 1955 sulfonylurea derivatives such as **tolbutamide** (Orinase) were available as antidiabetic agents. Guanidine derivatives also produced hypoglycemia, and phenethylbiguanide (Phenformin) represents another type of antidiabetic drug. When

$$CH_3-C_6H_4-\underset{\underset{O}{\|}}{\overset{\overset{O}{\|}}{S}}-\overset{H}{N}-\overset{\overset{O}{\|}}{C}-\overset{H}{N}-CH_2CH_2CH_2CH_3$$

Orinase (Tolbutamide)

$$C_6H_5-CH_2CH_2-\overset{H}{N}-\overset{\overset{NH}{\|}}{C}-\overset{H}{N}-\overset{\overset{NH}{\|}}{C}-NH_2$$

Phenformin (Phenethylbiguanide)

administered orally these two drugs exhibit different properties. Orinase stimulates the secretion of insulin from the beta cells, while Phenformin stimulates the oxidation of glucose by the peripheral tissues. Because of suspected cardiovascular difficulties with Tolbutamide dosage; its recommended use has recently been restricted to those cases in which diet and insulin are not effective.

GOUT

Gout is a chronic disease that is characterized by an increased level of uric acid in the blood, hyperuricemia, acute episodes of gouty arthritis, and degenerative changes in the joints. The disease is seen predominantly in men and the initial stages occur in their forties. In the early days, gout appeared to be more prevalent among the aristocracy, and at present, professors and clinicians exhibit a greater incidence of the disease; however, even vegetarians and lower economic groups show a predilection for the disease. A diet rich in protein and glandular meats containing nucleoproteins and more frequent visits to a physician or clinic may explain the increased occurrence in the upper income groups.

In the treatment of the disease an attempt is made to decrease the level of uric acid in the blood and in the body stores. Salicylates in the form of aspirin have long been used as a uricosuric agent, but fairly high doses and prolonged treatment are required. Drugs with a more potent action often produced toxic hepatitis. **Benemid,** a derivative of benzoic acid, was found to inhibit the enzymes responsible for the reabsorption of uric acid by the kidney. It is a safe drug that lowers serum uric acid by about 50 per cent within two to four days and maintains the reduced level as long as therapy is continued. A more recent therapeutic agent is **allopurinol** that inhibits the enzyme xanthine oxidase and thus reduces the formation of uric acid from its immediate precursors, hypoxanthine and xanthine. Treatment with allopurinol maintains the blood

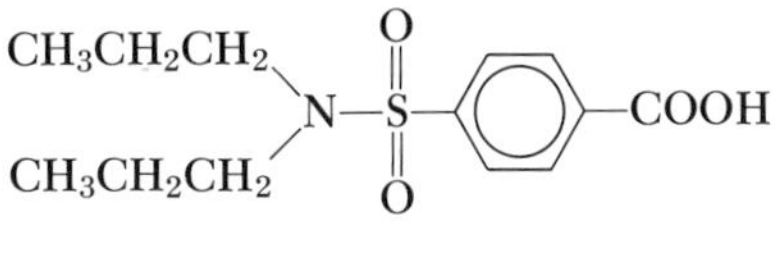

Benemid

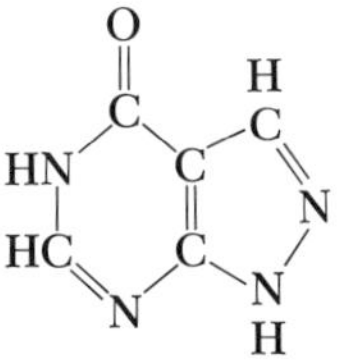

Allopurinol

uric acid level at a normal value and lowers the body pool of uric acid by constant excretion through the kidney.

CANCER

Cancer is a general term used to describe rapid multiplication of cells and increased growth of certain tissues in the body. Cancers are also called *tumors* and *neoplasms* and are classed as benign when they may be removed by surgery without reoccurrence, and as malignant when they spread to other parts of the body and exhibit reoccurrence of growth after surgery. *Leukemia* is a cancer of leucocytes or white blood cells, *carcinoma* involves epithelial cells, and *sarcoma* is a tumor of muscle or connective tissue. Considerable research effort and money has been expended in extensive studies on cancer in recent years. Tumors have been transplanted in experimental animals and produced by *carcinogenic agents* to aid in the study of the metabolism of cancer cells and possible therapeutic agents. Although some chemical compounds, such as 6-mercaptopurine derivatives, show promise in the treatment of leukemia, surgical removal of tumors and irradiation with X-rays and radioactive isotopes remain the major mode of attack on cancer. The drug **6-mercaptopurine** is thought to be converted first to hypoxanthine and then to ribonucleotides (see p. 409). It also is believed to suppress the biosynthesis of purines within the cell. These actions interfere with the production of RNA and DNA in the tumor cells and exert a cytotoxic action, especially on bone marrow and intestinal epithelial cells. The as yet unsolved problem is to discover a drug or type of radiation that will selectively kill cancer cells without damage to the normal cells in body tissues. A very active carcinogenic agent, **dimethylbenzanthracene,** and an antileukemic agent, 6-mercaptopurine possess the following structures.

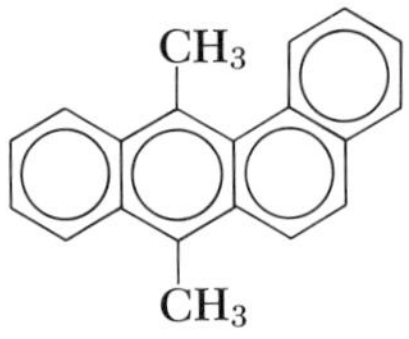

Dimethylbenzanthracene

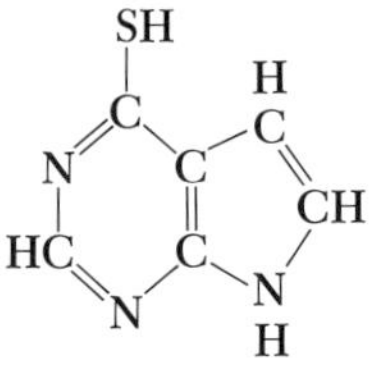

6-Mercaptopurine

Questions

1. Explain the biochemical basis of a specific disease.

2. Why is aspirin used in so many proprietary drug preparations? Explain.

3. Outline the main uses of antihistaminic drugs.

4. Compare meprobamate and reserpine on the basis of their chemical structure, potency, and mode of action.

5. Isocarboxazid is an example of a monoamine oxidase inhibitor. Describe the usefulness of the drug and compare its action with that of reserpine.

6. Give two examples of a hallucinogen drug. Should the use of these drugs be controlled by law? Explain.

7. Discuss the chemical nature and action of L-dopa in Parkinson's disease.

8. Explain the differences and similarities of sulfa drugs and antibiotics such as penicillin.

9. Give an example of a steroid drug used in the treatment of arthritis and explain its action.

10. Describe the mechanism of action of the antifertility drugs.

11. Explain the action of one oral insulin substitute and its advantages.

12. Compare the mechanism of action of benemid and allopurinol in the treatment of gout.

13. Why is 6-mercaptopurine used in the treatment of leukemia?

14. Compare the chemical structure of allopurinol and 6-mercaptopurine. Should allopurinol be effective in the treatment of leukemia? Explain.

Suggested Reading

Barron, Jarvick, and Bunnell: The Hallucinogenic Drugs. Scientific American, Vol. 210, p. 29, 1964.
Berelson, and Freedman: A Study in Fertility Control. Scientific American, Vol. 210, p. 29, 1964.
Braun: The Reversal of Tumor Growth. Scientific American, Vol. 213, p. 75, 1965.
Collier: Aspirin. Scientific American, Vol. 209, p. 96, 1963.
Frei, and Frereich: Leukemia. Scientific American, Vol. 210, p. 88, 1964.
Gates: Analgesic Drugs. Scientific American, Vol. 215, p. 131, 1966.
Grinspoon: Marihuana. Scientific American, Vol. 221, p. 17, 1969.
Himwich: The New Psychiatric Drugs. Scientific American, Vol. 193, p. 80, 1955.
Linder: The Health of the Americal People. Scientific American, Vol. 214, p. 21, 1966.
Marijuana Program Advances at NIMH. Chemical & Engineering News, July 6, 1970, p. 30.
Nichols: How Opiates Change Behavior. Scientific American, Vol. 212, p. 80, 1965.
Smith: Death from Staphylococci. Scientific American, Vol. 218, p. 84, 1968.
Weeks: Experimental Narcotic Addiction. Scientific American, Vol. 210, p. 46, 1964.

INDEX

DATE DUE